Amputation

Amputation

Surgical Practice and Patient Management

Edited by

G. Murdoch DSC MB ChB FRCS Ed

Emeritus Professor of Orthopaedic and Traumatic Surgery, University of Dundee, UK; formerly Visiting Professor, University of Strathclyde, UK; Past President of the International Society of Prosthetics and Orthotics workshops

and

A. Bennett Wilson Jr BSME

Associate Professor, Department of Orthopaedics and Rehabilitation, Medical School, University of Virginia, USA, (Retired)

Butterworth-Heinemann
Linacre House, Jordan Hill, Oxford OX2 8DP
A division of Reed Educational & Professional Publishing Ltd

 A member of the Reed Elsevier plc group

OXFORD   JOHANNESBERG   BOSTON
MELBOURNE   NEW DELHI   SINGAPORE

First published 1996

British Library Cataloguing in Publication Data
A catalogue record for this book is available from the British
Library

Library of Congress Cataloguing in Publication Data
A catalogue record for this book is available from the Library of
Congress

ISBN 0 7506 0843 9

ISBN 0 7506 2377 2
Butterworth-Heinemann International Edition

Composition by Genesis Typesetting, Rochester, Kent
Printed in Great Britain by Bath Press plc, Bath, Avon

Contents

Contributors

R. Baumgartner
Director, Universitatsklinik fur Technische Orthopadie und Rehabilitation, Munster, Germany

J.H. Bowker
Medical Director, Department of Orthopaedics and Rehabilitation, University of Miami School of Medicine, Miami, USA

D.N. Condie
Area Rehabilitation Engineer, Tayside Rehabilitation Engineering Services, Dundee, UK

H.J.B. Day
Hon. Specialist, Rehabilitation Medicine, Withington Hospital, Manchester, UK

W.G. Dykes
National Centre for Training and Education in Prosthetics and Orthotics, University of Strathclyde, Glasgow, UK

B. Ebskov
Consultant Orthopaedic Surgeon, Herlev Hospital, University of Copenhagen, Denmark

L. Ebskov
Herlev Hospital, University of Copenhagen, Denmark

J.E. Edelstein
Director, Program in Physical Therapy, College of Physicians & Surgeons of Columbia University, New York, USA

W.H. Eisma
Professor, Department of Rehabilitation, University Hospital Groningen, Groningen, The Netherlands

F.A. Gottschalk
Department of Orthopaedic Surgery, Southwestern Medical School, Dallas, USA

N.A. Govan
National Centre for Training and Education in Prosthetics and Orthotics, University of Strathclyde, Glasgow, UK

R.O. Ham
Medical Engineering and Physics, Kings College Hospital, London, UK

D.K. Harrison
Physicist, Vascular Laboratory, Ninewells Hospital and Medical School, Dundee, UK

P.E. Holstein
Department of Thoracic and Vascular Surgery, Bisperbjerg Hospital, Copenhagen, Denmark

J. Hughes
Director, National Centre for Training and Education in Prosthetics and Orthotics, University of Strathclyde, Glasgow, UK

G.A. Hunter
Consultant Orthopaedic Surgeon, Sunnybrook Health Sciences Center, North York, Canada

S.M.H.J. Jaegers
Department of Anatomy and Embryology, University of Groningen, The Netherlands

A.S. Jain
Consultant Orthopaedic Surgeon, Dundee Royal Infirmary, Scotland, UK

S.K. Jain
Senior Adviser in Surgery, Prosthetics and Orthotics, Artificial Limb Centre, Pune, India

J.S. Jensen
Formerly Department of Orthopaedic Surgery, University of Copenhagen, Denmark

E. Lyquist
Former Director, SAHVA, Copenhagen

P.T. McCollum
Consultant Vascular Surgeon, Ninewells Hospital and Medical School, Dundee, UK

L.J. Marks
Rehabilitation Consultant, Disablement Service Centre, Harrow Health Authority, Stanmore, UK

E.G. Marquardt
Professor Emeritus, Department of Orthopaedics, University of Heidelberg, Germany

J.W. Michael
Director, Department of Prosthetics and Orthotics, Duke University Medical Center, Durham, USA

G. Murdoch
Emeritus Professor of Orthopaedic and Traumatic Surgery, University of Dundee; formerly Visiting Professor, University of Strathclyde; Past President of the International Society of Prosthetics and Orthotics Workshops

G. Neff
Abt Technische Orthopadie und Rehabilitation der Freien, Universitat Berlin, Germany

G.E. Omer Jr
Chairman Emeritus, Department of Orthopaedics and Rehabilitation, University of New Mexico, Albuquerque, USA

B.M. Persson
Senior Consultant, Department of Orthopaedics, Lasarettet, Helsingborg, Sweden

P.D. Poonekar
General Surgeon in association with the Department of Orthopaedics and Rehabilitation, University of Miami School of Medicine, Miami, USA

K.C. Rankin
Consultant Orthopaedic Surgeon, Royal Infirmary, Glasgow, UK

K.P. Robinson
Consultant Surgeon, Queen Mary's University Hospital, London, UK

P.K. Sethi
Orthopaedic Surgeon, Vivekanand Marg. Jaipur, India

I.P. Torode
Consultant Orthopaedic Surgeon, Paediatric Consulting Centre, Royal Children's Hospital, Parkville, Victoria, Australia

A. Bennett Wilson Jr
Associate Professor, Department of Orthopaedics and Rehabilitation, Medical School, University of Virginia, USA (retired)

Preface

The content of the book is firmly focused on the surgeon and is comprehensively referenced. There is inevitably significant repetition both in the content and in the references. This has been seen as essential: examination candidates, reviewers and departmental heads are likely to read the entire book but most will refer to those parts specific to their surgical interest at the time. It is hoped that even in these circumstances the reader will also take the time to read those offerings relevant to most amputations such as principles, peri-operative management, rehabilitation etc. No responsible surgeon about to perform an amputation should do so without some understanding of the biomechanical and prosthetic factors involved in the patient's post-surgical future. To further this notion, a brief account of this essential information accompanies the surgical and medical management information displayed.

1

Introduction

G. Murdoch and A. Bennett Wilson Jr

The first record of an amputation performed deliberately, and as a direct result of trauma, was that of Herodotus (Loeb, 1924), who in 484BC reported that Hegesistratus, a Persian soldier confined in stocks, freed himself by cutting off his foot and replacing it later with a wooden one. Hippocrates (Le Vay, 1990), in the 4th century BC, reported the use of ligatures but the technique was lost during the Dark Ages, to be reintroduced by Ambroise Paré, a French military surgeon in 1529, when he was only 19 years old (Paré, 1564, 1951). Paré's use of ligatures replaced the mutilating practice of crushing the stump or immersing it in boiling oil to control bleeding, thus providing Paré the opportunity to fashion stumps that were more functional and easier to fit. He followed the course of his patients to the extent of designing artificial limbs for them and supervising their fabrication by a locksmith whom he referred to as 'le petit Lorraine'. Paré is also credited with introducing elbow disarticulation in 1536. Thus, it is only fitting that Paré be designated as 'the founder of modern principles of amputation'.

Amputation using Paré's methods received a big boost after Morel (Garrison, 1963) introduced the tourniquet in 1674 and amputations became commonplace in Europe, on those occasions when a large blood vessel was severed, an open fracture was present, or a joint was injured badly.

The first reported instance of a successful hip disarticulation was that performed in 1774 by William Kerr of England (Loon, 1957), and it was not unusual for this radical procedure to be used during the French Revolution and Napoleonic Wars.

The rate of survival after amputation before the introduction of antiseptic techniques by Lord Lister in 1867 (Guthrie, 1945) and the later development of the aseptic technique was extremely low. Another important milestone was the development of the Syme procedure involving disarticulation at the ankle (Syme, 1843 a, b, 1844, 1845 a b, 1846, 1857). James Syme, an Edinburgh surgeon, while attending a lecture by Lisfranc in 1822, recognized that the Chopart amputation, essentially a disarticulation at the mid-tarsal joint with a long plantar flap, was the one amputation that resisted 'hospital diseases' so common in surgery at that time (Fourcroy, 1792). We now know that because disarticulation results in the least damage to the vascular system there is less opportunity for bacteria to enter the circulatory system. Disarticulation at both ankle and knee had a profound effect in reducing the mortality rate of amputation surgery.

The Civil War in the United States (1861–1864) resulted in a huge number of amputations and amputees and although the literature reveals a great deal of activity in the development of artificial limbs, there seems to have been little innovation in surgical techniques.

It can be seen that through the years major wars have given impetus to developments in amputation surgery and artificial limbs. Because of World War II, government-supported research and development programmes of considerable size were started in the United Kingdom, Germany, Russia and the United States (US). Progress in the US programme was rapid, probably because in that country very little attention, especially from the scientific aspect, had been paid to the subject from the turn of the century. Teams of orthopaedic surgeons, engineers and prosthetists were formed in various parts of the country and assigned parts of the overall task to improve the quality of life of amputees. Emphasis was placed on the military veteran, but the US Congress specifically stated in the enabling legislation that all results of the programme were to be made available to the non-military ageing population as well. The overall programme was coordinated by the Committee on Prosthetics Research and Development (CPRD) of the National Academy of Sciences. Surgical techniques were considered as much a part of the programme as development of devices because it was soon recognized that the function of the prosthesis could be no better than the function provided by the stump.

From the beginning, workers in Canada were invited to participate in the US programme of work. Largely through the International Committee on Prosthetics and Orthotics of the International Society for the Rehabilitation of the Disabled (ISRD), and now styled Rehabilitation International, leaders from the US, Europe and Australia, following very stimulating and effective exchanges of information, produced improvements in surgical and prosthetic practice in many parts of the world.

The availability of new materials for the design and fabrication of artificial limbs often has an influence on surgery. The Syme ankle disarticulation amputation had fallen out of favour well before World War II for two reasons: the prostheses were ungainly and except in the US, Canada and Scotland, surgeons when they did carry out ankle disarticulation either failed to follow Syme's directions or tried to introduce their own modifications. With the use of the new plastic laminates, prostheses that were stronger, lighter and greatly improved in appearance were developed in Canada and the US. Through an education programme and the increasing commercial availability of the materials, Syme's amputation was revived and is used widely today.

The history of knee disarticulation is much the same. When workers at the University of California used polycentric joints in an attempt to replicate the movement of the centre of rotation of the human knee, Erik Lyquist of the Orthopaedic Hospital, Copenhagen (OHC) recognized the potential of placement of the knee unit in the shank of the knee disarticulation prosthesis, thus solving the long-standing problem of how to control the shank without the use of bulky 'outside' knee joints. The introduction of the OHC knee joint in the early 1970s and its later clones paved the way for the increasing number of knee disarticulation and long transfemoral amputations as carried out routinely today.

Myoplasty and osteoplasty techniques in lower limb stumps pioneered in Germany by Bier (1893), Bunger (1905), Ertl (1949), Mondry (1952) and Dederich (1956) were studied in the US and Scotland in the late 1950s but it was not until Marian Weiss of Poland delivered a lecture in Copenhagen in 1963 describing his attempts to measure the results of myoplastic surgery in the revision of stumps using EMG signals, and mentioning, almost as an aside, that the patient walked the day after surgery using a rigid dressing and temporary pylon in the manner described by Berlemont (1961) that myoplasty and osteoplasty were given the attention that they deserved.

Programmes in many parts of the US and Europe were initiated to study the so-called immediate post-surgical fitting technique. The results were impressive. The use of the rigid dressing was probably the biggest factor in the success, but it is difficult to measure the part played by the amount of attention to detail and analysis of the various factors involved in patient care when these techniques are used. Most surgeons were reluctant to use myoplasty and osteoplasty techniques in peripheral vascular cases. Fortunately, the effect of the immediate fitting programme was to give strong emphasis to the programmes within the US and Europe with the singular message of saving the knee joint in as many peripheral vascular cases as possible. Until then surgeons in training were taught that when infection was present amputation should be performed above the knee to ensure healing by first intention.

By means of a cooperative effort coordinated by CPRD in the US, the ratio of three AK amputations to one BK were reversed by the early 1970s. Equally centres in Scotland and in one or two centres elsewhere in Europe achieved the same objectives and have sustained the same excellence.

During the Vietnam War, Phillip Deffer demonstrated the possibilities of the Ertl osteomyoplasty techniques in transtibial amputations. Regrettably, these techniques are still not used except in a few specialist centres.

As an outgrowth of the immediate fitting programme and the work of Paul Brand at Carville, Louisiana with leprosy patients, surgeons at Rancho Los Amigos Hospital began a programme intended to avoid amputation when infection is confined to the foot. When this proves to be impossible, all practical length is saved, by débridement and application of a rigid dressing and later use of special shoes. The acceptance of these practices has taken longer than it should but through the efforts of John Bowker in America and Holstein and Jernberger in Europe these techniques are now routine in many areas.

By the late 1960s the International Commission on Prosthetics and Orthotics under the leadership of Knud Jansen had become increasingly active in promoting research, development, education and clinical practice worldwide in prosthetics, orthotics and related surgery to such an extent that it sought an independent stance. Thus, with the concurrence of the ISRD, the International society for Prosthetics and Orthotics (ISPO) was founded during a seminar on Prosthetics in Denmark in 1970 as an independent organization based on professional membership.

Despite a paucity of funds, ISPO achieved an early impact and has become stronger with each passing year. In addition to organizing a World Congress every 3 years, publishing a technical journal, *Prosthetics, Orthotics International*, ISPO conducts seminars, conferences, instructional courses and workshops dealing with various aspects of prosthetics and orthotics as appropriate and as time and resources permit.

In spite of the advances that have been made in amputation surgery and postoperative care during the past few decades, the Society realized that use of the improved techniques was still woefully lacking even in many of the most sophisticated surgical circles. The major reason for this shortcoming is that most of the amputations are done by surgeons as a life-saving measure, who see only a few patients a year, have little contact with prosthetists and do not realize the importance of the stump as an organ of locomotion and its relation to the ability of the amputee to obtain maximum use of the prosthesis. To begin the rectification of this condition, the ISPO organized a conference held in 1990 at Strathclyde University, Scotland with the object of arriving at a consensus concerning the best practices for amputation at all levels. The participants, comprising some 50 recognized authorities in the field, prepared themselves for the debates by undertaking a rigorous scrutiny of the surgical and prosthetic literature, published during the previous 20 years. This work was completed before the conference where teams consisting of a surgeon and a prosthetist presented their appreciation of the literature relevant to the specific levels of amputation and offered value judgements. Each presentation was discussed at length in plenary session before and after detailed discussion in panel meetings. The report of the conference is intended to provide the basis for texts and other instructional material that will eventually be used in the training of surgeons who are likely to perform amputations. The results of the conference provided the inspiration for the publication of this book.

References

Berlemont, M. (1961) Notre expérience de l'appareillage précoce des amputés des membres inférieurs aux establishments. Helio-Marins de Berck. *Ann. Méd. Phys.*, **4**, 213.

Bier, A. (1893) On the plastic formation of load bearing stumps following amputation at the tibia (Über plastiche Bildung tragfähiger stumpf nach Unterschenkel amputationen) *Arch. Klin. Chir.*, **46(1)**, 90–96.

Bunger, R. (1905) Technique for achieving weight bearing diaphysial stumps without osteoplasty. *Brun's Beitr. Klin. Chir.*, **47**, 808–827.

Dederich, R. (1956) Myoplastic stump revision. *Z-blatt f. Chir.*, **81**, 1194–1206.

Ertl, J.R. (1949) Uber Amputationsstumpfe. *Chirurg.*, **20**, 218–224.

Fourcroy, A.F. (1792) *La Médecine Eclairée par les Sciences Physiques*. Tom. 4 Paris. Chez Buisson [contains (pp. 85–88) the first description of Chopart's method of partial amputation of the foot. This is in the form of a note by Lafiteau: 'Observation sur une amputation partielle du pied'].

Garrison, F.H. (1963) *An Introduction to the History of Medicine*. Saunders, Philadelphia, p. 275.

Guthrie, D. (1945) in *A History of Medicine* (ed. Guthrie, D.), Nelson, pp. 321–336.

Le Vay, D. (1990) *The History of Orthopaedics*. Parthenon Publishing Group, Carnforth, pp. 20–41.

Lisfranc, J. (1815) *Nouvelle Méthode Operatoire pour l'Amputation Partielle du Pied dans sons Articulation Tarso-métatarsienne: methode précédes des Nombreuse Modifications qu'a Subies Celle de Chopart*. Gabon, Paris.

Lib. IX 37 (Loeb Classical library edition translated by Godley, A.D.) (1924) Vol. 4 Heinemann pp 202–205.

Loon, H.E. (1957) The past and present medical significance of hip disarticulation. *Artificial Limbs*, **4(2)**, 4–21.

Mondry, F. (1952) The muscular above-knee and below-knee stump. *Chirurg.*, **23**, 517–519.

Paré, A. (1564) *Dix Livres de la Chirurgie, avec le Magasin des Instrumens Necessaires à Icelle*. Jean Le Royer, Paris.

Paré, A. (1951) *The Apologie and Treatise of Ambroise Paré . . . with many of his Writings upon Surgery* (ed. Keyes, G.) Falcon Educational Books, London.

Syme, J. (1843a) *Edinburgh Monthly Journal of Medical Science*, **3** No. XXVI, 93.

Syme, J. (1843b) Ibid, **3** No. XXVIII, 274.

Syme, J. (1844) Ibid, **4** No. XLIV, 647.

Syme, J. (1845a) Ibid, **5** No. LIII, 341.

Syme, J. (1845b) *Monthly Journal of Medical Science*, **5** No. LIII, 337.

Syme, J. (1846) Ibid, **6** No. LXVIII, 81.

Syme, J. (1857) *Lancet*, **2**, 384, 394, 480.

Section 1

Principles of amputation

2

General principles of amputation

G. Murdoch

Amputation is a mutilation attended by not only physical and functional loss but often severe psychological trauma with the body image altered and distorted. Whether by accident or design, the more proximal the loss is, the more there is a progressive reduction in the ability to move, work and play and even, in some circumstances to survive. Our first responsibility must be to avoid amputation; all possible alternatives must be explored, evaluated and rejected only when the evidence points to amputation as the best solution for the patients' plight.

Introduction

Once the decision to amputate has been made efforts should be made to achieve the most distal level consistent with the causal condition and a well healed non-sensitive stump, as there is less functional loss and less energy is required to ambulate when wearing a prosthesis. In reaching a decision various factors (Murdoch, 1967) must be considered, including

pathology:
anatomy at the proposed level of amputation;
surgery (the management of the tissues);
available prosthetic replacement(s);
personal factors such as age, sex, occupation etc.

Once these considerations have been determined, the distilled goals for the surgeon, the patient and the other members of the clinic team are ablation

and reconstruction. Ablation of the limb or a part thereof implies removal of all that is necessary to eliminate the pathology and at the same time permit primary or secondary healing. Reconstruction requires the creation of the best possible stump in terms of both motor and sensory function and in doing so, provides for prosthetic substitution and restoration of function. To achieve those goals the amputating surgeon must understand functional limb physiology and the nature of prosthetic replacement in order to produce a functional organ of locomotion. In particular, there must be a basic understanding of the nature of the socket, which has a fixed volume and, with the possible exception of the 'flexible' socket (Kristinsson, 1988) a fixed shape. The volume is a derivative of the surgery and oedema control, given that the socket volume properly reflects stump volume and that in turn depends on the time interval between casting and fitting and on the skill of the prosthetist. The shape of the socket is dependent on the biomechanical realities of force transference from stump to prosthesis and *vice versa* and on the necessary deformations to avoid harmful levels of pressure and shear. That understanding will, in turn, depend on a basic knowledge of human locomotion and prosthetic gait.

Tissue management

The prime requirement of the surgery itself is gentle handling of the tissues. This is particularly true in the case of the *skin* where the specification is to produce a mobile, painless, well healed scar. The flaps should be mapped out with care to ensure that they are as broad based as possible. To avoid tapering of the skin and to permit good abutment in wound closure the knife should cut into the skin at right angles to the surface. Wound closure should be achieved without tension. The modern prosthetic socket is such that split skin and full thickness grafts can be tolerated. These procedures can be especially valuable in preserving stump length, particularly in the growth period.

Treatment of the *nerve* is predicated on the inevitable formation of a neuroma composed of nerve and scar tissue. The neuroma is painful in response to pressure, stretching or other physical manipulation. Cautery has been used in the past with a recent revival in interest with respect to digital nerves. Nerve ends have been injected with a variety of substances, buried in bone or enclosed in impervious materials such as silicone. The consensus view is that the nerve should be pulled down gently, ligated, divided cleanly and allowed to retract proximally into healthy muscle (Dellon *et al.*, 1984). In this way bleeding from the central vessel is prevented and the neuroma does not become entangled in the end scar. Similar treatment should be applied to smaller nerves such as the saphenous and sural nerves in the transtibial amputation. Failure to do so may result in the adherence of the neuroma to the skin scar and produce extremely painful episodes.

The primary objective in management of the *blood vessels* is effective haemostasis; the vessels should be isolated and securely ligated before division. The use of cautery should be limited as far as possible. Troublesome bleeding from a bone end is best controlled by sustained pressure and bone wax is rarely required (Hicks and McClelland, 1980)

In the case of *muscle* tissue one should retain the maximum bulk of functioning muscle to ensure strength, shape, circulation and proprioception. In order to maximize strength, control function and maintain stump shape distal muscle stabilization is essential. Various methods of muscle stabilization have been described and advocated including 'myofascial closure' (Burgess, 1968), 'myoplasty' (Dederich, 1967) 'myodesis' (Burgess, 1968a, b), myopexy (Weiss, 1969) and tenodesis (Gottschalk, this volume). Over the past 100 years the notion of myoplasty has been promoted mainly by German

surgeons. The original intention was to provide a 'pad' over the end of the stump. This cannot, in fact, be achieved but the work of Dederich (1967) shows that it is beneficial to attach the muscles to the end of the stump in some way. He showed in a series of stump revisions that the vascular supply to the end of the stump was significantly improved following this procedure. Danish workers Hansen-Leth and Reimann (1972) confirmed these findings in laboratory experiments using female rabbits. In transfemoral amputations it can be difficult to achieve myoplasty in the manner recommended by Dederich as the muscle loops produced can be difficult to stabilize, permitting emergence of the femoral end and a disfigured and less functional stump. Both Burgess (1968) in his myodesis and Murdoch (1968) in a procedure devised for the transfemoral amputation recommend fixation to the bone end via drill holes, at least in the case of some of the muscles. Sawamura (1988) has recommended a similar approach for the ankle disarticulation (Syme). The length at which you divide a muscle at amputation must be a matter of judgement. It is very difficult to make firm rules but one's objective should be to suture the muscles in slight tension.

The consensus view is now firmly in favour of myodesis, and even more favourable, tenodesis when it is feasible to perform. A notable example is fixation of the adductor magnus employing its tendinous attachment to the adductor tubercle and/or linea aspera in the transfemoral amputation.

There are two principal requirements in management of the cut end of the *bone*. One is to ensure that the cut edge is smoothly contoured over any aspect which is likely to be capable of being brought into contact with the socket wall. The classic situation is seen in the need for contouring the anterior distal end of the tibia (Murdoch, 1967). This is best achieved by starting with a 30–45° bevelling cut. If using a power tool the bone should be kept cooled with water to prevent thermal necrosis (Marsden, 1977). All edges of the cut surface now presented should be carefully filed and sculptured so that there are only smooth rounded bone edges in contact with the skin of the stump and socket wall. This can be readily checked by gently pulling the skin of the anterior flap over the anterior distal end and palpating with the other hand; any irregularity, however small, will be immediately apparent and can be remedied there and then. Similar bevelling and smoothing can be performed on the fibula at its postero-lateral aspect and at transfemoral level the anterior aspect of the femur should be carefully filed and smoothed. When it can be performed the medulla should be closed by a flap of periostium. This is always possible in the child and in the young adult but in many dysvascular cases the periostium is found to be thin and very friable. Time spent in careful preparation of the bone end is always worthwhile; otherwise it may cause serious problems for the prosthetist and untold misery for the patient.

Two-stage amputation

In certain situations it may be necessary to perform amputation in two or more stages. The initial procedure is designed to provide adequate drainage and once infection is controlled a second procedure, possibly at a higher level, may be performed. If the amputation is to be completed at the same level the tissues and bone end should be prepared, adequate skin assured and the wound left open. Dependent on tissue response secondary closure is performed after an interval. This approach is especially applicable in traumatic amputations when the wound is contaminated.

When revised amputations are deemed necessary it is usually to smooth the bone, sometimes after shortening, and to effect muscle stabilization. Revision amputation is never justifiable if the only complaint is pain.

Wound closure

In closing the wound opposing tissues are sewn under physiological tension, i.e. neither too tight nor too loose. Dead spaces should be eliminated. Careful abutment of the skin edges is essential to ensure wound healing. 'Blanket' sutures should be avoided. Burgess (1968) recommends avoidance of subcutaneous sutures in case they should compromise skin circulation. Interrupted sutures (Baker et al., 1977; de Cossart et al., 1983) are usually preferred to a continuous suture, although the latter has proved successful in the author's hands. Browse (1973), concerned about ischaemia of the small mass of tissue enclosed within the single suture, recommended interrupted intradermal sutures alternating with adhesive paper

strips to contain subcutaneous fat. Wound drainage is usually required and closed suction the preferred technique (Roon et al., 1977). The use of Penrose drains can be attended by an unwelcome incidence of infection (Tripses and Pollack, 1981; Kacy et al., 1982).

Stump environment

The options available include a stump free of all dressings, a variety of soft dressings, rigid dressing or controlled environment treatment. Few surgeons are brave enough to leave the stump totally free, although Vaucher in Geneva (personal communication) swears by it. Soft dressings are favoured by some and certainly permit easy inspection and free movement. In the case of the transfemoral amputation a soft dressing is preferable provided it is held in place by, for example, vertical strips of adhesive bandage. An encircling bandage is fraught with danger. Muller and Vetter (1954) point out that 'sustained pressures above 25 mmHg are potentially harmful'. Husni, Ximenes and Hamilton (1968) showed evidence that pressures greater than 15–20 mmHg applied to the popliteal fossa caused a tourniquet pressure effect even when the whole limb was bandaged. Johnson (1972) stated bluntly that 'no dressing over 10 mmHg should be left on overnight'. Pressures greater than those quoted are easily produced. Indeed, Isherwood, Robertson and Rossi (1975), in a study employing both skilled and unskilled bandagers, demonstrated that pressures as high as 130 mmHg were produced by operators in both groups. The rigid dressing applied with care and normally using Plaster of Paris is the preferred technique (Mooney et al., 1971; Baker et al., 1977; Barber et al., 1983) for knee disarticulation, transtibial and ankle disarticulation (Syme). There is less pain and less oedema; it provides excellent protection and in doing so encourages the patient to move about. Muscle 'setting' can be performed routinely and at the same time the cast prevents flexion contracture. The cast can be removed and replaced if bleeding or infection is suspected. It should also be renewed if there is significant reduction in stump/leg volume with a resultant loose cast. Those unused to handling Plaster of Paris bandages can learn the technique quickly using a helpful patient with a mature stump.

In the case of the transtibial amputation the rigid cast should be retained for 3 weeks (changed, if necessary) when the sutures can be removed.

Pre-operative phase

In an elective procedure in the younger patient there is no reason why the details of what will happen to them following surgery cannot be outlined. They should be given information regarding what pain they are likely to have and how this will be overcome, how long they are likely to be in bed, when the drain is likely to be removed, when dressings or plaster are likely to be changed, when they can expect to have the stitches removed from the wound and when it is likely that they will be fitted with their first prosthesis. The patient should have some understanding of the likely functional loss and to what extent this can be compensated by the prosthesis which will ultimately be provided. In furtherance of this, it is often useful for the patient to talk to someone, preferably of the same age and sex, who has undergone the same procedure and been fitted with a similar prosthesis. A discussion should be developed outlining the extent to which he will be able to undertake his previous employment or alternatively what retraining will be required. He should have some understanding of how well he will be able to negotiate the physical obstacles of life. Whatever communication is necessary and pertinent to the patient's needs should be established with his family, employer, and those persons involved in his social welfare.

In addition to evaluating any systemic or local infection by wound and blood cultures and determining sensitivities to antibiotics, additional baseline information should be obtained. If the haemaglobin is low, some (Bonhontos *et al.*, 1974; Bailey *et al.*, 1979) advocate withholding blood transfusion until the operation itself in order to avoid increasing the blood viscosity and decreasing flow. There are two further useful indicators as to the capacity for wound healing. These are, first, the serum albumin level which should be 3.5 g per decilitre or above and second, that the total lymphocyte count should be at least 1500 per mm^3 (Dichaut *et al.*, 1984). Smoking should be discouraged both pre-and post-operatively (Robinson, 1972).

The affected part of the limb is isolated within a plastic bag extending to just above the affected area and sealed to the skin. The limb should not be shaved because of the increased risk of infection (Seropian and Reynolds, 1962; Cruse and Foord, 1973; Hamilton *et al.*, 1977; Court-Brown, 1981; Fairclough *et al.*, 1987) but the skin is prepared with povidone–iodine on the day(s) prior to operation and by compress from groin to protected foot for 30 min preceding operation.

Penicillin 0.5 M units is given intramuscularly 2 hours before operation and continued for 3 days to obviate infection by *Clostridium welchii*.

Protective perineal pads are clearly in order.

Intra-operative phase

Anaesthesia may be local, regional, spinal or general. Both local and regional anaesthesia are feasible certainly at transtibial level or below and especially in the patient who is extremely ill or with severe cardio-pulmonary disease. Low spinal anaesthesia (Dale and Capps, 1959; Mann and Bissett, 1983) is, for the majority, the management of choice as it has no harmful effect on the pulmonary system, the patient suffers less pain and requires less analgesics. The reduced bleeding is welcomed by the operating surgeon. Equally, the nursing staff find their role less demanding as the patients do not suffer from sickness or nausea and are much less drowsy and restless. Blood pressure can be controlled by fluid administration and the use as required of vasopressors or alternatively vasodilators. General anaesthesia in the healthy young adult is safe and effective.

The *positioning* of the patient is important and should be carefully considered according to the level of amputation. For example, many consider the prone position is the best for a knee disarticulation procedure as it provides the best access to the relevant anatomy and dissection can proceed unimpeded in both extension and flexion of the knee.

Post-operative phase

Our first concern must be the control of pain. This presents in varying degrees. The avascular patient has, in all probability, suffered chronic pain for long periods, so that surgery when required, often urgently, paradoxically comes as a relief, and post-operative pain is minimized. On the other hand, surgery carried out after due deliberation as part of a planned programme of events, and also surgery following trauma, can produce a greater degree of post-operative pain. A variety of drugs may be used – from cyclomorphine to the simpler forms of analgesia – and preferably should be given on demand and not by the clock.

It is vitally important to control pain (Troup, 1988), but it is equally important to minimize the use of drugs. Very often one finds that simple drugs like mefenamic acid and paracetamol effectively control post-operative pain after the first 48 h. Continued heavy sedation can do nothing but harm, interfering with the patient's ability to increase activity towards final mobilization. Activity itself reduces the level of pain, particularly when directed at early active movement of the stump.

The next task is to encourage the patient to be as mobile as soon as possible. The patient should be out of bed sitting with the stump elevated on the second post-operative day. On the third or fourth day the physiotherapist will begin the process of 'standing' between parallel bars and ambulation with the aid of crutches. Much will depend on the individual's demonstrated balance, strength and proprioception. Early walking aids such as the pneumatic post-amputation mobility (PPAM) aid (Redhead, 1983) can be very useful, especially if a rigid cast is employed. In children and the young adult it may be possible to institute an Immediate Post-Operative Fitting (IPOF) regime (Burgess et al., 1969). If control of stump volume is strict it should be possible to proceed to a definitive prosthesis within 3 or 4 weeks. In the case of the transfemoral amputation when using a modular prosthetic system it is possible to develop a final prescription over the course of 7–10 days. Throughout this period the patient will be involved in a programme of exercise to promote balance, control of the prosthesis and strength in preparation for increasing mobility. Instruction in donning and doffing and taking care of the prosthesis is provided along with training in toileting, coping with stairs and the other obstacles of living. On discharge home the patient must be assured of how to seek help in the case of difficulty and admitted to a well organized follow-up system to monitor progress especially with regard to stump volume and any need for socket change.

Success in rehabilitation of the amputee depends not only on good surgery at the correct level but primarily on a confident clinic team (Murdoch et al., 1988) ensuring comprehensive assessment, realistic goal setting, careful attention to technical detail by all members of the team, coordination of activities and sound leadership.

References

Bailey et al. (1979) Pre-operative haemoglobin as predictor of outcome of diabetic amputations. *Lancet*, **July**, 168–170.

Baker, et al. (1977) The healing of below-knee amputations: a comparison of soft and plaster dressings. *Ann. J. Surg.*, **133**, 716–718.

Barber et al. (1983) A prospective study of lower limb amputations. *Can. J. Surg.*, **26**, 339–341.

Bonhontsos et al. (1974) The influence of haemoglobin and platelet levels on the results of arterial surgery. *Br. J. Surg.*, **61**, 984–986.

Browse, N.L. (1973) Choice of level of amputation in ischaemic arterial disease. *Scand. J. Clin. Lab. Invest.*, **31**, Suppl. 128, 249–252.

Burgess, E.M. (1968a) The stabilization of muscles in lower extremity amputations. *J. Bone Joint Surg.*, **50A**, 1486–1487.

Burgess, E.M. (1968b) The below-knee amputation. *Bull. Prosthet. Res.* **10(9)**, 19–25.

Burgess, E.M., Romano, R.L. and Zettl, J.H. (1969) *The Management of Lower Extremity Amputations.* TR10–6, August. US Government Printing Office, Washington.

de Cossart et al. (1983) The fate of the below-knee amputee. *Ann. R. Coll. Surg. Engl.*, **65**, 230–232.

Court-Brown, C.N. (1981) Preop skin depilation and its effect on post-op wound infection. *J. R. Coll. Surg. Edinb*, **26**, 238–241.

Cruse, P.J.E. and Foord, E. (1973) A five years prospective study of 23,649 surgical wounds. *Arch. Surg.*, **107**, 206.

Dale, W.A. and Capps, W. (1959) Major leg and thigh amputations: ten year survey and results. *Surgery*, **46**, 333–342.

Dederich, R. (1967) Technique of myoplastic amputations. *Ann. R. Coll. Surg.*, **40**, 222–227.

Dellon, A.L., Mackinnon, S.E. and Pestronk, A. (1984) Implantation of sensory nerve into muscle: preliminary clinical and experimental observations on neuroma formation. *Ann. Plast. Surg.*, **12**, 30–40.

Dichaut et al. (1984) Nutritional status: importance in predicting wound healing after amputation. *J. Bone Joint Surg.*, **66A**, 71–75.

Fairclough J. et al. (1987), *J. R. Coll Surg Edinb*, **32**, 76–78.

Hamilton, H.W., Hamilton, K.B. and Low F.K.J. (1977) Pre-operative hair removal. *Can. J. Surg.*, **28**, 269–274.

Hansen-Leth, C. and Reimann, I. (1972) Amputations with and without myoplasty on rabbits with special reference to the vascularisation. *Acta. Orthop. Scand.*, **43**, 68–77.

Hicks, L. and McClelland, R.N. (1980) Below-knee amputation for vascular insufficiency. *Ann. Surg.* **46**, 239–243.

Husni, E.A., Ximenes, J.O.C. and Hamilton, F.G. (1968) Pressure bandaging of the lower extremity. *JAMA*, **206**, 2715–2718.

Isherwood, P.A., Robertson, J.C. and Rossi, A. (1975) Pressure measurements beneath below-knee amputation stump bandages, elastic bandaging, the Puddifoot dressing and a pneumatic bandaging technique compared. *Br. J. Surg.*, **62**, 982–986.

Johnson, H.D. (1972) Mechanics of elastic bandaging. (Correspondence) *BMJ*, **3**, 767–768.

Kacy *et al.* (1982) Factors affecting the results of below-knee amputation in patients with and without diabetes. *Surg. Gynaecol. Obstet.*, **155**, 513–518.

Kristinsson, O. (1988) The flexible above-knee socket. In: *Amputation Surgery and Lower Limb Prosthetics* (ed. G. Murdoch), Blackwell Scientific Publications, Edinburgh.

Mann, R.A.M. and Bisset W.I.K. (1983) A comparison of spinal analgesia and general anaesthesia in the elderly. *Anaesthesia*, **38**, 1185–1191.

Marsden, F.W. (1977) Amputation: surgical technique and post-operative management. *Aust. N.Z. J. Surg.*, **47**, 384–392.

Mooney, V. Harvey, J.P., McBride, E. and Snelson, R. (1971) Comparison of post-operative stump management: plaster vs. soft dressings. *J. Bone Joint Surg.*, **53A (2)**, 241–249.

Muller, E.A. and Vetter, K. (1954) The effect of pressure loads upon blood supply to the skin. *Arbeitsphysiologie*, **15**, 295–304.

Murdoch, G. (1967) Levels of amputation and limiting factors. *Ann. R. Coll. Surg. Eng.*, **40**, 204–216.

Murdoch, G. (1968) Myoplastic techniques, *Bull. Pros. Res.*, **10–9**, 4–13.

Murdoch, G. (1970) The surgery of the below-knee amputation. In: *Prosthetic and Orthotic Practice* (ed. G. Murdoch) Edward Arnold, London, pp. 45–60.

Murdoch G. *et al.* (1988) The Dundee Experience. In: *Amputation Surgery and Lower Limb Prosthetics* (ed. G. Murdoch), Blackwell Scientific Publications, Edinburgh.

Redhead, R.G. (1983) The early rehabilitation of lower limb amputees using a pneumatic walking aid. *Prosthet. Orthot. Int.*, **7**, 88–90.

Robinson, K. (1972) Long posterior flap myoplastic below-knee amputation in ischaemic disease: review of experience in 1967–1971. *Lancet*, **1**, 193–195.

Roon *et al.* (1977) Below-knee amputation: a modern approach. *Ann. J. Surg.*, **134**, 153–158.

Sawamura, S. (1988) Syme's amputation – a review and a modified surgical-prosthetic approach. In: *Amputation Surgery and Lower Limb Prosthetics* (ed. G. Murdoch), Blackwell Scientific Publications, Edinburgh.

Seropian, B. and Reynolds, B.M. (1962), Wound infection after pre-operative depilatory versus razor preparation. *Am. J. Surg.*, **104**, 251–254.

Tripses, D. and Pollack, E.W. (1981) Risk factors in healing of below-knee amputation: appraisal of 64 amputations in patients with vascular disease. *Ann. J. Surg.*, **141**, 718–720.

Troup, I.M. (1988) Pre-operative and post-operative care: stump environment. In: *Amputation Surgery and Lower Limb Prosthetics* (ed. G. Murdoch), Blackwell Scientific Publications, Edinburgh.

Vaucher, J. personal communication.

Weiss, M. (1969) Physiologic amputation, immediate prosthesis and early ambulation. *Pros. Int.*, **3**, No. 8, 38–44.

3

Biomechanics and prosthetics

J. Hughes

Introduction

The provision of a prosthetic replacement for part of a missing limb is an exercise in restoring function with maximum attainable comfort while providing an appearance of normality. It is also intended that the prosthetic solution will so far as possible minimize the effort required of the individual in its use. There is a strong motivation for the lower limb amputee to use the prosthesis and using modern fitting techniques, materials and components there is every possibility of a good result. However, just as it is true that the prosthesis must be properly designed and constructed to obtain an optimum result it is also true that the stump must be fashioned in such a way that it gives best access to preferred prosthetic techniques.

The act of walking is typified by a series of events which occur in a rhythmic, repetitive pattern. Predictable rotations take place at the hip, knee and ankle joints, powered and controlled by the muscles which act across each joint. When part of the limb and one or more joints are replaced by a prosthesis a new system is created. Prosthetic joints are not powered and some of the movements of the normal joints, such as flexing and extending of the knee during weight-bearing are no longer possible. As a consequence, prosthetic walking always involves an adapted form of gait and an adaptation in the use of the remaining musculature.

At every level of amputation, in addition to the loss of anatomical joints, a new joint must be formed between the prosthesis and the body. The stump/socket interface will experience changing patterns of force corresponding to gravity and inertia effects. This is similar to the situation which exists between the foot and shoe of the normal individual but whereas the foot provides a large plantar area well adapted to the application of load, most stumps have little end-bearing capability and only at ankle and knee-disarticulation levels can any significant use be made of normal 'load-bearing' tissue. The proper design of both elements of the interface, the socket and the stump will be crucial to the achievement of comfortable function.

Biomechanics is concerned with the study of forces and their effects on the human body. In considering the design and construction of the stump for each level of amputation an understanding of both the biomechanics and the prosthetic considerations is essential. Every chapter in this volume dealing with amputation levels has a brief description of the biomechanics and prosthetic considerations applicable to that level. The intention of this chapter is to provide an explanatory background of general principles to facilitate an understanding of the detailed information on specific levels.

Normal locomotion

Normal locomotion is a very individual activity and each individual has a walking pattern which is an easily recognizable characteristic. However, although small individual variations occur, there are certain common elements in the normal gait

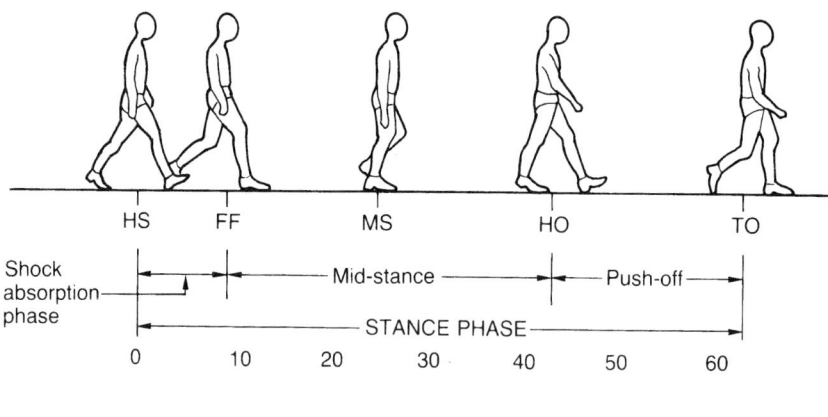

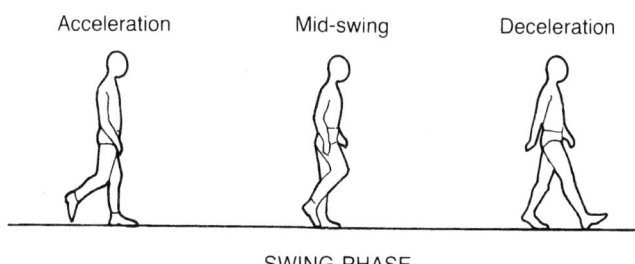

Fig. 3.1. The gait cycle.

cycle. The gait cycle is the name given to the period during walking from the time one foot comes into contact with the ground until the next ground contact with the same foot (Fig. 3.1). The gait cycle itself is composed of two phases – the stance phase when the foot is in contact with the ground and the swing phase when the foot is moving forward in space preparatory to contacting the ground and commencing a new gait cycle.

The stance phase may be further subdivided into a number of significant events – heel strike when the heel comes into contact with the ground, foot flat as the sole of the foot comes into contact with the ground, mid-stance when both ankles are in apposition, heel-off when the heel loses contact with the ground and toe-off which signifies the end of stance phase. The period between heel strike and foot flat is known as the shock absorption phase; that between heel off and toe off as the push-off phase. The stance phase accounts for about 60% of the gait cycle at normal walking speeds.

The swing phase is divided into three segments. The first is acceleration immediately after toe off

when the leg is being actively accelerated forward; the second is mid-swing when both ankles are in apposition, corresponding to mid-stance on the other leg; the third is deceleration when the leg is being slowed down preparatory to placing the foot to commence another stance phase.

Biomechanics of the gait cycle

The above actions take place under the combined effects of gravity and inertia. To enhance the understanding of the locomotion process it is necessary to understand something of joint biomechanics. Joints are activated by the muscles which act across them. The requirement for muscle activity, however, is not dictated by the movement involved but by the applied force system. This may be appreciated by considering the simplified representation of the elbow joint in Fig. 3.2.

Force can have two effects – translation and rotation. For the force system to be in equilibrium

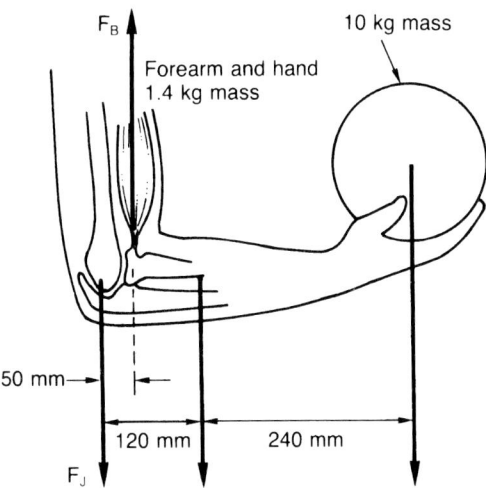

F_B

10 kg mass

Forearm and hand
1.4 kg mass

50 mm

120 mm

240 mm

F_J

Fig. 3.2. Force effects at the elbow joint.

both of these effects must balance. The rotational effect of the force is known as the moment of the force and is equal to the magnitude of the force multiplied by the perpendicular distance from its line of action to the point of rotation.

In the example shown, the mass of the forearm and hand (assumed to be 1.4 kg) and the mass held in the hand (10 kg) are both producing turning effects or moments tending to extend the elbow. If the arm is held in equilibrium the biceps muscle must produce a flexion moment equal to the extension moment to resist the extension effect. Noting that:

F_B = Force in biceps
F_J = Force in elbow joint
A mass of 1 kg exerts a force of 9.81 Newtons due to gravity.

It may be seen that:

$$F_B \times 50 = (9.81 \times 1.4 \times 120) + (9.81 \times 10 \times 360)$$

Thus F_B, the force in the biceps muscle is 739 Newtons.

If the translational effects are then considered it can be seen that the forearm and hand and the mass held in the hand exert forces downward of (1.4 + 10) 9.81 Newtons. Biceps exerts a force upward of 739 Newtons. For equilibrium there must be a force in the joint acting downward on the radius and ulna

$$F_J = 739 - (11.4 \times 9.81) = 627 \text{ Newtons}$$

This simple example (Hughes and Jacobs, 1979) illustrates some important principles. When force actions are exerted on the body they tend to produce turning effects or moments at joints. These turning effects are counterbalanced by turning effects or moments due to pull in appropriately placed muscles. Muscle lines of action are frequently much closer to the joint centres than the external actions and hence, to produce equal turning effects or moments, much larger forces are required in the muscle. These large muscle forces, again because the muscle is acting very close to the joint, produce correspondingly large joint forces.

The requirement for muscle activity in the situation depicted in Fig. 3.2 is determined by the applied force system. If from the configuration shown, the elbow was very slowly extended *or* flexed so that the acceleration was very low and therefore inertia effects might be considered negligible the same force system would apply and as a consequence biceps activity would be required to control either the extension or the flexion of the elbow. This confirms that it is the force system and not the movement involved which determines the requirement for muscle activity.

In considering normal locomotion from a perspective of prosthetic replacement it is necessary to understand the implications of the changing configuration of the leg during walking, the changing force actions and the consequent requirement for muscle activity. These may be easily understood with reference to the simplified diagrams depicted in Figs. 3.3–3.11 (Hughes and Jacobs, 1979: reproduced with permission from *Prosthetics and Orthotics International*). In the diagrams of the stance phase, only the reaction force on the foot by the ground is shown. There are, of course, forces due to the mass of the leg and to the inertia effects, but during stance these effects are relatively insignificant in relation to the ground reaction force and for the purpose of determination of muscle activity may be ignored.

Fig. 3.3   Heel strike
            Position of reaction force:
                Anterior to hip causing flexion
                moment

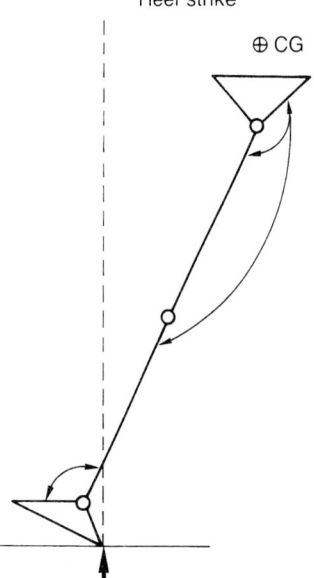

Fig. 3.3. Heel strike.

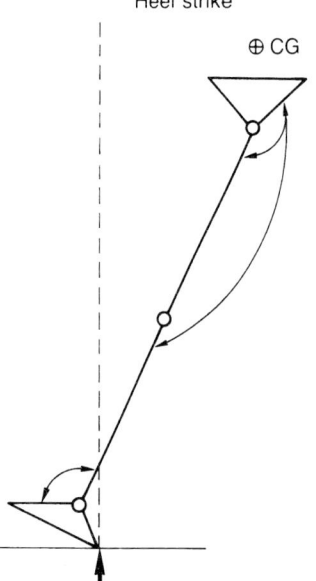

Fig. 3.5. Foot flat.

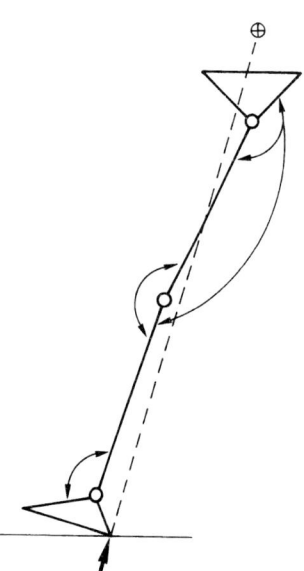

Fig. 3.4. Shortly after heel strike.

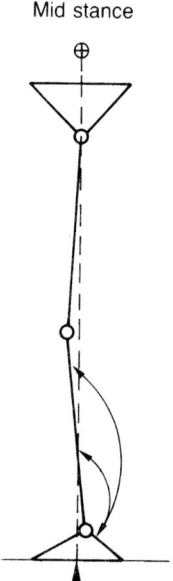

Fig. 3.6. Mid-stance.

Heel off

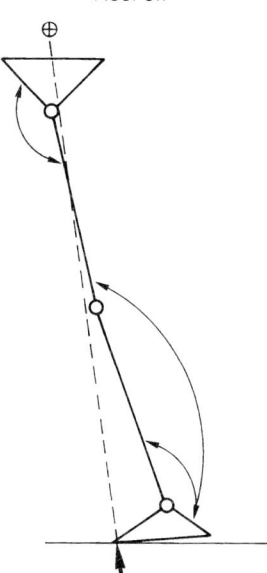

Fig. 3.7. Heel off.

Acceleration

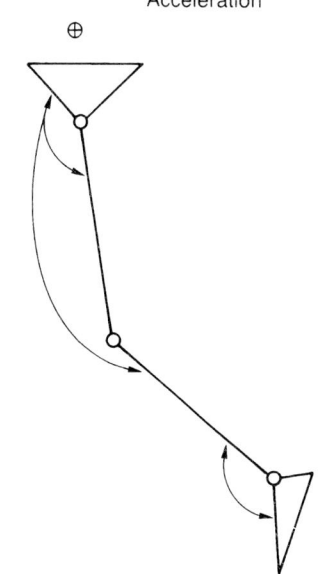

Fig. 3.9. Acceleration.

Toe-off

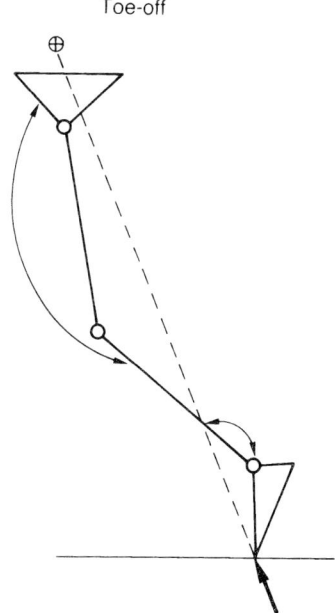

Fig. 3.8. Toe off.

Mid swing

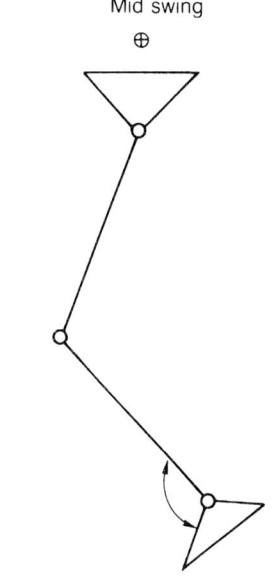

Fig. 3.10. Mid-swing.

Deceleration

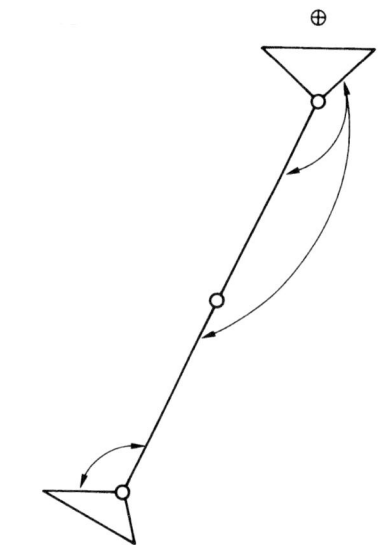

Fig. 3.11. Deceleration.

Anterior to knee causing extension
moment
Posterior to ankle causing
plantarflexion moment
Joint position and muscle activity:
Hip flexed to 25°; gluteus maximus
and hamstrings active in preventing
further flexion
Knee in full extension. Extension
moment overcome by action of
hamstrings which initiate flexion.
Ankle in neutral position,
beginning to plantarflex.
Plantarflexion controlled by action
of the pre-tibial muscles.

Fig. 3.4   Shortly after heel strike
Position of reaction force:
Anterior to hip causing flexion
moment
Posterior to knee causing flexion
moment
Posterior to ankle causing
plantarflexion moment
Joint position and muscle activity:
Hip held in 25° flexion by action
of gluteus maximus and
hamstrings.
Knee in 5° flexion, continuing to
flex

Rate of flexion controlled by
quadriceps
Ankle in 5° plantarflexion and
continuing to plantarflex under
control of the pretibial muscles.

Fig. 3.5   Foot flat
Position of reaction force:
Anterior to hip causing flexion
moment
Posterior to knee causing flexion
moment
Posterior to ankle causing
plantarflexion moment
Joint position and muscle activity:
Hip in 25° flexion beginning to
extend due to action of gluteus
maximus and hamstrings.
Knee reaches 15° flexion and
continues to flex to 20° shortly
after foot flat then begins to
extend. Quadriceps are active in
controlling flexion.
Ankle in 10° of plantarflexion.
Plantar flexion moment reduces as
reaction force moves towards the
joint centre and pretibial muscle
activity falls off.

Fig. 3.6   Mid stance
Position of reaction force:
Through hip joint, no moment
Posterior to knee causing flexion
moment
Anterior to ankle causing
dorsiflexion moment
Joint position and muscle activity:
Hip in 10° flexion, continuing to
extend
Knee reaches 10° flexion,
continues to extend. Quadriceps
action has fallen off, suspected that
soleus is active in controlling
flexion.
Ankle in 5° dorsiflexion and
continuing to dorsiflex controlled
by calf group of muscles.

Fig. 3.7   Heel off
Position of reaction force:
Posterior to hip causing extension
moment
Anterior to knee causing extension
moment

Anterior to ankle causing
dorsiflexion moment
Joint position and muscle activity:
Hip reaches 13° flexion then begins
to flex. Iliacus and psoas major are
active.
Ankle reaches 15° dorsiflexion then
plantarflexes with powerful
contraction of calf muscles.

Fig. 3.8   Toe off
Position of reaction force:
By toe-off the reaction force is
insignificant in magnitude as majority
of weight is on the other foot.
Joint position and muscle activity:
Hip in 10° extension contrives to
flex with activity of rectus femoris.
Knee flexed to 40° continuing to
flex.
Ankle has reached 20°
plantarflexion with contraction of
calf muscles which become
inactive immediately after toe-off.

Fig. 3.9   Acceleration
Joint position and muscle activity:
Hip in 10° extension and flexing as
hip flexors accelerate the limb
forward.
Knee in 40° flexion and continuing
to flex under pendulum action.
Ankle in 20° plantarflexion,
beginning to dorsiflex under action
of pre-tibial group of muscles.

Fig. 3.10   Mid-swing
Joint position and muscle activity:
Hip flexed to 20° and continuing to
flex
Knee reaches 65° flexion then
starts to extend under pendulum
action.
Ankle reaches neutral position and
held there by slight activity of the
pre-tibial muscles.

Fig. 3.11   Deceleration
Joint position and muscle activity:
Hip reaches 25° flexion and is
restrained by gluteus maximus and
the hamstrings.
Knee in full extension, restrained
by the hamstrings.
Ankle still held in neutral position
by action of pretibial muscles.

It is noted that at the end of deceleration the leg
is positioned for the following heel strike and all
the muscles necessary for counteracting the ground
reaction force are already active. The joint positions indicated in the above figures are of course
approximate and will vary from individual to
individual, with speed of walking and with terrain.
They are however typical of normal level walking.
The same technique of visualizing joint positions,
and force, actions is invaluable in analysing
different ranges of activities and conditions. Consideration of the above examples also helps to
explain the problems of amputee gait.

Amputee gait

It has already been said that the amputee has an
adapted gait which also involves an adaptation in
the use of the remaining musculature. In fact, the
amputee's gait is a process of controlling and
adjusting force patterns so that the forces and the
effort required of the musculo-skeletal system are
within his or her capability and the pressures on the
stump within acceptable limits. The object of
prosthetic design is to facilitate this process.

The problems of joint design are demonstrated
by the different situations encountered firstly
between mid-stance and heel off (Fig. 3.6–7) and
then between heel off and toe off (Fig. 3.7–8).
Throughout both periods the line of action of the
force passes anterior to the ankle tending to
dorsiflex it and as a consequence activity is
required of the calf muscles, the plantarflexors.
However, before heel-off the ankle is dorsiflexing
under the external force action whereas following
heel-off the ankle is plantarflexing. This represents
a significant difference in terms of prosthetic
replacement. Before heel-off the calf muscles are
active but are lengthening; after heel-off the
muscles are active and are shortening. In the
former situation the muscles are providing a
resistance to the movement which the ground
reaction is tending to produce (dorsiflexion due to
the moment of the force). This action may be
substituted in a prosthetic sense by a mechanism
which provides a resistance to dorsiflexion as
shown in Fig. 3.12 where dorsiflexion of the ankle
is resisted by the moment due to the force
generated by compression of the dorsiflexion
'bumper'. The situation after heel-off where the

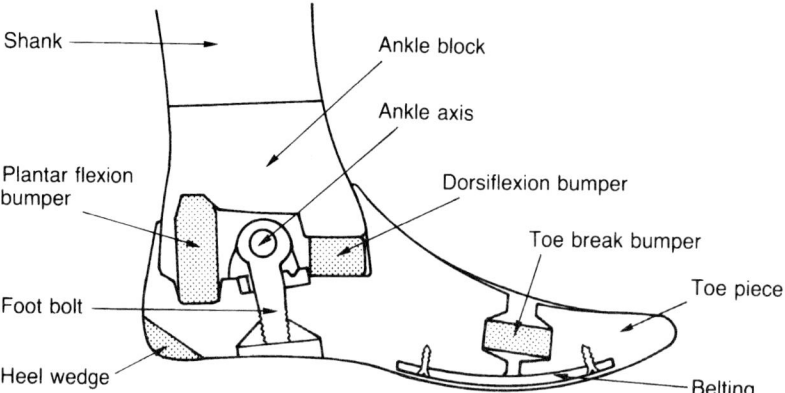

Shank

Ankle block

Ankle axis

Plantar flexion
bumper

Dorsiflexion bumper

Toe break bumper

Toe piece

Foot bolt

Heel wedge

Belting

Fig. 3.12. Conventional foot mechanism.

muscle is active but shortening is quite different in terms of prosthetic replacement. In the first part of this period the dorsiflexion 'bumper' has been compressed and therefore due to the stored energy may apply force while plantarflexion is taking place. However, beyond the neutral position, duplicating the situation where the muscle is active but shortening would require an external energy input. This can be visualized in Fig. 3.12 where to replace the action in question the dorsiflexion 'bumper' would require to lengthen beyond its uncompressed length while producing force. To date, no mechanism utilizing external power input to the prosthetic leg has evolved so, of the situations visualized above, one can to some extent be duplicated in the case of the resistive moment but not in the case of the 'active' moment. It can be seen from this that prosthetic joint mechanisms used at the ankle can only duplicate some of the lost function and so an adaptation in both gait and muscle activity will be required. Exactly the same logic may be applied to similar situations at the hip and knee. Resistive moments, that is where muscles would be active but lengthening, may to some extent be provided by prosthetic joint mechanisms but not 'active' moments which would require power input.

The line of action of the ground reaction force in relation to the joint is highly significant in the case of a prosthesis. Considering the situation at Foot-flat illustrated in Fig. 3.5, it can be seen that the ground reaction force passes posterior to the knee joint tending to flex it and flexion is taking place. This requires the activity of the quadriceps group.

This is a situation where, in a prosthesis, a resistive moment could be provided to duplicate this action, and indeed is in the case of so-called stabilized knees. However, a more common solution would be to arrange to alter the line of action of the force so that it passes just anterior to the knee and to provide the knee mechanism with an 'extension stop', which does not permit hyperextension, thus providing a stable situation. It should be noted that in biomechanical terms this is the same solution as the stabilized knee provides, where the force action passes posterior to the knee, a resistive moment is being provided at the joint to counteract the moment due to the ground reaction force. The extension stop is, however, a mechanically simpler solution in that it does not involve the complexity of providing a moment which resists flexion at parts of the cycle where this is required but not at others where it is not. Such a resistance would not be helpful, for example, in the phase between heel-off and toe-off.

The prosthetist in designing and aligning the prosthesis is attempting to provide an optimum force situation for the amputee in terms of his ability comfortably to exert and control the forces in the socket and in his musculo-skeletal system. Again the different aspects of the alignment process may be appreciated with reference to the situation at foot flat which was just considered (Fig. 3.5). The relationship of the line of action of the ground reaction force relative to the knee can obviously be changed physically by moving the knee joint in an antero-posterior direction. So if it is required to arrange for the reaction force to pass anterior to the knee, the knee may simply be moved in a posterior direction. However, referring to the figure it can be seen that if the foot were plantarflexed the point of application of the reaction force would be moved in an anterior direction and the line of action would therefore also tend to pass anterior to the knee. The skilled prosthetist in aligning the prosthesis must be aware of all of the biomechanical implications. The situation is not of course a simple one as different limb configurations will alter the amputee's gait and as a consequence the forces due to inertia and the resulting ground reaction. The amputee may also at any time alter the ground reaction force in magnitude or direction by adopting an altered gait pattern. A grossly abnormal gait which the amputee uses to relieve pain in a socket is a control of gravity, inertia and the ground reaction force to change the force pattern in the socket.

Socket design

The general procedure which is followed in designing any socket is the same. The areas which are available to provide the support function are identified and consideration is then given to the other forces which will be generated during

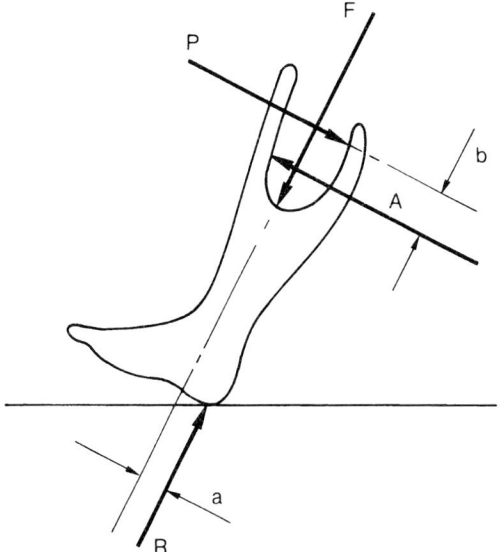

Fig. 3.13. Transtibial prosthesis – force actions.

walking and other normal activities. In essence, the 'joint' between prosthesis and stump is treated in biomechanical terms in the same way as an anatomical joint. However, the rotational effect of the ground reaction force within the socket is resisted, not by muscle activity but by force actions applied by the stump tissues. Figure 3.13, a simplified representation of a transtibial prosthesis shortly after heel strike, demonstrates the force effects which might be generated in a socket with end bearing. The mass of the prosthesis and inertia effects are ignored in this consideration.

From equilibrium conditions the end bearing load F must
equal the ground reaction R assumed to be parallel.
However, they are off-set by a distance, a.

There is therefore a turning effect or moment which is equal to Fa. Forces P and A also assumed to be parallel and from equilibrium requirements of equal magnitude must be generated and produce a turning effect to balance moment Fa. Hence Pb = Fa.

This is a simplified representation of a complex situation. It illustrates the main point, however, that although the socket must transmit the support force, the changing magnitude and line of action of the ground reaction force with respect to the socket will from equilibrium considerations lead to changing but predictable force patterns in the socket. It is the control of the application of these forces to the stump which will determine the comfort of the socket. The most important factor in this respect will be the avoidance of excessive pressure. Pressure is directly proportional to the force applied and indirectly proportional to the area of application. Thus every effort must be made to minimize force while at the same time maximizing the area over which the force is applied. It must also be recognized that some areas of the stump will tolerate pressures well while others will be sensitive. Successful prosthetic fitting embodies all of these concepts.

Frequently incorrect reference is made to the lever arm of the stump with confusion between the turning effects of the forces in the socket and the turning effects at the joint proximal to the socket. The above example (Fig. 3.13) should clarify this. In this situation which shows a transtibial amputee, shortly after heel strike, the line of action of the floor reaction force passes posterior to the knee tending to flex it. The flexion moment is resisted by the action of quadriceps whose lever arm is unaffected by the amputation and independent of stump length. On the other hand, the turning effect in the socket is resisted by the moment Pb and here the length of the stump is important as the greater the length, the greater can be dimension b and hence the lower the stabilizing force P. A longer stump length does therefore have two effects in regard to pressure application. Firstly, as shown above, it reduces the magnitude of the stabilizing forces but also because of its greater length provides a greater area for application of the reduced force. Both effects tend to reduce the pressure on the stump with a consequent effect on comfort.

This perhaps exemplifies the interaction between socket design and stump design. On the one hand, the greater length of stump may be seen to have a beneficial effect in regard to pressure reduction and hence comfort. On the other hand, there are other prosthetic considerations which will limit the extent to which length is desirable – it will be necessary to ensure adequate space is available for the prosthetic foot/ankle complex and excessive length with the typically slightly flexed socket of the transtibial prosthesis would lead to cosmetic problems due to the stump protruding in the postero-distal aspect of the calf.

The stump and socket are after all two halves of the one interface. The socket is designed to apply forces to the stump in the most comfortable way. However, it is essential that the stump is in the same way designed with a view to the forces to which it will be subject and to the optimization of this process. Stumps which are too short and inadequately padded, areas which are intolerant of pressure and muscles which cannot properly exert forces about the involved joint will all tend to reduce the effectiveness of the prosthetic solution.

Designing both the prosthesis and the stump involves the optimal use of presently available techniques. With a good stump and a well-designed prosthesis a good functional, cosmetic solution is possible. An essential element in the integrated process is a real understanding of the biomechanics.

Reference

Hughes, J. and Jacobs, N. (1979) Normal human locomotion. *Prosthet. Orthot. Int.*, **3**, 4–12.

4

Epidemiology

B. Ebskov and L. Ebskov

Lower limb amputation continues to represent a major health and socio-economic problem. Denmark with its 5 100 000 uniform population (ethnicity, education, religion etc.) is a well defined area to perform epidemiologic studies. The lifestyle, the typical population-related trends with an increasing fraction of elderly people, and the quality level of the health services is comparable to other developed western countries.

This chapter aims to present the nationwide epidemiology of lower limb amputation in Denmark for the period 1980–1990.

Materials and methods

The Danish National Health Board (NHB) established the National Patient Register (NPR) in 1976 and decreed a duty for all general hospitals to submit standardized information on all patients admitted. The reliability of the figures, which represent full national coverage, has been found adequate for epidemiological studies.

The Danish Amputation Register (DAR) was established in 1972 for the purpose of collecting data on major upper and lower extremity amputations in Denmark. In the first years the collecting of data was dependent on voluntary reports, but since 1978 the DAR has had access to NPR, and in this contribution information on amputation is based on data from NPR.

Diagnoses in the LPR are recorded according to the WHO International Classification of Diseases (ICD). Besides information on diagnosis, the following information is available in the NPR: surgical procedures (based upon the classification of the NHB); age and sex; information regarding hospital, department and geography; the date and type of admission and discharge including death.

Trends in lower limb amputation

Based upon the ICD codes, the lower limb amputations can be divided in a number of aetiology groups, i.e. non-diabetic vascular insufficiency; diabetes; trauma; malignant bone and soft tissue neoplasia; congenital malformation; infection and miscellaneous conditions (venous ulcer, pseudarthrosis, other neoplasms including metastases). Figure 4.1 shows the total number of amputations in 1980 and 1990 in the different aetiology groups. The total number of major amputations has decreased by about 27%. The non-diabetic vascular insufficiency (NDMVI) amputations and the diabetic amputations are responsible for the decrease. The percentage representing these aetiological groups shows a minor increase for the NDMVI amputations (from about 64% to 66%) and a more pronounced decrease for the diabetic amputations (from 27% to 22%). The NDMVI and the diabetic amputations represent about 89% of all amputations. The percentage for the other aetiology groups shows no systematical changes.

Corrections of the figures according to the increasing number of elderly people, as well as an

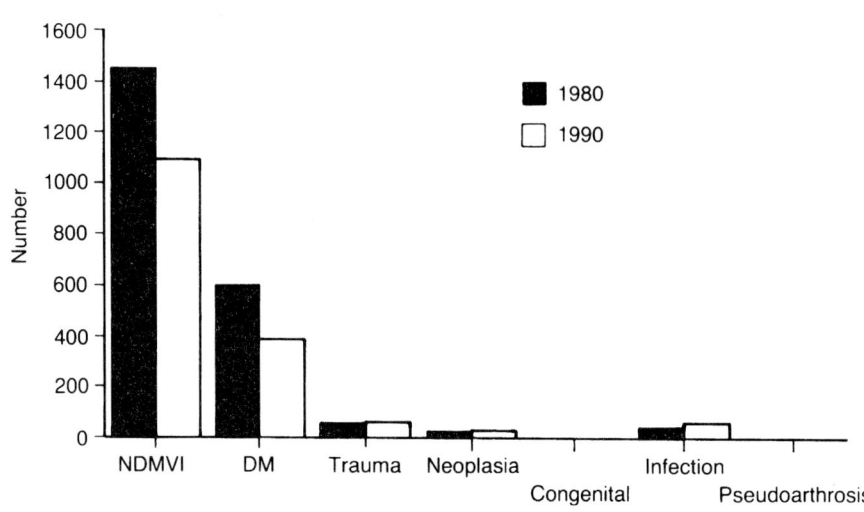

| 1980 | 1.446 | 601 | 61 | 31 | 1 | 50 | 1 |
| 1990 | 1.093 | 390 | 63 | 34 | 3 | 69 | 1 |

Aetiology

Fig. 4.1. Annual number of major lower limb amputations in 1980 and 1990. The amputations are divided in seven aetiological groups. Any amputations impossible to place in one of the seven specific aetiological groups are classified as miscellaneous (about 60 per year).

increase of the diabetic population (about 7% in the period under study) have not been attempted.

The epidemiology for the major amputation groups, i.e. NDMVI, DM, trauma and neoplasia, will be discussed in detail.

Non-diabetic vascular insufficiency amputations

The annual number of amputations is shown in Fig. 4.2. After a minor increase in the period 1980 to 1984 the number of amputations decreases significantly. A thorough discussion of this phenomenon can be found in Chapter 22.

The mean age for the NCO amputations is 73 years old, for females 76, and 72 for males, respectively. The overall sex ratio (male: female) is 1:0.7. Figure 4.3 shows the age distribution. The age-specific amputation rate was significantly higher for males in all age groups.

The intranosocomial mortality defined as death while staying in the department where the surgical intervention was carried out, is about 13%. The

relative mortality increases with age irrespective of sex from about 10% in the age group 60–69 to about 20% in the age group 80–89.

In addition, the mortality rate is related to the level of amputation, with 17% in the transfemoral (TF) group; 12% in the knee disarticulation (KD) group and 9% in the transtibial (TT) group.

Long-term survival is described in a separate section. Figure 4.4 shows the distribution of amputation level. A longitudinal study of the period 1980–1990 shows some systematic changes; the relative fraction of TF amputations seems to decrease, and the share of KD amputations increases from about 2% to 8%.

The average duration of stay in hospital is 46 days, but is dependent on the destination upon discharge: discharge to home, 60 days; to other department/hospital, 36 days and to nursing home 40 days.

Diabetic amputations

Diabetic amputations have been considered a serious health problem of increasing significance.

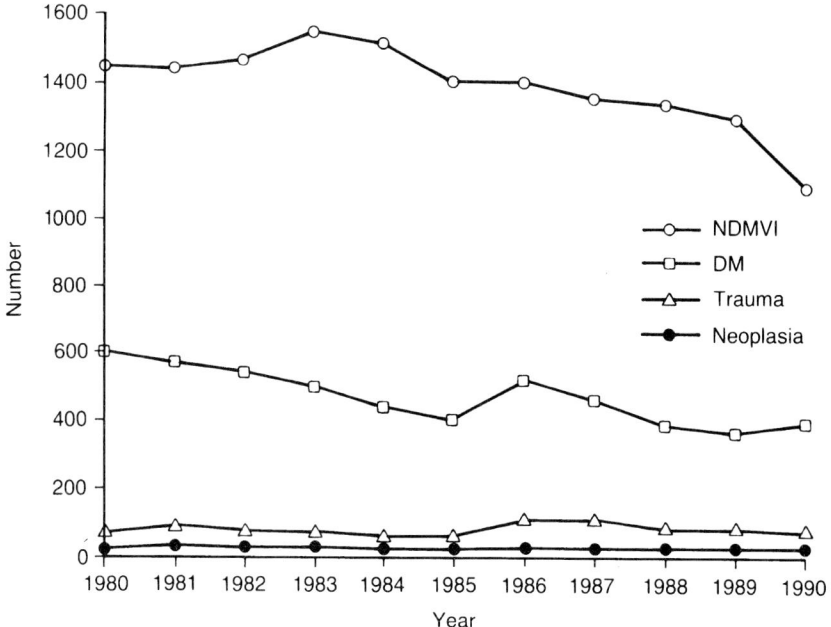

Fig. 4.2. Annual number of major lower limb amputations in the four dominating aetiological groups, i.e. non-diabetic vascular insufficiency (NDMVI), diabetes (DM), trauma and neoplasia, in the period 1980 to 1990.

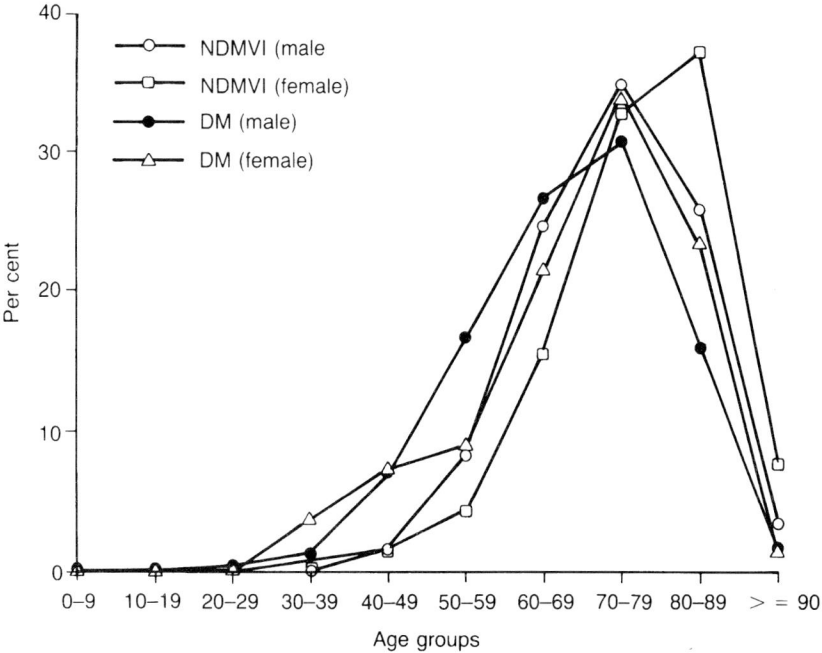

Fig. 4.3. Age distribution for the NDMVI and the DM amputations in relation to sex.

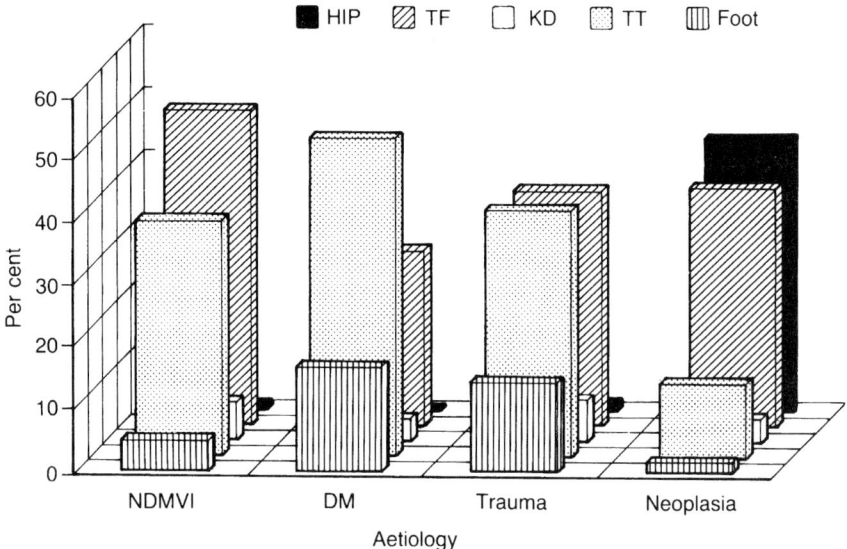

Fig. 4.4. Distribution of amputation level in the four dominating aetiological groups, i.e. NDMVI, DM, trauma and neoplasia.

It had been assumed previously that approximately 2% of the non-insulin dependent diabetics and about 12% of the insulin dependent diabetic patients will eventually undergo lower limb amputation. Pecoraro *et al.* (1990) describe a 'causal pathway to amputation' for diabetic patients: neuropathy → minor trauma → ulceration with or without infection → failure to heal because of vascular insufficiency → amputation.

Prevention of one or more of the steps in the causal pathway can reduce the risk of amputation. A considerable effort to prevent late diabetic complications, including amputation, has been made in Denmark. Besides improvements in the medical treatment, the introduction of vascular surgery (Chapter 22) and a well arranged and publicly supported podiatric service is considered to be of importance. The epidemiology of the diabetic lower limb amputation is probably influenced by these factors.

Figure 4.2 shows the number of diabetic amputations in Denmark for the period 1980–1990.

For the diabetic population there is a fairly steady decrease, with a tendency to level off during the last 2 years of the period. It should be noted that during the same period the diabetic population

has increased 6–8%, but no correction for this change has been attempted.

The mean age of the diabetic is about 71.3 years old. The mean age for males was 69.9 and for females 73. The age distribution is shown in Fig. 4.3. The male to female ratio was 1:0.9. Analysis of insulin dependent (IDDM) and non-insulin dependent (NIDDM) diabetics shows a similar pattern of sex and age.

The intranosocomial mortality is about 10%. Similar to the NDMVI amputations, the relative mortality increases with age irrespective of sex from about 6% in the age group 60–69 to about 15% in the age group 80–89.

Mortality and level of amputation is related. The mortality rate in the group amputated at foot level is 4.5%; for the TT group 9%; the KD group 14.3% and the TF group 18.2%.

Figure 4.4 shows the distribution of amputation level for the diabetic amputations. We have observed changes in the level of amputation in diabetic patients during the decennium 1980–1990 similar to the NDMVI amputations.

The number of hospital days for diabetic amputees decreases from 55 in 1980 to 37 in 1989.

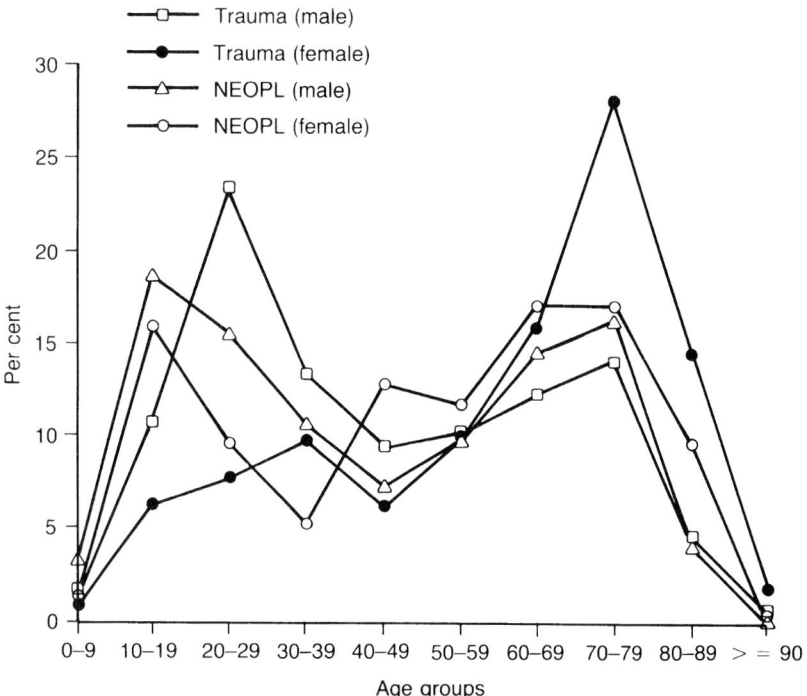

Fig. 4.5. Age distribution for the trauma and the neoplasia amputations in relation to sex.

Traumatic amputations

In the period 1980–1990 about 750 traumatic lower limb amputations (1.4 per 100 000 population per year), have been recorded in Denmark. Figure 4.2 shows the annual number of amputations.

The mean age was 49.4 years, for males 44.8 and for females 58.8. Figure 4.5 shows the age distribution. The male to female ratio is 2:1.

The average of stay in hospital was 45.8 days for men and 54.5 days for women.

Figure 4.4 shows the distribution of amputation level. Sex and level of amputation (TT and TF) are significantly related. The only apparently systematic change in the distribution of level during the period 1980–1990 is the increase of the knee disarticulation (KD) group from below 5% to about 10% by the end of the period.

The male group has a significantly lower in-hospital mortality (3.8%) than the female group (6.3%). The in-hospital mortality is significantly related to the level of amputation; 75% of all the

amputees who died during the hospitalization period were amputated at the TF level.

Neoplasia amputations

Malignant bone and soft-tissue neoplasia are a relatively infrequent cause of amputation (Fig. 4.1). Figure 4.2 shows the annual number of amputations in the period 1980–1990.

The mean age is 47.6 years, for females 50.6 and 45.4 for males. The sex ratio male to female is 1:0.75.

The age distribution is seen in Fig. 4.5.

The average stay in hospital is about 35 days with no difference between male and female.

The distribution of level is seen in Fig. 4.4. The level of amputation and in-hospital mortality is related; 90% of the in-hospital mortality is found in the TF group (including hip disarticulation and trans-pelvic amputation). The in-hospital, mortality is about 4% with no difference between males and females. Sixty-eight percent of the amputees were discharged to home, the remaining 32% were discharged to another hospital or department, nursing home etc. or died during hospitalization (4%).

Long-term mortality

All Danes (8691 patients) who underwent a primary amputation for NDMVI (6745) or diabetes (1946) between 1982 and 1987 were followed through 1989, and all deaths during this period were recorded for the purpose of establishing long-term survival rates.

The mortality of arteriosclerotic and diabetic amputees depends upon age at primary amputation, the level of amputation and the aetiology. The mortality increases significantly with increasing age at the primary amputation, irrespective of level of amputation.

The average annual death rate increases with more proximal levels of amputation irrespective of the age at primary amputation. The hazard is significantly higher in amputees with concomitant diabetes.

The average annual death rate for a 73-year-old Danish male is 5% whereas the corresponding death rate for an arteriosclerotic TF amputee is 19.1%, and for a diabetic TF amputee 25.4%.

The mortality rises sharply with time; thus only one out of five (20%) diabetic amputees, and two out of five (38%) NDMVI amputees are alive 5 years after the primary amputation.

Risk of reamputation

In a study from the year 1980, 2029 primary amputees were followed over 48 months in order to analyse the risk of ipsilateral reamputation and contralateral amputation. The computations were carried out for NDMVI and diabetic amputees individually, but since no statistically significant differences were found the results are given for the entire group.

Ipsilateral reamputation was carried out on 10.4, 14.8 and 16.5% of all amputees within 1, 2 and 3 months, respectively, after the primary amputation on the same limb. After 6 months almost one out of five had been reamputated, but extremely few reamputations were carried out at a later stage during the period of observation; in other words, if the limb does 'survive' the first half year after the primary amputation, it stands a very good chance

of permanent survival. This is not so with contralateral amputation, which was recorded in 11.9, 17.8, 27.2 and 44.3% of the primary amputees after 1, 2, 3 and 4 years, respectively.

One might say that there is an ongoing 'race' between ipsilateral reamputation, contralateral amputation and death, in the sense that 4 years after primary amputation only 20% of all amputees were still alive without any further amputation having been performed.

Future perspectives

Presently work is going on to generate a model which will permit us to approximate the total population of lower extremity amputees alive in Denmark. This seems realistic since we know the annual 'input', i.e. the number and category of primary amputations, and further the 'loss', i.e. the mortality over a rather long span of time. This model may be utilized to prognosticate the probable future need for amputation related resources, i.e. hospital services, rehabilitation facilities, prostheses, wheelchairs, social services and hence the economical demands as well as future education needs within the relevant vocations. An annual update is quite feasible and would serve the dual purpose of validating the model and adjusting the ensuing prognoses of 'needs' in the widest sense of the word. Furthermore, such a model would presumably yield a tool for cost-benefit analyses, e.g. regarding the profitability of vascular surgery and other activities aimed at reducing the risk of amputation.

Acknowledgement

The present study was generously supported by grants from the Krista and Viggo Petersen Foundation, for which the authors are deeply grateful.

References

Ebskov, B. (1977) Fruhergebnisse des Danischen Amputations-registers. *Orthopadische Praxis*, **13**, 430–433.

Ebskov, B. (1986) The Danish Amputation Register 1972–1984. *Prosthet. Orthot. Int.*, **10**, 40–42.

Ebskov, L. (1991a) Lower extremity amputations for vascular insufficiency. *Int. J. Rehab. Res.*, **14,** 59–64.

Ebskov, L. (1991b) Epidemiology of lower limb amputations in diabetics in Denmark (1980 to 1989). *Int. Orthop.*, **15,** 285–288.

Ebskov, L. (1992) Level of lower extremity amputation in relation to etiology. *Prosthet. Orthot. Int.*, **16,** 163–167.

Ebskov, L. and Ebskov, B. (1989) Mortality following primary major amputation on the lower extremity. *Proceedings ISPO VI World Congress*, 12–17 Nov. 1989. Kobe, Japan.

Ebskov, B. and Josephsen, P. (1980) Incidence of reamputation and death after gangrene of the lower extremity. *Prosthet. Orthot. Int.*, **4,** 77–80.

Ebskov, L. (1993) Major amputations for malignant melanoma. *J Surg Oncol.*, **52**, 89–91.

Pecoraro, R.E., Reiber, G.E. and Burgess, E.M. (1990) Pathways to diabetic limb amputations. *Diabetes Care*, **5,** 11.

5

Peri-operative care

A. S. Jain

The role of meticulous surgery in the speedy and effective rehabilitation of the amputee is unquestionable but all other aspects of care such as patient assessment, level selection in the pre-operative phase, positioning of the patient on the table and establishing the stump environment at the time of the operation; and, post-operatively the execution of a carefully coordinated plan by the clinic team are of significant importance.

Pre-operative care

The *first imperative* is to determine the level of amputation which will serve the patient best. The key decision centres on the knee joint – should the level be a transfemoral (TF), a knee disarticulation (KD) or a transtibial amputation (TT).

The consensus view, strengthening as time has passed, is that salvage of the knee joint is of prime importance for a successful rehabilitation (Pedersen, 1968; Murdoch *et al.*, 1988). Despite this advocacy there remains a disappointingly high proportion of transfemoral in contrast to transtibial amputation world wide. The advantages of the transtibial procedure is well described in the literature. Waters *et al.* (1976) assessed the energy consumption and demonstrated that the more proximal the amputation the more energy required to walk and, of course, significantly more than a normal, able individual.

Anderson *et al.* (1967) were able to demonstrate that the transtibial amputee appeared to be more physically active than the transfemoral amputee. Other advantages were quoted:

(a) Improved balance.
(b) Ease of donning and doffing of the prosthesis.
(c) The bony contours demand and ensure accurate fitting.

Level selection

Clinical assessment

Assessment in the office, consulting room or hospital ward in the case of the possible candidate or declared amputee is essentially no different from any assessment in other medical circumstances but, it has to be said, there is an extra dimension and significant considerations in respect of the dysvascular patient. These include:

(a) the skin condition;
(b) joint contracture;
(c) patient's general condition;
(d) patient's mental state;
(e) personal factors, e.g. age, sex, occupation, hobbies, etc.

These considerations are used with a view to setting the goals for the future rehabilitation of the patient in the light of the level selection. If a patient is clearly mentally and physically incapable of walking with a prosthesis then a decision may be taken to perform a transfemoral amputation irrespective of other indications. Such a patient will require a wheelchair. In that circumstance a transtibial amputation can cause problems with the development of a flexion contracture. Skin conditions such as ulceration, eczematous rash, scarring, skin grafts, etc. may serve to dictate the level of amputation for a successful prosthetic management.

Ancillary methods of assessment

Selection of the level of amputation in the case of tumour is usually determined by the state of the pathological tissue, the nature of the tumour and other indicators of malignancy including radiography, magnetic resonance imaging, biopsy, computer aided tomography and various biological investigations.

Similar circumstances exist in the case of infection as the main causal condition requiring amputation, mainly based on bacteriological investigations. The majority of lower limb amputations in the industrial nations are performed for peripheral vascular disease. A variety of ancillary investigations have been developed over the past 20 years for patient assessment and level selection. These investigations vary considerably in their value and several are continuing to be developed.

Detailed information can be found elsewhere in this volume (Chapter 20).

In Dundee our practice allied to clinical assessment is based on 27 years of experience and is as follows:

(i)   Systolic pressure measurements.
(ii) Skin blood flow measurement.
(iii) Infrared thermography.
(iv) Where arteriograms have been obtained, to assess the possible value of vascular surgery, the presence of a patent profunda femoris artery makes a successful transtibial amputation more likely.

Systolic blood pressure measurements

The use of the Doppler technique for estimating systolic blood pressure measurements (Fig. 5.1) has gradually become widely accepted since first adopted (Dean *et al.*, 1975) as a standard method for the non-invasive evaluation of the patient with peripheral vascular disease. While less than precise at other levels, a blood pressure measurement in the thigh of over 70 mmHg strongly suggests that a successful transtibial amputation is likely (Spence *et al.*, 1988).

Skin blood flow measurement

A quantitative measurement of skin blood flow at the precise level of operation is an added asset to

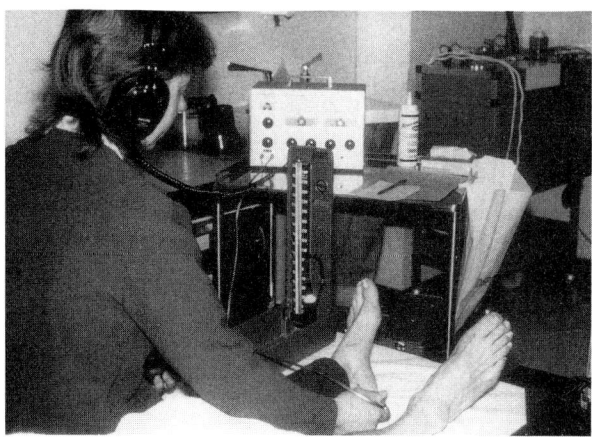

Fig. 5.1. The Doppler technique.

predict wound healing. ^{125}I–4 iodoantipyrine has been used for this purpose and has been shown to provide a good indication of tissue viability, especially below the knee (Spence and Walker, 1984).

Infrared thermography

Infrared thermography has been used in the assessment of peripheral vascular disease for the last 25 years and was suggested as a possible means of selecting amputation levels by Lloyd-Williams (1964). Recently, high resolution digital systems became available and when linked to a microcomputer the outcome from these systems can be quantified. It is now possible to obtain an accurate calibrated thermal map of the limb which provides highly acceptable assessment of an ischaemic limb. The great advantage of the ther-

mographic mapping of the skin is the ability to design the skin flaps (McCollum et al., 1985), thus obtaining good healing and at the same time preserving the knee joint (Fig. 5.2).

Clinical management

The aim of pre-operative management is to prepare the patient in the best way for surgery to reduce morbidity and mortality. While the level is often pre-determined in trauma or tumour cases, in the case of dysvascularity with or without diabetes the surgeon's skills, knowledge and experience, albeit with the aid of ancillary methods of assessment, can improve the TF:TT ratio (Murdoch et al., 1988).

Over 80% of amputations in industrial societies are performed for peripheral vascular disease with or without diabetes. The majority of these patients

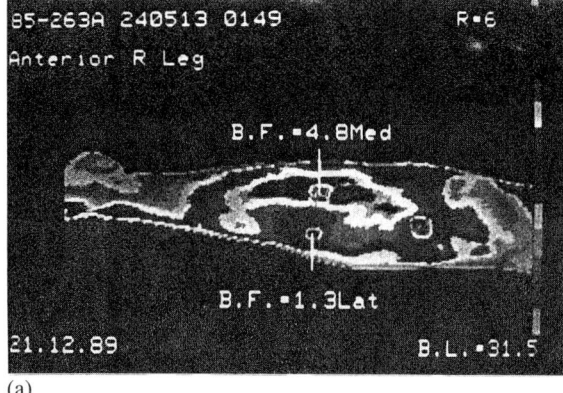

(a)

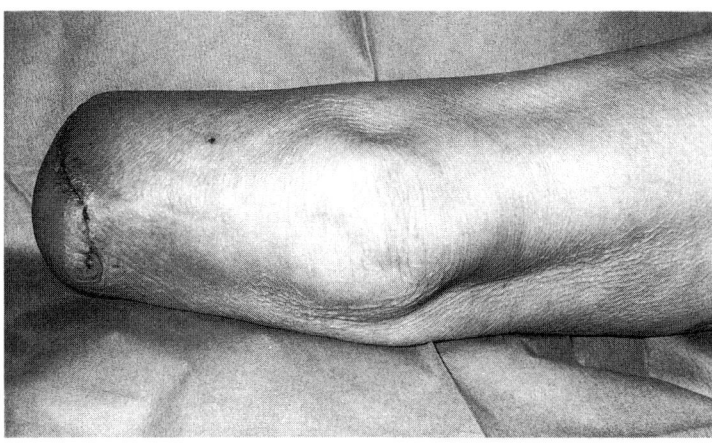

(b)

Fig. 5.2. (a) Medial to lateral thermal gradient. (b) Healed stump and medially based flap.

also suffer from concurrent medical problems and these increase the morbidity and mortality if not managed adequately. A review of the literature suggests that mortality varies from under 2% to 30% within 1 year of surgery. In most cases surgery is not required immediately and patients should be prepared at least 48 h prior to surgery. Every case should be considered carefully on its own merits; however, the management can be divided into two categories:

(a) *General*: medical, physical and mental preparation.
(b) *Local*.

General medical management

It is important to optimize the patient's general condition. This involves careful comprehensive assessment including biochemical and bacteriological analyses. The medical condition in these patients leads to disturbed blood chemistry which requires appropriate treatment.

Conditions such as heart failure, chest infection, diabetes, hypertension and urinary tract infection require adequate treatment to reduce the surgical risk.

Recent work has also indicated that blood chemistry estimations such as blood viscosity, plasma protein and white cell count can give some indication of wound healing. A serum albumin level above 3.5 g/dl and a total leucocyte count (TLC) over 1500 per mm^3 appear to predict satisfactory wound healing. The prevention of infection is always an important factor and it is no different in amputation surgery.

Appropriate prophylactic antibiotic cover should be provided. Three-dose broad spectrum antibiotic cover is most commonly used in surgical practice. Additionally, penicillin pre-operatively is used for prophylaxis for clostridium infection.

As some of these patients are at high risk of developing deep venous thrombosis, prophylactic cover should be provided by one or other form of heparin therapy.

Physical preparation

All patients who are to become amputees require sufficient upper body strength to use crutches, sticks and even wheelchairs. For this purpose it is important that where possible they be geared physically for all of these activities. Donning and doffing of a prosthesis requires extra energy.

Mental preparation

Counselling has an important role in amputation surgery. In the toxic, dysvascular patient it is often difficult to counsel at the pre-operative stage, but in cases of tumour, trauma, infection and other conditions patients must be counselled with sensitivity and care.

In dysvascular patients it is the relatives who most require counselling. Apprehension, anxiety and ignorance lead to psychological problems in the amputee and the relatives. It is the task of the clinic team who look after the amputee to reduce these problems; communication and assurance helps in this regard. Counselling can be done in various ways.

One valuable approach in dealing with this problem, especially in cases of tumour and infection is 'matching patient counselling'. This is counselling by other patients of the same sex and age group who have undergone the same level of amputation.

Local Factors

Important local factors which influence surgery include:

extent of gangrene;
degree of infection;
condition of adjacent tissues;
degree of arterial impairment;
severity of pain;
line of demarcation;
joint contractures, etc.

Preparation of the limb is important as many of patients have ulcers, discharging sinuses or frank gangrene. The operative area should be cleaned twice a day for 48 h prior to surgery with appropriate antiseptic solutions such as colloidal iodine or Hibitane and then a bandage impregnated with the antiseptic solution is applied. Various publications have shown that pre-operative shaving of hair increases the risk of infection (Seropian, 1962; Cruse, 1973; Hamilton, 1977; Court-Brown,

1981). The consensus of opinion is clearly that pre-operative shaving is harmful. However, if a surgeon feels very strongly that it is preferable then shaving should be carried out immediately prior to surgery. Any discharging sinus or ulcer should be isolated and encapsulated in a polythene bag to avoid any contamination.

Intra-operative management

Anaesthesia

Little has been written by amputating surgeons regarding anaesthesia for amputation. The various types of anaesthesia available to the amputating surgeon are the same as for any surgeon, i.e. local, regional, spinal and general anaesthesia. In summary, the choice of anaesthesia depends on the condition of the patient, the available special skills and expertise of the anaesthetist and, where applicable, patient choice. However, low spinal anaesthesia seems to give good post-operative pain relief and less morbidity (Mann and Bisset, 1983).

Positioning of patient

The author prefers to carry out all his amputations with the patient supine. A sandbag is placed under the limb proximal to the site of operation. Some surgeons prefer to carry out the disarticulation of the knee procedure with the patient in the prone position; however, the choice of positioning of the patient on the table is dictated by the experience of the surgeon, comfort and common sense.

Tourniquet

No tourniquet should be used in patients with peripheral vascular disease. In other conditions a pneumatic tourniquet should be used. It is placed around the thigh except in high above-knee amputations where it is not feasible.

The tourniquet should be inflated after the limb has been exsanguinated with an Esmarch bandage, except in the case of tumour involvement.

General principles of surgery

Meticulous surgery plays an important part in wound healing, especially with respect to tissue viability.

The tissues should be handled very carefully and there should be minimal trauma by instrumentation and retraction. It must be emphasized that the surgery should be carried out through remaining viable tissue.

Skin

The skin flaps should be mapped out prior to surgery to ensure adequate skin coverage. It is preferable to have longer flaps to begin with rather than shorter flaps. Extra skin can be trimmed at the end of the surgery if need be.

Bone

Cutting of bone is also important, especially in the transtibial amputation. Many surgeons use electric saws to cut the bone which always produces some bone necrosis. One way to avoid bone necrosis is to use a Gigli saw. Once the bone is cut it should be smoothed or sculptured to avoid rough edges.

Muscles

Muscles carry blood from the deep arterial supply to the skin. It is generally accepted that preservation of muscle is important and it should not be dissected from the deep fascia or the skin. Muscles also provide padding for the bone and an anchorage for myodesis. Sometimes if the stump is bulky those muscles that do not contribute can be excised and filleted.

The surgeon's first and main objective is to produce a stump which will proceed to primary uncomplicated wound healing. The stump should be capable of good muscular control and be able to withstand the pressures experienced at the socket/stump interface. Stump control is usually achieved by total motor function of the muscles and proprioceptive feedback coming from these functioning muscles. To achieve this the techniques of myoplasty and myodesis are employed.

Nerves

Various methods have been described for handling nerves and probably every surgeon has his own. The important factor in nerve handling is to prevent painful neuroma formation. The consensus of opinion is that the nerve should be pulled down gently and divided and ligated so it can retract into the underlying muscles, thus preventing the migration of axons and bleeding from the central vessel.

Vessels

There is very little which can be added regarding vessels in amputation surgery apart from the need to secure ligation.

Drainage

Drainage is an important part of the surgery to prevent haematoma formation, thus helping in the prevention of infection and oedema formation.

Post-operative care

The aim of post-operative care is integrated rehabilitation of the amputee to the optimum functional level. The important factors in post-operative management are the control of pain, swelling and infection. In terms of patient management the activity can be seen to consist of general care and stump care.

General care

(i)   Control of pain.
(ii) Prevention of oedema.
(iii) Prevention of infection.
(iv) Prevention of deep venous thrombosis.
(v)   Care of concurrent medical problems.
(vi) Biochemical monitoring.

Control of pain following surgery is vitally important for the patient's comfort and in gaining patient confidence. Spinal anaesthesia helps in this respect for several hours following surgery. The patient should be adequately covered for analge-

sics for pain for at least 48 h after operation. This will make the patient more comfortable and cooperative for the rehabilitation process. If surgery is carried out properly then patients with rest pain prior to surgery will require very little post-operative analgesia. However, patients who require amputation for trauma or tumour and infection will require narcotic administration for pain control.

A recent approach is the placement of a small silastic catheter at the time of surgery within or alongside the nerve (Malawer *et al.*, 1991). The catheter is led out of the cast or the dressing so that a local anaesthesia can be administered by a standard infusion pump at the rate of 2–3 ml/h over the first 72-h period. Most patients are managed by an appropriate dose of narcotic administration, usually every 3–6 h or on demand. This is more convenient, suitable and easy to manage.

For *control of infection* a three-dose intravenous antibiotic regime should be sufficient for most cases unless infection is the over-riding factor in the amputation. A broad spectrum antibiotic should be used for cover.

For *prevention of deep venous thrombosis* in high risk patients a 5-day course of subcutaneous heparin seems to reduce the incidence.

The next task is to make the patient as mobile as soon as possible. This requires adequate *care of concurrent medical conditions* and improvement in upper body strength. In the past, post-surgical patients were often kept in bed rest for prolonged periods, with significant deconditioning occurring after a few days. This situation has changed dramatically and now the aim is to encourage patients to be up and about as soon as possible. From the first post-operative day the patient should start by sitting out of bed with the stump elevated to the level of the chair seat. By the next day, the patient should attend the physical therapy department for ambulation in parallel bars and for upper body exercises.

The greatest impetus to early mobilization was the introduction of the immediate post-operative fitting of the prosthesis (IPOF) (Burgess and Romano, 1968).

There was a time when these devices were fitted in theatre but now it has been sensibly modified and the consensus is that mobilization on any device should be avoided for at least 7–10 days to allow the tissues to settle and the wound to recover from surgery. In this regime as soon as the patient is able and the wound is satisfactory, early mobilization should commence with the help of a walking aid.

Ensure that communication with the members of the clinic team is at its best by supplying appropriate aids such as hearing aids and properly prescribed spectacles.

Biochemical monitoring should be maintained. It may be that at this stage diabetes can be confirmed by plasma glucose levels and glucose tolerance testing. The known diabetic requires special care – the absence of an infected foot and the increasing post-operative activity and general well-being cause marked changes in plasma glucose levels, thus altering the need for insulin and oral hypoglycaemic drugs. Frequently, insulin is required at the time of surgery, but it is usually possible to change to oral hypoglycaemic drugs during the post-operative phase.

Whilst blood transfusion may have been contra-indicated in the preoperative phase, it may now become necessary. Even a moderate drop in haemoglobin level induces tiredness and lethargy, which interfere with the patient's rehabilitation.

Stump care

Like the wound of any other surgery all fresh amputees require wound protection of some sort. Stump environment has been the subject of considerable debate. The ideal goal is to achieve a comfortable, effective, functional, stable stump which will be able to cope with the socket of the prosthesis.

Pain and swelling are factors which delay the production of such a stump. Swelling is usually the main reason and any system designed or technique applied has to take this factor into account. Ideally, the stump environment should provide relief of pain, reduction of swelling and prevention of infection.

The various methods available are:

1. soft dressing;
2. rigid dressing;
3. controlled environment treatment;
4. stump shrinkers;
5. bandaging.

Soft dressing

The advantage of a soft dressing is that it allows easy access to the wound for inspection and for regular motion of the stump. It does not offer any protection to the wound from trauma, nor does it prevent joint contractures. The use of soft dressings has been associated with delay in prosthetic fitting. The only situation where a simple soft dressing is used is in the transfemoral amputation because of difficulty in applying a rigid dressing. A simple elastoplast adhesive dressing used in vertical strips without encircling bandaging is sufficient for the transfemoral amputation (Fig. 5.3).

Rigid dressing

Rigid dressing in the form of a plaster cast was advocated by Burgess *et al.* (1969) and subsequently evaluated by Mooney *et al.* (1971), and still remains the most favoured (Fig. 5.4).

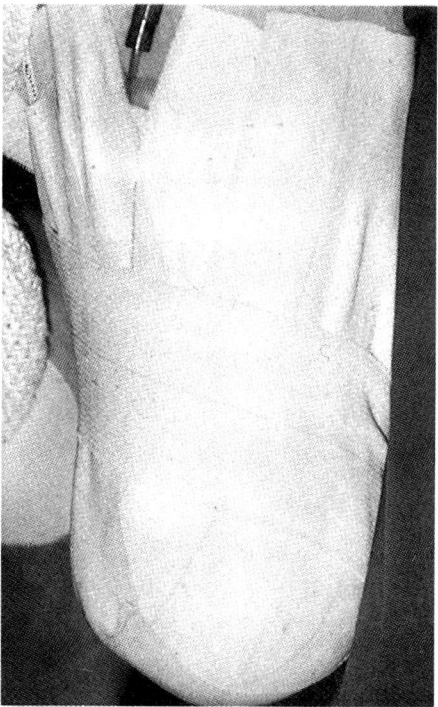

Fig. 5.3. An elastoplast, adhesive strip dressing provides an above-knee stump with protection and support, without producing an adverse pressure gradient.

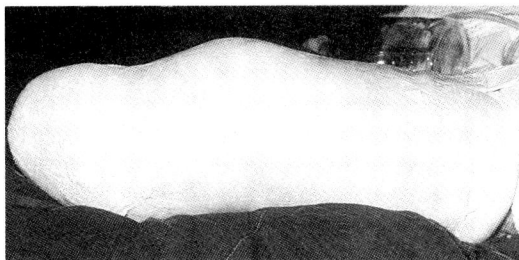

Fig. 5.4. Shows rigid (plaster-cast) dressing for below-knee stumps.

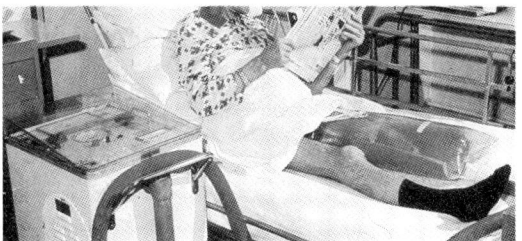

Fig. 5.5. Rapid healing of stumps below the knee is possible using a sterile, controlled-environment air dressing.

Experience suggests that the following are the main advantages of this method:

(a) Pain is diminished because the stump is not handled for repetitive bandaging and is well protected.
(b) Oedema is prevented.
(c) Wounds heal more quickly.
(d) The risk of infection is minimized because the rigid dressing is applied in theatre under aseptic conditions.
(e) The patient is free to move about and can begin early walking training using such aids as the pneumatic amputation mobility aid (PPAM-aid) or a Tulip device or where applicable, an 'immediate post-operative fitting' regime.

A rigid plaster cast dressing is ideal but care is essential in its application. However, the technique can be learned with 1 or 2-h practice. Sutures are not removed until 3 weeks post-operatively following knee disarticulation or transtibial amputations. Elderly patients have poor skin nutrition and require this length of time for a sound collagen bond to develop.

For stumps generated by ankle disarticulation (Syme amputation), a rigid dressing is again ideal as it assists in maintaining an optimum position for the heel pad.

Controlled environment treatment

This form of treatment was developed by Redhead and Snowdon (1978). It consists of an air dressing which creates an 'ideal environment' to obtain rapid healing of the stump (Fig. 5.5).

The stump is enclosed in a *sterile* transparent plastic container with a control system. This treatment is used for 5–7 days and thereafter a rigid dressing is applied. Experience shows that the use of this treatment is fast decreasing.

The ventilated compression system (Flowtron Air Treatment) (Jain, 1992) utilizes the principle of ventilated intermittent compression to reduce oedema and assist in recovery of the stump (Fig. 5.6). Flowtron Air comprises a PVC garment with

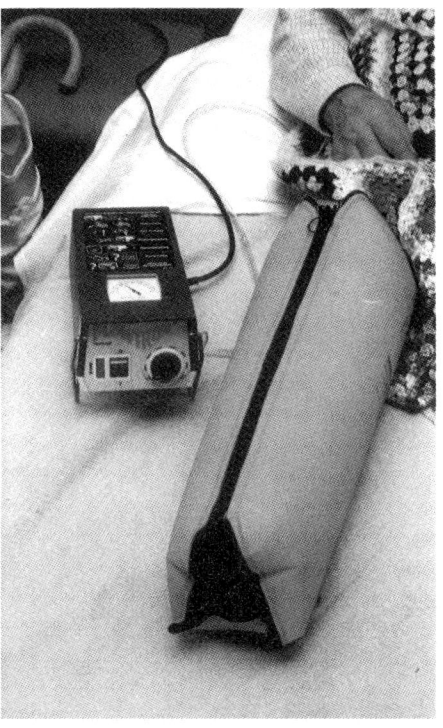

Fig. 5.6. Once the amputation wound has healed, recovery of below-knee stumps can be encouraged using a Flowtron air system. Alternating inflation and deflation cycles encourage fluid movement and joint mobility.

inner and outer chambers, an air pump, providing alternating inflation and deflation cycles to encourage fluid movement and joint mobility. A constant flow of filtered air (particle filter) reduces sweating and resulting tissue maceration. The advantages of the system are:

(a) Helps in wound recovery.
(b) Reduces tissue oedema.
(c) Relieves pain.
(d) Allows joint mobility.
(e) Helps in reduction of flexion contracture of knee joint.

The disadvantage is that it is not a sterile system so it can only be used after the wound has healed or over a dressing.

Two-way stretch amputation stump shrinkers

Recently these stump shrinkers have been used in place of bandaging to control stump oedema (Fig. 5.7). It is essential in using this system to have available the whole range of sizes and to use the appropriate size of both dimensions, length and circumference, otherwise the top of the shrinker can curl over and lead to distal congestion.

It should also be emphasized that as the swelling reduces the size of the shrinkers should be changed accordingly.

The shrinkers appear to have some advantages over bandages:

(a) Provide good skin tolerance for the sensitive post-operative stump.
(b) Provide pressure gradient from distal end of the stump proximally.
(c) Elasticity allows changes to be tolerated without discomfort.

Bandaging

Bandaging of the stump should only be instituted after the wound is securely healed (Fig. 5.8). There can be no doubt that a properly applied bandage is beneficial, provided the pressure produced does not exceed intravascular hydrostatic pressure and in particular does not produce an adverse pressure gradient. This can only be done by trained and experienced staff who are aware of the risk involved. Later the patient or a caring relative can be trained to undertake bandaging once the stump is more stable. Bandaging helps to reduce the stump volume and maintain a good stump shape.

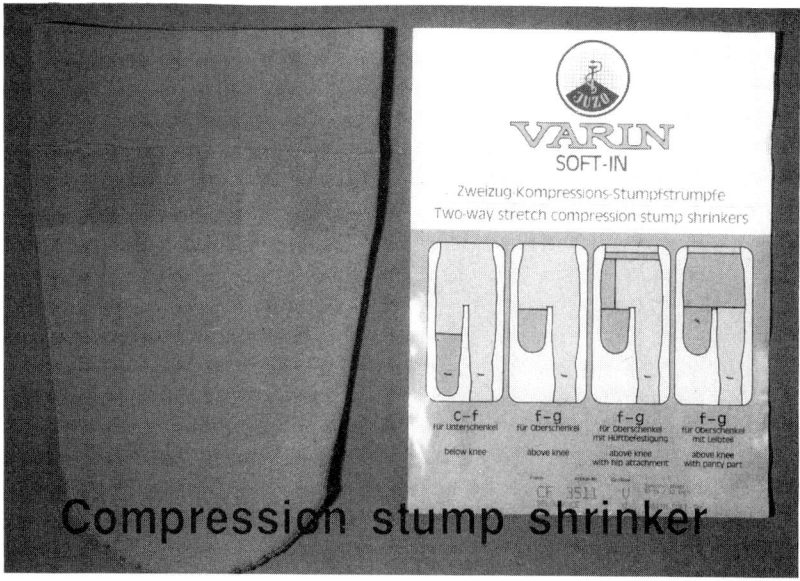

Fig. 5.7. Stump shrinkers.

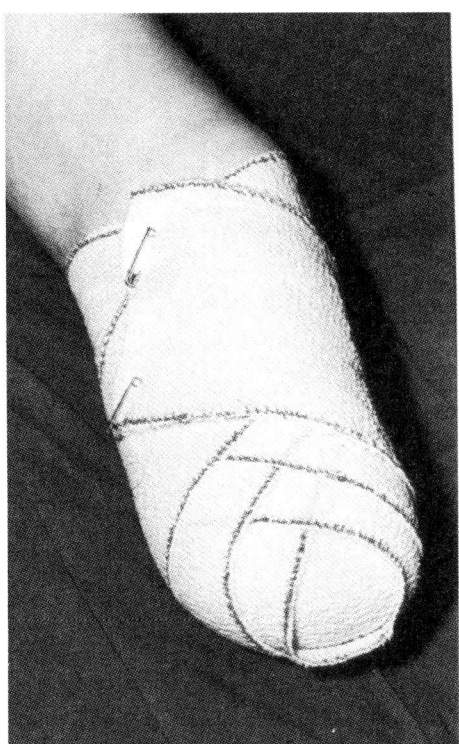

Fig. 5.8. Properly applied bandage on a below-knee stump.

Management of overall locomotor system

People who have lost one or both legs rely considerably on their upper limbs. The importance of regular coordinated exercises to improve the muscle tone, power and stamina cannot be over-emphasized. It also has subsidiary benefits such as improved respiratory function and overall circulation. Early ambulation not only requires good upper limb function, but clearly good control and strength of the stump.

The physiotherapist works with the patient from the first post-operative day and directs attention to both these factors. Therapists, both physical and occupational, play an important role in training the patient how to sit on a chair, to rise from a chair, to rise from the floor, to ascend and descend stairs, to take off and put on the artificial leg and to perform other related activities.

Continuing care

The patient's progress should be closely monitored on return home and arrangements made to check stump volume, socket fit, any possible change in prosthetic prescription and above all establish lines of communication should the patient need help at short notice.

References

Anderson, A.D., Cummings, V., Levine, S.L. *et al.* (1967) The use of lower extremity prosthetic limbs by elderly patients. *Arch. Phys. Med. Rehabil.*, **48**, 533.

Burgess, E.M. and Romano, R.L. (1968) The management of lower extremity amputees using immediate post-operative prostheses. *Clin. Orthop.*, **57**, 137–146.

Burgess, E.M., Romano, R.L. and Zettl, J. (1969) *Management of Lower Extremity Amputations: TRIO-6.* US Government Printing Office, Washington, D.C.

Court-Brown, C.N. (1981) Preop skin depilation and its effect on post-op wound infection. *J. R. Coll. Surg. Edinb.*, **26**, 238–241.

Cruse, P.J.E. and Foord, E. (1973) A five years prospective study of 23,649 surgical wounds. *Arch. Surg.*, **107**, 206.

Dean, R.H., Yao, J.S.T., Thompson, R.G. *et al.* (1975) Predictive value of ultrasonically derived arterial pressure in determination of amputation level. *Am. Surg.*, **41** 731–737.

Hamilton, H.W., Hamilton, K.B., and Low, F.K.J. (1977) Pre-operative hair removal. *Can. J. Surg.*, **28**, 269–274.

Jain, A.S. (1992) Management of lower limb amputation. *Update*, **April**, 616–628.

Lloyd-Williams, K. (1964) Pictorial heat scanning. *Phys. Med. Biol.*, **9**, 433–456.

McCollum, P.T., Spence, V.A. and Walker, W.F. (1985) Circumferential skin blood flow measurements in the ischaemic limb. *Br. J. Surg.*, **72**, 310–312.

Malawer, M.M., Buch, R., Khurana, J.S. *et al.* (1991) Postoperative infusional continuous regional analgesia (PICRA): a technique for relief of postoperative pain following major extremity surgery. *Clin. Orthop.*, **266**, 227–237.

Mann, R.M. and Bisset, W.K. (1983) Anaesthesia for lower limb amputations: a comparison of spinal analgesia and general anaesthesia in the elderly. *Anaesthesia*, **38**, 1185–1191.

Mooney, V., Harvey, J.P., McBride, E. *et al.* (1971) Comparison of post-operative stump management: plaster v soft dressings. *J. Bone Joint Surg.*, **53A**, 241–249.

Murdoch, G., Condie, D.N., Gardner, D. *et al.* (1988) The Dundee experience. In: *Amputation Surgery and Lower Limb Prosthetics* (eds G. Murdoch and R.G. Donovan), Blackwell Scientific Publications, Edinburgh, pp. 440–457.

Pedersen, H.E. (1968) The problems of the geriatric amputee. *Artificial Limbs*, **12**, (part 2), i–iii.

Redhead, R.G. and Snowden, C. (1978) A new approach to the management of wounds of the extremities: controlled environment and its derivatives. *Prosthet. Orthot. Int.*, **2**, 148–156.

Seropian, B. and Reynolds, B.M. (1962) Wound infection after pre-operative depilatory versus razor preparation. *Am. J. Surg.*, **104**, 251–254.

Spence, V.A. and Walker, W.F. (1984) The relationship between temperature isotherms and skin blood flow in the ischaemic limb. *J. Surg. Res.*, **36**, 278–281.

Spence, V.A., McCollum, P.T. and Walker, W.F. (1988) Recommendations for the objective determination of the level of limb viability. In: *Amputation Surgery and Lower Limb Prosthetics* (eds G. Murdoch and R.G. Donovan), Blackwell Scientific Publications, Edinburgh, pp. 9–20.

Waters, R.L., Perry, J., Antonelli, D. *et al.* (1976) Energy cost of walking amputees: the influence of level of amputation. *J. Bone Joint Surg.*, **58A,** 42–46.

Section 2

The tibia

6

Transtibial amputation

J. H. Bowker

Introduction

It can be fairly said that the transtibial (below-knee) amputation is the most important one in the lower limb because it is the most proximal level at which near-normal function is available to a wide spectrum of lower limb amputees. Although transtibial amputation has had strong supporters in the past (Bickel, 1943), more often its detractors held the day. An example is the statement of Homans (1969) that 'amputation below the knee can almost never be expected to offer a healthy stump'. With widely varying opinions as to the desirability and efficacy of transtibial amputation and the acceptability of a higher amputation in cases of failure at the lower level, surgeons have struggled with the conflicting demands of wound healing and preservation of function.

While it is true that primary healing occurs more readily at the transfemoral (above-knee) level, it is far from certain. In two series totalling 277 transfemoral and 174 transtibial cases, the failure rates for primary healing were as follows: 28% and 34% for transfemoral cases and 35% and 51% for transtibial cases, respectively (Dale and Capps, 1959; Boontje, 1980). In an attempt to find the place of transtibial amputation, two groups decided to do amputations at this level on virtually all patients who had intact, uninfected skin at the proposed operative site and a functioning knee joint, without regard to the presence or absence of pulsation at the popliteal level (Bickel, 1943; Baker et al., 1977). Although the percentage of wound healing in these two studies was virtually identical, the two papers came to entirely different conclusions as to the acceptable percentage of re-amputation to the transfemoral level. In Bickel's opinion, a failure rate of 27% for transtibial amputations in cases of absent popliteal pulse was acceptable. Baker et al. (1977), on the other hand, decided that a 26% failure rate was too high to continue to offer the option of transtibial ablation to their patients.

A broader look at healing rates for transtibial amputation also helps to put the question in perspective. Eight reported series, each with at least 50 patients, were combined for an overall view of primary and secondary healing rates and rate of failure at the transtibial level. There was a total of 942 cases. Seventy per cent healed primarily and 16% secondarily for a total healing rate of 86% and a failure rate of 14% (Kendrick, 1956; Ecker and Jacobs, 1970; Chilvers et al., 1971; Moore et al., 1972; Castronuovo et al., 1980; Fleurant and Alexander, 1980; DeCossart et al., 1983; Yaramenko and Andruhova, 1986). Murdoch et al. (1988), in a carefully controlled prospective study of 81 transtibial amputations, noted a 93% healing rate. Seven stumps which failed to heal primarily did so after a local 'wedge' resection of necrotic tissues.

In the series just noted, no attempt was made to separate diabetics with or without ischaemia from those with ischaemic disease alone. To illustrate more clearly the differences between these groups, four series which distinguished them were combined. Of 194 diabetics, 92% healed. In contrast, only 75% of 188 patients with purely ischaemic disease healed (Lim *et al.*, 1967; Cranley *et al.*, 1969; Roon *et al.*, 1977; Kacy *et al.*, 1982). This showed that diabetic patients, even with ischaemia, had significantly higher healing rates following transtibial amputation. Two additional series (Hoar, 1962; Fearon *et al.*, 1985), each with 100 diabetics, reported healing rates of 99% and 90%, respectively. Thus, the notion that diabetics should have a primary transfemoral amputation for a foot lesion should be laid to rest.

Comparative mortality rates must also be included among the relative merits of transtibial versus transfemoral amputation. Three studies made careful note of this factor (Lim *et al.*, 1967; Boontje, 1980; Barber *et al.*, 1983). Their combined mortality for transtibial amputation, at 9.5%, was one-third of the 29.7% they reported for transfemoral amputation. In other reports, Bickel (1943) had a transtibial mortality rate of 0.02% while Smith (1956) stated that mortality for transfemoral amputation in New York city during the 1930s ranged from 40% to 90%.

As Loon (1961) noted, relative emphasis in amputation surgery has undergone many changes, both as new techniques have become available and as new goals have been appreciated. To the two traditional stages of amputation, ablation and reconstruction, a third, rehabilitation, has now been added. With preservation of ambulatory function now a major goal, comparative reports on prosthesis usage for transtibial and transfemoral amputees become valuable when considering amputation level. Four papers (Bickel, 1943; Smith, 1956; Paloschi and Lynn, 1967; Barber *et al.*, 1983) reported that transfemoral prosthesis usage varied from 10% to 64% with an average usage of 26.5%. In contrast, an average 73.5% transtibial prosthesis usage rate with a range of 31–100%, was noted in 13 papers (Bickel, 1943; Smith, 1956; Hoar, 1962; Block and Whitehouse, 1963; Perry, 1963; Murray, 1965; Lim *et al.*, 1967; Chilvers *et al.*, 1971; Robinson, 1972; Roon *et al.*, 1977; Barber *et al.*, 1983).

Prosthesis usage following transtibial amputation for trauma was examined by Purry and Hannon (1989). Twenty-five unilateral traumatic amputees were queried regarding function and lifestyle. Twenty-one wore their prosthesis more than 13 h a day. In regard to functional activities, 18 could walk a mile if necessary and 21 drove cars. Various sports were played by 18. The most striking finding was that 21 regarded themselves as minimally or non-disabled.

In ischaemic conditions, unilateral transtibial amputation may be followed by amputation of the opposite limb as vascular disease progresses. Thornhill *et al.* (1986) noted an interval of 23 months, on average, between transtibial amputations in 80 patients. Harris *et al.* (1988) showed a risk of contralateral limb loss of 10% per year. With sufficient longevity, patients often face the prospect of loss of the other leg. Their chances of ambulation as bilateral transtibial amputees were reported by several authors (Smith, 1956; McCollough, 1972; Roon *et al.*, 1977; Thornhill *et al.*, 1986). Pooling their data on 137 patients showed that 77% of bilateral transtibial amputees were able to utilize prostheses. A larger single series of 110 cases showed the same general result with 68% successfully fitted (Murdoch *et al.*, 1988).

With the relationship of knee preservation to ambulatory function quite clear, what level should be elected when the patient is thought to have no prospect of walking? Barber *et al.* (1983) and Castronuovo *et al.* (1980) questioned the value of knee preservation in these circumstances. Persson (1974), however, advocated preservation of the knee joint even in patients unable to walk so that they might use the stump for purposes of sitting and turning in bed. His only requirement was that knee joint contracture not exceed 30°. Perry (1963) stated that the transtibial level was not useful to a wheelchair-bound patient. He then contradicted himself by advocating saving one knee joint in bilateral amputations due to its usefulness in bed mobility and sitting balance.

As new information has become available on the efficacy of transtibial amputation, the reported ratios of transtibial to transfemoral amputation have changed to favour the transtibial level. Two studies spanning the 1930s through the 1950s (Bickel, 1943; Dale and Capps, 1959) had an averaged TT:TF ratio of 1:2.6. Harris *et al.* (1988) reported a 1964 ratio in their institution of 1:6. Data from 1970 to 1983 gleaned from six papers showed a definite reversal to an average TT:TF ratio of 2:1 (Boontje, 1980; Rizzo and Matsumoto, 1980; Kacy *et al.*, 1982; Fearon *et al.*, 1985; Keagy *et al.*, 1986; Harris *et al.*, 1988). Of special interest was the reversal reported by Harris *et al.* (1988) from their 1964 TT:TF ratio of 1:6 to 3:1 during 1978–1982. Aggressive deliberate policy changes can have a profound effect. At Dundee, just prior to the organization of their Limb Fitting Centre in 1966, the transtibial:transfemoral ratio was 1:2. By 1973, the ratio was more than reversed to 3:1, indicating the value of a programmatic approach.

Roon *et al.* (1977) set down as the objectives of a successful amputation programme several of the areas of inquiry which were the goals of the consensus conference held by the International Society for Prosthetics and Orthotics (ISPO) in 1990. Their first objective was to reduce healing failures by accurate determination of healing potential. The second was to facilitate rehabilitation by performing the most distal amputation that will heal. Their third objective was to shorten hospitalization by immediate postoperative rehabilitation. To these, ISPO added an attempt to reach consensus on the best surgical techniques. In this regard, the author has attempted to find areas of agreement in each area under study as well as the areas of disagreement which may change with time.

Causal conditions

As the general population has aged, trauma is no longer the leading cause of lower limb amputation; it has been vastly overtaken by peripheral vascular disease. Smoking appears to be a risk factor related to this increase. Stewart (1987), in a review of 51 male lower limb amputees, found a significantly highly incidence of smokers as compared to the general population (82.4% vs. 55%). Rush *et al.*, (1981) reported that 58% of his 110 transtibial amputees were smokers. In diabetes mellitus, the great majority of amputations seem to be related to various types of foot injury secondary to peripheral neuropathy, with often minor damage to the foot providing a portal for infection. In diabetics, infection may be difficult to combat at the tissue level due to the decrease in leucocyte activity in the hyperglycaemic state (Bagdade *et al.*, 1970), The patients often continue to walk on an infected foot due to the loss of deep pain, thereby rapidly spreading the infection along tissue planes. Neuropathic arthropathy, often initiated by presumably minor trauma, may also lead to amputation once the foot and ankle skeletal structure is severely damaged. Although diabetics seem to develop atheromatous disease at an earlier age than the general population, it is at times difficult to distinguish, in causation of gangrenous changes, the relative importance of the atheromatous changes seen in larger vessels and the more peripheral small vessel disease common in diabetes (Dwars *et al.*, 1989).

The population of diabetics appears to be growing. Whether this is due to earlier detection, to increased longevity related to better treatment or to other factors is conjectural. It is certain, however, that a greater percentage of lower limb amputations is being done in diabetics in recent decades. In 17 studies published between 1961 and 1988, an average 52% of patients had diabetes mellitus as the primary or secondary causal factor in amputation with a range of 30–75% (Harris P.D. *et al.*, 1961; Perry, 1963; Paloschi and Lynn, 1967; Cranley *et al.*, 1969; Chilvers *et al.*, 1971; Moore *et al.*, 1972; Robinson, 1972; Baker *et al.*, 1977; Murdoch, 1977; Roon *et al.*, 1977; Termansen, 1977; Alter *et al.*, 1978; Boontje, 1980; Castronuovo *et al.*, 1980; Cumming *et al.*, 1987; Harris *et al.*, 1988). In contrast, an earlier study showed diabetes as a factor in only 16% of cases (Kendrick, 1956). The precipitating cause of amputation may be gangrene, infection or intractable claudication (Boontje, 1980).

Level selection

Selection of the transtibial level as the correct one in an individual case has several facets. In cases of trauma, the exact length of tissue capable of playing a part in the preservation of a transtibial stump is usually predetermined by the accident and treatment to that point. Tumour surgery demands that adequate margins free of disease are the surgeon's first concern, with preservation of limb length a secondary one. In dysvascular cases, the surgeon must first determine if the limb may be salvaged by reconstructive vascular surgery, either entirely or with limited loss at toe, ray or transmetatarsal level after successful revascularization. Secondly, it should be determined whether a transtibial stump would have a reasonable chance of healing. Thirdly, an amputation level that will heal, be durable and functionally optimal should be chosen.

Level selection as currently practised is certainly multifactorial. Many studies have tried to oversimplify the problem by using only one criterion, basing success or failure solely on that. The fact remains, however, that while both clinical evaluation and objective laboratory measurements of vascularity are reasonably predictive of success or failure at high and low ends of the measurement spectrum, there remains an intermediate grey zone of unpredicability. The inability to reach a consensus on the best test or tests for level selection, as reflected in the literature, has clearly shown that the best test, which may not yet exist, would be that which predicts failure with 100% accuracy, correctly guiding the surgeon away from that level (Cheng, 1982; Lepantalo et al., 1987). This would avoid the possible imposition of higher levels of amputation on those patients who are deemed capable of healing at the transtibial level, but were eliminated by strict application of criteria which include a built-in failure rate for reasons undetermined by the chosen study method. If failure then occurred, factors other than tissue blood flow could be sought, such as poor nutritional status, infection, suboptimal surgical technique or poor post-operative stump management.

The more traditional methods of level selection will be considered in this paper. Unfortunately, in practice, level selection is often somewhat idiosyncratic based on the attitudes and prejudices of the surgeon and the prosthetist regarding the level under consideration. This is attested to by the varying ratios of transtibial to transfemoral amputations performed in similar institutions in various parts of the world and indeed, various parts of the same country.

Even with the development of more sophisticated tests, many surgeons have depended on factors that can be easily evaluated by sight and touch, including skin texture and warmth, colour of the foot elevated and dependent, hair growth and the presence of indolent ulcers, tissue necrosis, gross infection or lymphangitis (Block and Whitehouse, 1963; Perry, 1963; Pedersen et al., 1964; Lim et al., 1967; Ecker and Jacobs, 1970; Boontje, 1980; Castronuovo et al., 1980; DeCossart et al., 1983; Fearon et al., 1985).

In regard to peripheral pulses, if they can be easily felt, they are usually there. If they cannot be easily felt, however, they may still be present but obscured by oedema, obesity or hypotension. At any rate, a significant number of transtibial amputations will heal despite the absence of palpable pulsation at any given level, including superficial femoral. In a series of 113 transtibial amputations, Roon et al. (1977) found that 57% of stumps healed with only an aortic pulse palpable; the addition of a femoral pulse increased success to 81%. With popliteal or pedal pulses palpable, all transtibial stumps in their series healed. Six papers were examined in the aggregate regarding healing rate and presence of a palpable popliteal pulse. Although 65% of these combined patients had no popliteal pulse, 82.5% healed at the transtibial level (Bickel, 1943; Harris et al., 1961; Block and Whitehouse, 1963; Eraklis and Wheeler, 1963; Chilvers et al., 1971; Barber et al., 1983). These findings point out the difficulty in evaluation of collateral circulation development.

Kacy et al. (1982) found that healing of a transtibial stump in the absence of a femoral pulse did not occur in the presence of pre-existing cellulitis of the foot. In contrast, 85% of those with neither cellulitis nor a femoral pulse healed. In the presence of a femoral pulse, however, much less sensitivity to cellulitis was found; 82% with cellulitis and 100% without, healed. Baker et al. (1977) emphasized the difficulty of assessing collateral circulation, which is certainly not available to the palpating finger.

Persson (1974) evaluated arteriography in regard to determining the potential for transtibial amputation healing. He found that it bore little correlation to healing potential, on a par with reliance on palpable pulses. Arteriography is now used chiefly to determine the feasibility of vascular reconstruction (Barber et al., 1983).

One test that many surgeons have relied on is the trial skin incision (Bickel, 1943; Burgess, 1968; Brodie, 1970). The presumption is that if the skin bleeds at the proposed level of amputation, then it should heal at the that level; if it does not, one should immediately move to a higher level. Kendrick (1956) noted no correlation between healing of a trial skin incision and healing potential. The basic question of how distally the initial trial skin incision should be made remains unaddressed. Thus, until very recently, amputation level selection has been something of an art. More recently, various tests designed to give more objective measurements of limb and tissue blood flow have been devised. For a detailed discussion of these methods, the reader is referred to Chapter 20.

Once a decision has been made to amputate at the transtibial level, a secondary and equally important choice must be made as to the exact length to be retained (Fig. 6.1). The shortest useful

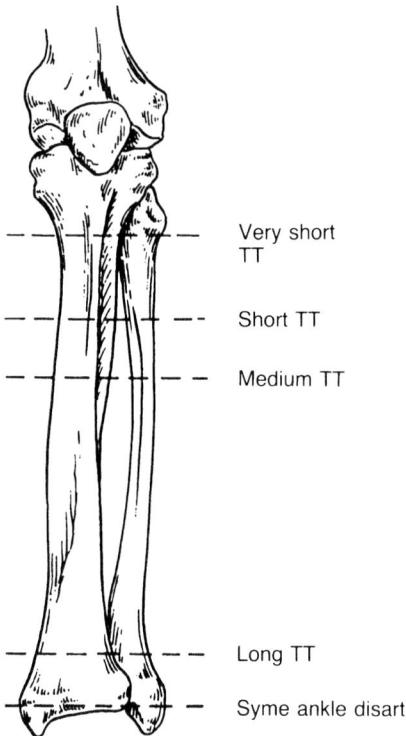

Fig. 6.1. Levels of transtibial amputation. (Fig. 199–1B, page 5124, from Epps, C.H. Amputation of the Lower Limb in Surgery of the Musculoskeletal System, 2nd Edition, (Ed. C. McC Evarts). Churchill Livingston, New York, 1990).

stump must include the tibial tubercle to preserve knee extension by the quadriceps (Lim *et al.*, 1967). Flexion at this level is adequately provided by the semimembranosus and biceps femoris.

Beyond universal agreement as to this shortest possible functional level, the ideal length for durability and optimal prosthetic function has not been scientifically determined. This is exemplified by the opinion of Brodie (1970) that the surgeon should maintain as much length as possible so long as it is no more than 12.5 cm. The amputation method advocated by Burgess (1968), which produces a cylindrical stump, effectively limits stump length to approximately 15 cm, since the leg begins to taper beyond that point. Marsden (1977) recommends limiting stump length to 15 cm on the basis that the prosthetist will have less trouble fitting a prosthesis. Thus, there appears to be solid support for a standard transtibial segment of

approximately 15 cm measured from the knee joint line. This consensus figure was obtained by averaging the recommendations found in 18 papers (Bickel, 1943; Harris, 1944; Smith, 1956; Hoar, 1962; Block and Whitehouse, 1963; Eraklis and Wheeler, 1963; Burgess, 1968; Moore *et al.*, 1968, Brodie, 1970; Condon and Jordan, 1970; Hunter-Craig *et al.*, 1970; Baker *et al.*, 1977; Alter *et al.*, 1978; Castronuovo *et al.*, 1980; Fleurant and Alexander, 1980; Hicks and McClelland, 1980; Harris, 1987).

Although little has appeared in the literature to contradict this apparent consensus, there are a number of opinions, expressed over several decades, which cast doubt on this certainty. Harris (1944), although recommending a short transtibial amputation, noted that a long stump is stronger

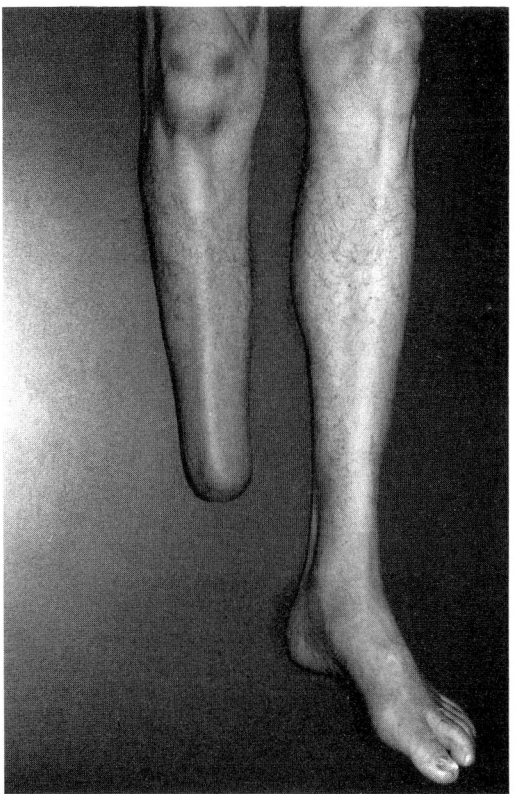

Fig. 6.2. Long right transtibial amputation in 15-year-old male following severe foot trauma. He is a very active athlete. Reprinted with permission from AAOS Atlas of Limb Prosthetics, Second Edition (Eds. J.H. Bowker and J.W. Michael), Mosby-Year Book. St. Louis, 1992.

than a shorter one. Despite this functional advantage, he recommended the short stump due to the complications from wearing prostheses with plug-fit sockets and thigh corsets available at that time. McCollough (1972), although not specifying what he considered the optimal length, flatly stated that the longer the stump obtained, the better the gait (Fig. 6.2). This position is supported by a study showing that transtibial amputees with longer stumps require less energy to ambulate (Gonzalez et al., 1974). In regard to another important factor in length selection, Eraklis and Wheeler (1963) recommended that transtibial stump length should not be absolute, but related to the individual's height.

Pre-operative care

The pre-operative management of prospective amputees has several very important aspects. These are to a large extent related to the reason for amputation. Patients undergoing amputation for trauma, while usually young and healthy, frequently have concomitant injuries due to other skeletal parts, soft tissue or viscera. A careful evaluation to rule out injury to areas other than the affected limb is therefore essential.

When dysvascularity related to peripheral vascular disease and/or diabetes mellitus leads to amputation, the presence of associated disease must be assumed. Barber et al. (1983) found that 76% of 70 patients coming to transtibial amputation had various other degenerative diseases. Smith (1956) advised special attention to control of congestive heart failure, electrolyte imbalance, dehydration, hypertension and diabetes for optimum results. Rapid pre-operative treatment of congestive heart failure, bronchitis and diabetes was advocated by Robinson (1972).

In cases of diabetes, the infection which often leads to the need for amputation has often totally disrupted diabetic control. Since the control of infection and control of blood sugar are interdependent, both must be approached at the same time for optimum effect. Icing of a necrotic/infected limb to control local and systemic effects of the process is controversial. Kendrick (1956) suggested its use in selected cases, while Pedersen et al. (1964) condemned this practice, stating that following icing, a transfemoral amputation is

unavoidable. Instead, he advocated prompt drainage of abscesses, followed by appropriate antibiotics and bedrest. A wide range of bacteria is associated with foot infections in diabetics. They may include gram positive, gram negative, aerobic and anaerobic organisms, occasionally singly, but more often in various combinations. Hoar (1962) found Staphylococcus aureus, Streptococcus haemolyticus, and Proteus vulgaris to be most common. Fearon et al. (1985) cultured more than 15 different bacteria in their series of diabetic gangrene cases.

Systemic infection secondary to wet gangrene or infections independent of the foot problem must also be controlled. Particularly, one should look for genitourinary and pulmonary infections. In addition to evaluating the infection by aerobic and anaerobic wound and blood cultures and determination of sensitivities to antibiotics, assessments as to wound healing potential are in order. These include obtaining the serum albumin level as an indicator of nutritional status; ideally this should be 3.5 g/dl or above. Also, the total lymphocyte count should be at least 1500 per mm^3 as an assurance of immunocompetence. If these are abnormal, one may expect difficulties with primary wound healing (Dickhaut, 1984).

Anaesthesia

Little has been written by amputation surgeons regarding anaesthesia for amputation. Dale and Capps (1959) divided debilitated patients into two groups, suggesting spinal anaesthesia for poor risk and local for high risk patients. In cases appropriate for local anaesthesia, the agent, without adrenaline is injected along the proposed incision line with infiltration of deeper tissues as necessary. Special attention is given to injection of nerves, especially the posterior tibial, prior to any manipulation and section. Mann and Bissett (1983) did a randomized study of 60 elderly patients undergoing lower limb amputation, using spinal anaesthesia for one group and general for the other. Twenty-six per cent were transfemoral cases with the rest at the transtibial or Syme ankle disarticulation level. Excellent results with both methods were attributed to meticulous peri-operative management. There was a definite preference for spinal anaesthesia noted by the surgeons because of less intra-operative bleeding and by the nurses for the smoother post-operative course. For those patients with severe cardiopulmonary compromise in whom speedy anaesthesia is not critical, a sciatic–femoral block has proven very effective.

In summary, the choice of anaesthesia depends on the condition of the patient, the special skills and experience of the anaesthetist and the patient's choice provided he or she is fit to receive any type of anaesthetic.

Incision

The placement and measurement of flaps, no matter which type is favoured, must be correctly related to the cross-sectional area of the leg at the bony level selected. Otherwise, the surgeon may have to further shorten the bone to avoid skin closure under tension, or alternatively, excise redundant soft tissue. Skin incisions should be made at 90° to the skin surface to avoid having portions of the flaps unsupported by subcutaneous tissue and hence more difficult to close accurately and also more liable to necrosis.

In the recent literature, the long posterior myofasciocutaneous flap with a short anterior flap appears to be most commonly adhered to (Boontje,

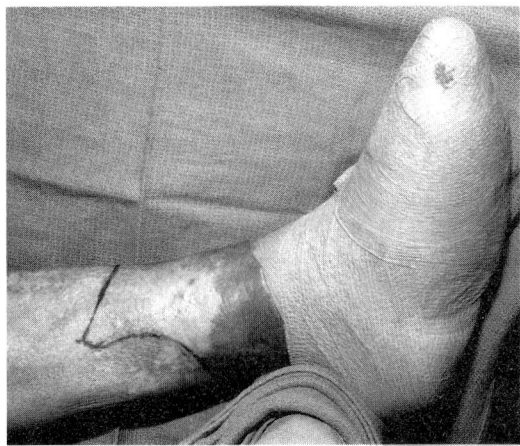

Fig. 6.3. Operative outline of long posterior myofasciocutaneous flap with short anterior flap as adapted for use in long transtibial amputation. Reprinted with permission from AAOS Atlas of Limb Prosthetics, Second Edition (Eds. J.H. Bowker and J.W. Michael), Mosby-Year Book, St Louis, 1992.

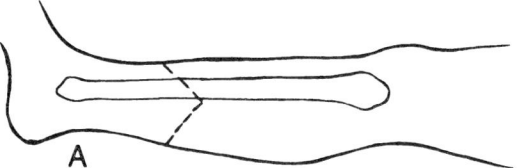

Fig. 6.4. Diagram of equal anterior and posterior flap design for transtibial amputation. (Fig. 199–9A). Epps, C.H. Amputation of the Lower Limb in Surgery of the Musculoskeletal System, 2nd Edition (Ed. C. McC. Evarts), Churchill Livingstone, New York, 1990).

1980; DeCossart *et al.*, 1983; Harris, 1987; Yaramenko and Andruhova, 1986) (Fig. 6.3). The major impetus for the acceptance of this configuration may be attributed to the educational efforts of Burgess (1968). It must be noted, however, that a long posterior myofasciocutaneous flap technique was reported in 110 amputations done between 1930 and 1942 by Bickel (1943). Prior to the popularization of the long posterior myofasciocutaneous flap, equal anterior and posterior fasciocutaneous flaps were perhaps most common (Eraklis and Wheeler, 1963) and some continue to report favourably on this method (Fearon *et al.*, 1985) (Fig. 6.4). Moore *et al.* (1972) reported an improvement in healing rate from 76% to 89% since changing from a circular incision, closed transversely, to a long posterior myofasciocutaneous flap.

Another configuration is that of equal medial and lateral (sagittal) myofasciocutaneous flaps (Figs 6.5a,b,c). The advantages of this approach were outlined by Persson (1974). He stated that these flaps were less likely to become necrotic for two reasons: firstly, the approach automatically reduces the amount of poorly vascularized anterior skin and secondly, the resultant flaps are widely based and short. These sagittal flaps also allow the skin to be more easily cut proximal to any anterior or posterior necrotic skin present, helping to preserve bony length. He also feels that a side-to-side myoplasty covers the bone better and provides good spontaneous drainage. In his series, Persson (1974) compared 58 patients with sagittal flaps to 40 with equal anterior and posterior flaps. With the sagittal technique, 74% had primary healing and 10% were revised to the transfemoral level, compared to 41% and 39%, respectively, for the 'conventional' group. Termansen (1977) compared

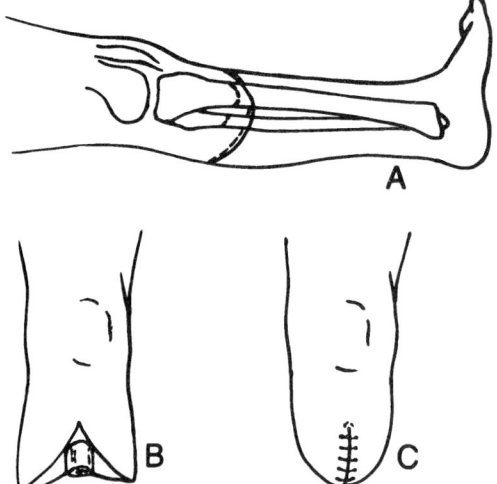

Fig. 6.5. (a) Medial view of sagittal myofasciocutaneous flaps related to level of bone section in short transtibial amputation. (b) Anterior view of same. (c) Midline closure of sagittal flaps (above are Figs. 199–11A,B,C, from Epps, C.H., Amputation of the Lower Limb in Surgery of the Musculoskeletal System, 2nd Edition (Ed. C. McC. Evarts). Churchill Livingstone, New York, 1990).

41 cases of sagittal myofasciocutaneous flaps to 47 with long posterior myofasciocutaneous flaps. The total healing rates of 58% and 59%, respectively, as well as primary and secondary healing rates, were virtually identical. Another configuration is a long anterior fasciocutaneous flap designed to place the incisional scar posterior to the tibia to avoid scar trauma from the prosthesis (Hoar, 1962; Block and Whitehouse, 1963). For a description of skew flap techniques, the reader is referred to the contributions of A.S. Jain and K.P. Robinson in this volume.

Singer *et al.* (1988) reported a novel reconstructive approach in five transtibial amputations for trauma which gives end weight bearing capability. The criteria are strict and the surgery is precise. Distal tibial bone loss must be extensive enough to preclude reconstruction, while the posterior tibial nerve and foot should be basically intact. The heel pad and sole are dissected free from the skeletal foot with the posterior tibial nerve in continuity. The posterior tibial nerve is folded into the soft tissues, a posterior tibial-popliteal arterial anastomosis is done and the heel pad is sutured over the stump end.

Treatment of bone

All articles that mention the treatment of bone recommend bevelling of the tibial crest (Eraklis and Wheeler, 1963; Burgess, 1968; Moore *et al.*, 1972). While most authors make no mention of the acuity of the bevel, several suggest 45° (Bickel, 1943; Lim *et al.*, 1967; Harris, 1987). Roon *et al.* (1977) report that a bevel between 45° and 60° is adequate, while Hicks and McClelland (1980) prefer a bevel of 30°. Bevelling combined with careful smoothing of the bone edges will help prevent damage to the skin which lies between hard bone and a firm prosthetic socket (Fig. 6.6). If the bone is cut with a power saw, it should be cooled with water to prevent thermal necrosis (Marsden, 1977). The only other physical treatment of the tibia mentioned is drilling of the bone to effect a myodesis. Hicks and McClelland (1980) decry the use of bone wax to control bone bleeding because of its tendency to provoke a foreign body reaction.

There is considerable variability in the literature in regard to the preferred length of the residual fibula. Suggestions range from 1 to 3 cm shorter than the tibia (Eraklis and Wheeler, 1963; Burgess, 1968) (Fig. 6.7). Removing sharp edges with a file was mentioned by Brodie (1970), but only one instance was found of bevelling the fibula

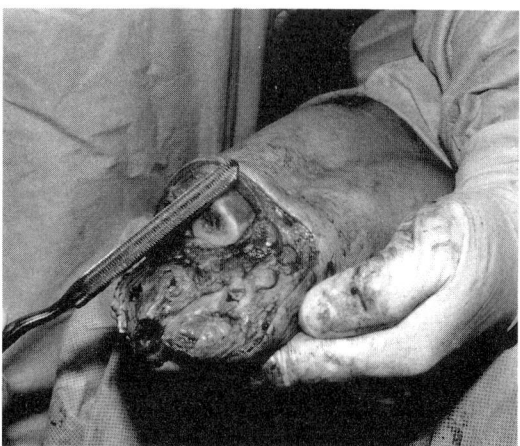

Fig. 6.6. Final contouring of anterior tibial level with bone file. Reprinted with permission from AAOS Atlas of Limb Prosthetics, Second Edition, (Eds. J.H. Bowker and J.W. Michael) Mosby-Year Book, St. Louis, 1992.

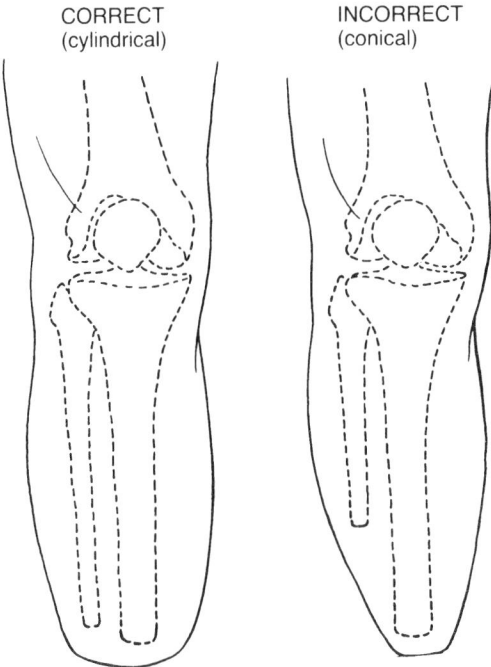

CORRECT
(cylindrical)

INCORRECT
(conical)

Fig. 6.7. Contrast of cylindrical transtibial stump (minimal fibular shortening) with conical stump (excessive fibular shortening) (Fig. 12–14, page 126 in Lower Extremity Amputation (Eds. W. Moore and J. Malone), W.B. Saunders, Philadelphia, 1989).

(Cumming *et al.*, 1987). In our own practice, a fibular bevel facing posterior-lateral seems to decrease complaints of soft tissue impingement during prosthetic use. Management of the short fibula and the Ertl osteomyoplasty are fully covered in A.S. Jain's contribution on special situations in transtibial amputation.

Treatment of muscle

Since muscle is considered to carry blood from the deep arterial supply of the leg to the skin, it is generally accepted that preservation of muscle, especially if it is not dissected from the crural fascia or skin, will enhance healing of an amputation stump. Continuous myofascial closure over the cut bone end should also prevent adherence of the skin scar to bone. For these reasons, the most commonly advocated method of treating muscle is the creation of a long posterior myofasciocuta-

neous flap (Burgess, 1968). Several authors advocate reducing the bulk of the posterior muscle by excising the posterior tibial, toe flexor and peroneal muscle bellies to the level of the cut tibia, leaving only the gastrocnemius and perhaps a tapered portion of the soleus (Bickel, 1943; Robinson, 1972; Marsden, 1977; Murdoch, 1977; Yaramenko and Andruhova, 1986; Harris, 1987) (Fig. 6.8). This approach is in sharp distinction to the statement of Block and Whitehouse (1963) that muscle closed over bone is unlikely to survive. Condon and Jordan (1970) recommended excising any ischaemic or necrotic muscle found during transtibial amputation, but reported a healing rate of less than 50% in these circumstances.

With a long posterior myofasciocutaneous flap, bone coverage may be obtained by myoplasty or

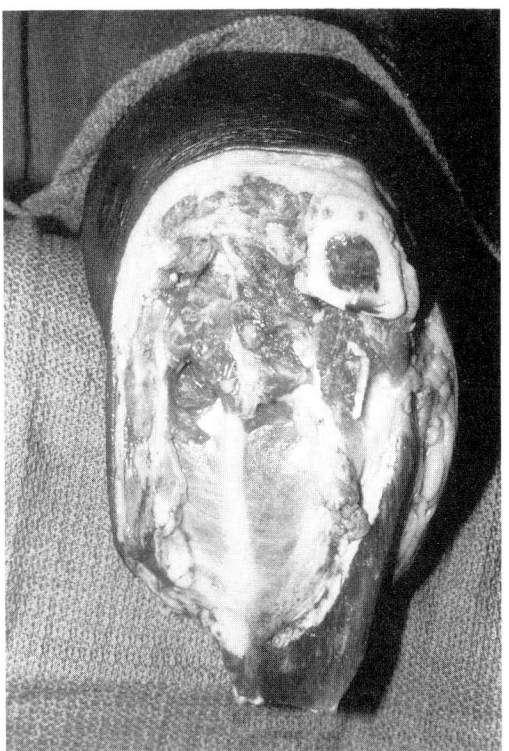

Fig. 6.8. Intra-operative photograph of long transtibial stump, following excision of deep posterior calf muscle. This reduces distal bulkiness while preserving the gastrocnemius and tapered soleus for distal padding. Note myodesis drill holes in anterior tibia. Reprinted with permission from AAOS Atlas of Limb Prosthetics, Second Edition (Eds. J.H. Bowker and J.W. Michael), Mosby-Year Book, St. Louis, 1992.

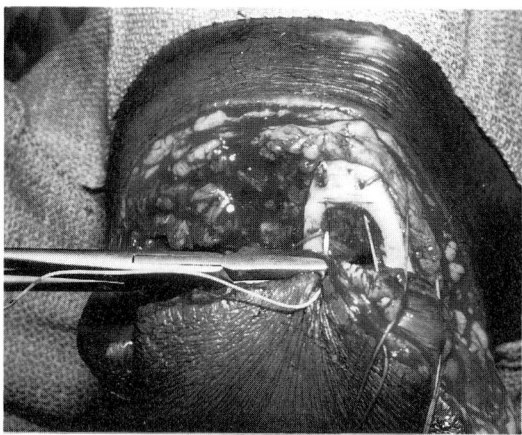

Fig. 6.9. Intra-operative photograph of myodesis, in which the posterior myofascia and posterior and anterior investing fascia of the leg are being sutured to the tibia through drill holes. Reprinted with permission from AAOS Atlas of Limb Prosthetics, Second Edition (Eds. J.H. Bowker and J.W. Michael), Mosby-Year Book, St. Louis, 1992.

by myodesis with direct attachment of muscle to bone through two or more drill holes (Fig. 6.9). There is no clear preference of one over the other noted in the literature in regard to amputation in non-ischaemic patients. Burgess (1969) and Marsden (1977) specify that myodesis is contraindicated in ischaemic disease and recommend myoplasty only. In the case of sagittal myofasciocutaneous flaps, only myoplasty is done (Persson, 1974).

Complete removal of all muscle tissue leaving a long posterior fasciocutaneous flap was reported by Kendrick (1956). Cranley *et al.* (1969) elected equal sagittal fasciocutaneous flaps. Long anterior flaps devoid of muscle tissue were used to place the wound scar posteriorly by Block and Whitehouse (1963) and Hoar (1962).

Treatment of nerve

In those series in which nerve management is discussed, the consensus is that the nerves should be carefully sought out, gently drawn distally, sharply cut and allowed to retract into the soft tissues where the inevitable neuroma will, hopefully, be protected from trauma during prosthetic gait (Murdoch, 1977; Alter *et al.*, 1978). The nerves to be specifically sought out during transtibial amputation include the superficial peroneal, the saphenous, the deep peroneal, the sural and the posterior tibial. Several authors feel that the posterior tibial nerve has sufficient intrinsic vascular supply to warrant ligation (Smith, 1956; Burgess, 1968; Condon and Jordan, 1970; Marsden, 1977; Hicks and McClelland, 1980).

Many attempts have been made at the time of amputation to inhibit neuroma formation by traumatizing the proximal cut end. These include crushing, cauterization, injection with phenol or absolute alcohol and application of nitrogen mustard. Burgess (1968) advised against such approaches in favour of simple division following mild traction on the nerve. Dellon *et al.* (1984) compared transected nerve ends buried in muscle with those simply left in their bed, both in humans and in baboons. Those buried in muscle showed little enlargement and an orderly fascicular pattern on cross section. Those left *in situ* showed the scar fixation, fibrous enlargement and disorganized cross sectional pattern typical of classic neuromata.

Treatment of fascia

It is generally accepted that the crural fascia should be cut at the same level as the skin and subcutaneous tissue (Gillis, 1965). It should never be separated from subcutaneous tissue or from the muscle to prevent damage to small perforating vessels serving the skin (Brodie, 1970). Fascia that is stripped from the underlying muscle should be excised to avoid dense scarring that may prevent re-establishment of these vessels (Loon, 1961). In closing a myofasciocutaneous flap, whether by myoplasty or by myodesis, care should be taken to ensure that the crural fascia is indeed found and captured in the sutures to assure maximum initial wound strength.

Skin closure

To avoid failure of primary skin healing in transtibial amputation, especially in dysvascular cases, surgeons have two worthwhile approaches

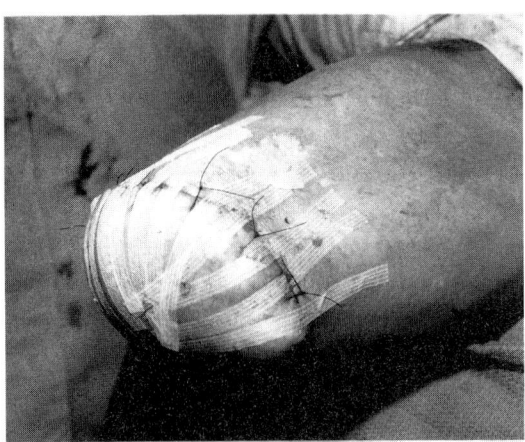

Fig. 6.10. Damage to skin flaps in ischaemic cases can be limited by use of widely spaced sutures alternated with adhesive paper strips. Reprinted with permission from AAOS Atlas of Limb Prosthetics, Second Edition (Eds. J.H. Bowker and J.W. Michael), Mosby-Year Book, St. Louis, 1992.

open. One is judicious level selection and the second, equally important, is careful skin closure. Gentle handling of the skin throughout surgery, including avoidance of skin forceps, is strongly recommended by several authors (Kendrick, 1956; Hoar, 1962; Perry, 1963; Chilvers *et al.*, 1971; Termansen, 1977; Alter *et al.*, 1978).

Precise anatomical approximation (Block and Whitehouse, 1963) and avoidance of skin closure tension (Harris, 1987) also contribute significantly to primary healing. The compromise of skin circulation can also be averted by avoidance of subcutaneous suture (Burgess, 1968) and the use of interrupted rather than continuous skin suture (Baker *et al.*, 1977; DeCossart *et al.*, 1983). Damage to the skin may be further limited by use of widely spaced sutures alternated with adhesive paper strips to contain subcutaneous fat (Robinson, 1972; Alter *et al.*, 1978) (Fig. 6.10).

Stump environment

All fresh amputations require wound protection of some sort. The issues beyond that are basically two: 1) are there real benefits to be gained from a rigid versus a soft dressing and, if so, what are they? 2) what is the need, if any, to look at the stump wound on a daily basis?

The advantages of a soft dressing are that it allows easy access to the wound for inspection and for regular motion of the stump, with or without the guidance of a therapist. In distinction to a rigid dressing, it does not offer any protection to the wound from trauma nor does it prevent knee flexion contracture if the patient is not disposed to move the knee on a regular basis (Eraklis and Wheeler, 1963). Mooney *et al.* (1971) randomly assigned 182 transtibial amputees to one of three stump management groups: soft dressing, rigid dressing or rigid dressing with pylon and foot attached. The use of soft dressings was associated with the greatest delay in prosthetic fitting. Wounds protected in a rigid dressing alone healed more rapidly than those subjected to early weight bearing on a pylon. The use of a pylon did not decrease the time to achieve independent prosthetic use. Barber *et al.* (1983) did a randomized study of soft dressings versus rigid circular dressings with early touch-down weight bearing in 70 patients. Their conclusions were that the rigid dressing resulted in less pain, improved sense of well-being and enhanced prosthetic fitting progress. In another series, Baker *et al.* (1977) compared similar groups and found that hospital time was halved for casted patients to only 7 days. Barnet *et al.* (1976) and Hunter-Craig *et al.* (1970) did not recommend rigid dressings due to logistical difficulties in getting them applied in non-specialized centres.

A posterior plaster splint keeps the knee straight so long as the splint is not broken and the wrapping is firm. Since problems with a simple splint occur so often, it is better to make a strong posterior hemicylinder by removing the anterior half of a full cast if the surgeon's goal is regular wound inspection and knee motion. On the other hand, a complete circular plaster or fibreglass cast is easy to apply, offers excellent protection from knee flexion contracture and trauma and will control post-operative oedema (Fig. 6.11). The protection of a cast is especially needed against falls while the patient is learning to manage a walker or crutches. If the cast tends to slip off a rather conical lower limb, it can be further secured with the aid of a waist belt. Knee flexion contracture may also be prevented by the use of light skin traction of from 2 to 5 lb for up to 2 weeks (Smith, 1956; Cranley *et al.*, 1969). The obvious disadvantage is delay in patient mobilization.

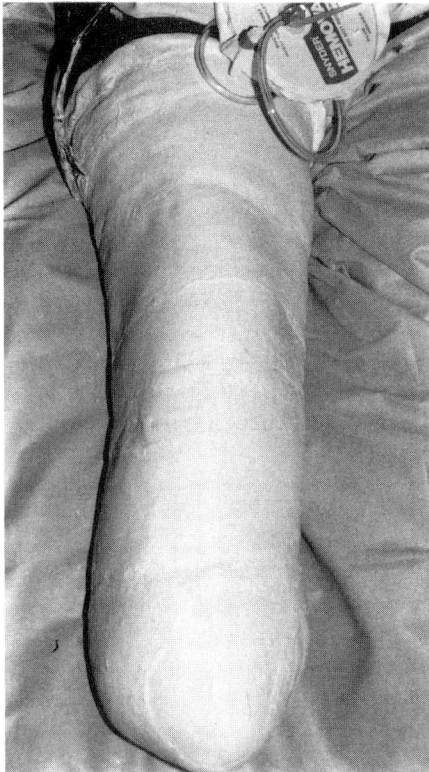

Fig. 6.11. A lightweight plaster of Paris cast has been applied with the knee in full extension. The patella is heavily padded to avoid a pressure ulcer. Drain tube exiting proximally will be removed without disturbing cast. Reprinted with permission from AAOS Atlas of Limb Prosthetics, Second Edition (Eds. J.H. Bowker and J.W. Michael), Mosby-Year Book, St. Louis, 1992.

Another aspect of stump environment to be considered is the use of wound drainage. The majority of surgeons reporting series with long posterior flaps drained their amputation wounds to prevent haematoma accumulation. Roon *et al.* (1977) recommended suction drainage, but only in those cases with active bleeding. Two authors reported major problems with the use of Penrose drains. Kacy *et al.* (1982) noted a 55% complication rate in drained wounds compared with 16% in those not drained. Tripses and Pollak (1981) found a 46% infection rate in those with open drainage, compared to 20% in those undrained. The equal sagittal flap method may not require drainage devices due to the inferior placement of the wound

which encourages spontaneous drainage of haematoma (Persson, 1974; Alter *et al.*, 1978).

In conclusion, while it may reassure the surgeon to look at the wound daily, it is certainly not necessary. The only time the wound needs to be evaluated is if the patient develops unexplained fever or excessive drainage. One might be concerned about developing loss of knee flexion range, but this can easily be overcome by weekly change of cast at which time the knee is fully ranged and the wound inspected. Most wounds, even in dysvascular patients, will be well healed at 3 weeks, at which time the cast can be discontinued and the sutures removed. In cases of trauma, earlier suture removal may be possible.

Immediate post-operative care

The issue of first concern to most patients post-operatively is pain control. Most patients will require narcotic administration, usually every 3–6 h or by means of an on-demand machine. After a maximum of 5 days, the patient should be switched to oral narcotics or even less potent drugs. In this way, habituation should not occur. A newer approach is the placement of a small silastic catheter at the time of surgery, within or alongside the posterior tibial nerve sheath. The catheter is led out of the cast so that a local anaesthetic (bupivacaine, 0.25–0.5%) can be administered by a standard infusion pump at the rate of 2–3 ml/h over the first 72 h post-operatively (Malawer *et al.*, 1991). Any of the above measures plus avoidance of stump dependency should give good relief.

Peri-operative intravenous antibiotics should be sufficient for most cases unless infection was an overriding factor in the amputation. In this case, one or more antibiotics, chosen from organism sensitivities, should be continued for 2–5 days post-operatively (Ecker and Jacobs, 1970; Robinson, 1972). Further need for antibiotics can be determined by direct evaluation of the wound. Deep breathing to prevent atelectasis should be encouraged. Various types of blow-bottles and incentive inspiration devices are effective if used regularly.

The next task is to make the patient mobile as soon as possible. In the past, post-surgical patients were often kept at bedrest for prolonged periods, with significant deconditioning occurring after just a few days. This situation has changed dramatically from the time when 2 weeks of post-amputation bedrest was recommended (Bickel, 1943). On the first post-operative day, the patient should be out of bed sitting with the stump elevated to the level of the chair seat. By the next day, at the latest, the patient should be sent to the physical therapy department to begin ambulation in the parallel bars. This is followed by the use of crutches or walker as conditioning progresses.

Perhaps the greatest impetus to early mobilization was the introduction of the immediate post-operative prosthesis and its more commonly used descendant, the rigid post-operative cast (Weiss, 1966; Moore et al., 1968; Brodie, 1970; Mooney et al., 1971; Fleurant and Alexander, 1980). If the patient has had an immediate postoperative prosthesis applied, limited weight bearing on the stump can start almost immediately, provided the patient demonstrates sufficient strength, balance, proprioception and cognition to accurately determine the weight applied. Fearon et al. (1985) recommended no walking for 5–7 days, probably because the posterior splint he used was unlikely to stay in place during ambulation as reliably as a circular cast. Adjuncts to mobility training include positioning such as prone lying to prevent hip flexion contracture (Moore et al., 1972) and knee range of motion if a soft dressing is elected (Barber et al., 1983). Since dysvascular patients may heal skin wounds somewhat more slowly than healthy patients, Robinson (1972) wrote that sutures should be left in place for 3 weeks.

The cost of hospital stay has become a major issue in some countries. In the past, patients remained in the hospital or rehabilitation centre following surgery until the wounds had healed and they had been fitted with a prosthesis and thoroughly trained in its use. In many areas, this is no longer financially feasible. In the United States, for example, transtibial amputees are often discharged from the hospital 4–5 days after surgery unless they have failed to achieve their maximum level of independence in transfers and one-legged ambulation. In that case, they stay until these goals have been achieved or abandoned as unrealistic. Further care, including prosthetic fitting, is accomplished on an out-patient basis. Hospitalization for prosthetic gait training can still be justified, however, in cases of marked deconditioning, bilateral concomitant lower limb amputations or distance from the centre.

Early fitting type of prosthesis

The ultimate early fitting type of prosthesis is that applied on the operating table at the conclusion of the amputation. Weiss (1966) is credited with popularizing this technique which was first described by Berlemont (1961). Burgess et al. (1967) instigated an intensive evaluation of this technique in 1964 by a number of amputation surgeon-prosthetist teams at Veterans hospital and university centres throughout the United States.

The most commonly used immediate post-operative prosthesis (IPOP) consists of a rigid plaster cast carefully moulded to a patella-tendon-bearing configuration to which a pylon and foot are attached. In most cases, the patient starts with touch-down weight bearing on the amputated side within 24–48 h after surgery (Burgess, 1968; Condon and Jordan, 1970; Mooney et al., 1971; Moore et al., 1972; Murdoch, 1977; Roon et al., 1977). The amputated side is used for balance with only minimal weight bearing allowed initially. This activity begins on the parallel bars with progression to crutches or walker. Patients must be carefully selected for this procedure in that they should have good strength, balance, proprioception and cognition to avoid putting excessive amounts of weight on the stump too soon. Because of problems with wound dehiscence related to these factors, Cohen et al. (1974) and Harris (1987) have stopped using immediate fitting, while Murdoch (1977) makes a plea for individualized prescription of this technique. Pneumatic prostheses can also be used for immediate or early fitting and have the advantage of easy access to the wound for inspection as the patient increases weight-bearing (Marsden, 1977; Cumming et al., 1987). Because of the virtual impossibility of regulating weight bearing, immediate post-operative prostheses are not fitted to patient with concomitant bilateral transtibial amputation (McCollough, 1972).

Another approach is to use a rigid dressing for its environmental advantages until the wound is deemed safe for touch-down weight bearing. The length of time for the wound to achieve this condition is defined differently in various series. Harris (1987) is willing to apply a temporary prothesis at 7–10 days if the wound is adequate at the first cast change. Other authors suggest a 2–3 week delay before beginning partial weight-bearing (Eraklis and Wheeler, 1963; Robinson, 1972; Marsden, 1977; Barber et al., 1983; Fearon et al.,

1985). The surgeon must make the decision whether this activity can be undertaken safely on an out-patient basis or must be done in hospital. The financial and social difficulties of maintaining patients in hospital for this purpose can be a deterrent to its use.

Summary

The task of the author has been to determine the consensus on various aspects of transtibial amputation as reflected in the literature. The advantages of transtibial amputation over transfemoral amputation in relation to rates of healing, revision, mortality, prosthetic rehabilitation, as well as functional advantages are corroborated. It was noted that the prevalence of diabetes has grown to the point where it is involved primarily, in relation to peripheral neuropathy, or secondarily together with peripheral vascular disease in at least 50% of all lower limb amputations.

Since fully 75% of dysvascular patients coming to amputation have one or more significant associated disease processes, pre-operative management must be intensive, rapid and skilled to avoid intra-operative and immediate post-operative complications. If feasible, the patient should meet the team members who will be working with them as prosthetic care proceeds.

While amputation level selection is often pre-determined in trauma or tumour, in cases of dysvascularity with or without diabetes, the surgeon often has considerable latitude in selecting a suitable amputation level. This should be based on clinical and laboratory estimates of healing, durability and functional value of the resulting stump. In this regard, rehabilitation as the goal of amputation surgery must become a part of the thinking of all amputation surgeons so that the patient gets the benefit of the lowest possible useful level.

The prime requisite of amputation surgery is gentle handling of tissues with the use of myo-fasciocutaneous flaps whenever possible. Post-operative casting is generally recommended to protect the soft tissues and prevent knee contracture. Mobility should resume the day after surgery to prevent deconditioning. Early prosthetic weight-bearing has value if closely monitored.

In closing, this review certainly supports the need for both standard methods of reporting and for long-term true outcome analyses based on prospective protocols, if progress is to accelerate in the areas of amputation surgery and prosthetics.

References

Alter, A.H., Moshein, J., Elconin, K.B. and Cohen, M.J. (1978) Below-knee amputation using the sagittal technique: a comparison with the coronal amputation. *Clin. Orthop.*, **131**, 195–201.

Bagdade, J.D., Nielsen, K., Root, R. and Bulger, R. (1970) Host defense in diabetes mellitus: The feckless phagocyte during poor control and ketoacidosis. *Diabetes*, **19**, 364.

Baker, W.H., Barnes, R.W. and Shurr, D.G. (1977) The healing of below-knee amputations: A comparison of soft and plaster dressings. *Am. J. Surg.*, **133**, 716–718.

Barber, G.G., McPhail, N.V., Scobie, T.K., Brennan, M.C.D. and Ellis, C.C. (1983) A prospective study of lower limb amputations. *Can. J. Surg.*, **26**, 339–341.

Barnet, A.J., Twist, E. and Balfe, A. (1976) Lower limb amputation in general hospital: A comparative review. *Med. J. Aust.*, **2**, 14–18.

Berlemont, M. (1961) Notre expérience de l'appareillage précoce des amputés de membre inférieur aux établissements. Heliomarins de Berck. *Ann. Med. Phys.* **4**, 213.

Bickel, W.H. (1943) Amputation below the knee in occlusive arterial diseases. *Surg. Clin. North Am.*, **23**, 982–994.

Block, M.A. and Whitehouse, F.W. (1963) Below-knee amputation in patients with diabetes mellitus. *Arch. Surg.*, **87**, 682–689.

Boontje, A.H. (1980) Major amputations of the lower extremity for vascular disease. *Prosthet. Orthot. Int.*, **4**, 87–89.

Brodie, I.A.O. (1970) Lower limb amputation. *Br. J. Hosp. Med.*, **4**, 596–604.

Burgess, E.M. (1968) The below-knee amputation. *Bull. Prosthet. Res.*, **10**, 19–25.

Burgess, E.M. (1969) The below-knee amputation. *ICIB*, **8**, 1–22.

Burgess, E.M., Traub, J.E. and Wilson, Jr, A.B. (1967). *Immediate Post Surgical Prosthetics in the Management of Lower Extremity Amputees, TRIO.5.* Prosthetic and Sensory Aids Service, Veterans Administration, Washington, DC.

Castronuovo, J.J., Deane, L.J., Deterling, R.A., O'Donnell, T.F., O'Toole, D.M. and Callow, A.D. (1980). Below-knee amputation. Is the effort to preserve the knee joint justified? *Arch. Surg.*, **115**, 1184–1187.

Cheng, E.Y. (1982) Lower extremity amputation level: selection using noninvasive hemodynamic methods of evaluation. *Arch. Phys. Med. Rehab.*, **63**, 475–479.

Chilvers, A.S., Briggs, J., Browse, N.L. and Kinmonth, J.B. (1971) Below-and through-knee amputations in ischaemic disease. **Br. J. Surg.**, **58**, 824–826.

Cohen, S.I., Goldman L.D., Salzman, E.W. and Glotzer, D.J. (1974) The deleterious effect of immediate postoperative prosthesis in below-knee amputations for ischaemic disease. *Surgery*, **76**, 992–1001.

Condon, R.E. and Jordan, P.H. (1970) Below-knee amputation for arterial insufficiency. *Surg. Gynecol, Obstet*, **130**, 641

Cranley, J.J., Krause, R.J., Strasser, R.S. and Hanfer, C.D. (1969) Below-the-knee amputation for arteriosclerosis obliterans: with and without diabetes mellitus. *Arch. Surg.*, **98**, 77–80.

Cumming, J.G.R., Jain, A.S., Walker, W.F., Spence, V.A., Stewart, C. and Murdoch, G. (1987) Fate of the vascular patient after below-knee amputation. *Lancet*, **Set. 12**, 613–615.

Dale, W.A. and Capps, W. (1959) Major leg and thigh amputations. *Surgery*, **46**, 333–342.

Daly, M.J. and Henry, R.E. (1980) Quantative measurement for skin perfusion with xenon-133. *J. Nucl. Med.*, **21**, 156–160.

De Cossart, L., Randall, P., Turner, P. and Marcuson, R.W. (1983) The fate of the below knee amputee. *Ann. R. Coll. Surg. Eng.*, **65**, 230–232.

Dellon, A.L., MacKinnon, S.E. and Pestronk, A. (1984) Implantation of sensory nerve into muscle: Preliminary clinical and experimental observations on neuroma formation. *Ann. Plast. Surg.*, **12**, 30–40.

Dickhaut, S.C., DeLee, J.C. and Page, C.P. (1984) Nutritional status: Importance in predicting wound-healing after amputation. *J. Bone Joint Surg.*, **66A**, 71–75.

Dwars, B.J., Rauwerda, J.A., van den Brock, T.A.A., Hollander, W.D., Heidendal, G.A.K. and Van Rij, G.I. (1989) A modified scintigrafic technique for amputation level selection in diabetics. *Eur. J. Nucl. Med.*, **15**, 38–41.

Ecker, M.L. and Jacobs, B.S. (1970) Lower extremity amputation in diabetic patients. *Diabetes*, **19**, 189–195.

Eraklis, A. and Wheeler, B. (1963) Below-knee amputations in patients with severe arterial insufficiency. *N. Engl. J. Med.*, **269**, 938–943.

Fearon, J., Campbell, D.R., Hoar, C.S., Gibbons, G.W., Rowbotham, J.L., and Wheelock, F.C. (1985) Improved results with diabetic below-knee amputations. *Surgery*, **120**, 777–780.

Fleurant, F.W. and Alexander, J. (1980) Below knee amputation and rehabilitation of amputees. *Surg. Gynecol. Obstet.*, **151**, 41–44.

Gillis, L. (1965) Amputations. *Br. J. Surg.*, **52**, 821–826.

Gonzalez, E.G., Corcoran, P.J. and Reyes, R.L. (1974) Energy expenditure in BK amputations: correlation with stump length. *Arch. Phys. Med. Rehab.*, **55**, 111–119.

Harris, J.P., Page, S., Englund, R. and May, J. (1988) Is the outlook for the vascular amputee improved by striving to preserve the knee? *J. Cardiovasc. Surg.*, **29**, 741–745.

Harris, P.D., Schwartz, S.I. and DeWeese, J.A. (1961) Midcalf amputation for peripheral vascular disease. *Arch. Surg.*, **82**, 71–73.

Harris, R.I. (1944) Amputations. *J. Bone Joint Surg.*, **26**, 626–634.

Harris, W.R. (1987) Below-knee amputation: a technical note. *Can. J. Surg.*, **30**, 392–393.

Hicks, L. and McClelland, R.N. (1980) Below-knee amputations for vascular insufficiency. *Am. Surg.*, **46**, 293–243.

Homans, J. (1939) *Circulatory Diseases of the Extremities.* Macmillan, New York.

Hoar, C.S. and Torres, J. (1962) Evaluation of below-the-knee amputation in the treatment of diabetic gangrene. *N. Engl. J. Med.*, **266**, 440–443.

Hunter-Craig, D., Vitali, M. and Robinson, K.P. (1970) Long posterior=flap myoplastic below-knee amputation in vascular disease. *Br. J. Surg.*, **57**, 62–65.

Kacy, S.S., Wolma, F.J. and Flye, M.W. (1982) Factors affecting the results of below knee amputation in patients with and without diabetes. *Surg. Gynecol. Obstet.*, **155**, 513–518.

Keagy, B.A., Schwartz, J.A., Kotb, M., Burnham, S.J. and Johnson, G. (1986) Lower extremity amputation: The control series. *J. Vasc. Surg.*, **4**, 321–326.

Kendrick, R.R. (1956) Below-knee amputation in arteriosclerotic gangrene. *Br. J. Surg.*, **44**, 13–17.

Lepantalo, M., Isoniemi, H. and Kyllonen, L. (1987) Can the failure of a below-knee amputation be predicted? *Ann. Chirurg. Gyn.*, **76**, 119–123.

Lim, R.C., Blaisdell, F.W., Hall, A.D., Moore, W.S. and Thomas, A.N. (1967) Below-knee amputation for ischaemic gangrene. *Surg. Gynecol. Obstet.*, **125**, 493–501.

Loon, H.E. (1961) Below-knee amputation surgery. *Artif. Limbs*, **6**, 86–99.

Malawer, M.M., Buch, R., Khurana, J.S., Garvey, T. and Rice, L. (1991) Postoperative infusional continuous regional analgesia (PICRA): a technique for relief of postoperative pain following major extremity surgery. *Clin. Orthop.*, **266**, 227–237.

Mann, R.A.M. and Bisset, W.I.K. (1983) Anaesthesia for lower limb amputation: a comparison of spinal analgesia and general anaesthesia in the elderly. *Anaesthesia*, **38**, 1185–1191.

Marsden, F.W. (1977) Amputation: surgical technique and postoperative management. *Aust. NZ. J. Surg.*, **47**, 384–392.48.

McCollough, N.C., Jennings, J.J. and Sarmiento, A. (1972) Bilateral below-the-knee amputation in patients over fifty years of age: results in 31 patients. *J. Bone Joint Surg.*, **54A**, 1217–1223.

Mooney, V., Harvey, J.P., McBride, E. and Snelson, R. (1971) Comparison of postoperative stump management: plaster vs. soft dressings. *J. Bone Joint Surg.*, **53A**, 241–249.

Moore, W.S., Hall, A.D. and Wylie, E.J. (1968) Below knee amputation for vascular insufficiency. *Arch. Surg.*, **97**, 886–893.

Moore, W.S., Hall, A.D. and Lim, R.C. (1972) Below the knee amputation for ischaemic gangrene: Comparative results of conventional operation and immediate postoperative fitting technique. *Am. J. Surg.*, **124**, 127–134.

Murdoch, G. (1977) Amputation surgery in the lower extremity. *Prosthet. Orthot. Int.*, **1**, 72–83.

Murdoch, G., Condie, D.N., Gardner, D., Ramsay, E., Smith,

A., Stewart, C.P.U., Swanson, A.J.G. and Troup, I.M. (1988) The Dundee experience. In: *Amputation Surgery and Lower Limb Prosthetics* (Ed. G. Murdoch). Blackwell Scientific Publications, Edinburgh.

Murray, D.G. (1965) Below-knee amputations in the aged: evaluation and prognosis. *Geriatrics*, **20**, 1033–1038.

Paloschi, G.B. and Lynn, R.B. (1967) Major amputations for obliterative peripheral vascular disease with particular reference to the role of below-knee amputation. *Can. J. Surg.*, **10**, 168–171.

Pedersen, H.E., LaMont, R.L. and Ramsey, R.H. (1964) Below-knee amputation for gangrene. *S. Med. J.*, **57**, 820–825.

Perry, T. (1963) Below-knee amputations. *Arch. Surg.*, **56B**, 110–114.

Persson, B.M. (1974) Sagittal incision for below knee amputation in ischaemic gangrene. *J. Bone Joint Surg.*, **56B**, 110–114.

Pohjolainen, T. and Alaranta, H. (1988) Lower limb amputations in Southern Finland. *Prosthet. Orthot. Int.*, **12**, 9–18.

Purry, N.A. and Hannon, M.A. (1989) How successful is below-knee amputation for injury? *Injury*, **20**, 32–36.

Rizzo, R.L. and Matsumoto, T. (1990) Above vs. below knee amputations: a retrospective analysis. *Int. Surg.*, **65**, 265–267.

Robinson, K.P. (1972) Long-posterior-flap myoplastic below-knee amputation in ischaemic disease: review of experience in 1967–1971. *Lancet*, **Jan. 22**, 193–195.

Roon, A.J., Moore, W.S. and Goldstone, J. (1977) Below-knee amputation: a modern approach. *Am. J. Surg.*, **134**, 153–158.

Rush, D.S., Huston, C.C., Bivins, B.A. and Hyde, G.L. (1981) Operative and late mortality rates of above-knee and below-knee amputations. *Am. Surg.*, **47**, 36–39.

Singer, D.I., Morrison, W.A., McCann, J.J. and Renney, J.T.G. (1988) The fillet foot for end weight-bearing cover of below knee amputations. *Aust. NZ J. Surg.*, **58**, 817–823.

Smith, B.C. (1956) A twenty year follow-up in fifty below knee amputations for gangrene in diabetics. *Surg. Gynecol. Obstet.*, **103**, 625–630.

Stewart, C.P.U. (1987) The influence of smoking on the level of lower limb amputation. *Prosthet. Orthot. Int.*, **11**, 113–116.

Termansen, N.B. (1977) Below-knee amputation for ischaemic gangrene. *Acta. Orthop. Scand.*, **48**, 311–316.

Thornhill, H.L., Jones, G.D., Brodzka, W. and VanBockstaele, P. (1986) Bilateral below-knee amputations: experience with 80 patients. *Arch. Phys. Med. Rehabil.*, **67**, 159–163.

Tripses, D. and Pollak, E.W. (1981) Risk factors in healing of below-knee amputation: Appraisal of 64 amputations in patients with vascular disease. *Am. J. Surg.*, **141**, 718–720.

Weiss, M. (1966) *Report of Workshop Panel on Lower Extremity Prosthetic Fitting*. National Academy of Sciences, Committee on Prosthetics Research and Development.

Yaramenko, D. and Andruhova, R.V. (1986) Below-knee amputation in patients with vascular disease and prosthetic fitting problems. *Proshet. Orthot. Int.*, **10**, 125–128.

7

Biomechanics and prosthetic management

N. A. Govan

In prosthetic management at the transtibial level, there is general acceptance of the biomechanical analysis of Radcliffe (1961) and almost all such amputees are fitted with a patellar-tendon-bearing prosthesis of one design or another. The primary goal is to produce a prosthesis which permits virtually unhindered function of the knee on the amputated side. To achieve this requires a total contact socket in which the largest possible area of the stump shares in the distribution of the weight support and control forces generated between stump and socket when the amputee walks. Certain areas of the stump are of particular importance in weight support. The sloping surfaces of the proximal tibia are loaded, particularly on the medial side, this being counterbalanced by pressure against the shaft of the fibula. Pressure on the patellar ligament and on the popliteal area has a part to play although the importance of this weight support has been over-emphasized by the very name of the prosthesis. In the classical socket there is an indentation distal to the patella. If this indentation is made too deep a pressure sore on ligamentum patellae will result and indeed some prosthetists can produce a successful socket without this patella-tendon bar. Where possible the distal end of the socket should share in the weight support but a very careful assessment of the pressure bearing capability of the stump end is necessary before attempting to load this area.

Where reliefs have to be built into the socket in pressure sensitive areas, the aim is to reduce the pressure in these areas rather than to create spaces within the socket. During rectification of the positive plaster model an addition to, for example, the area of the head of the fibula should be restricted to a very few millimetres. Protection of the tibial crest within the socket is considered necessary by some prosthetists. Others would contend that the tibial crest is not over sensitive to pressure within the socket and that pressure on the medial aspect of the tibia and in the area of the antero-lateral muscle compartment creates a wedge effect which adequately protects the tibial crest.

The loads applied to the stump include not only the forces concerned with weight support but also the forces necessary for achieving stability of the stump within the socket. When considering the spreading of the load over the largest possible area to minimize pressure it becomes evident that the longer the stump the larger is the area presented. Some surgeons will always attempt to produce the longest stump possible in the circumstances. Below the musculo-tendinous junction of the gastrocnemius the stump is largely tendinous and in cold climates may become uncomfortably sensitive. Equally, many prosthetists find that very long transtibial stumps present those problems of over-sensitivity and present problems when aligning the prosthesis in an optimal manner and avoid producing a prosthesis with an unsightly bulge on the medial aspect of the prosthesis distally. Most of the stumps encountered by prosthetists will be of medium to short length, and the other way in which the area of loading of the stump can be increased is by extending the socket proximally. The use of the soft tissue areas above the femoral condyles, or the use of these areas coupled with the

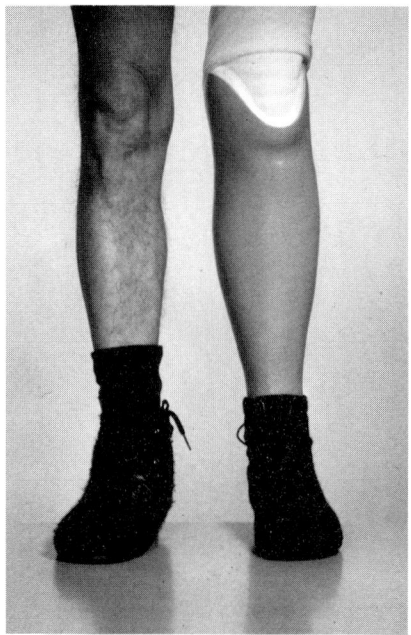

Fig. 7.1. Supracondylar suspension using Pelite.

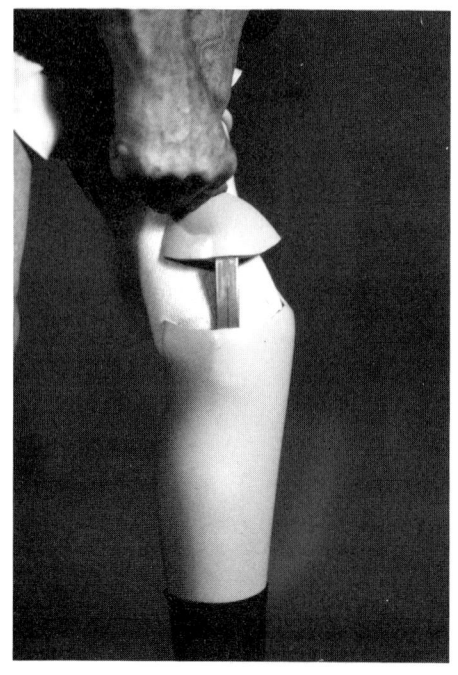

Fig. 7.2. The removable brim of Fillauer.

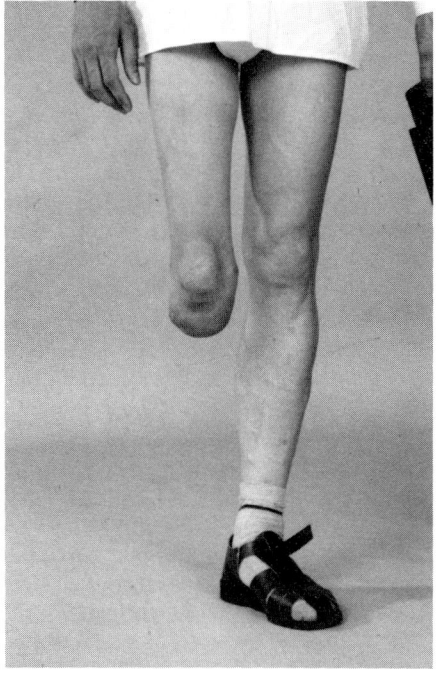

(a)

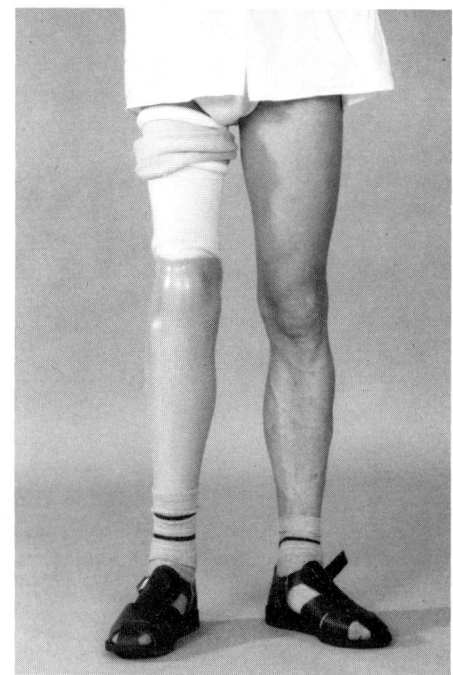

(b)

Fig. 7.3. (a,b) The use of supra-patellar suspension with a short stump.

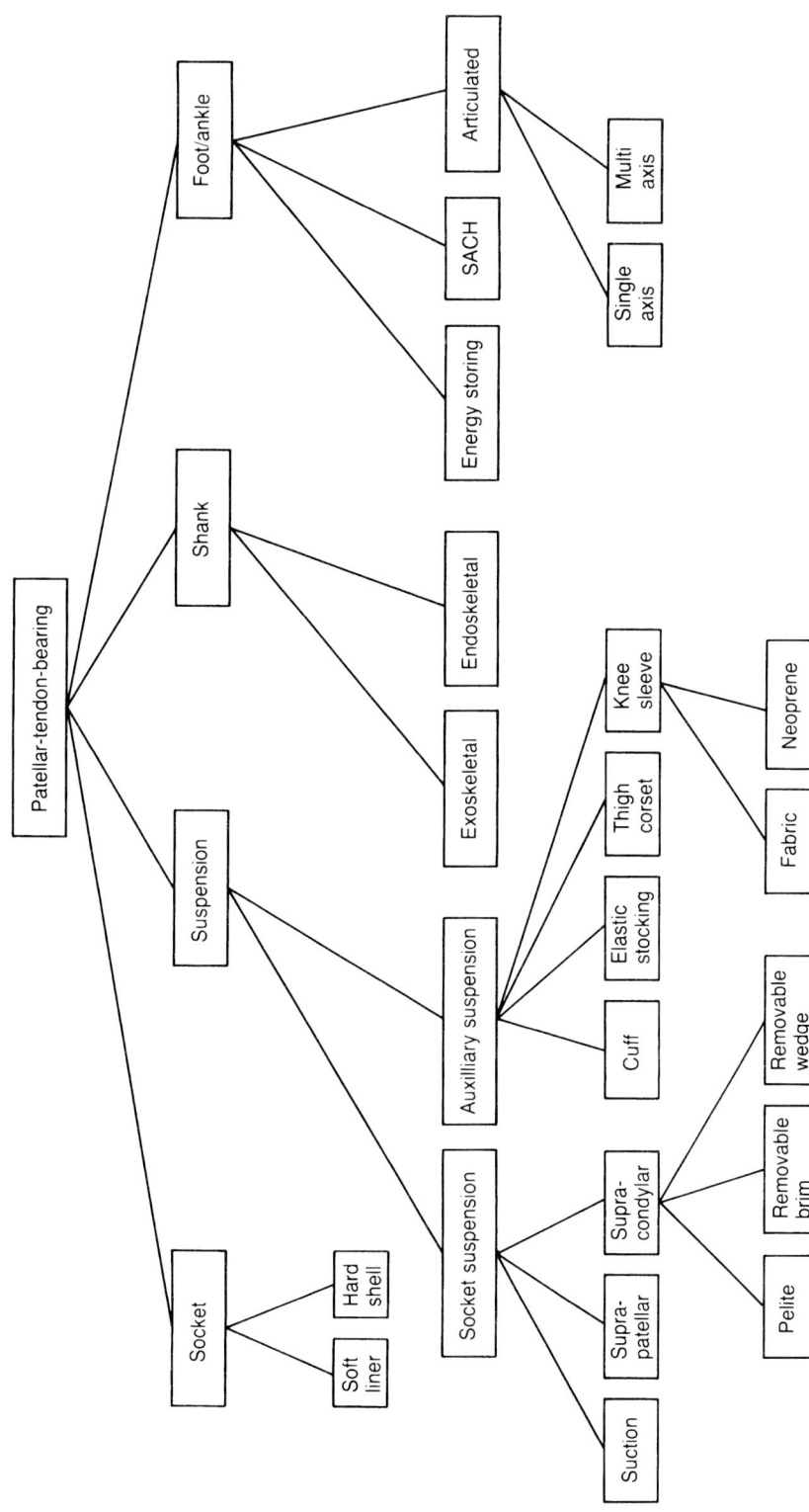

Fig. 7.4. Schema indicating the various prescription options.

supra-patellar area for suspension of the prosthesis also offers a larger area on which to apply the forces required for medio-lateral stability of the prosthesis, thus reducing pressure. The most extreme example of such an increase in area is by the addition of a thigh corset. This is sometimes prescribed when the stump is very short or the knee unstable or when there is some combination of problems such as those related to the patient's lifestyle or occupation.

The procedure which produces a transtibial stump offering the prosthetist the best opportunity for ideal prosthetic fitting is the osteomyoplasty. As a result of the bone bridge developed between tibia and fibula and the secure muscle fixation obtained, a stump with a very stable shape and volume is achieved. Furthermore, in the hands of an experienced prosthetist a measure of end bearing can be incorporated in the socket.

The other factor which influences the distribution and magnitude of these forces is the alignment of the prosthesis. An alignment with a foot placed medial to the socket leads to an increased medial/proximal and lateral/distal force distribution at mid-stance, thereby ensuring that good use is made of the medial-tibial flare. Conversely, lateral displacement of the foot reduces the stabilizing forces. An optimum alignment aims to achieve a natural looking walking base with acceptable pressure.

There now exists an extensive range of variants of the original patellar-tendon-bearing socket and a large range of options are available to prescribers. These variants and options are summarized in the table which shows one way of thinking about a possible prescription in terms of the socket, suspension, the choice of an endoskeletal or exoskeletal prosthesis and a decision as to which foot/ankle assembly to incorporate.

Most sockets have a soft inner liner (Fig. 7.1). A hard shell socket is less bulky, is probably cooler and can be just as comfortable as one with a liner. It does, however, offer less scope for socket adjustment.

Suspension systems can be considered as auxiliary suspension, as in the case of the original patellar-tendon-bearing prosthesis with its cuff suspension, or soft suspension which is provided by the socket itself. At all levels of amputation, self suspending sockets give a more efficient suspension than is the case with auxiliary suspension (Fig. 7.2). Suction sockets such as the Iceross (Kristinsson, 1993) or 3S (Fillauer et al., 1989) give patients good suspension. Supra-patellar suspension is particularly good when there is a need to prevent knee hyperextension (Fig. 7.3). When the appearance of the prosthesis is the prime consideration, a combination of a hard shell socket, elastic stocking suspension and an exoskeletal prosthesis is hard to beat. Some prescription decisions may be largely dictated by the working practices of the prosthetic centre in question.

As at other amputation levels, the wide selection of prosthetic feet available offers prescribers options to suit all requirements.

The success of the patellar-tendon-bearing type of prosthesis is evidenced by the great variety of stump types which prosthetists can successfully fit. To make the best of that success, the surgeon should strive for a well fashioned stump of an acceptable length and the clinic team responsible for prescription decisions needs to work hard at determining the very best prescription for each case that presents (Fig. 7.4).

References and further reading

Fillauer, C.E., Pritham, C.H. and Fillauer, K.D. (1989) The evolution and development of the silicone suction socket (3S) for below knee prosthesis. *J. Prosthet. Orthot.*, **2**, 92–103.

Kapp, S. and Cummings, D. (1992) The transtibial amputation – prosthetic management. In: *Atlas of Limb Prosthetics*. American Academy of Orthopaedic Surgeons, Mosby, St Louis, pp. 453–478.

Kristinsson, O. (1993) The ICEROSS concept: a discussion of a philosophy. *Prosthet. Orthot. Int.*, **17**, 49–55.

Radcliffe, C.W. (1961) The biomechanics of below knee prostheses in normal, level, bipedal walking. *Artif. Limbs*, **6(2)**, 16–24.

Radcliffe, C.W. and Foort, J. (1961) *The Patellar-Tendon-Bearing Below Knee Prosthesis*. University of California, Berkeley and San Francisco.

Wilson, A.B. Jr. (1989) *Limb Prosthetics*, 6th Edition. Demos Publications, New York. pp. 45–55.

8

Skew flap method of transtibial amputation

K. P. Robinson

The skew flap transtibial amputation method is designed to make the best use of the available blood supply at below knee level (Fulford, 1967; Termansen, 1977; Haertsch, 1981; Robinson, 1991), to provide a stump which is shaped to the ideal for fitting with a patellar tendon bearing (PTB) prosthesis with a scar that avoids the bone ends and which incorporates a myoplasty to retain the flexor function of the gastrocnemius remnant. The flaps, based on every aspect of the transtibial stump, have been described in the past and sagittal flaps are known to have been used since 1839 (Bourgery). There are a number of series indicating the advantages of this flap design in wound healing and function (Robb *et al.*, 1965; Tracy, 1966; Termansen, 1977; Alter *et al.*, 1978; Towne *et al.*, 1979; McCollum *et al.*, 1985). With conventional sagittal flaps the anterior part of the incision, and therefore the scar, overlies the anterior crest of the tibia where pressure cannot be avoided when wearing a PTB socket. By rotating the line of the incision by 15° the scar is brought 2 cm lateral to the bone, midway between the tibia and fibula bone ends and the flaps are, as a result, skewed and the scar removed from the point of high pressure. This skew flap cannot be formed in the myocutaneous method, and therefore the flaps are separated from the underlying muscles preserving the tissue plane carrying the cutaneous blood supply (Haertsch, 1981). This allows the gastrocnemius and soleus muscle remnant to be formed into a long posterior flap which can be utilized as a myoplasty within the skewed skin and subcutaneous tissue flaps. The muscle flap is folded over the bone ends, closing the open medullary cavity of the tibia which is thought to be an important factor in the blood supply of the residual bone. In practice, very little separation of the skin from the muscle flap is necessary and no significant perforating arteries have been observed crossing between the surface of the muscle and the superficial fascia.

Method

The patient is prepared for surgery by cleansing the limb with povidone–iodine solution and enclosing any infected and necrotic tissue in an impermeable bag, sealed to the skin. The perineum is isolated with a large cotton wool pad, to avoid contamination with clostridial spores from the bowel, and the limb is wrapped in dry, sterile towels. A preliminary rectal swab may identify clostridial carriers but, in any case, prophylactic antibodies, usually metronidazole and cefotaxime, are recommended. The operation can be conducted under regional or general anaesthesia.

The patient is placed supine on the operating table with a pad to prevent excess pressure on the heel of the opposite leg. If the operation is being performed for diabetic gangrene without large vessel arterial disease, or in a patient with normal blood supply, a pneumatic tourniquet is placed in position on the thigh but only inflated if an unacceptable degree of blood loss is encountered.

At this stage, the skin flaps can be marked on the skin with a waterproof marking pen (Fig. 8.1). A level of bone section 10–15 cm below the tibial plateau appears to be ideal. The shortest transtibial stump that can be accepted allows 3 cm of stump below the flexor tendons when the knee is at 90° of flexion. When the level of bone section has been marked a line is drawn around the circumference. A point is drawn on the skin 2.5 cm lateral to the subcutaneous crest of the tibia, this being the point of anterior intersection of the two flaps. A measuring tape, the length of the circumference, is folded in half and used to locate the posterior intersection of the flaps. The half length of tape is passed around the circumference line to reach the point opposite to the mark on the anterior tibial compartment. The tape is then folded into four to mark the centre of each flap and the same quarter circumference can then be marked below the circumferential line to indicate the length of each flap. A 2 cm proximal extension is drawn over the anterior tibial compartment to allow easy access to the line of bone section. When the skin flaps are fully drawn, the leg can be prepared and draped to isolate both foot and perineum. The thigh may be supported on an up-turned bowl or block to allow easy access to the operation site.

The skin flaps are cut along the previously marked lines. The skin and superficial fat are divided down to the deep fascia. The long and short saphenous veins are identified and ligated. The saphenous nerve and sural nerve are carefully separated, pulled down and divided under tension so that they will not become incorporated into the ligatures securing the veins. The sural nerve artery

and the saphenous nerve artery will usually require diathermy to obtain haemostasis. Retraction of each side of the anterior scar exposes the anterior tibial compartment where the muscles are divided transversely until the anterior tibial artery and vein are exposed (Fig. 8.2); these are secured with artery forceps, divided and ligatured with fine catgut. The peroneal muscles are similarly divided and the musculo-cutaneous nerve cut under traction. The interosseous membrane is divided at the same level and the periosteum raised around both tibia and fibula. To divide the bones a cantilever blade powersaw is ideal. Good retraction with a double hook retractor allows access to the fibula first which is divided 2 cm proximal to the line of the tibial division. Similar retraction should expose the tibia which is cut with a smooth curve; this can easily be achieved with the power saw, but if a hand saw has to be used, the anterior bevel is cut prior to the vertical cut. Saline irrigation is important throughout the bone cutting to disperse the heat from the power saw and to remove bone, dust and any fragments. A bone hook in the distal tibia medullary cavity steadies the lower leg, allowing the tibialis posterior muscle to be exposed and carefully cut transversely revealing the peroneal and posterior tibial neurovascular bundles. A plane of cleavage can easily be found between the tibia and fibula and the tibialis posterior and also between the tibialis posterior and the gastrocnemius and soleus muscle mass. In this plane will be found the peroneal and posterior tibial arteries with the venae comitans and the posterior tibial nerve. Sometimes the soleus has a more extensive insertion into the fibula and sharp dissection will

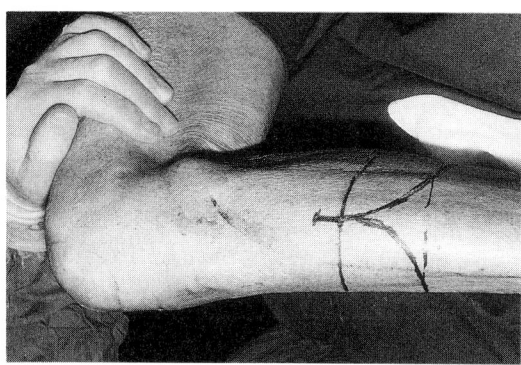

Fig. 8.1. The leg prior to amputation with the skin incision marked.

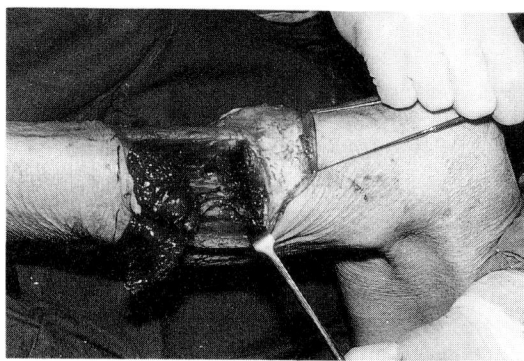

Fig. 8.2. The skin and deep fascia reflected, the anterior tibial compartment opened and the muscles divided to expose the anterior tibial vessels and nerve.

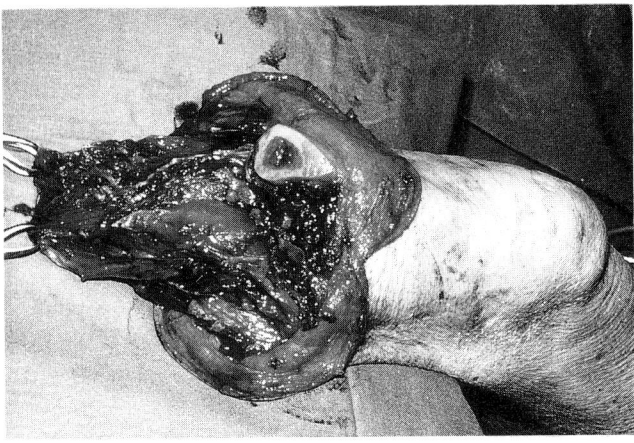

Fig. 8.3. The completed transection with the long muscle flap of gastrocnemius before it is trimmed to shape and the bone end filed to a smooth contour.

then be required to enter this surgical plane. While this part of the dissection is being performed digital pressure on the popliteal vessels will minimize any bleeding and as soon as the lower tibial remnant can be lifted clear, the peroneal and posterior tibial vascular bundles are sought and temporarily clamped with a large haemostat, while the gastrocnemius/soleus mass is dissected out of the amputated leg and cut across 15 cm distal to the line of bone section to free the specimen.

The posterior tibial nerve is extricated from the vessel clamps and the posterior tibial and the peroneal arteries are individually ligated with catgut, as are their venae comitans. The posterior tibial nerve is drawn down and cut across with a scalpel, allowing it to retract deep into the stump. The remainder of the gastrocnemius and soleus muscle masses are held out flat with tissue forceps and then thinned by cutting with a Symes amputation knife to taper the muscle flap down to the bare aponeurosis at its distal limit (Fig. 8.3). In addition, some tissue must be removed from both medial and lateral sides of the posterior muscle flap, to within even a few centimetres proximal to the level of bone section, to ensure a slender stump. In the course of this part of the procedure several soleal sinuses and vessels will be cut and these can only satisfactorily be controlled by under-running with a 'Z' or 'N' suture of 2/0 catgut. A suction drain tube can now be placed across the bone ends and brought out through the skin well above the scar on the lateral aspect of the leg. At this stage further attention is directed at the bone ends. The fibula is

smoothed with a fine bone rasp to remove any sharp corners or irregular surface and a more detailed shaping of the tibia is undertaken with a coarse bone rasp. First, the anterior curve is fashioned to remove the sharp prominence of the tibial crest and to round the posterior aspect of the bone and then the whole circumference is rounded to bring the overall contour as near to a hemisphere as the contour of the tibia will permit.

After ensuring that any bone fragments and filings are washed away a myoplasty is constructed, providing a new insertion for the gastrocnemius/soleus muscle into the anterior tibial periosteum and fascia. The muscle flap closes the tibial medullary cavity and provides soft tissue cover for the bone end (Fig. 8.4). The muscle mass

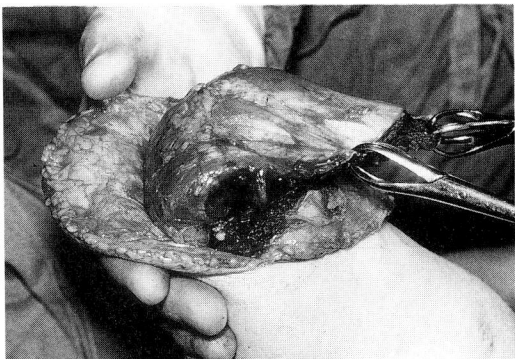

Fig. 8.4. The muscle flap folded over the bone end before it is trimmed to the correct length and attached to the anterior tibial fascia and periosteum.

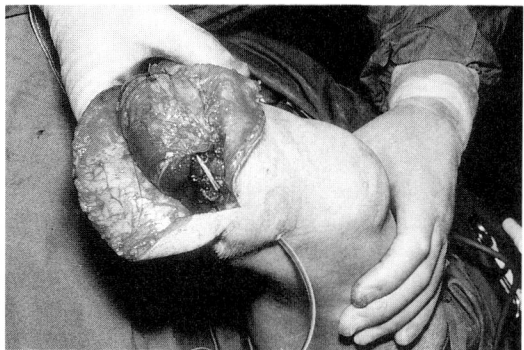

Fig. 8.5. The gastrocnemius soleus mass sutured in position with a Redivac drain in place prior to skin closure.

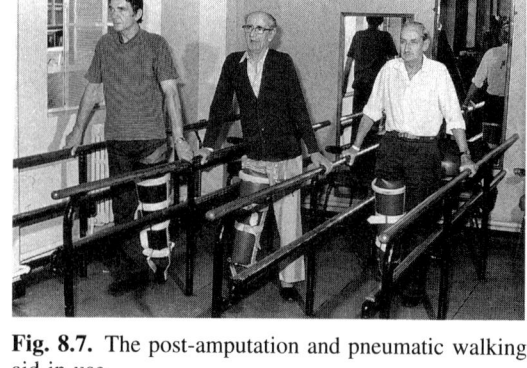

Fig. 8.7. The post-amputation and pneumatic walking aid in use.

must be trimmed to fit the outline of the periosteum and fascia and then the posterior muscle flap is secured with a No. '1' chromic catgut continuous suture to minimize the amount of foreign material left in the wound (Fig. 8.5). When secure and dry the skin falls into place and can be sutured without any tension using 2/0 or 3.0 nylon or prolene sutures interspersed with 5 mm Steristrips. There is some redundancy of skin but this is readily taken up during the suturing, especially if the halving technique of suture placement is adopted. The drain is secured and the lightest dressing of fluffed gauze must be applied and secured with a net stocking or a loose spiral elastic bandage which must not exceed 4 inches or 10 cm in width.

The drain can be removed after 48 h and its contents cultured (Fig. 8.6). Prophylactic antibiotics are usually continued for 3 days. The patient is allowed to stand with assistance on the

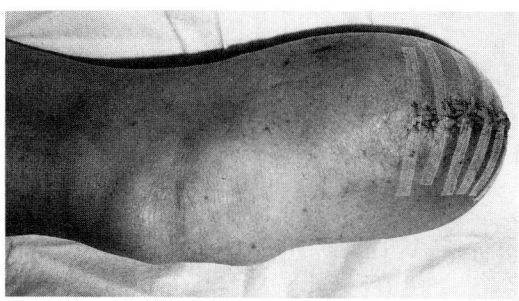

Fig. 8.6. Completed amputation stump shown 48 h after operation with skin sutures and steristrips in place.

second post-operative day and, in favourable cases, the pneumatic walking aid can be used at this time. More usually the pneumatic walking aid is used on the fifth post-operative day under supervision by physiotherapists with a pressure of 40 mm Hg. The pneumatic walking aid is not worn for more than 1 h at a time and always under supervision. Initially the walking is only in the parallel bars, although later the use of a walking frame, tetrapods or sticks can be allowed (Fig. 8.7). The cast for the first resin socket can be taken when all the oedema and swelling has resolved; this occurs usually well before the sutures are removed on the 21st day. Walking in the pneumatic walking aid has a beneficial effect on the overall volume of the stump from the time at which a prosthesis can be used. The first PTB socket can be worn between the 21st and 28th post-operative days, although in patients with non-vascular problems 10–14 days is not unusual (Fig. 8.8). Any swelling that remains a problem can be resolved by placing the limb in the controlled environment chamber and subjecting it to an intermittent positive pressure. Any unexplained pain suggests that there is continuing ischaemia in the stump and that ultimately wound breakdown may necessitate a re-trimming procedure or re-amputation to a higher level.

The choice of transtibial level for amputation must be carefully made to ensure that the vascularity and oxygenation are adequate for wound healing and that secondary surgery or re-amputation will be avoided. The part of the leg causing rest pain must be eliminated by the amputation and the operation must be conducted through warm,

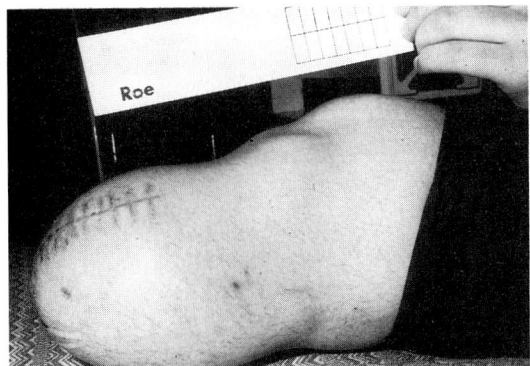

Fig. 8.8. Healed stump following removal of sutures at a time when the patient had commenced activity with the PTB below knee prosthesis.

well perfused, normally coloured skin and soft tissue. To aid clinical judgement there are a wide range of techniques to aid the process of level selection (Chapter 20). The single most readily available test is undoubtedly Doppler derived systolic pressure measurement in the lower thigh. A pressure of more than 70 mmHg virtually assures wound healing at transtibial level.

The transtibial amputation should not be selected for patients who will not be able to walk. A wheelchair user may allow the stump to be dependant while sitting and this can rapidly lead to knee flexion and swelling of the end of the stump which can ulcerate and herald secondary surgery. Certainly if the patient cannot cooperate and allows the knee to flex, ulceration will be inevitable. Similarly, a patient who already has an established fixed flexion deformity of the knee is unlikely to reverse this and both these factors are a contraindication to the transtibial amputation. Flexion deformities of up to 45° may be accepted as it is possible to incorporate this degree of flexion in the socket of the prosthesis.

Results

One hundred and eighty-three transtibial amputations have been performed by this technique. Twelve of these have been for orthopaedic indications and the longest follow-up is 9 years. The incidence of delayed wound healing, re-amputation and mortality are exactly comparable with those in published series of transtibial amputation by other techniques. A series of 35 skew flap amputations (Harrison *et al.*, 1987) suggests that healing may be obtained by this technique in patients who might not heal using a long posterior flap. However, the major benefit is in the reduction of time in which walking activity in the pneumatic walking aid can be achieved on the third to fifth post-operative day. The stump is ready for casting for a PTB socket between the seventh and twenty-first post-operative day in most patients allowing early discharge from hospital and rapid rehabilitation with cost saving and a reduction in the deterioration that patients experience while waiting for the supply of an artificial limb. The use of bandages, stump shrinkers and plaster dressings is eliminated. In the long term, there is a minimum incidence of complications associated with this stump design. Two female patients with obesity have required soft tissue reduction utilizing the same flap design. One patient developed abduction of the fibula which required its later resection but soft tissue ulceration has been greatly reduced as a problem.

References and further reading

Alter, A.H., Moshein, J., Elconin, K.B. and Cohen, M.J. (1978) Below knee amputation using the sagittal technique: a comparison with coronal amputation. *Clin. Orthop.*, **131**, 195–201.

Bek, K.M. (1982) Sagittal technique for leg amputation for vascular insufficiency. *Ugeskr. Laeger.*, **144**, 1470.

Bourgery, J.M. (1839) *Traite Complet de l'Anatomie de l'Homme.* Tome 6, Paris, pl. 83.

Buckley, C.B., Stonebridge, P.A. and Prescott, R.J. (1991) Skew flap versus long posterior flap in below knee amputations. *J. Vasc. Surg.*, **13(3)**, 423–427.

Dellon, A.L. and Morgan, R.F. (1981) Myodermal flap closure of below knee amputation. *Surg. Gynaecol. Obstet*, **153**, 383.

Fulford, G.A., (1967) *Symposium on Amputation Surgery.* Queen Mary's Hospital, Roehampton.

Haertsch, P.A. (1981) The blood supply to the skin of the leg: a post mortem investigation. *Br. J. Plastic Surg.*, **34**, 470.

Harrison, J.D., Southworth, S. and Callum, K.G. (1987) Experience with the 'skew flap' below knee amputation. *Br. J. Surg.*, **74**, 930.

McCollum, P.T., Spence, V.A., Walker, W.F. and Murdoch, G. (1985) A rationale for skew flaps in amputation surgery. *Prosthet. Orthot. Int.*, **9**, 95.

Persson, B.M. (1981) Lower leg amputation with sagittal section in vascular diseases–a study of 692 patients. *Beitr. Orthop. Traumatol.*, **28**, 656.

Pilgard, H.K. (1985) Muscle blood flow after amputations. Increased flow with medullary plugging. *Acta. Orthop. Scand.*, **56**, 500.

Robb, H.J., Jacobson, L.F. and Jordan, P. (1965) Midcalf amputation in the ischaemic extremity: the lateral and medial flap. *Arch. Surg.*, **91**, 506.

Robinson, K.P. (1982) Skew flap myoplastic below knee amputation: a preliminary report. *Br. J. Surg.*, **69**, 554–557.

Robinson, K.P. (1991) Skew flap below knee amputation. *Ann. R. Coll.* Surg. **73(3)**, 155–157.

Robinson, K.P. (1992) Skew flap below knee amputation. *Proceedings of The World Congress*, I.S.P.O.

Tremansen, H.B. (1977) Below knee amputation for ischaemic gangrene: Prospective randomized comparison of a transverse and sagittal operative technique. *Acta Orthop. Scand.*, **48**, 311.

Towne, J.D. and Condon, R.E. (1979) Lower extremity amputation for ischaemic disease. *Adv. Surg.*, **13** 199.

Tracy, G.D. (1966) Below knee amputation of ischaemic gangrene. *Pac. Med. Surg.*, **74** 251.

Yamanaka, M. and Kwong, P.K. (1985) The side to side flap technique in below knee amputation with long stump. *Clin. Orthop. Rel. Res.*, **201**, 75–79.

9

Salvage of the knee joint

A. S. Jain

It is now accepted that salvage of the knee joint in elderly dysvascular patients is of prime importance for successful rehabilitation. Realization of the critical importance of the knee joint in the gait cycle has been highlighted by the patients who have hip arthrodesis or ankle arthrodesis. On the other hand, fusion of the knee, functioning as it does between the two long lever arms of femur and tibia, critically compromises limb function. Mechanical knee function replacement has proved to be, and continues to be, an elusive engineering challenge.

Transtibial amputation is the most favoured level of amputation in all cases whatever the pathology, but it is a critical level especially in the elderly dysvascular patient. The advantages of a transtibial amputation are:

1. Retention of the knee joint.
2. Reduction of energy consumption during walking.
3. Ease of donning and doffing of the prosthesis.
4. Improvement in gait.
5. The possibility of the patient returning to original occupation and lifestyle.

The over-riding need to retain a functional knee joint when possible has placed increasing emphasis on the technique of transtibial amputation surgery. Along with increased and better knowledge of limb viability and level selection, more and more knee joints can be saved by various surgical procedures. The long posterior flap technique is discussed elsewhere but there are other complementary procedures which may prove useful in retaining the knee joint.

Long posterior flap technique

This is the most popular technique and gives excellent results. It was first described by Bickel and Gormley in 1943 and then popularized by Kendrick (1956), Burgess and Romano (1968), Robinson (1972), Murdoch (1975) and Bowker (1990). Extensive clinical studies have demonstrated excellent results using this technique but there are times when it may not be feasible and other modified techniques can be deployed to salvage the knee joint.

It is important to emphasize that level selection is vitally important and goes hand in hand with the surgical technique. The selection of amputation level should not be based solely on the successful application of the long posterior flap technique. There are other complementary techniques available which may yet salvage the knee joint and provide commensurate function. These techniques are listed here:

1. Short stump.
2. Wedge resection.
3. Excision of fibula.
4. Medially based skin flap (skew flap).
5. Osteomyoplasty.
6. Combination of procedures including skin graft and musculo-cutaneous flaps, etc.

Short stump

It has always been thought that a satisfactory transtibial stump should measure between 15 and

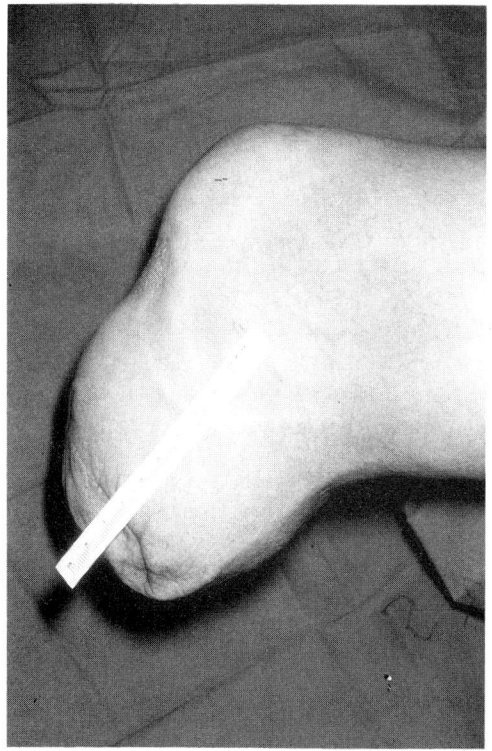

20 cm in length but with increasing advances in technology of prosthetics it is possible to fit a stump measuring 6–10 cm in length (Fig. 9.1). It is important for the amputating surgeon to know that the responsible prosthetist is able to fit the shorter stump, thereby saving the knee joint by a deliberate team effort of consultation.

Wedge resection

This is a salvaging procedure for an infected/non-healing ulcer on a transtibial stump. The technique was first described by Murdoch (1977) and subsequently its success was confirmed by Kronberg and Netz (1984) and Hadden *et al.* (1987). The principle of the technique is to excise the necrotic, infected or devitalized tissue *en bloc* from skin to bone using the full diameter of the stump and without disturbing the involved tissues, thus avoiding further spread of infection. This technique leads to slight shortening of the stump but secures primary healing of the ulcer (Fig. 9.2).

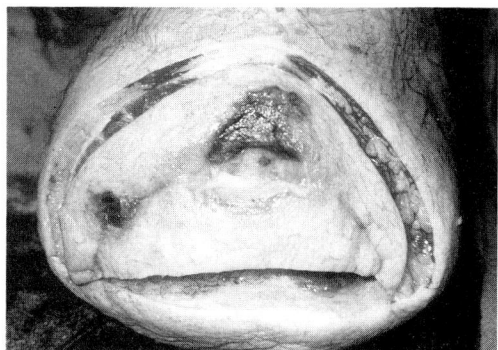

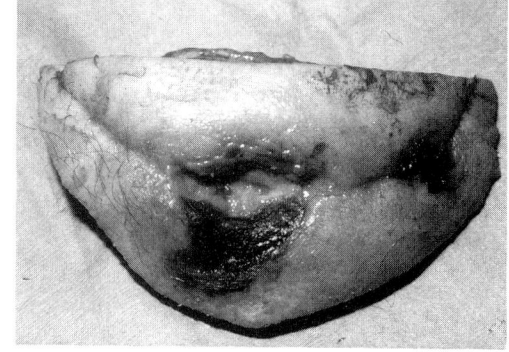

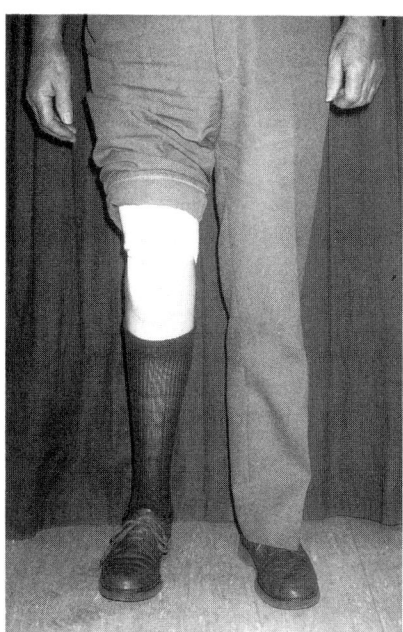

Fig. 9.1. Patient with a short stump fitted successfully.

Fig. 9.2. Elliptical incision across the unhealed area and excision of the total wedge.

Excision of fibula

As highlighted previously the preservation of even a short tibial stump is of great functional importance to the amputee. The need for a proprioceptive sense of a stump for a proper forward thrust of the leg is self-evident. The amputating surgeon may well encounter difficulty in skin coverage in some short stumps due to ischaemic problems or lack of tissue due to trauma, etc. In the short trans-tibial stump the prosthetist has considerable difficulty in transferring force action to the lateral aspect of the stump to balance those forces readily absorbed on the smooth curved surface of the medial aspect of the tibia because of the unsecured fibular remnant, the lack of underlying muscle and the unwelcome presence of the remnant of the lateral peroneal nerve. Excision of the fibula (fibulectomy) gives that latitude to gain extra skin to provide adequate

cover and to emancipate the lateral curved surface of the tibia for force transference (Spira and Steinbach, 1933) (Fig. 9.3). The fibula has to be excised from the upper tibio-fibular joint. It must be emphasized that the lateral ligament which is attached to the fibula, if not sutured to the periosteum of the tibia, can lead to some valgus deformity of the knee at a later date. If the ridge of the tibia at the tibio-fibular joint is prominent then it should be smoothed.

Medially based skin flap (skew flap)

Experience with vascular assessment over the years has shown that there is a medial to lateral gradient in the skin blood flow of dysvascular patients. There are some patients where only medial skin blood flow at the below-knee level is sufficient and compatible with wound healing, whereas the lateral blood flow is below the predicted level of wound healing, thus making those patients as an 'at risk' group for wound healing. In this group a medially based flap may possibly save the knee joint.

The surgical technique does not differ from the standard long posterior flap technique apart from the skin flaps which are mapped out according to the skin blood flow patterns. Generally the base of the main flap is some 30–70° medially to that of a standard long posterior flap. The flap is marked such that it is around 12–15 cm long and takes in approximately half the circumference of the leg. The lateral skin incision passes horizontally around the leg just at the level of bone section (Fig. 9.4).

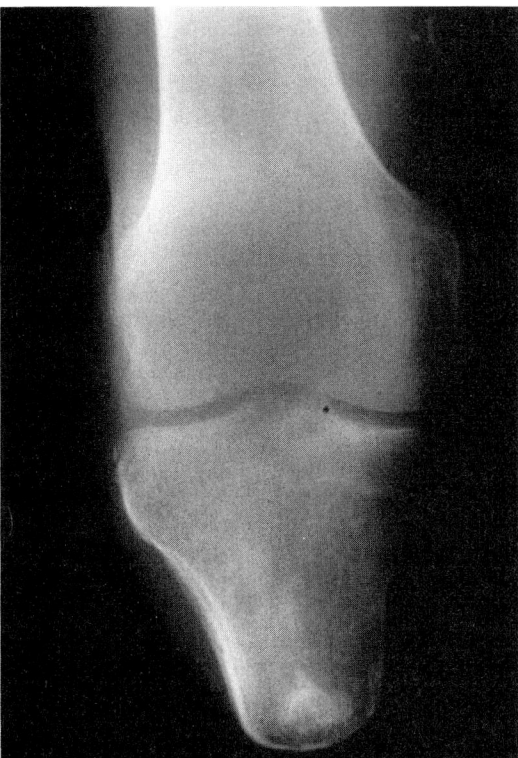

Fig. 9.3. Short tibial stump following excision of fibula, smoothing of margins of tibio-fibular joint permitting force transference to lateral curved surface of tibia.

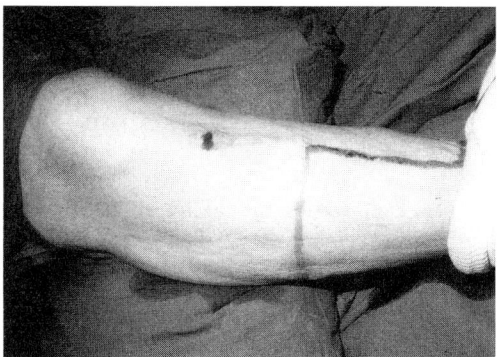

Fig. 9.4. Lateral transverse incision with medially based flap.

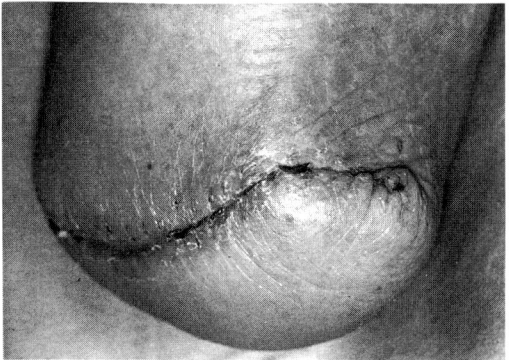

Fig. 9.5. Well healed stump.

When the flap has been marked the skin, soft tissue and bone are divided as described by Kendrick (1956). To obtain comfortable closure considerable trimming of muscle is required. This is the most demanding stage of the technique and differs a little from the classical description. The soleus is sacrificed and the gastrocnemius, deep fascia and skin are left as the major components of the flap. The flap should be closed without tension over the sculptured end of the tibia giving a uniform round appearance (Fig. 9.5).

Osteomyoplasty

Ertl (1949) devised a procedure which produced a stump with end bearing properties. This was done by joining osteoperiosteal flaps from the tibia and fibula to produce the formation of a solid bony bridge. This method has been employed in the American military (Deffer, 1970), but to a much lesser extent in the civilian population. Marsden (1977) also recommended it as a useful technique in the young, especially in traumatic cases.

The author uses this procedure mainly in young patients where there is enough skin and good quality bone is present. This method is not employed in vascular disease because the procedure is more elaborate and has an added risk when the level of amputation itself is difficult to assess and the patient's survival is limited. The purpose of osteomyoplasty is to provide a durable, possibly end-bearing strong stump which is more useful in the young adult since it provides better locomotor capacity. The surgical technique is as described by Murdoch (1970). The level of amputation is

usually just above the musculo-tendinous junction of the calf muscles and this, of course, will vary depending on the total length of the tibia. Vertical incisions are made on the antero-lateral and postero-medial aspects of the stump distally from a point about 1 inch above the anticipated level of bone section. They are carried down the limb sufficiently far to expose 3 to 4 inches of tibia beyond the level of ultimate bone section. At the distal end these incisions are joined by a circular incision. The two flaps thus formed are elevated subcutaneously to ensure that the deep fascia and muscle aponeurosis remain undissected. A vertical cut is then made through the deep fascia of the limb just lateral to the anterior tibial crest avoiding the periosteum. A further vertical incision is made through the deep fascia overlying the fibula. The whole of the antero-lateral group of muscles including the peroneals is then elevated by sharp dissection from the distal part of the operative field from the bed formed by the tibia with its overlying periosteum, the interosseous membrane and the fibula, to a point just proximal to the level of bone section. The posterior muscle flap is treated in a similar manner, the deep fascia being incised just medial to the subcutaneous surface of the tibia. Again after division of the muscles at the distal end of the wound the whole muscle flap is elevated by sharp dissection from the bed formed by the tibia, interosseous membrane and fibula to a point just proximal to the level of the bone section.

Up to this point the two bones remain intact. The fibula is divided with a Gigli saw at the same level required for the tibia along with the whole of the attached interosseous membrane. Two vertical incisions are then made in the periosteum of the tibia so that roughly equal osteo-periosteal flaps can be elevated. The antero-medial flap comprising at least that periosteum covering the subcutaneous surface of the tibia is elevated with a medium-sized gouge so that small flakes of bone remain securely attached to the parent periosteum and this elevation proceeds to a point half an inch above the anticipated level of bone section. The same procedure is employed in elevation of the postero-lateral flap. Only now can the tibia be divided at the prescribed level and this is best done with a reciprocating power saw. The antero-distal end of the divided tibia should then be carefully sculptured.

Before formation of the osteo-periosteal bridge the posterior tibial nerve and vessels should be

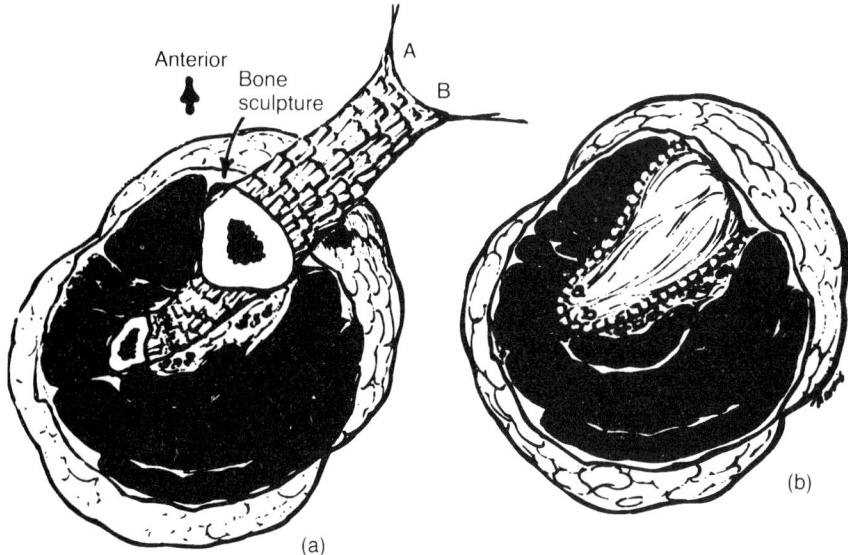

Fig. 9.6. Osteomyoplasty. Sketch (a) shows the postero-lateral osteo-periosteal flap reflected laterally and sutured to periosteum of fibula. The antero-medial osteo-periosteal flap displayed demonstrates attached osseous chips. It is brought over the end of the tibia after bone sculpture (sketch b) and sutured to the periosteum of the fibula (note A,B), forming a firm tube. (From Artificial Limbs, **6,** No. 2, June 1962, pp. 90–91.)

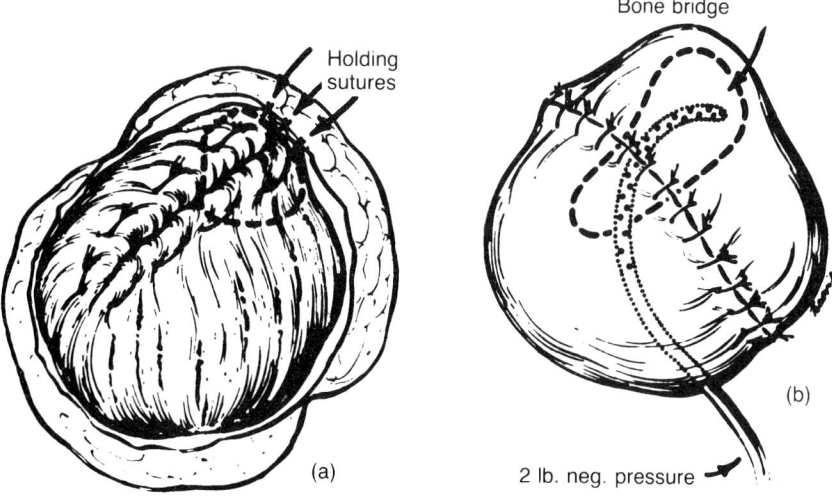

Fig. 9.7. Osteomyoplasty. Sketch (a) shows antero-lateral group of muscles sutured under tension to the posterior muscles. Note the important anchor sutures securing muscle groups to the base of the osteo-periosteal bridge and the periosteum of the tibia. Sketch (b) indicates the relationship of the final skin suture at right angles to the muscle suture and osteo-periosteal bridge. Closed suction drainage is employed. (From Artificial Limbs, **6,** No. 2, June, 1962, p. 91.)

secured, isolated and divided and the posterior tibial nerve divided cleanly under light tension. The postero-lateral osteo-periosteal flap is then reflected upon itself and, after trimming, sutured to a small cuff of periosteum elevated for this purpose from the fibula at the time of its division. The antero-medial osteo-periosteal flap is then brought over the prepared end of the tibia and sutured to the lateral part of the fibular periosteal cuff after trimming to length. Suture of the two flaps is now completed between the tibia and fibula, thus forming a rather firm osteo-periosteal tube bridging the two bones (Fig. 9.6).

Attention is then directed towards the two muscle flaps with their intact aponeurosis and deep fascial coverings. Briefly, they are cut to length, trimmed, contoured and sutured under some tension over the periosteal tube. Discussion could be developed as to how best this could be done. In practical terms trial shortening of the flaps carried out with tension maintained on the muscle flaps by some suitable instrument permits the procedure to be done effectively and the retention of the deep fascia ensures secure fixation. The posterior flap will require considerable trimming, usually necessitating excision of the deep posterior muscles and most of the soleus. The gastrocnemius itself will probably require lateral and medial trimming. The remaining vessels and nerves should be picked up as this trimming proceeds. One important point of technique must be underlined and that is the need to secure both groups of muscles by anchoring sutures to the periosteum of the tibia and to the base of the osteo-periosteal bridge (Fig. 9.7). This prevents dislocation laterally. The two long skin flaps are now carefully tailored and sutured. It will be noted that the skin scar will now lie at approximately right angles to the muscle scar. Closed suction draining is recommended and it is occasionally advisable to use one drain deep to the muscle suture and one in the subcutaneous space. The dressing of the would and its subsequent management is as for the posterior flap operation.

A combination of above procedures and inclusion of skin grafts, musculocutaneous rotation flaps, etc.

The surgeon should use all his skills and every available procedure in his armamentarium in order to save the knee joint (Fig. 9.8). In the traumatic

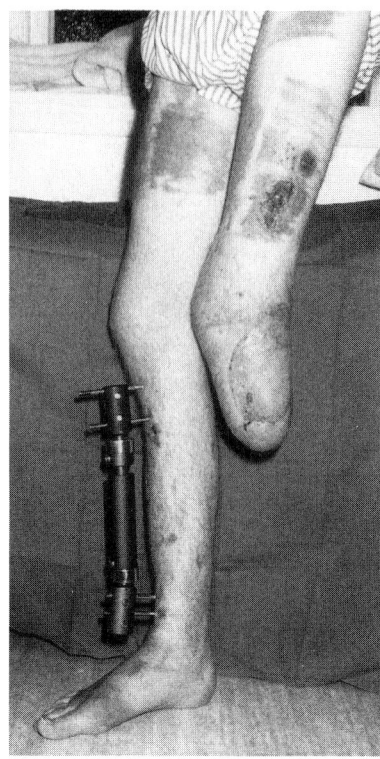

Fig. 9.8. Salvage of the knee joint by myocutaneous flap and skin graft.

group it is always useful to use skin grafts, musculocutaneous flaps, rotation flaps and sometimes the distal part of the limb, i.e. the heel pad and os calcis (Hallock, 1991) to save the knee joint.

It is important to conclude that the difference between a trans-tibial amputee and a trans-femoral amputee can be so extreme that one can return to all his previous activities and the other may end up in long-term care. In view of all the physical, economical and psychological factors, every attempt should be made to salvage the knee joint.

References

Bickel, W.H. and Gormley, (1943) Amputations below the knee in occlusive arterial diseases. *Surg. Clin. North Am*, **23**, 982–994.

Bowker, J.H. (1990) Transtibial (below-knee) amputation. In: *Report of ISPO Consensus Conference on Amputation Surgery* (eds. G. Murdoch, N.A. Jacobs and A. Bennett

Wilson), International Society for Prosthetics & Orthotics, Copenhagen, pp. 10–25.

Burgess, E.M. and Romano, R.L. (1968) The management of lower extremity amputees using immediate post-surgical prostheses. *Clin. Orthop.*, **57**, 137–146.

Deffer, P.A. (1970) More on the Ertl osteoplasty. Newsletter, *Amputee Clinics,* **2** (1), 7.

Ertl, J. (1949) Uber Amputationsstumpf. *Chirurg.*, **20**, 218.

Hadden, W., Marks, R., Murdoch, G. *et al.* (1987) Wedge resection of amputation stumps: A valuable salvage procedure. *J. Bone Joint Surg*, **69**, 306–308.

Hallock, G.G. (1991) Isle of palm and sole fillet flaps. *Ann. Plast. Surg.*, **26(6)**, 514–519.

Kendrick, R.R. (1965) Below knee amputation for ischaemic gangrene. *Br. J. Surg.* **44**, 13–17.

Kronberg, M. and Netz, P. (1984) Wedge resection in infected below-knee amputation stumps. *Acta Orthop. Scand.* **55**, 106–107.

Marsden, F.W. (1977) Amputation: surgical technique and postoperative management. *Aust. N.Z. J. Surg.* **47**, 384–392.

Murdoch, G. (1970) The surgery of below-knee amputation. In *Prosthetic and Orthotic Practice* (ed. G. Murdoch), Arnold, London, pp. 45–60.

Murdoch, G. (1975) Research and development within surgical amputee management. *Acta Orthop. Scand.* **46**, 526–547.

Murdoch, G. (1977) Amputation surgery in the lower extremity. *Prosthe. Orthot. Int.*, **1**, 72–83.

Robinson, K.P. (1972) Long posterior flap myoplastic below knee amputation in ischaemic disease: a review of experience in 1967–1971. *Lancet*, **1**, 193–195.

Spira, E. and Steinbach, T. (1973) Fibulectomy and Resection of the peroneal nerve for 'short tibia stumps'. *Acta Orthop. Scand.*, **44**, 589–596.

Section 3

The ankle

10

The Syme ankle disarticulation

A. S. Jain

James Syme, Professor of Clinical Surgery in the University of Edinburgh, first performed an amputation at the ankle joint in 1842 on a 16-year-old boy who suffered from caries of a tarsal bone. Since then this operation has been used extensively throughout the world and has been favoured by some and condemned by others. The favourable reports have come from Scotland and North America; the unfavourable reports mainly from English surgeons (Langdale-Kelham and Perkins, 1942). It is important to note that the unfavourable reports were probably due to failure to adhere to the precise surgical technique. Most of these operations were not in fact Syme amputations as described, as the techniques used included excision of the distal 2 inches of the tibia, i.e. a supramalleolar excision. The credit or the discredit goes to Elmslie (1942) and then others followed (Warren *et al.*, 1955). It must be emphasized that the procedures described were not true Syme amputations but were nevertheless identified as such and popularized as 'Syme's' operation. Inevitably the bad results following these techniques resulted in condemnation of the true Syme operation.

There have been numerous publications on this subject. Some are of a purely historical nature, providing a description of the Syme amputation and presenting a few cases. Others are mainly review papers written by famous people in the field of amputation surgery but have very little further to contribute.

The subject can be usefully examined under the following headings:

1. Pre-operative assessment/level selection.
2. Surgical technique.
3. Assessment of the stump.

Pre-operative assessment/level selection

The Syme amputation is the main functional level as it produces an end bearing stump and allows the patient to walk without a prosthesis in an emergency or on nocturnal visits to the toilet. It has, however, proved to be very difficult to identify those factors which are critical to level assessment.

Silbert and Haimovici (1954) felt that there was a requirement for careful pre-operative level selection to obtain satisfactory results. The following important factors were identified:

1. Every case should be considered carefully on its own merits.
2. Local factors.
3. Patient's general condition.

The following local factors were deemed important:

1. Extent of gangrene.
2. Degree of infection.
3. Condition of adjacent tissue.
4. Degree of arterial impairment.
5. Severity of pain.
6. Line of demarcation.

On the basis of this clinical experience it was concluded that a gangrenous toe inevitably requires amputation above the ankle whereas gangrene of the tip of the toe has a favourable result following removal of the digit. The authors also suggested that gangrene of the anterior one-third of the foot required a transtibial amputation as the Syme amputation was unlikely to be viable. This conclusion emphasizes the critical role of assessment in amputation surgery.

In diabetic patients Baddeley and Fulford (1964) looked at the role of arteriography for level selection. The arteriograms were performed on the table prior to surgery under anaesthesia. The factors scrutinized were:

1. Collateral efficiency.
2. Degree of the run of flow.

Sixty-seven lower limb amputations were performed; of these 48 healed giving an overall success rate of 71.6%. Only two Syme amputations were performed and both healed.

Syme, Swanson, Fish et al. (1982) is the only randomized study to be found on review of the literature and is concerned with the healing rate. There were two groups of patients.

Group 1 – where conventional clinical criteria were used for level selection.
Group 2 – Doppler ultrasound blood pressure measurement was used for level selection.

The results showed that patients in Group 1 had a significantly higher rate of revision surgery, whereas Group 2 showed an improvement in the rate of primary healing. Although this article highlights the importance of ancillary methods of level selection it failed to give precise indicators.

Malone, Leal, Moore et al. (1981) suggest that xenon isotope clearance for skin blood flow measurements is the 'gold standard' in level selection. Thirteen Syme amputations were performed out of 76 lower limb amputations studied by this method. Out of 13 cases 12 healed primarily and one required further surgery. They stated that a blood flow of more than 2.2 ml/100 g/min at the proposed level of surgery predicts satisfactory healing. In the case of the Syme amputation recorded levels ranged from 3.1–22.0 ml/100 g/min.

Dickhaut, Delee, and Page et al. (1984) published a significant paper which underlined a different and significant aspect of assessment of wound healing, viz., the nutritional status of the patient. Two factors were examined:

1. Albumin level.
2. Total lymphocyte count.

Twenty-three Syme amputations were carried out using the criteria identified by Wagner (1992) viz.

(a) Patient must be a potential prosthetic user.
(b) Heel pad must be free of open lesions.
(c) Doppler arterial pressure at ankle must be above 70 mm Hg.
(d) Ischaemic index (ankle/brachial ratio) should be more than 0.45%.
(e) No gross pus at amputation site.
(f) No ascending lymphaginitis or cellulitis.
(g) No gas in tissues.
(h) Intra-operative bleeding should occur in skin flaps within 3 min of tourniquet release.

Out of 23 Syme amputations, 10 healed. All of these had a serum albumin level over 3.5 g/dl and a total leucocyte count (TLC) over 1500 per mm^3. The other 11 patients had positive Wagner criteria but with a low albumin and TLC. Of this number, two patients only had satisfactory wound healing and the other nine failed to heal. This highlights that nutritional assessment plays an important part in predicting wound healing, certainly in the Syme procedure.

Surgical technique

The technique can be discussed under three headings:

(a) Original technique.
(b) Two-stage Syme amputation.
(c) Modified Syme amputation.

Original technique

There are many descriptions of the surgical procedure and all emphasize the importance of good technique.

The important publications are:

(a) James Syme's original paper of 1843 and a further four papers in the same decade.

(b) Harris (1956) made a classic contribution concerning the total end bearing quality of the stump and the importance of the elastic adipose tissue of the heel pad. It highlights the following problems which are encountered:

(i)   Misplaced heel flap.
(ii)   Sloping cut surface of tibia.
(iii)   Too small cross-sectional area of the cut end causing localized pressure leading to callus formation.
(iv)   Unstable or 'wobbly' heel pad.
(v)   Tender heel pad.
(vi)   Neuroma of posterior tibial nerve.
(vii)   Marginal gangrene of heel flap.
(viii) Vascular insufficiency.

Indications for the Syme amputation

(i)   Severe injury to foot.
(ii)   Intractable infection of foot, e.g. pyogenic, TB, blastomycosis, 'Madura' foot.
(iii)   Deformities of the foot.
(iv)   Selected cases of peripheral vascular disease.
(v)   Diabetic foot.
(vi)   Deficiencies and deformities present at birth.
(vii)   Frostbite/immersion foot as a result of thrombosis of the vessels due to cold.
(viii) Neurological problems, e.g. ulcers due to leprosy, diabetes and syringomyelia.
(ix)   Malignant disease of the foot.

Positioning of patient

Surgery is carried out with the patient lying supine with the heels projecting over the edge of the operating table. A sandbag should be placed a few inches proximal to the ankle joint.

Tourniquet

With the exception of peripheral vascular disease and diabetes all other conditions requiring the Syme amputation are carried out under tourniquet.

Incision

Figure 10.1 shows the two points which are joined with slight concavity proximally. The lateral incision is extended from these two points downwards viz. the tip of the lateral malleolus and medially a point about 2 cm posterior to the tip of the medial malleolus. The plantar incision is completed across the sole with slight convexity towards the toes.

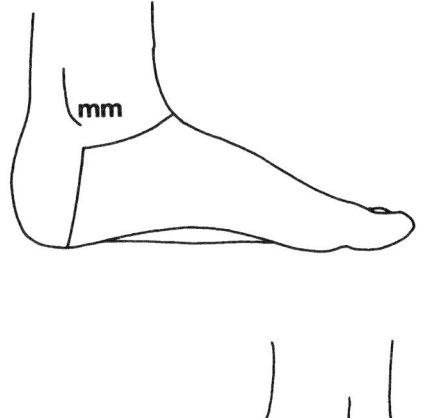

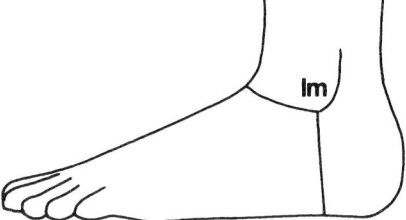

Fig. 10.1. Incisions joining lateral malleolus (LM) to point 2 cm posterior to tip of medial malleolus (MM). Plantar incision completed across sole.

Procedure

The joint is opened by dividing all structures in deepening the dorsal incision. The anterior tibial neurovascular bundle is identified and dealt with in the normal surgical fashion. The internal saphenous vein should be ligated to avoid post-operative oedema.

The foot is fully plantar flexed, exposing the articular surface of the talus (Fig. 10.2). This procedure can be facilitated by using a bone hook to pull the talus anteriorly. The knife is

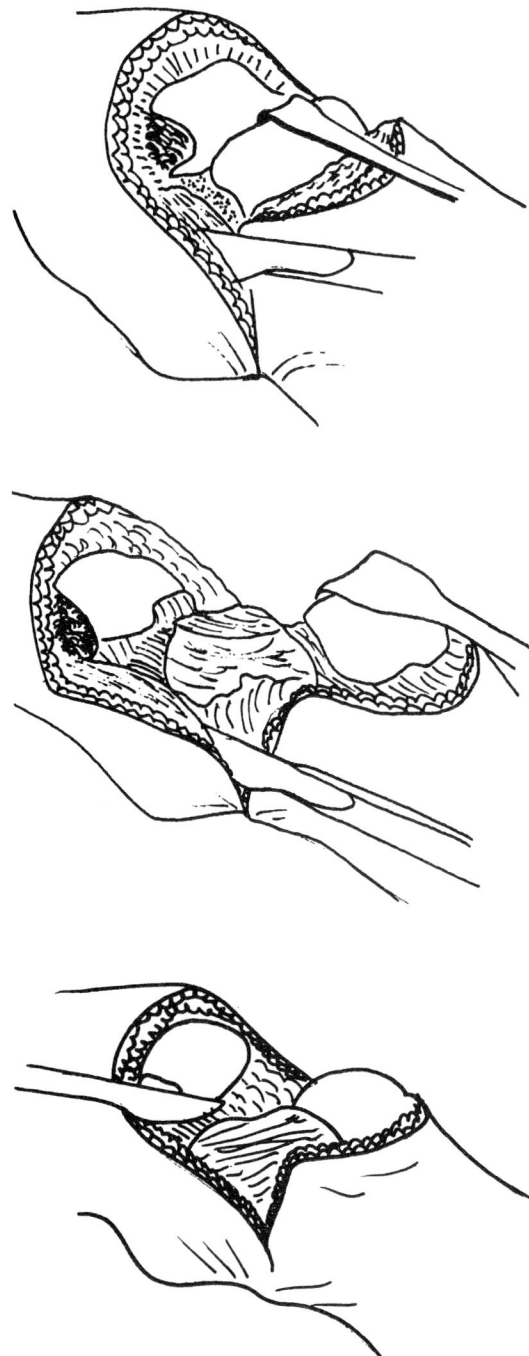

passed down on either side to divide the collateral ligaments. This allows further plantar flexion of the foot. By subperiosteal dissection the posterior capsule of the ankle joint is stripped from the os calcis. All dissections should be carried out close to the bone of the os calcis in all directions. The tendo-Achilles is divided and its distal attachment is stripped off the back of the os calcis. The foot is further plantar flexed till dissection meets the plantar dissection which should be deepened directly to the os calcis so that the os calcis is finally peeled out of the heel flap as cleanly as a boiled egg from its shell. This technique should leave an intact fibro-fatty heel pad with its special compartmented structure undisturbed. All the long tendons are cut short. The important factor in the technique is that one should not seek out the posterior tibial neurovascular bundle since accidental division of these vessels leads to ischaemia of the heel pad.

The tibial articular surface is prepared ideally by a saw cut. The heel pad is carefully turned upwards and backwards. Both malleoli and the distal quarter-inch of the tibia is exposed. A thin sliver of tibia with both malleoli is removed using a saw. The saw cut should be in the horizontal plane parallel to the ground irrespective of any deformity, if present, of the tibia. The thinnest possible amount of tibia should be removed allowing the largest possible cross-sectional area of tibia and fibula. If a tourniquet has been used it should be removed at this stage and perfect haemostasis secured.

Wound closure

Once haemostasis is secured the wound is closed by bringing the heel flap up to the anterior surface of the tibia. When closing the wound particular attention should be paid to the disparity in size, shape and thickness between the heel flap and the anterior skin margins. In most instances absorbable sutures can be used to oppose the subcutaneous tissues. Occasionally in vascular disease skin sutures alone are advisable. The skin is sutured with fine non-absorbable sutures (Fig. 10.3).

Drainage is essential and the author prefers the use of closed suction drainage.

Fig. 10.2. (a) Foot is fully plantar flexed exposing articular surface of tibia. (b) Knife passed down either side of talus permitting division of collateral ligaments and by subperiosteal dissection, removal of the posterior capsule and (c) division of Achilles tendon.

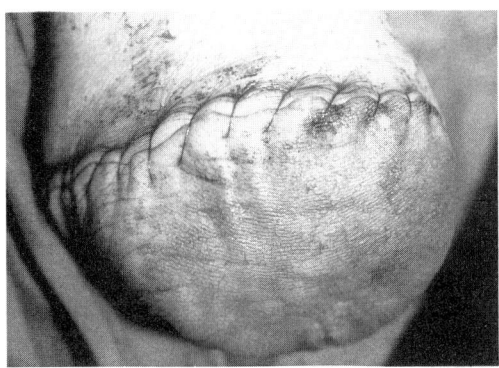

Fig. 10.3. Heel flap in position and suture completed.

Heel flap fixation

Most writers have emphasized the importance of heel flap fixation but the methods vary. Harris prefers to avoid transfixion of the stump and uses external fixation by Elastoplast strips applied from behind, forward and from side to side. It is also important to note that elimination of dead space and any haematoma is as important as fixation of the heel pad. The author prefers to transfix the heel pad in non-vascular cases, with a transfixing pin, ('K' wire or a thin Steinman pin) until the plaster cast is applied. Following application of the plaster cast the fixation pin is removed as the plaster takes over the stabilization. In vascular cases it is better to use some sort of broad 'steristrips' prior to plaster cast application since any damage to the skin in the heel pad area due to the transfixing pin may lead to skin necrosis. Following surgery, the stump is enclosed in a rigid dressing.

Pavot (1973) describes the pure disarticulation of the ankle joint without any excision of articular cartilage equating with the Syme amputation (stage one) as carried out by Wagner (1981). This is described as a definitive procedure. Twenty-seven patients underwent disarticulation of the ankle. Some 23 cases were war related injuries and four were diabetics. In 25 patients the stumps healed satisfactorily and the patients were discharged with prostheses. There is no record of their subsequent progress but two patients with diabetic lesions did not heal and required further surgery.

The two-stage Syme amputation

There have been reports in the literature of two-stage Syme amputations giving better results than the one-stage procedure. This technique was described by Hulnick, Highsmith and Bouton (1949) and a paper from Spittler, Brennan and Payne (1954) describes the results in 36 patients who were fitted with prostheses. Thirty patients were still using their prostheses at the time of review and two required higher amputation because of osteomyelitis. Sixteen patients were manual labourers and returned to work and four military personnel also went back to their original duties.

Sarmiento and Warren (1969) reported a 50% revision rate to higher levels following the Syme amputation.

Hunter (1975) and Shelswell (1954) strongly recommended that the Syme amputation should not be performed on a pulseless foot.

Wagner (1981), in a significant publication regarding the two-stage Syme amputation performed on some 500 cases, describes that technique and indications in detail but regrettably does not report any results in detail. He purported to contradict the statement made by Alldredge and Thomson (1946) that the Syme amputation should not be performed in patients with diabetes, vascular disease or neuropathy but the statement is made without conviction and is not backed by scientific evidence.

Pinzur, Jordan and Rana (1982) described the results of nine patients treated by the two-stage Syme amputation on Wagner's criteria. Seven patients had good healing and were fitted with a prosthesis. One required revision of a Syme amputation at the same level and the other required a transtibial amputation.

Modified Syme amputation

Various modifications have been proposed based on the Syme technique but with the objective of obtaining a less bulbous end and providing a more cosmetic appearance once fitted.

Wilder (1964) stated that in such a technique:

(a) The articular surface is retained.
(b) Protruding tibial and fibular malleoli are removed.

This publication emphasizes the importance of the cartilaginous weight bearing surface as being invaluable. The advantages of the technique are listed as:

(a) Stump tolerates end pressure.
(b) Long lever arm of the leg provides increased control of prosthesis.
(c) Ability to walk without prosthesis.

The last two advantages are common to all techniques relating to the Syme amputation.

Five cases were reported without any further assessment or results.

The classical technique was further modified by Sarmiento, Gilmerre and Finnieston (1966) in order to reduce the circumference by rongeuring of the metaphyseal flare of the distal tibia and bevelling of the distal fibula. This leads to a slightly wider distal bony stump than the diaphyseal portion of the leg. The authors contended that the eventual reduction of stump circumference is approximately one-third without interfering with the end bearing properties.

Further modifications were proposed once again by Sarmiento (1972). This time the reduction in circumference was obtained by osteotomy of the tibial and fibular malleoli to narrow the distal diameter: nearly one half-inch of the distal end of the tibia was trimmed. It was also mentioned that further reduction could be obtained by reducing the flares. Fifty-three cases had a Syme amputation performed in one or other of these two modifications. Surgery was performed on one group of 38 patients in the initial period of the technique and the second group of 15 in the later phase of the technique. In the first group when the technique was just being established 19 patients required reamputation; 15 to transtibial and four to transfemoral levels. In the second group when the technique was further modified only two required re-amputation.

These modified techniques are not amputations as described by Syme and later endorsed by Harris and in the author's view should be classified as long transtibial amputations.

Mathur, Piplanik and Majid (1985) describe their modification as follows:

(a) Smaller heel flap.
(b) Tibia and fibula are cut half-an-inch above the articular surface.
(c) Heel pad firmly secured to the cut end of bone by suturing to periosteum.

Sawamura (1989) described the advantages of tenodesis as:

(i) Prevention of muscle atrophy
(ii) Provision of better circulation.
(iii) Better proprioception.
(iv) Augmented suspension by isometric contraction of muscles.

In his series 44 patients had the Syme procedure carried out using this modified technique. Fifteen of the 44 had tenodesis performed. Twenty-one other patients who had the Syme amputation carried out in other hospitals were also reviewed, making the total 65. Out of the 65 patients, four died and six required further revision to transtibial level. One conclusion was that there is always a risk of late breakdown of the Syme amputation in peripheral vascular disease cases.

Modifications by Mazet (1968) included the reduction of the medio-lateral diameter but the heel cup was managed much as in the classic Syme amputation.

Post-operative assessment of stump

Most assessments have been carried out based on the rate of wound healing or the re-amputation rate. Some papers have highlighted one or two other factors especially with respect to children, such as functional assessment, psychological assessment and physical evaluation. The paper by Murdoch (1976) also included a biomechanical study on end-bearing qualities of the stump. He reported on 67 cases of the Syme amputation of which 30% were performed for congenital deformities.

Three factors were examined:

(a) Mobility of heel pad grouped in four grades.
(b) Bone ends: 25 patients were deemed satisfactory.
(c) Biomechanical considerations using a dynamometer specifically designed to measure the end bearing capacity of the stump fitted with a prosthesis: these tests demonstrated a range from 34% to 90% end bearing.

Herring (1986) reported on 21 patients with 23 Syme amputations. These were mainly children under the age of 5 years. The evaluation consisted of physical examination and psychological function testing. The physical examination was in the form of a 50-yard dash sprint and Cybex testing of the quadriceps and hamstrings by dynamometer. Psychological function was tested by the Piers Harris concept scale and the Wechster intelligence test. Other factors considered included the presence or otherwise of genu-valgum and the state of the heel pad.

A paper by Fergusson, Morrison and Kenwright (1987) also looked at the functional aspects of the Syme amputation, especially in the case of children who had amputation carried out for congenital deficiencies. They looked at factors such as the ability to walk on the bare stump, ability to run, walking distance, ability to use stairs, ability to walk on uneven ground, the need for a stick, ability to ride a bicycle or play sports and the effect on school and occupation.

Conclusion

Experience over the years has shown that the following conclusions can be made:

(a) The Syme amputation still remains one of the best amputations in the lower limb especially for trauma, infection and congenital deformities.
(b) The surgical technique has to be precise and should follow the principles and techniques as described by Harris.
(c) The selection of the Syme procedure remains difficult in diabetes and peripheral vascular disease. There is no single test which will give a good prediction of healing.

More than one factor has to be considered in level selection, for example:

(a) Clinical examination.
(b) Assessment of skin viability.
(c) Nutritional assessment.

A combination of these three factors can give a fairly good prediction and a likely satisfactory outcome.

Regarding the assessment of the stump and the results, the following points should be considered:

(a) Revision rate.
(b) Wound healing.

(c) Weight bearing quality of the stump.
(d) Stability of the heel pad.
(e) Functional assessment (walking capacity).
(f) Physical assessment (e.g. strength of muscles and 50-yard dash).

Ankle disarticulation is the term identified by the International Standards Organisation to describe amputations at or about the ankle joint and as such conforms to the terminology chosen for amputations in the lower limb. That being so, the amputation choice at this level is the *Syme* amputation described first some 150 years ago and remains today the procedure which gives the best results in terms of surgical, prosthetic and biomechanical aspects. Indeed its only drawback is the cosmetic one with respect to the female, where a transtibial amputation could provide the best solution in personal and social terms. This aspect is of great importance and should demand a clear presentation to the patient of the only negative element of the Syme amputation.

References

Alldredge, R.H. and Thompson, T.C. (1946) The technique of the Syme amputation. *J. Bone Joint Surg.* **28**, 415–426.

Baddeley, R.M. and Fulford, J.C. (1964) The use of arteriography in conservative amputations for lesions of the feet in diabetes mellitus. *Br. J. Sur.*, **51**, 658–663.

Dickhaut, S.C., Delee, J.C. and Page, C.P. (1984) Nutritional status: importance in predicting wound healing after amputation. *J. Bone Joint Surg.* **66A**, 71–75.

Elmslie, R.C. (1942) Amputations. In: *Modern Operative Surgery* (ed. H.W. Carson), Cassell & Co., Vol. **1**, pp. 123–160.

Fergusson, C.M., Morrison, J.D. and Kenwright, J. (1987) Leg-length inequality in children treated by Syme's amputation. *J. Bone Joint Surg,* **69B**, 433–436.

Harris, R.I. (1956) Syme's amputation: technical details essential for success. *J. Bone Joint Surg.*, **38B**, 614–632.

Herring, J.A. (1986) Syme amputation: an examination of the physical and psychological function in young patients. *J. Bone Joint Surg.*, **68A,** 573–578.

Hulnick, A., Highsmith, C. and Bouton, F.J. (1949) Amputation for failure in reconstructive surgery. *J. Bone Joint Surg.*, **31A,** 639–649.

Hunter G.A. (1975) Results of minor foot amputation for ischaemia of the lower extremity in diabetics and non-diabetics. *Can. J. Surg,* **18**, 273–276.

Langdale-Kelham, R.D. and Perkins, G. (1942) *Amputations and Artificial Limbs.* Oxford, London.

Malone, J.M., Leal, J.M., Moore, W.S. *et al.* (1981) The 'gold standard' for amputation level selection: Xenon-133 clearance. *J. Surg. Res.*, **30**, 449–455.

Mathur, B.P., Piplanik, C.L., and Majid, M.A. (1985) A new approach to the Syme's amputation and its prosthesis, *Orthot. Prosthet.*, **39**, (3), 47–52.

Mazet, R. (1968) Syme's amputation: a follow-up study of fifty-one adults and thirty-two children. *J. Bone Joint Surg.*, **50A**, 1549–1563.

Murdoch, G. (1976) Syme's amputation. *J. R. Coll.* Surg. Edinb., **21**, 15–30.

Pavot, A.P. (1973) Ankle disarticulation: a definitive type of amputation in adults. *Arch. Phys. Med. Rehabil.*, **54**, 307–310.

Pinzur, M.S., Jordan, C., and Rana, N.A. (1981) Syme's two-stage amputation in diabetic dysvascular disease. *IMJ*, **160**, (1), 23–27.

Sarmiento, A. (1972) A modified surgical-prosthetic approach to the Syme's amputation. *Clin. Orthop. Rel. Res.*, **85**, 11–15.

Sarmiento, A. and Warren, W.D. (1969) A re-evaluation of lower extremity amputations. *Surg. Gynecol. Obste,* **140**, 800–802.

Sarmiento, A., Gilmer, R.E. and Finnieston, A. (1966) A new surgical-prosthetic approach to the Syme's amputation: a preliminary report. *Artif. Limbs,* **10**, (1), 52–55.

Sawamura, S. (1989) A review and a modified surgical-prosthetic approach. In: *Amputation Surgery and Lower Limb Prosthetics* (eds G. Murdoch and R.G. Donovan), Blackwell Scientific Publications, Edinburgh, pp. 81–86.

Shelswell, J.H. (1954) Syme's amputation. *Lancet,* **2**, 1296.

Silbert, S. and Haimovici, H. (1954) Criteria for the selection of the level of amputation for ischaemic gangrene. *JAMA,* **155**, 1554–1558.

Spittler, A.W., Brennan, J.J. and Payne, J.W. (1954) Syme amputation performed in two stages. *J. Bone Joint Surg.*, **36A**, 37–42.

Syme, J. (1843) On amputation of the ankle joint. *Lond. Edinb. Month. J. Med. Sci.,* **2**, XXVI, 93.

Syme, J.F., Swanson, L., Fish, W.W. *et al.* (1982) The value of non-invasive predictors of healing in dysvascular amputations. *J. Bone Joint Surg.* **64B**, 259.

Wagner, W.F. (1981) The Syme amputation: surgical procedures. In: *Atlas of Limb Prosthetics: Surgical and Prosthetic Principles.* American Academy of Orthopaedic Surgeons, C.V. Mosby, St. Louis, pp. 326–334.

Wagner, W.F. (1992) The Syme ankle disarticulation: surgical procedures. In: *Atlas of Limb Prosthetics; Surgical, Prosthetic and Rehabilitation Principles.* American Academy of Orthopaedic Surgeons, 2nd edn. CV Mosby, St. Louis, pp. 413–422.

Warren, R., Thayer, T.R. and Achenback, H. (1955) The Syme amputation in peripheral arterial disease: a report of six cases. *Surgery,* **37**, 156–164.

Wilder, M.J. (1964) Modified Syme amputation. *Inter-Clin. Info. Bull.* **4** (4), 6–15.

11

Biomechanics and prosthetic management

N. A. Govan

There are well-known prosthetic advantages and disadvantages with regard to amputations at the ankle. The prosthetist's job is made easier by the amputees' ability to bear a large proportion of body weight directly through the stump end. It is, however, unfortunate if such amputations are incorrectly performed. The prosthetist is then faced with a long stump without its principal advantage of end bearing and is from time to time asked to fit prostheses to stumps which are less than ideal. The principal disadvantage of amputations at this level is the difficulty in providing the amputee with good cosmetic restoration. Some surgeons trim the malleoli to reduce the medio-lateral dimension, making a cosmetic result easier albeit at the cost of reduced end-bearing. However, end-bearing in a stump is rarely total. No individual would feel comfortable stomping about on their heels all day – even on heels which have not been surgically assaulted. The prosthesis for such amputations extends proximally up the stump and in order to make use of both medial and lateral tibial flares to support some of the amputee's weight in the same way as at transtibial level. Indeed, given the extremely atrophied nature of most ankle disarticulation stumps, it is a relatively straightforward task to shape the socket to support weight proximally (Radcliffe, 1961a). However, this proximal weight support should not be over-used with the objective of reducing end bearing. The normal procedure is to share weight support between the proximal socket and the distal end, with the distal socket taking the larger share. There is always

doubt as to a single prosthetist's ability to accurately apportion proximal and distal bearing because a very small change in the individual amputee's habits alters the situation as, for example, when a slightly thicker or thinner stump sock is donned.

Radcliffe stated a number of requirements for an ankle disarticulation prosthesis (Radcliffe, 1961b). There must be comfortable support of the body weight in the manner described. The socket should contain an end bearing pad of some shock absorbing material, carefully shaped to conform to the stump end. There should be firm support against the antero-proximal surface of the stump in respect of the point of push-off. It is noted that the wedge effect of the medial aspect of the tibia combined with the antero-lateral muscle compartment of the stump within the socket adequately provides this. A slight exaggeration of this wedging conformation by plaster removal during cast rectification also serves to protect the tibial crest, although a few millimetres of plaster build-up is usually added in the area of the tibial crest itself.

Support of the postero-proximal aspect of the stump at heel contact is achieved by slight flattening of the tissues in the popliteal fossa and this can be done at casting or during cast rectification.

As the amputee walks there is a continually changing location of the centre of pressure in relation to the distal end of the stump. A modern socket can be accurately shaped to cup the distal end in a protective embrace. The surgeon can contribute by ensuring that the cut surface of the tibial end is parallel to the ground to prevent the shear force displacing the stump end pad. For the same reason the correct placement and securing of the heel pad are necessary for optimal force transfer. Even in those stumps with a mobile heel pad – a difficult problem with more traditional sockets – it can be adequately held in the correct place providing it is properly positioned during the casting procedure (Murdoch, 1976).

Stabilization against torques about the long axis of the stump is attained by the combined effect of the anterior wedge shape described and the flattening of the popliteal tissues. In this way the triangular nature of the stump shape is emphasized in the socket shape itself.

Most muscles in the ankle disarticulation amputation atrophy, the extent depending on the distal attachment of the major muscles viz., the Achilles tendon and the antero-lateral group. There is of course one advantage in a stump in which muscles atrophy leaving a bulbous end (Fig. 11.1). This shape is ideal for suspension of

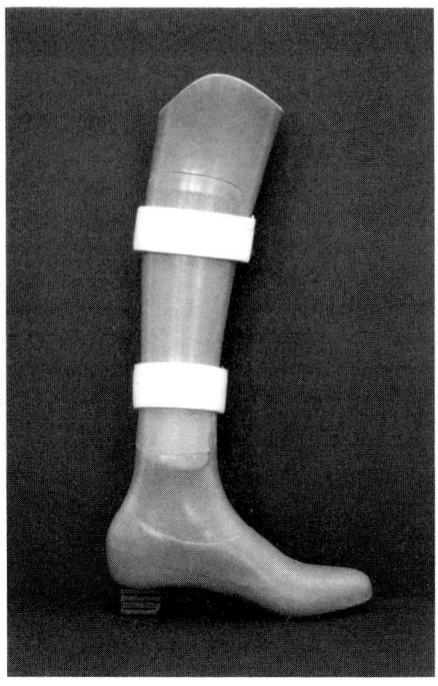

Fig. 11.2. Prosthesis with a medial panel.

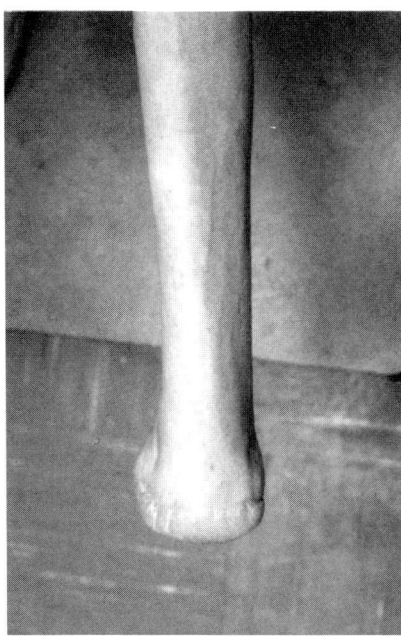

Fig. 11.1. The bulbous end, typical in an ankle disarticulation.

the prosthesis, although a method of donning and doffing has to be incorporated in the socket design (McLaurin, 1970). Where there is a small bulbous end, a hard shell plastic socket may be carefully made so that the amputee is able to push the stump and sock into the socket, requiring, in turn, a pull larger than the weight of the prosthesis plus shoe and normal inertial forces to remove it. Less elegant is a socket with a window cut, usually on the medial aspect, and covered with a panel secured by straps (Fig. 11.2). A bulkier socket results if a soft inner liner is used. The inside of the liner is contoured to the stump shape and the outside has almost parallel walls. A thin nylon sock is usually worn over the soft liner. The amputee then dons the soft liner over the stump sock and pushes both the liner and nylon sock into the hard shell. Because of the bulk of this type of socket it may be more appropriate for a stump with a small distal end, but this type of socket should also be considered when the strength of the socket is important as in the case of a heavy, active amputee.

A disadvantage of older types of prosthesis was that, with reduced space distal to the socket for the foot/ankle assembly, the function provided by the foot was often less effective than that in prostheses for more proximal amputation levels. Some of the modern feet, termed 'dynamic feet' are available for prostheses for distal amputation levels and overcome this disadvantage (Voner and Michael, 1992).

Amputations at the ankle provide good functional stumps with the advantage of end-bearing. Even when the amputation is done incorrectly, the prosthetist should be able to supply a useful and comfortable prosthesis albeit with difficulty.

References

McLaurin, C.A. (1970), The Syme type prosthesis. In: *Prosthetic and Orthotic practice* (ed. G. Murdoch), Arnold, London, Pp. 125–137.

Murdoch, G. (1976) Syme's amputation. *J.R. Coll. Surg. Edinb.*, **21**, 15–30.

Radcliffe, C.W. (1961a) *The Patellar Tendon Bearing Prosthesis*. University of California, Berkeley.

Radcliffe, C.W. (1961b) The biomechanics of the Syme prosthesis. *Artif Limbs* **6(1)**, 76–85.

Voner, R. and Michael, J.W. (1992) The Syme ankle disarticulation: prosthetic management. In: *Atlas of Limb Prosthetics*. American Academy of Orthopaedic Surgeons, C.V. Mosby, St. Louis, pp. 423–427.

Section 4

The foot

12

Partial foot amputations

R. Baumgartner

In lower extremity amputation level selection, the possibility of preserving part of the foot must be given priority. Partial foot stumps providing excellent functional results are feasible whatever the causal condition, including peripheral vascular disease.

A good foot stump normally permits full end bearing and, accordingly, is more physiological than any more proximal level of amputation. Even with a hind-foot stump, a patient is able to walk short distances without the need of a prosthesis, e.g. from bed to the toilet at night. Furthermore, the proprioceptive properties of the sole of the foot are preserved: even when the sensory functions are disturbed and neuropathy has gravely affected what remains of the foot, the stump is unsurpassed in facilitating rehabilitation. In dysvascular patients, preservation of part of the foot is particularly important, since sooner or later the opposite leg may well require amputation.

Patients with toe or metatarsal amputations are often better off without any prosthesis at all or at most with orthopaedic shoe adaptations. The need for physical therapy is reduced if required at all. The shorter the stump, the more the need for prosthetic replacement. The objection regarding poor cosmesis is no longer valid if the chosen surgical technique recognizes the needs of the prosthesis and modern prosthetic techniques are employed. Finally, the psychological aspects of the loss of one or even both feet also must be considered.

Amputation levels

The weight bearing area of the plantar surfaces and the triangle formed by both feet with the body's line of gravity in its centre is of basic importance in the biomechanics of human standing and locomotion. This 'basement' is reduced in area by every successive shortening. Asymmetrical amputations are a great help in preserving the weightbearing area and therefore constitute an important alternative to transverse sections (Fig. 12.1).

The following levels of amputation will be considered (Fig. 12.2):

(1)   Distal phalanx.
(2)   Disarticulation at proximal interphalangeal joint (PIP) (hallux only).
(3)   Disarticulation of the toes.
(4)   Distal metatarsals.
(5)   Proximal metatarsals.
(6)   Lisfranc (Lisfranc, 1815).
(7)   Chopart.
(8–10) Partial or total calcanectomy.
(11)   Calcaneo-tibial arthrodesis (Pirogoff, 1854; Boyd, 1939).
(12)   Syme amputation (Syme, 1843) plus fusion between scaphoid and cuboid bones and the anterior part of the tibia.
(13)   Syme amputation (see Chapter 10).

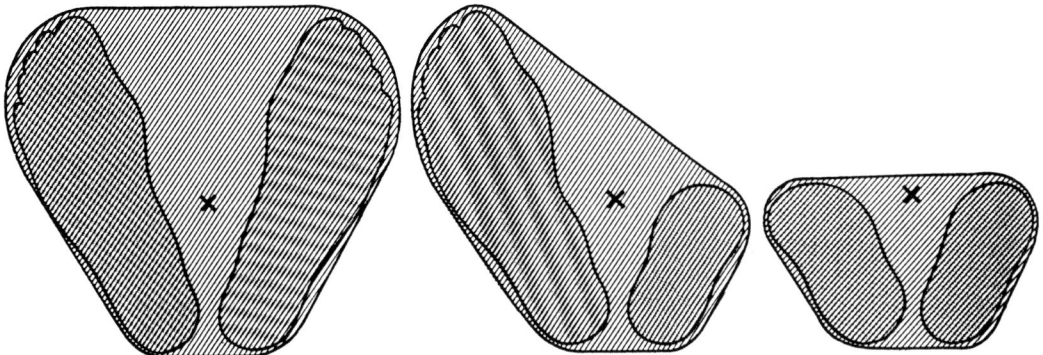

Fig. 12.1. Weight-bearing of plantar surfaces and the area formed by both feet – the 'basement' reduced by each successive shortening.

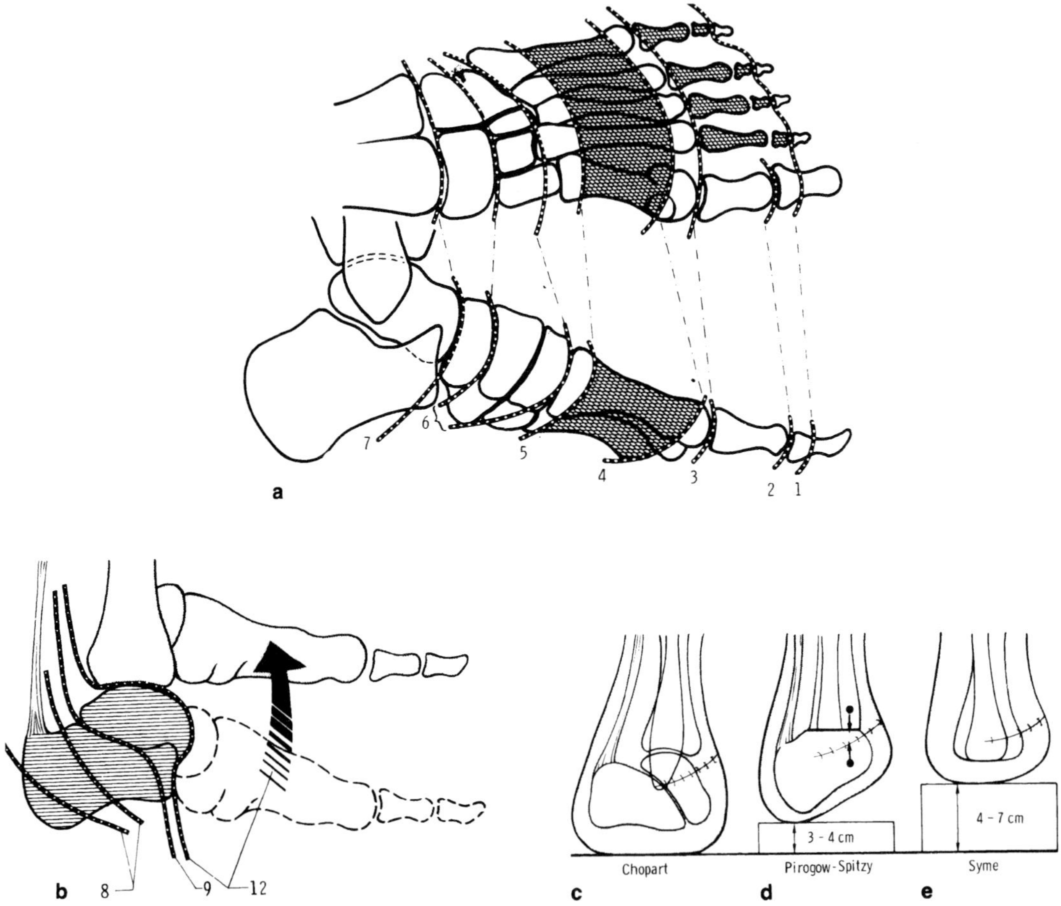

Fig. 12.2. Levels of amputation. (a) From distal phalanx to Chopart. (b) Partial or total calcanectomy; calcaneo-tibial arthrodesis; Syme amputation with fusion of scaphoid and cuboid bones to anterior part of tibia. (c) Chopart. (d) Pirogow-Spitzy. (e) Syme ankle disarticulation.

Operative technique

As in any amputation level, every effort should be made to preserve a maximum of tissue and still obtain a functional stump free of pain. In the foot, preservation of the plantar soft tissues is particularly important. Special attention should be paid to the stump ends, which towards the end of the stance phase, have to withstand high loads and shear forces. Gentle tissue handling and clean-cut skin edges are mandatory. The skin incision should always proceed down to the bone and dissection be kept strictly to the contours of the bone if the soft tissues are to be separated en bloc.

The proximal side of the saw blade is kept cool by saline solution as overheating of the bone leads to tissue necrosis. Tendons and nerves are cut transversely without damage to their stumps. In disarticulation, the cartilage of the retained articular surface is preserved unless the bone has to be shortened due to lack of soft tissues.

The bones of the stump end therefore must be carefully aligned and well rounded in both sagittal and transverse planes. They require to be covered with a full thickness plantar soft tissue flap. The flap is bevelled, tendons are cut at the level of the bone and nerves are shortened by 5–10 mm, ensuring their separation from the skin suture. The long plantar flap is now fashioned to fit the short dorsal one. The surplus of length in the plantar flap is distributed evenly along the dorsal flap (the so-called 'moccasin' technique). Deep interrupted skin sutures are usually sufficient so long as they avoid strangulation. As an alternative, closure by Steristrip is sometimes feasible or in case of infection the wound is left open until delayed primary closure. Open or closed drainage for 2–5 days is important in every foot stump.

In hind-foot stumps, muscle balance is extremely disturbed as the Achilles tendon continues to act in plantar flexion and supination whereas in contrast the dorsal flexors and pronators have lost their lever. This may result in stump contractures in flexion and supination which compromise function and make prosthetic fitting difficult. The risk of such contractures is greatest in the interval between surgery and wound healing. Neither plaster cast nor physical therapy are able to prevent this deformity. Therefore, external fixation joining the calcaneus and tibia is recommended for 2–4 weeks. This procedure also facilitates positioning and wound care (Fig. 12.3).

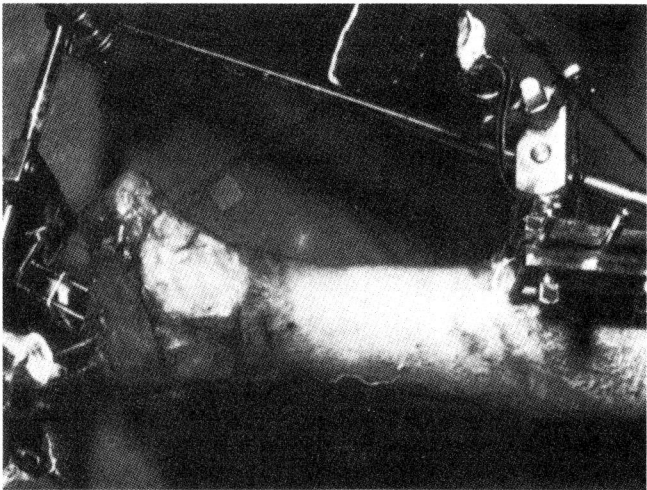

Fig. 12.3. External fixation ensuring stability between tibia and calcaneus.

Toes

Partial or total amputation of the distal phalanx of every toe is sufficient in localized dry gangrene. The bone is rounded and the plantar flap adapted without any tension and requiring only one suture if any (Fig. 12.4).

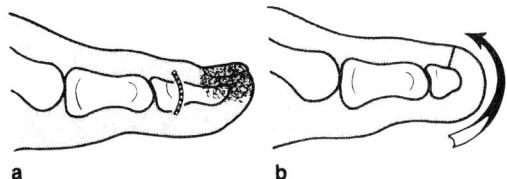

a   **b**

Fig. 12.4. Partial or total amputation of a distal phalanx in localized dry gangrene. Plantar flap requiring only one suture.

For digits II–V only, total disarticulation is the next level of amputation. Any attempted partial amputation of these toes leads inevitably to a stump contracted in dorsiflexion with consequent pain and problems in shoe fitting.

However, in the case of the hallux, partial amputations or disarticulation at the interphalangeal joint are most valuable. In disarticulation, the condyles of the proximal phalanx must be rounded off (Fig. 12.5).

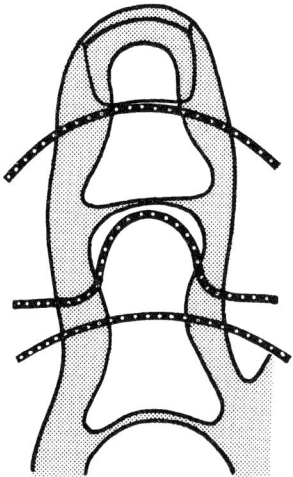

Fig. 12.5. In hallux partial amputations or disarticulation at interphalangeal joint apply.

Toe disarticulation

In toe disarticulation, a racquet-type incision is recommended, beginning at the dorsum of the metatarso-phalangeal articulation (Fig. 12.6). The incision is carried down to the proximal phalanx of the toe. The extensor tendons are cut and then by

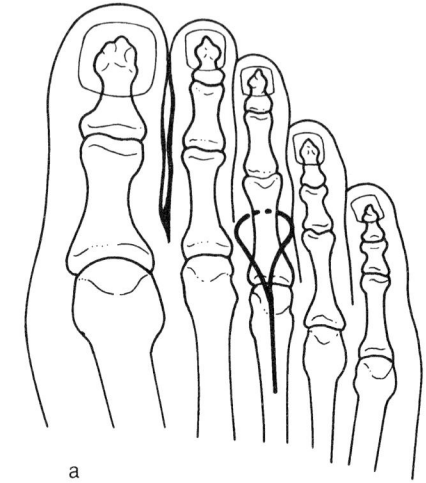

a

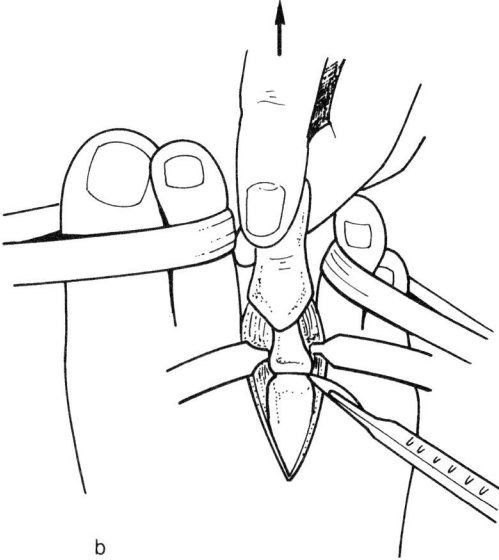

b

Fig. 12.6. The amputation. (a) Racquet incision on dorsum of metatarso-phalangeal joint. (b) Traction on toe permitting division of extensor and flexor tendons and collateral ligaments. Nerves shortened and divided, vessels coagulated.

traction and plantar flexion of the toe, the collateral ligaments and the plantar flexion tendons are divided, the latter at the level of the metatarsal head. The nerves are identified and shortened by 5–10 mm and the vessels are coagulated or ligated. The skin is brought together without tension. In dysvascular patients, no suture is necessary. In all cases wound drainage for 1–2 days is recommended.

In disarticulations of the marginal rays (I and V) the tubercle presenting outwards on the head of the metatarsal should be removed.

Trans-metatarsal amputations

The metatarsal bones should only be amputated through cancellous bone either at the head or preferably at the base. Amputations through the cortical diaphysis lead inevitably to bony resorption. These stumps become as sharp as a pencil with the risk of pain and perforating ulcers in the case of neuropathy. Partial amputations give excellent stumps as long as the first metatarsal or at least two minor rays are preserved.

However, the asymmetrical partial foot stump requires well designed foot orthotics in order to avoid overloading of the remaining rays.

In plantar ulcer due to neuropathy, it is sometimes possible to preserve the toes and resect only the distal two-thirds of one or all of the metatarsals (Figs. 12.7, 12.8). The functional result is the same as in amputation, but by preserving the toes, the patient does not become an amputee, will not suffer from stump and phantom pain and after 1–2 years, the toes shorten and retract to the bony section of the metatarsals and may even regain some function.

Plantar ulcers are simply cleansed, left open and used for drainage of the space created (Fig. 12.9). No suture is necessary.

In total amputation, the plantar flap is bevelled and, after insertion of a drain, shaped to the dorsal flap (Fig. 12.10). In the case of infection, open wound treatment and delayed primary closure should be considered.

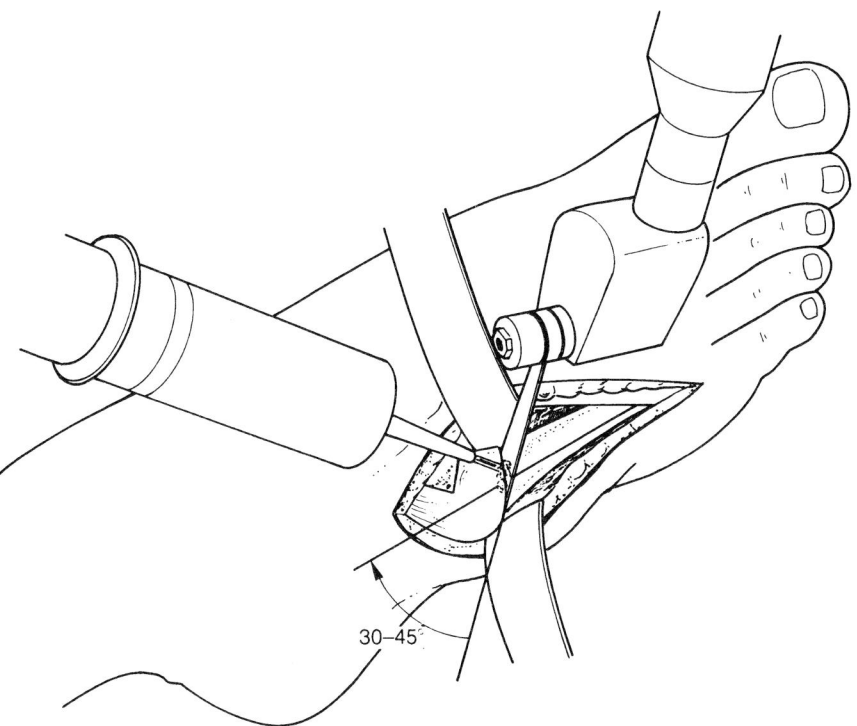

30–45°

Fig. 12.7. Removal of distal two thirds of fifth metatarsal via section through cancellous bone with 30–45° angle to ensure a sloping surface to present to shoe or walking surface.

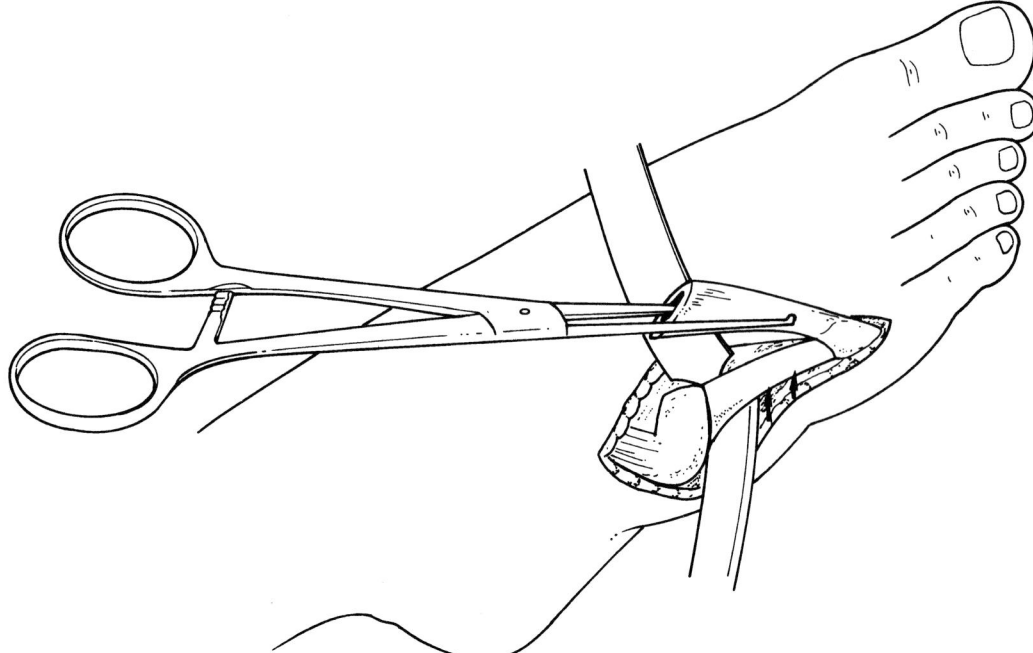

Fig. 12.8. Removal of distal two thirds of fifth metatarsal by traction and division of capsule and collateral ligaments. Function result is the same as amputation but with toe(s) preserved patient is not an 'amputee'.

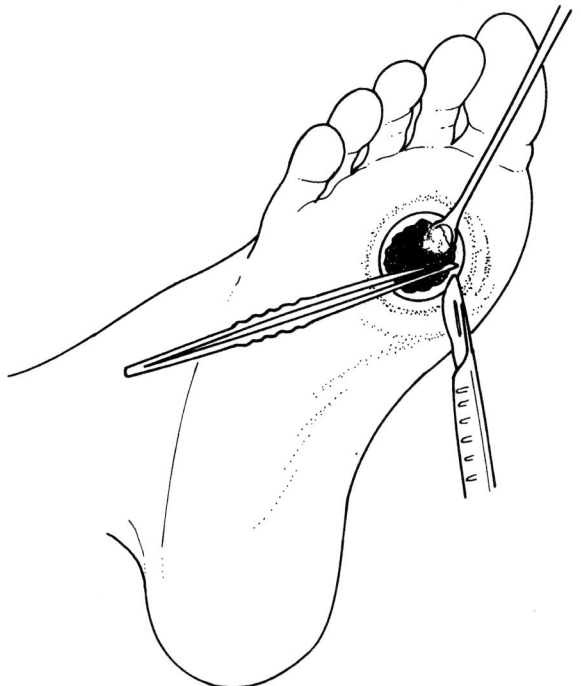

Fig. 12.9. Plantar ulcer. Open drainage.

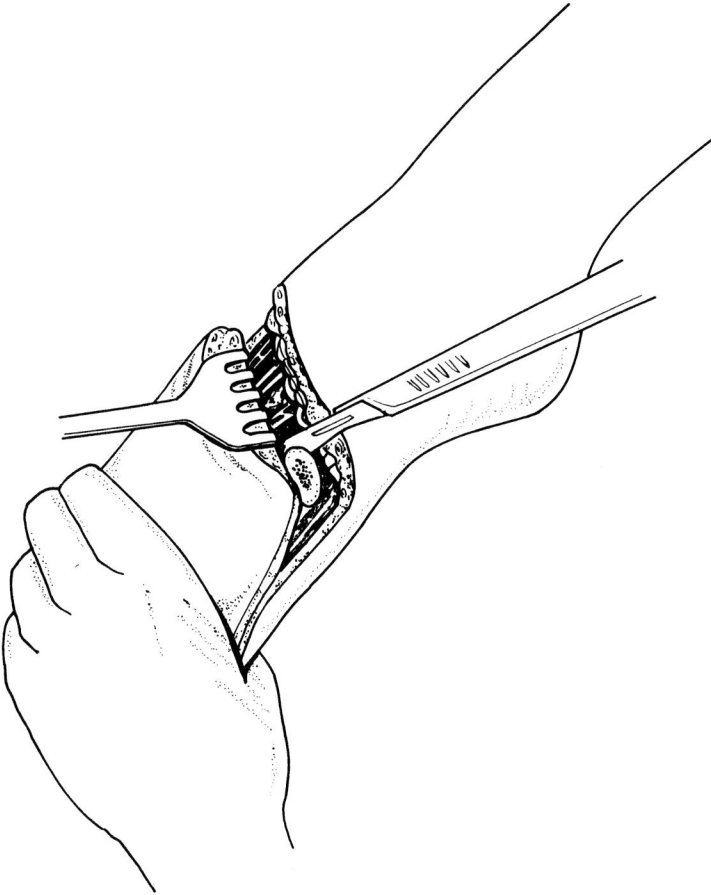

Fig. 12.10. Transmetatarsal amputation through cancellous bone at base of metatarsals. Plantar flap bevelled and shaped to dorsal flap.

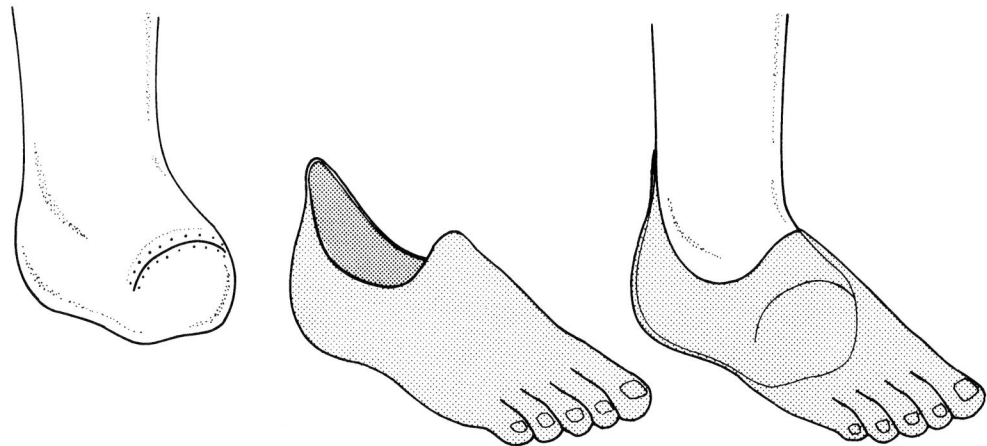

Fig. 12.11. Transmetatarsal amputation. Dorsal scar. Lightweight snugly fitting prosthesis of acrylic resin with carbon fibre reinforcements.

Employing a plaster cast with the stump in slight pronation and dorsiflexion, the patient is mobilized within a few days after surgery in a programme of gradually increasing weight bearing.

Prosthetic fitting should not restrict the range of motion of the hindfoot articulations. This can be achieved by means of a lightweight, snugly fitted prosthesis made of acrylic resin suitably reinforced with carbon fibre (Fig. 12.11).

Lisfranc procedure

Pure disarticulations of these joints do not result in good stumps because of the uneven length of the remaining articular surfaces: however, careful alignment of all rays, disregarding the anatomy of the articulations, may give excellent results. Care must be taken to preserve the plantar arteries which are very close to the second and third cuneiform bones.

The operative technique is the same as in metatarsal amputations. Asymmetrical procedures are also possible.

Hindfoot amputation

There are several techniques of hindfoot amputations, all of which may give excellent results:

(1) Chopart's disarticulation.
(2) Pirogoff–Boyd technique: calcaneo-tibial fusion.
(3) The Syme amputation: ankle disarticulation.
(4) Calcanectomy, partial or total.

These techniques vary widely in their effectiveness.

Chopart's disarticulation

This procedure should be seen as first choice. It preserves the full length of the lower limb and the ankle joint, even if its range of motion remains limited. In order to obtain a well shaped stump end, the calcaneus and the talus itself should be rounded off (Fig. 12.12).

Severe stump contractures in equinus and supination, if not prevented by external fixation, may require corrective surgery such as (Fig. 12.13):

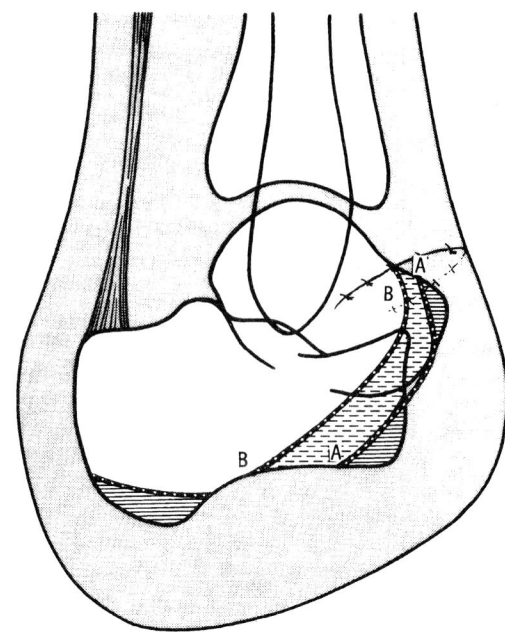

Fig. 12.12. Chopart's disarticulation. Talus and calcaneum bones rounded off.

(a) lengthening of the Achilles tendon;
(b) transfer of the anterior tibial tendon to the lateral border of the stump;
(c) wedge osteotomy and fusion of the subtalar joint.

Calcaneo-tibial fusion

This procedure is seen as second choice as the ankle joint is sacrificed and there will be a shortening of leg length by 3–4 cm. The calcaneus is placed in a physiological position, particularly in the sagittal and transverse planes and after resection of the hindfoot joints, fused to the distal end of the tibia. For better cosmesis of the prosthesis, the malleoli should be resected. A biomechanical advantage is also gained if the calcaneus is displaced forwards by 1–2 cm.

For fusion, internal or external fixation is possible. However, care should be taken not to perforate the soft tissues of the sole.

The author recommends the Charnley frame placed in the frontal plane as far forwards as possible. The tension of the Achilles tendon will further compress the bony surfaces (Fig. 12.14).

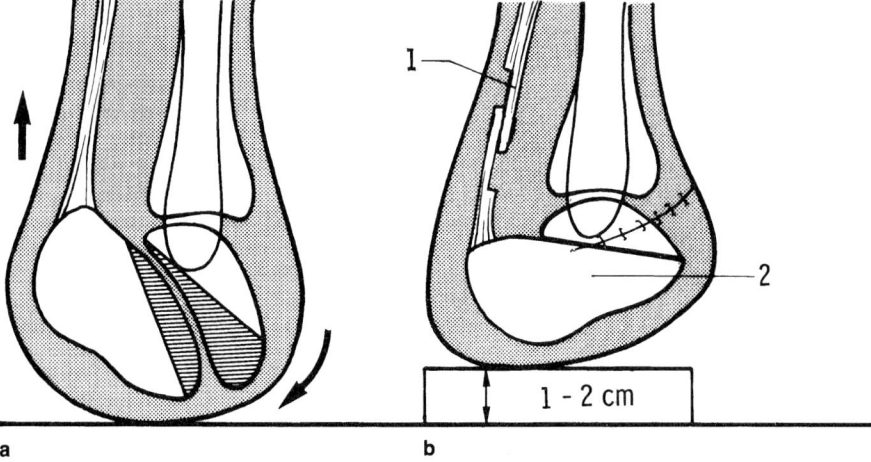

Fig. 12.13. (a) Severe stump contractures can develop. (b) Corrective surgery including (i) tendo achilles lengthening and (ii) fusion of subtalar joint.

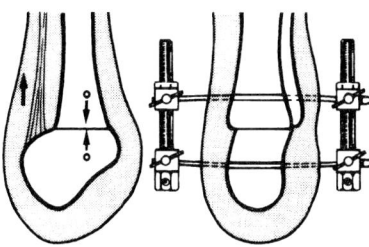

Fig. 12.14. Compression by anteriorly disposed Charnley frame and tension in tendo achilles.

External fixation is removed after about 6 weeks and with a walking plaster cast, bony consolidation continues for a further 6 weeks. The stump, however, only assumes its final shape after some 3–6 months or more.

The procedure is not recommended in peripheral vascular disease and neuropathy, as there is an increased risk of non-union, infection and gangrene.

The Syme amputation

The Syme amputation offers a reduced weight bearing surface and a shortening of the leg length of 6–8 cm. However, this level is still by far superior to any transtibial amputation. For details see Chapter 10.

Calcanectomy

In lesions limited exclusively to the calcaneus, e.g. after open infected fracture, partial or total calcanectomy with or without disinsertion of the Achilles tendon should be considered before proceeding to a transtibial amputation.

The aim is to remove the affected part of the calcaneum and by removing a sufficient amount of bone, providing the borders of the soft tissue ulcer the possibility of closing without tension. Skin grafts of any type are not recommended as they cannot withstand the mechanical strains in this area over a prolonged time.

With the patient in prone position, the incision begins at the ulcer and continues longitudinally and medially of the Achilles tendon. The first intention must be to preserve the insertion of the Achilles tendon but if it has to be sacrificed the functional result can still be excellent after hemi-calcanectomy (Fig. 12.15).

It has to be recognized that total calcanectomy causes a severe instability of the forefoot with important loss of function and therefore is not recommended unless the patient is prepared to use an ankle foot orthosis on a permanent basis.

In order to stabilize the forefoot it is possible to perform a fusion between the anterior part of the tibia and the scaphoid and cuboid bones. Once the calcaneus, the talus and the malleoli are

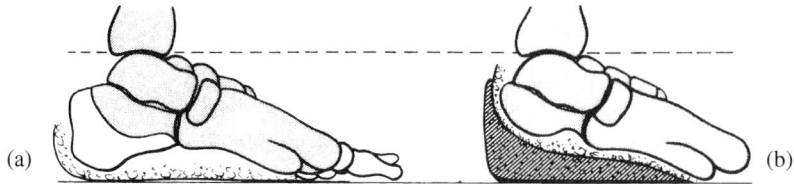

Fig. 12.15. Hemicalcanectomy. (a) Diagram showing amount excised and (b) Foot in adapted footwear.

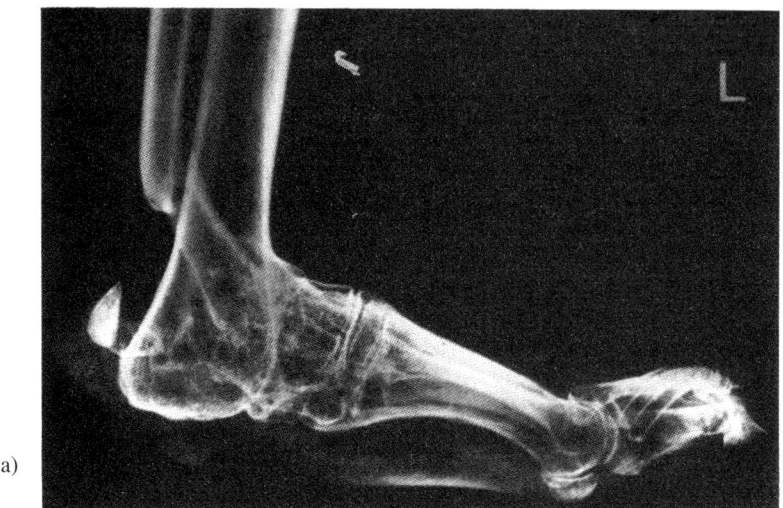

(a)

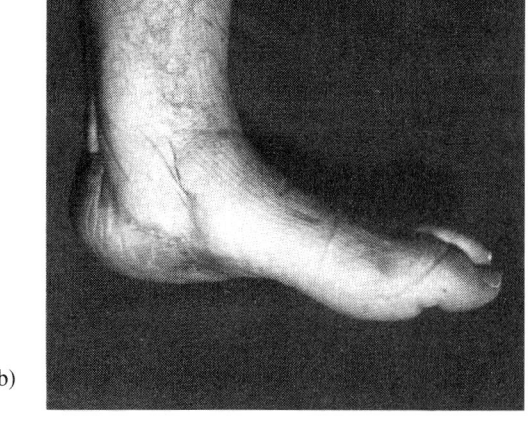

(b)

Fig. 12.16. (a) Satisfactory fusion of cuboid and scaphoid to anterior surface of tibia. (b) Appearance of foot underlining large weight bearing area and 'natural' appearance of foot.

resected the cartilaginous surfaces of the scaphoid and cuboid bones are excised. The cortical bone of the anterior part of the tibia is then removed in order to obtain a large surface of cancellous bone. External fixation joining the forefoot bones to the tibia is recommended as the method of choice (Fig. 12.16).

The shortening of the leg is comparable to that found in the Syme amputation but the advantage of the procedure is the resultant increased weight bearing surface and the fact that the patient is not seen to be an amputee.

References

Boyd, H.B. (1939) Amputations of the foot with calcaneo-tibial arthrodesis. *J. Bone Joint Surg.*, **21**, 997–1000.

Lisfranc, J. (1815) *Nouvelle Méthode Operatoire pour l'Amputation Partielle du Pied dans son Articulation Tarso-metatarsienne; Méthode Précédée des Nombreuses Modifications qu'a Subies celle de Chopart.* Gabon, Paris.

Pirogoff, N.I. (1854) Osteoplastic elongation of the bones of the leg in amputation of the foot. *Voyerno-med. J.*, **63,2 sect.**, 83–100 (Pirogoff's method of complete osteoplastic amputation of the foot. German translation, Leipzig, 1854).

Syme, J. (1843) On amputation at the ankle joint. *Lond. Edinb. Month. J. Med. Sc.*, **2**, XXVI, 93.

13

Biomechanical and prosthetic considerations

D. N. Condie

Introduction

When contemplating a partial foot amputation it is essential that the surgeon has a clear understanding both of the functions of the normal foot and of the consequences of the proposed surgical procedure in relation to prosthetic replacement

Normal foot function

The normal foot is an extremely complex structure, the detailed functions of which are still only partially understood. This discussion of the mechanics of normal foot function will be restricted to a brief consideration of the *load-bearing structure* and the *function of the foot joints* during normal walking.

Load-bearing structure

The foot is the means whereby the ground reaction forces generated during physical activities are transmitted to the body structure. During normal level walking these loads are directed initially onto the heel. The specially adapted fatty tissues of the heel pad are ideally suited to the absorption of the high forces generated at impact and during the subsequent loading of the limb. Once the foot is flat and until the heel leaves the ground as push-off is initiated, the supporting forces are shared between the heel and the fore-foot, with only a

small contribution from the lateral aspect of the midfoot.

This method of load transfer is commonly attributed to the 'arch structure' of the foot skeleton, even though it is now generally understood that its effectiveness is a function of a number of both structural and neuromuscular mechanisms.

Once the heel leaves the ground, the increased ground reaction force associated with push-off must be transmitted through the area defined by the metatarsal heads and the pulps of the toes. As body weight is transferred to the contralateral limb, this load falls and localizes on the plantar surface of the hallux.

Joint function

The functions of the joints of the foot have been the subject of endless investigation. Clearly, the ability of the foot to alter its shape and alignment are of considerable importance in adapting to variations in the slope of the walking surface. A more subtle but equally important role concerns the absorption of the rotation about the longitudinal axis of the lower limbs that occur during each stride.

Internal rotation of the entire lower limb, which is initiated during the swing phase, continues after heel contact until the foot is flat. During this phase the foot pronates about the subtalar joint axis, thereby maintaining the normal toe-out position of the foot. Elevation of the lateral margin of the foot, which is a consequence of this movement, is counteracted by supination of the forefoot. After the foot is flat, as the lower limb commences external rotation, the foot supinates about the subtalar joint axis to absorb this motion, thus avoiding slippage occurring between foot and ground. The associated elevation of the medial margin of the foot is in this instance counteracted by pronation of the forefoot. This action ensures that the ground reaction force is distributed across the full width of the fore-foot.

After the heel leaves the ground, external rotation of the limb continues; however, the subtalar joint now reverses its direction of motion to pronate in conjunction with the forefoot, thus transferring the area of support medially onto the first metatarsal head and finally the hallux as the foot loses contact with the ground.

A final word should be reserved for the role of the mid-tarsal joint. During the initial loading phase this joint acts in concert with the subtalar joint. Once subtalar supination commences, however,

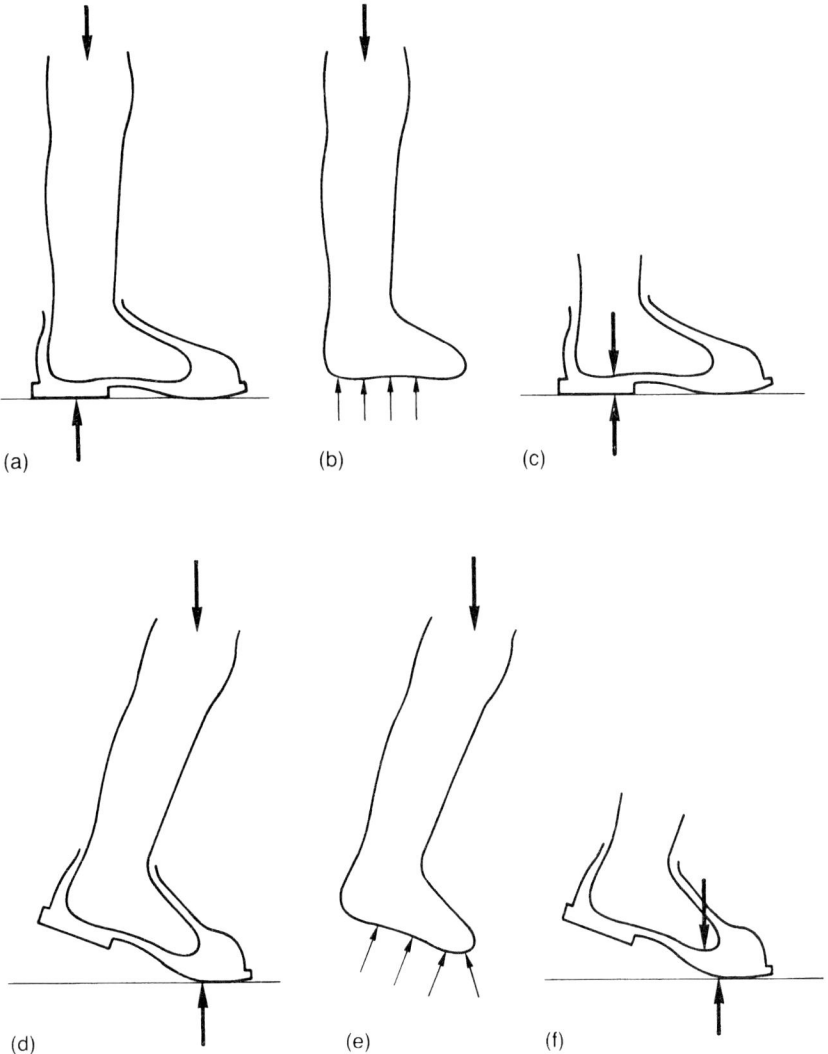

Fig. 13.1. Distal amputations. (a) Mid-stance, external forces; (b) Stump pressure distribution; (c) Socket forces; (d) Push off, external forces; (e) Stump pressure distribution; (f) Socket forces.

this joint locks and, by doing so, stiffens the long arch of the foot to prepare it for the higher dorsiflexion moment to which it is subjected after the heel leaves the ground.

The implication of amputation

All accepted partial foot surgical procedures leave the talo-crural and talo-calcaneal structures with the underlying heel pad intact. The capacity of the resulting stump to absorb the initial contact forces is therefore unimpaired.

In contrast, virtually all amputations, excepting that of the toes alone, will seriously disrupt the mechanisms responsible for the transmission of forces from mid-stance until toe-off.

Similarly, as the level of amputation moves proximally, firstly the supinatory/pronatory rotation of the forefoot, then the mid-tarsal joint

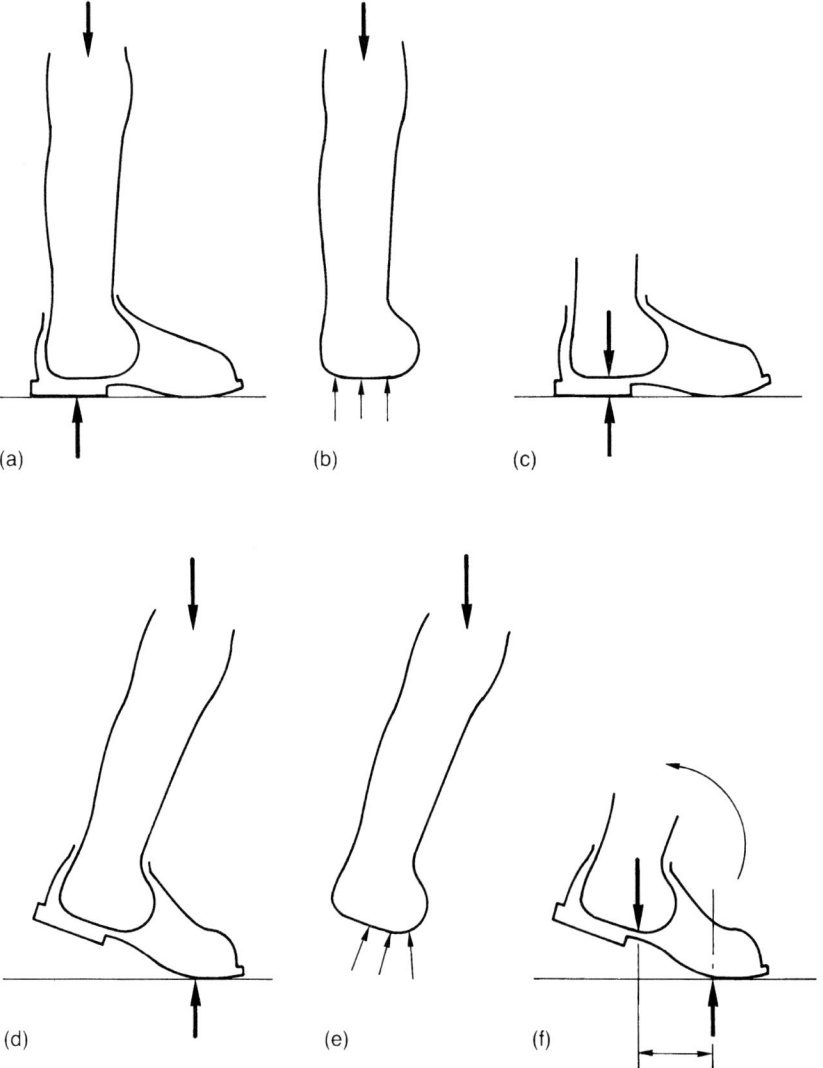

Fig. 13.2. Proximal amputations. (a) Mid-stance, external forces; (b) Stump pressure distribution; (c) Socket forces; (d) Push off, external forces; (e) Stump pressure distribution; (f) Socket forces (causing tendency to rotate dorsally).

function and finally the subtalar joint function will be compromised.

Modern prostheses are designed to compensate for these functional losses. It is vital, however, that the surgeon ensures that the stump is 'designed' to interface with the prosthesis in a manner which will enable optimal comfortable restoration of function.

Distal amputations

Trans-metatarsal amputation and ray amputations will all modify or eliminate the normal weight bearing properties of the fore-foot. Prosthetic solutions will require to redistribute this force across the remaining plantar surface of the stump (Fig. 13.1). In general, this may be successfully achieved by the use of an insole or insert in conjunction with appropriate shoe adaptations. Even with these patients, however, inevitably, during push-off, significant pressure will be applied to the end of the stump.

Prosthetic fitting will be considerably facilitated if proper attention is paid to the cut surfaces of bones, e.g. the metatarsal shafts, and if the suture line is positioned dorsally. There is little that the surgeon can do to alleviate the loss of forefoot mobility, the replacement of which depends on the skill of the prosthetist and his choice of materials and method of construction.

Proximal amputations

When the amputation is performed through the tarso-metatarsal joints (Lisfranc's procedure) and even more so at mid-tarsal level (Chopart's procedure) the replacement of the normal weight bearing mechanisms is a much more challenging task.

During mid-stance and even during push-off it is still possible to contemplate redistributing the ground reaction force across the plantar aspect of the stump (Fig. 13.2). However once the heel leaves the ground a second problem arises, with the 'offset' position of the socket force tending to cause it to rotate dorsally relative to the stump.

In order to prevent this occurring the prosthetist will require to 'lock' the prosthesis on to the stump by totally enclosing it in a socket which conforms closely to the underlying skeleton. This may take either the form of a 'Symes-type' of socket which will inevitably prevent all ankle and subtalar motion or preferably a 'slipper-type' of socket which will still allow free motion at both these joints.

The surgical implications of this situation are twofold. The tendency of the socket to rotate relative to the stump during push-off will inevitably give rise to a shearing effect at the stump–socket interface, no matter how intimate the socket fit. Experience has demonstrated that this effect is less likely to cause skin trauma if the suture line resulting from the amputation is located on the distal dorsal surface rather than on the end or even on the plantar surface of the stump.

Secondly, as has been previously mentioned, with the slipper type of socket, ankle and subtalar motion is still possible. In addition, the powerful calf muscles (soleus and gastrocnemius) have retained their normal attachment on the calcaneus.

Failure to secure the principal dorsiflexor (tibialis anterior) and the pronators (the peroneal muscles) will inevitably result in an equinovarus deformity of the stump. If, however, these muscles are fixed securely, it is possible to retain some degree of normal ankle and subtalar function, thereby enabling a more normal gait pattern to be achieved.

Section 5

The femur

14

Transfemoral amputation

F. A. Gottschalk and S. M. H. J. Jaegers

It is now well accepted that most patients with peripheral vascular disease and diabetes will benefit from a transtibial (below-knee) amputation where feasible and that those who have a transfemoral (above knee) amputation are sicker and are more at risk for complications (Harris *et al.*, 1988). Those patients are also deemed to have a decreased chance of adequate rehabilitation (Couch *et al.*, 1977). Patients with generalized medical problems are at greater risk for failure, and increased morbidity and mortality (Dale and Capps, 1959; Colt and Lee, 1972; Barber *et al.*, 1983; Christensen *et al.*, 1988). There are many articles on transfemoral amputation which are anecdotal and have no scientific evidence to prove or disprove some of the claims made. Most cite the incidence of amputation and the ratio of transfemoral to transtibial amputations (Dale and Capps, 1959; Colt and Lee, 1972; Barber *et al.*, 1983; Christensen *et al.*, 1988; Harris *et al.*, 1988).

The experienced surgeon is likely to bring good clinical judgement to bear on level selection. Even so, such ancillary methods as are available to predict viability of selected levels of amputation can aid the surgeon in the choice of procedure. Of those available, measurement of skin blood flow and thermography appear to provide the most valuable information (Chapter 20). The first cited is invasive and the second is expensive. The single most readily available test is undoubtedly Doppler derived systolic pressure measurement. There is considerable evidence to support the use of a cut-off figure of 70 mm Hg at the lower thigh for a successful transtibial amputation and conversely for a transfemoral.

Younger patients who have had a transfemoral amputation for trauma or neoplastic condition are more likely to become functionally independent because their problem is a more localized situation. This is in contrast to the older diabetic and peripheral vascular disease patients who have systemic involvement and are generally in poor health (Kay and Pennal, 1958).

The majority of articles which describe technique are review articles with very little scientific basis. The techniques seem to have been handed down from generation to generation and accepted as the standard. The notable controversy is still in terms of myodesis versus myoplasty. The proponents of myoplasty advocate anchoring of the antagonistic muscle groups to each other in such a way that they do not slide over the end of the cut bone. It would appear that the major disadvantage of this technique is an inability to restore satisfactory muscle tone in the residual muscle as described by Thiele *et al.* (1973). These authors noted that a reduction in muscle mass combined with inadequate mechanical fixation, as well as atrophy of the remaining musculature, were the major factors for the decrease in muscle strength detected in transfemoral amputees. The muscle groups that were most notably affected were the flexor, extensor, abductor and adductor muscles of the hip and these correlated directly with inadequate muscle stabilization.

Magnetic resonance imaging (MRI) provides an accurate method for determining the muscular changes and cortical atrophy of the femur that occur in transfemoral amputee (Jaegers and de Jongh, 1992). An MRI study of 12 male patients with unilateral transfemoral amputations due to trauma or osteosarcoma was done to evaluate the changes that occur in the stump, at least 2 years after amputation. Axial images were obtained from the iliac crest to the head of the fibula on the intact side. The contours of the muscles and bones of the hip and thigh were traced manually using a computer mouse and the data entered into a computer. An integrated three-dimensional software program enabled the data to be viewed as a three-dimensional graphical reconstruction, and from various angles.

The area and volume of the muscles and bone on the amputated and sound side were measured and calculated. The amount of atrophy of the muscles on the stump was expressed as a percentage of the intact limb. The more proximal the amputation, the greater the change in geometry of the muscles, and the more the atrophy. Muscles which had not been sectioned at the time of surgery were atrophied by up to 30%. These included the iliopsoas and gluteus medius and minimus. Although both the gluteus maximus and tensor fascia latae are not sectioned, they nevertheless lose part of their distal insertion via the iliotibial tract at the time of amputation. Both muscles proved to be retracted and atrophied when the ilio-tibial tract had not been re-anchored. Atrophy of the gluteus maximus and tensor fasciae latae was shown to be from 36° to 46%. Muscles which had been sectioned and attached by myoplasty or myodesis were markedly atrophied between 40% and 60%, but not as much as those where no re-attachment had been done. Despite the marked atrophy, no fatty degeneration was noted.

Adductor magnus showed the most atrophy of the adductor group, since it is most often sectioned at the time of surgery. In none of the subjects had it been adequately re-anchored although in four subjects a myoplasty had been performed by attachment to the iliotibial tract. Adductor longus was atrophied in those subjects where the amputation was at a higher level involving the belly of the muscle. In those subjects where adductor magnus and longus had not been re-anchored, an abduction contracture of the stump occurred in five despite loss of the distal attachment of the iliotibial tract. A partial attachment remains, however, via the intermuscular septum to the linea aspera (plates 1 and 2). Enlargement of the medullary cavity of the femur was observed on the cross-sectional MRI, although the overall cortical diameter of the bone remained unchanged.

Changes in muscle morphology after amputation are due to volume and geometric changes. Those most likely to change give rise to the flexed and abducted position of the residual femur, because the muscles have shortened in concert with the residual femur in an abnormal position (Plate 3). As a result, muscle function is altered and the deforming force compounded.

Maintenance of muscle function on the amputated side is dependent on adequate distal anchorage of the divided muscles. The residual muscles should be re-attached in as functional a position as possible in order to maintain the anatomical axis of the residual femur as compared to the intact side.

No research results are as yet available in terms of the prevention or handling of neuroma formation after sectioning of the nerve. Phantom sensation and phantom pain continue to be problems in amputee patients and there does not appear to be a satisfactory way to determine the result. However, a recent study describing placement of a small catheter in the sciatic nerve for local anaesthesia infiltration suggests that it is possible to decrease post-operative pain and phantom sensation (Malawer et al., 1991).

The most noticeable hiatus in terms of management of the transfemoral amputee is the lack of information in terms of a team approach to the patient. It would appear that most surgical procedures are done without thought to the subsequent function of the patient. Although many patients may undergo transfemoral amputation as a preterminal event because of severe medical problems, many of these patients will survive for an additional 3–5 years but will be significantly functionally impaired in terms of mobility (Dale and Capps, 1959; Couch et al., 1977; Dardik et al., 1988). No studies are available to evaluate attempts of reducing energy expenditure for walking in transfemoral amputees. A properly performed surgical procedure offers the patient a much better chance, functionally, than any other intervention. It also enhances prosthetic fitting by providing the prosthetist with a well contoured and functional stump. The biomechanics of the lower limb need to be considered by the surgeon at the time of performing the amputation. The procedure should be considered analogous to any other type of lower limb reconstruction in terms of functionally improving the patient and should be based on an understanding of the relevant functional mechanics.

Surgical procedures

Introduction

It is well accepted that transfemoral amputees require increased energy for walking (Waters *et al.*, 1976). Even amputees with no other medical problems are unable to achieve normal gait in terms of velocity, cadence or energy consumption. Older dysvascular amputees are at a further disadvantage in that they do not have the physical resources to expend the extra energy; often, they may not use a prosthesis at all, or be limited household walkers (Volpicelli *et al.*, 1983).

Despite improvements in prosthetic design and fabrication, the artificial limbs available are unable to provide a satisfactory replacement for the lost limb when poor surgery has been done and an inadequate stump has been created (Long, 1985; Sabolich, 1985). Too often, the procedure is performed without thought for the biomechanical principles or preservation of muscle function. Although one of the major goals of the surgery is primary wound healing, this can still be achieved together with the maintenance of biomechanical principles of lower limb function. In the majority of transfemoral amputees who have had a conventional surgical procedure, the energy expenditure will be 65% or more above that for level walking at a regular walking speed (Gonzalez *et al.*, 1974; Waters *et al.*, 1976). To minimize energy expenditure it is important to maintain as much length as possible when doing a transfemoral amputation. The longer the stump, the easier it is to suspend and align a prosthesis. The functional ability of the patient is also improved with a longer stump. In some circumstances the level of the amputation may be dictated by the prevailing local pathology. Even in those patients where prosthetic use is not considered, as long a stump as possible should be left. This provides a longer lever arm and may help with sitting and transfers.

Biomechanics

The normal anatomical and mechanical alignment of the lower limb have been well defined (Freeman, 1980; Maquet, 1980; Hungerford *et al.*, 1984) In two-legged stance the mechanical axis of the lower limb runs from the centre of the femoral head through the centre of the knee to the midpoint of the ankle, and at an angle of 3° from the vertical. The femoral shaft axis measures 9° from the vertical (Fig. 14.1). The normal anatomical alignment of the femur is thus in adduction, which allows the hip stabilizers and abductors to function normally and reduce the lateral motion of the centre of the mass of the body. A smoother and more energy efficient gait is thus observed.

In the case of most transfemoral amputees, both the mechanical and anatomical alignments are disrupted since the residual femur no longer has its natural anatomical alignment and the femoral shaft

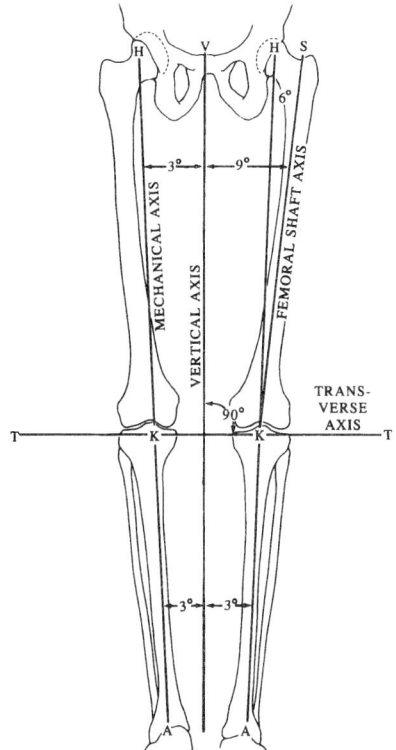

Fig. 14.1. Mechanical and anatomical axes of normal lower limbs (From Gottschalk, F., Kourosh, S., Stills, M. *et al.* J Prosthet Orthot 1989; 2: 94–102. Used with permission).

axis is in abduction compared to the sound limb. The abducted femur of the transfemoral amputee leads to a lurch to the side and higher energy consumption. The major portion of the adductor attachment is lost in conventional transfemoral amputations, because the adductor magnus insertion has an attachment to the medial aspect of the distal third of the femur. Once this attachment is lost, the femur swings into abduction because of the relatively unopposed action of the abductor system. This situation is established at the time of surgery when the surgeon sutures the residual adductors and the other muscles around the femur while the stump is in the abducted position.

As the attachments of the adductor muscles are lost, the moment arm becomes shorter. Thus, a smaller mass of adductor muscle would have to generate a larger force to hold the femur in its normal position. In conventional transfemoral amputations, the residual adductor muscles invariably have been re-attached to the femur with the femur in an abducted position, and they are thus unable to generate sufficient force to pull the femur over and maintain the normal anatomical position.

The adductor magnus has a moment arm with the best advantage as compared to the adductor longus and brevis (Fig. 14.2). Transection of the adductor magnus through the belly of the muscle at the time of amputation leads to a loss of muscle bulk and a reduction in the moment arm and a loss of up to an estimated 70% of adductor power (Gottschalk *et al.*, 1989). This combination results in overall weakness of the adductor force of the thigh and resultant abduction of the residual femur. In addition, loss of the extensor portion of the adductor magnus leads to a decrease in hip extension power and a greater likelihood of a flexion contracture.

It has been noted that a reduction in muscle mass at amputation, combined with inadequate mechanical fixation, as well as atrophy of the remaining musculature, are the major factors for the decrease in muscle strength detected in transfemoral amputees (James, 1973b; Thiele *et al.*, 1973) Most noticeable is a significant decrease in the strength of the flexor, extensor, abductor and adductor muscles of the hip, which correlated with inadequate muscle stabilization (James, 1973a).

The goal of surgery in transfemoral amputations should be the creation of a dynamic stump with good motor control and sensation. Preservation of

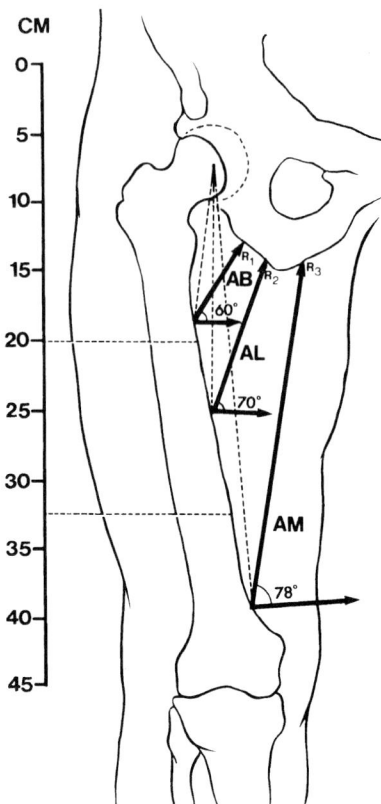

Fig. 14.2. Diagram of the moment arms of the three adductor muscles. Loss of the distal attachment of the adductor magnus will result in a loss of 70% of adductor strength. AB = adductor brevis; AL = adductor longus; AM = adductor magnus. (From Gottschalk, F., Kourosh, S., Stills, M. *et al.* J Prosthet Orthot 1989; 2: 94–102. Used with permission.)

the adductor magnus muscle is possible and helps maintain the muscle balance between the adductors and abductors. The retained muscle bulk allows the adductor magnus to maintain nearly normal muscle power and has a better advantage for holding the femur in a normal anatomical position. A stump with dynamic function will allow the amputee to function at a more normal level and use a prosthesis with greater care.

Several authors recommend transecting the muscles through the muscle belly at a point below the bony transection so they can retract to the level of the bony amputation. Although muscle stabilization is advocated as a means of controlling the femur, in actuality this is infrequently achieved,

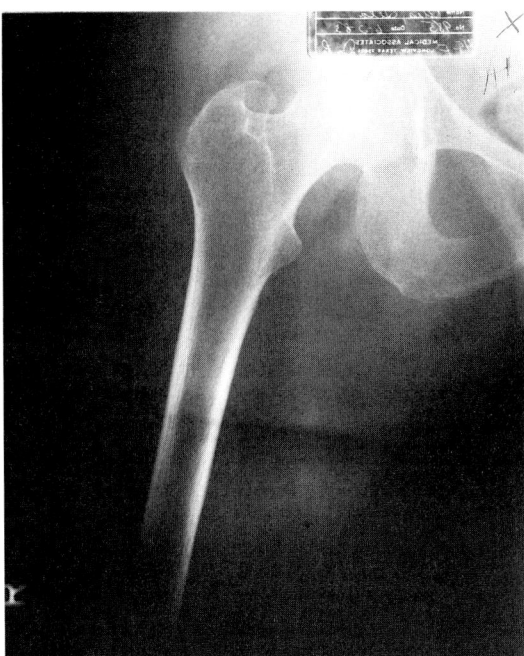

Fig. 14.3. Radiograph of abducted residual femur resulting from inadequate muscle stabilization. (From Gottschalk, F., Kourosh, S., Stills, M. *et al.* J Prosthet Orthot 1989; 2: 94–102. Used with permission).

since the remaining muscle mass will have retracted at the time of transection (Fig. 14.3). It is then difficult to re-establish the normal muscle tension, as recommended in the standard texts (Harris, 1981; Bohne, 1987; Burgess, 1989). A muscle-preserving technique is preferred, whereby the distal insertions of the muscles are detached from the bone with excision of excess muscle tissue once the myodesis has been done.

Indications for transfemoral amputation

Vascular disease

This is probably the most common cause for transfemoral amputation. Although this procedure is done less frequently than in the past, it is often necessary in those patients with very severe

vascular and diabetic disease who are deemed to have poor potential to heal at a lower level amputation (Jensen *et al.*, 1982; Harris *et al.*, 1988; McCollum *et al.*, 1988) The majority of these patients have widespread systemic manifestations of the disease which will often compromise their postoperative rehabilitation. Their physical reserve is often insufficient for them to become prosthetic users (Kay and Pennal, 1958). Patients with combined diabetes and vascular disease tend to be on average some 10 years younger than those patients with purely vascular problems (Christensen *et al.*, 1988; Pohjolainen and Alaranta, 1988)

Trauma

The majority of patients who require a transfemoral amputation for trauma are generally younger. Most often, the indication for the surgery will be severe soft tissue, vascular, neurological and bone injury. The maximum stump length should be retained. It is mandatory to do at least a two-stage reconstruction, leaving the wound open at the initial stage (Harris, 1944). Where feasible, skin grafts may be appropriate with a view to secondary soft tissue expansion. Fractures of the femur should be stabilized by appropriate means. The orientation of skin flaps is not critical, but closure should be without tension.

Infection

Amputation for severe soft tissue infection or osteomyelitis should be done as a two-stage procedure with antibiotic cover dependent on the type and nature of infection. In some situations, the placement of antibiotic impregnated methacrylate beads is useful for controlling local infection. All infected tissue must be excised.

Tumours

The level of amputation is determined by the type and location of the tumour. The principles of tumour eradication need to be considered, while at the same time, preserving as long a stump as possible. Preservation and restoration of function are important factors.

Technique

In general, a tourniquet is not used for the majority of transfemoral amputations. When ischaemia is not a factor, a sterile tourniquet can be placed proximal to the site of surgery and released prior to myodesis to allow proper setting of muscle tension. The skin flaps should be marked out prior to the skin incision. Anterior flaps are fashioned longer than posterior flaps so that its length is equivalent to the diameter of the thigh at the level of amputation. A long medial flap in the sagittal plane is also acceptable and is the preferred flap (Fig. 14.4). The skin flaps should be made longer than may be thought necessary to avoid having to shorten the bone further.

Once the major vessels have been isolated, they should be ligated and cut at the proposed level of bone section. The major nerves should be dissected 2–4 cm proximal to the bone cut and sectioned with a new sharp blade. An epidural catheter may be placed in the sciatic nerve to administer local anaesthetic intermittently postoperatively to reduce pain and phantom sensation (Malawer et al., 1991).

Muscles should not be sectioned until they have been identified. The quadriceps should be detached just proximal to the patella to retain some of its tendinous portion. The adductor magnus is detached from the adductor tubercle by sharp dissection and reflected medially to expose the femoral shaft. It may be necessary to detach 2–3 cm of the adductor magnus from the linea aspera. The smaller muscles should be transected approximately an inch longer than the proposed bone cut.

The femur is exposed just above the supracondylar level and cut with a power saw using an oscillating blade approximately 12 cm proximal to the knee joint (Fig. 14.5). The blade should be cooled with saline. Two or three small drill holes are made on the lateral cortex of the distal femur 1–1.5 cm from the cut end. One or two additional holes may be made anteriorly and posteriorly. The cut edges of the femur should be smoothed with a rasp prior to muscle suture.

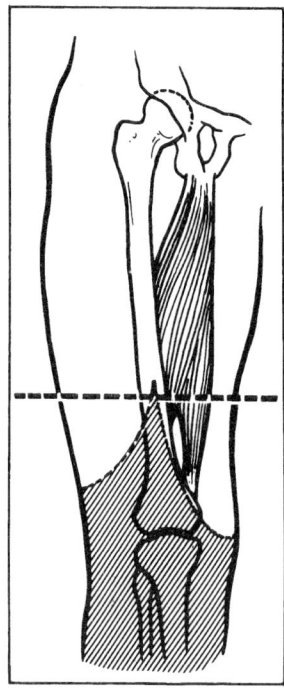

Fig. 14.4. Diagram depicting the proposed skin flaps and level of bone section.

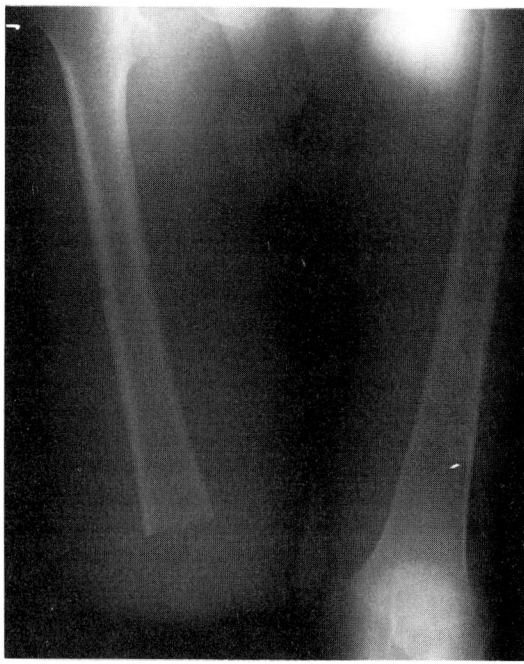

Fig. 14.5. Radiograph of residual femur held in normal anatomic alignment following adductor myodesis.

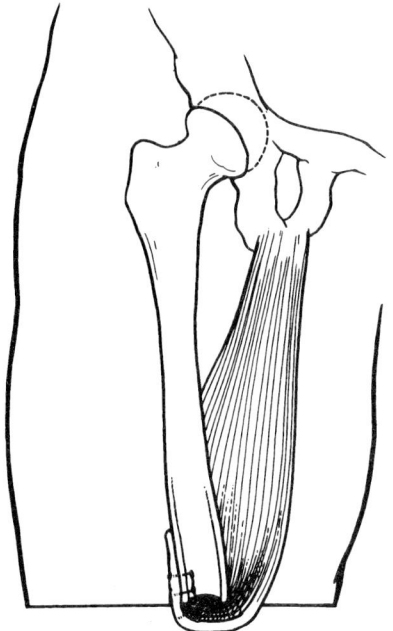

Fig. 14.6. Diagram depicting attachment of the adductor magnus to the lateral part of the femur.

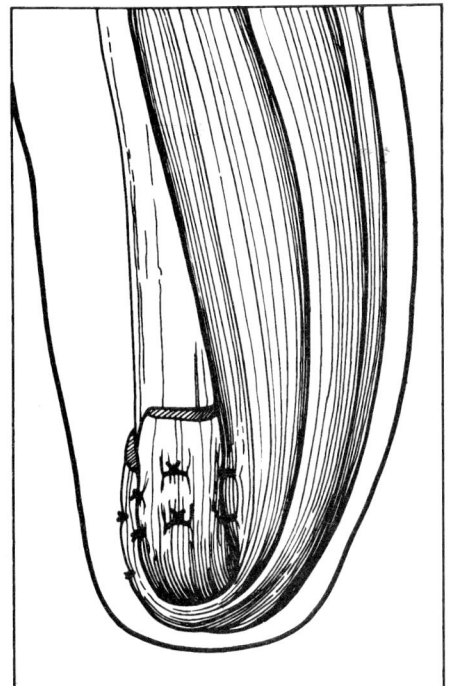

Fig. 14.7. Diagram to show the attachment of the quadriceps over the adductor magnus.

The adductor magnus tendon is sutured with non-absorbable or absorbable long-lasting suture material to the lateral aspect of the residual femur via the drill holes (Fig. 14.6). Prior to securing the sutures, the femur is held in maximum adduction, while the adductor is brought across the cut end of the femur while maintaining its tension. Additional anterior and posterior sutures are placed to prevent the muscle from sliding forwards or backwards on the end of the bone.

Once the adductor magnus has been anchored, the quadriceps is sutured to the posterior femur, via the posterior drill holes (Fig. 14.7). The hip should be in extension when this is done to prevent creating a hip flexion contracture. The remaining hamstring muscles are then anchored to the posterior area of the adductor magnus. The fascia of the thigh is then sutured as dictated by the skin flaps. Subcutaneous stitches are used to approximate the skin edges and fine nylon sutures are used to close the skin. Figure 14.5 shows a post-operative radiograph with the femur held in adduction.

Post-operative stump care

The stump should be wrapped with an elastic bandage applied as a hip spica, with the hip extended. Although rigid dressings control the oedema and stump position better than soft dressings, they are cumbersome to apply and do not offer any great advantage in the long term in transfemoral amputations. A well applied elastic bandage will not slip off the limb.

Another method of controlling the swelling and reducing discomfort is to apply a stump shrinker with a waist belt. The shrinkers are made of a one- or two-way elastic material which applies even pressure from distal to proximal. The waist belt helps prevent the shrinker from slipping off the stump. Post-operative pain may be controlled by administering local anaesthetic through a previously placed catheter. This is done for 3–4 days and then discontinued (Malawer *et al.*, 1991).

General post-operative care

While the wound on the stump is healing, the patient should be mobilized in a wheelchair, between parallel bars, and upper body exercises

started. In addition, conditioning of the contralateral leg should take place. Most often the
sutures can be removed at around two weeks.
During this time the patient will have been
wrapping the stump or using a shrinker. A
temporary plastic adjustable prosthesis can now be
fitted and gait training started. The patient needs to
have sufficient upper body strength to use crutches
or a walking aid. Flexion contractures should be
prevented from occurring by correct positioning of
the patient in bed as well as muscle strengthening
exercises. By using aggressive rehabilitation techniques in a motivated patient, return to walking can
be accomplished in a short time. Those patients
who do not have the physical or mental ability to
participate in a rehabilitation programme designed
to teach prosthetic use will be better off using a
wheelchair. Transfer training is important in this
regard.

References

Barber, G.G., McPhail, N.V., Scobie, T.K., Brennan, M.C.D.
and Ellis, C.C. (1983) A prospective study of lower limb
amputations. *Can. J. Surg.*, **26**, 339–341.

Bohne, W.H.O. (1987) Above knee amputation. In: *Atlas of
Amputation Surgery* ... Thieme Medical Publishers, New
York, pp. 86–90.

Burgess, E.M. (1989) Knee disarticulation and above-knee
amputation. In: *Lower Extremity Amputation* (ed. W. Moore
and J. Malone) Saunders, Philadelphia.

Christensen, K.S., Falstie-Jensen, N., Christensen, E.E. and
Brochner-Mortensen, J. (1988) Results of amputation for
gangrene in diabetic and non-diabetic patients. *J. Bone Joint
Surg.*, **70A**, 1514–1519.

Colt, J.D. and Lee, P.Y. (1972) Mortality rate of above-the-knee
amputation for arteriosclerotic gangrene: a critical evaluation. *Angiology*, **23**, 205–210.

Couch, N.P., David, J.K., Tilney, L., Nicholas, L. and Crane, C.
(1977) Natural history of the leg amputee. *Am. J. Surg.*, **133**,
469–473.

Dale, W.A. and Capps, W. (1959) Major leg and thigh
amputations (ten-year survey of results). *Surgery*, **46**,
333–342.

Dardik, H., Kahn, M., Dardik, I., Sussman, B.I. and Ibrahim,
I.M. (1982) Influence of failed vascular bypass procedures on
conversion of below-knee to above-knee amputation levels.
Surgery, **91**, 65–69.

Freeman, M.A.R. (1980) The surgical anatomy and pathology
of the arthritic knee. In: *Arthritis of the knee* (ed. M.A.R.
Freeman) Springer, New York, pp. 32–33.

Gonzalez, E.G., Corcoran, P.J., and Peters, R.L. (1974) Energy
expenditure in below-knee amputees: correlation with stump
length. *Arch. Phys. Med. Rehab.*, **55**, 111–119.

Gottschalk, F., Kourosh, S., Stills, M., McClennan, B. and
Roberts, J. (1989) Does socket configuration influence the
position of the femur in above-knee amputation? *J. Prosthet.
Orthot.*, **2**, 94–102.

Harris, J.P., Page S, Englund, R. and May, J. (1988) Is the
outlook for the vascular amputee improved by striving to
preserve the knee? *J. Cardiovasc. Surg.*, **29**, 741–745.

Harris, R.L. (1944) Amputations. *J. Bone Joint Surg.*, **26**,
626–635.

Harris, W.R. (1981) Principles of amputation surgery. In:
*Amputation Surgery and Rehabilitation: the Toronto Experi-
ence* (ed. J.P. Kostuik) Churchill Livingstone, New York, pp.
37–49.

Hungerford, D.S., Krackow, K.A. and Kenna, R.V. (1984) *Total
Knee Arthroplasty*. Williams & Wilkins, *Baltimore*, pp.
36–39.

Jaegers, S.M.H.J. and deJongh, H.J. (1992) The morphology of
the muscles around the hip joint of unilateral above-knee
amputees based on MR scans. *Proceedings of 7th World
Congress of the International Society of Prosthetics and
Orthotics*, p. 261.

James, U. (1973a) Maximal isometric muscle strength in healthy
active male unilateral above-knee amputees with special
regard to the hip joint. *Scand. J. Rehab. Med.*, **5**, 55–66.

James, U. (1973b) Unilateral above knee amputees, a clinico-
orthopaedic evaluation of healthy active men, fitted with a
prosthesis. *Scand. J. Rehab. Med.*, **5**, 23–24.

Jensen, J.S., Mandrup-Poulson, T. and Kransnik, M. (1982)
Wound healing complications following major amputations
of the lower limb. *Prosthet. Orthot. Int.*, **6**, 105–107.

Kay, G.D. and Pennal, G.F. (1958) Rehabilitation of the elderly
amputee. *Can. J. Surg.*, **2**, 44–51.

Long, I.A. (1985) Normal shape – normal alignment (NSNA)
above-knee prosthesis. *Clin. Prosthet. Orthot.*, **9**, 9–14.

Malawer, M., Buch, R., Khurana, J., Garvey, T. and Rice, L.
(1991) Postoperative Infusional Continuous Regional
Analgesia. *Clin. Orthop.*, **266**, 227–237.

Maquet, P. (1980) *Biomechanics of the Knee*. Springer, New
York, p. 22.

McCollum, P.T., Spence, V.A. and Walker, W.F. (1988)
Amputation for peripheral vascular disease: the case for level
selection. *Br. J. Surg.*, **75**, 1193–1195.

Pohjolainen, T. and Alaranta, H. (1988) Lower limb amputations in Southern Finland 1984–1985. *Prosthet, Orthot, Int.*,
12, 9–18.

Sabolich, J. (1985) Contoured adducted trochanteric-controlled
alignment method (CAT-CAM): introduction and basic
principles. Clin. *Prosthet. Orthot.*, **9**, 15–26.

Thiele, B. James, U. and Stalberg, E. (1973) Neurophysiological studies on muscle function in the stump of above-
knee amputees. *Scand. J. Rehabil. Med.*, **5**, 67–70.

Volpicelli, L.J., Chambers, R.B. and Wagner, F.W. (1983)
Ambulation levels of bilateral lower-extremity amputees. *J.
Bone Joint Surg.*, **65A**, 599–604.

Waters, R.L., Perry, J., Antonelli, D. and Hislop, H.J. (1976)
Energy cost of walking of amputees: influence of level of
amputation. *J. Bone Joint Surg.*, **58A**, 42–46.

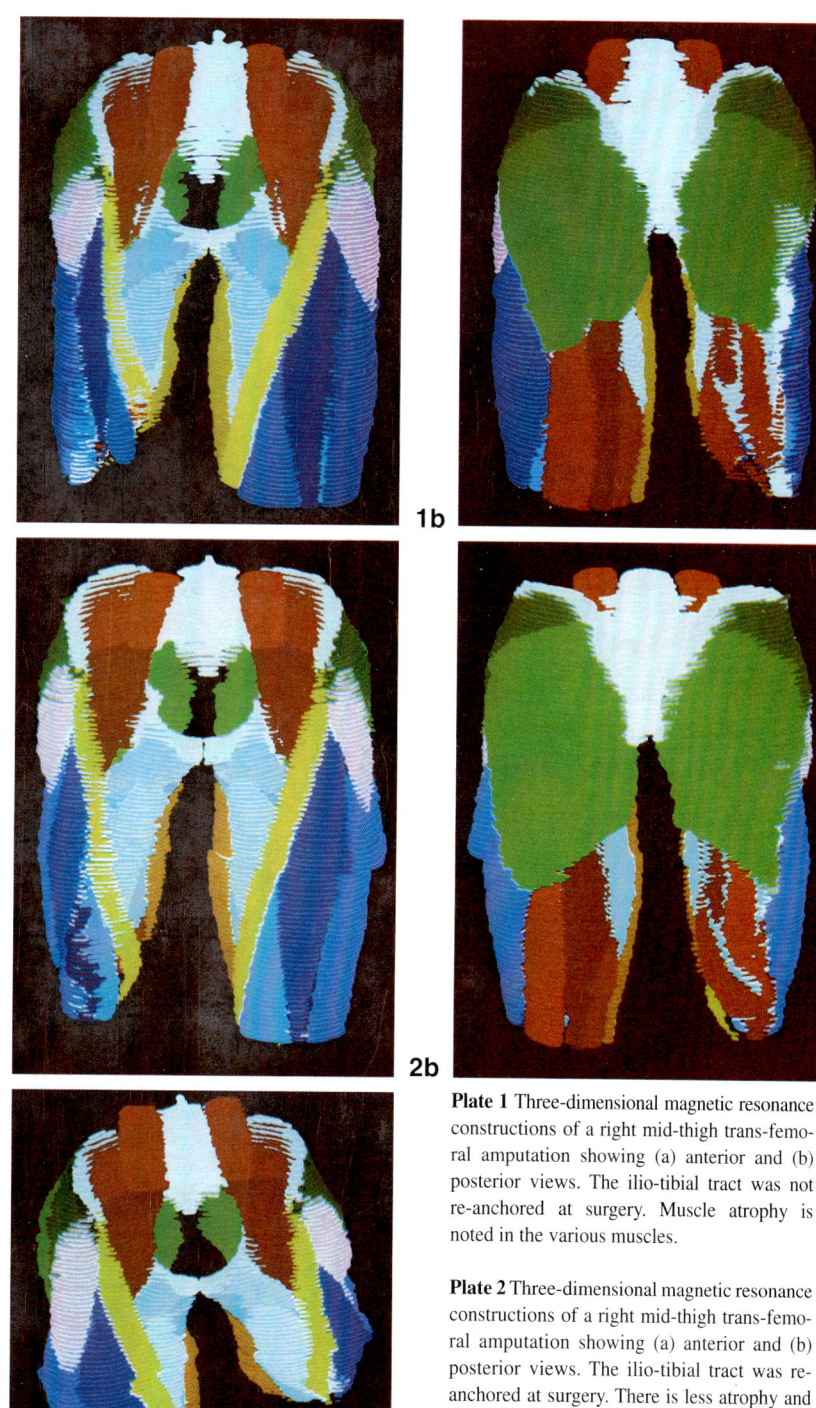

Plate 1a

1b

Plate 2a

2b

Plate 3

Plate 1 Three-dimensional magnetic resonance constructions of a right mid-thigh trans-femoral amputation showing (a) anterior and (b) posterior views. The ilio-tibial tract was not re-anchored at surgery. Muscle atrophy is noted in the various muscles.

Plate 2 Three-dimensional magnetic resonance constructions of a right mid-thigh trans-femoral amputation showing (a) anterior and (b) posterior views. The ilio-tibial tract was re-anchored at surgery. There is less atrophy and better maintenance of muscle position compared to Plate 1.

Plate 3 Anterior view of a three-dimensional magnetic reconstruction of a left short trans-femoral amputation with an abduction contracture. There is a loss of muscle mass as well as altered muscle/stump position.

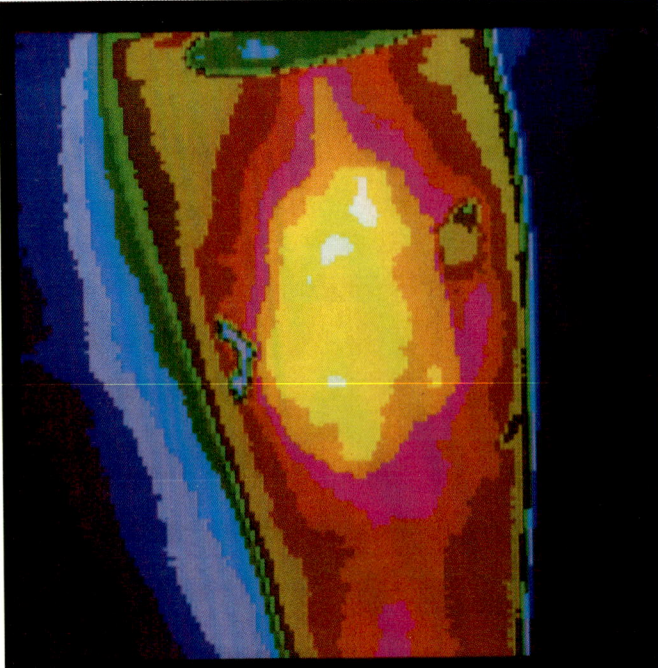

Plate 4 Infrared thermoscan of the forearm of a subject with an intradermal injection of tuberculin, demonstrating the hyperaemic response obtained three days later in this reaction. The dramatic increase in the blood flow is seen as yellows and whites in contrast to the cooler surrounding skin.

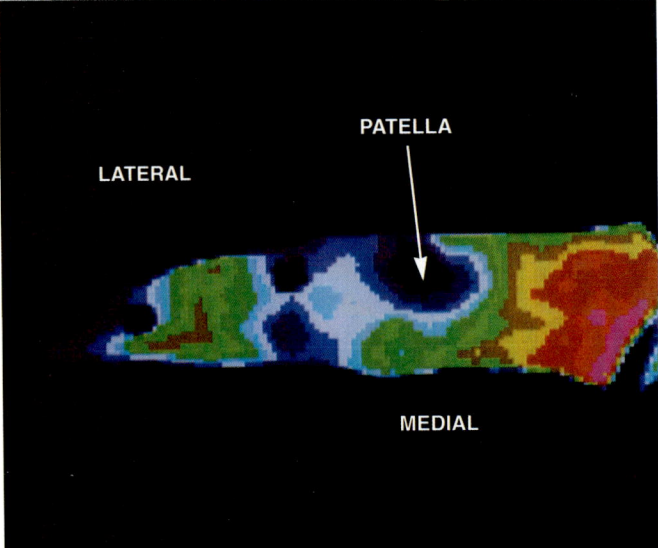

Plate 5 Infrared thermoscan of the right leg of a pre-amputation patient. The patella is seen as a black area in the mid-section with the thigh to the right and the foot (unseen because of severe ischaemia) to the left. There is a clear thermal gradient at a level 10cm distal to the tibial tubercle and a transverse thermal gradient, suggesting a better blood flow in the medial aspect of the leg.

15

Prosthetic requirements and transfemoral biomechanics

J. W. Michael

Introduction

Creating an effective prosthesis for the trans-femoral amputee is a difficult task. In part, this is because the passive mechanical replacement of the formerly active functions of the knee, ankle and foot is technically challenging. Limitations in both current prosthetic technology and in the scientific understanding of the complexities created by transfemoral amputation are also factors. Finally, the transfemoral level is the only amputation in which the major weight bearing structure (the pelvis) is proximal to the propulsive structure (the thigh).

Weight bearing

One fundamental requirement for any lower limb prosthetic socket is to provide comfortable, stable support for the body mass during single limb support. For the transfemoral amputee, the pelvis supplies the majority of the weight bearing areas.

Throughout the world, there are a large number of successful variations in the configuration of transfemoral sockets. Figure 15.1 illustrates the different load-bearing surfaces utilized by the older style quadrilateral design and the more contemporary ischial containment type. Many additional configurations can be identified in the literature (Chodera, 1970). Although each variant has both

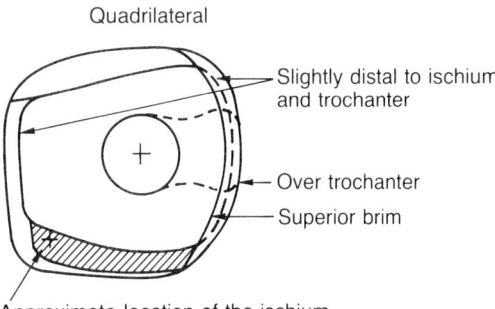

Quadrilateral

— Slightly distal to ischium and trochanter

— Over trochanter
— Superior brim

Approximate location of the ischium with weight bearing area cross hatched

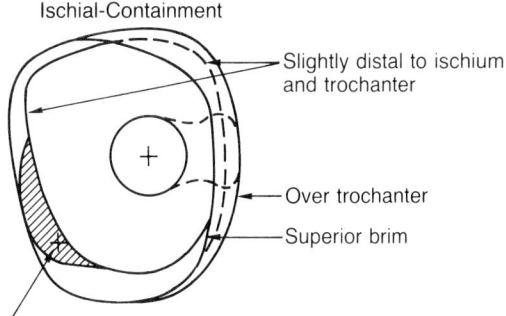

Ischial-Containment

— Slightly distal to ischium and trochanter

— Over trochanter
— Superior brim

Approximate location of the ischium with weight bearing area cross hatched

Fig. 15.1. Commonly encountered transfemoral socket configurations include the 'quadrilateral' type (top) and the 'ischial containment' type (bottom). Although the cross-hatched weight bearing areas differ somewhat, many variants achieve good clinical results. (Reprinted with permission from Michael, 1990, p. 87, Fig. 7.)

advantages and disadvantages, the only consensus which has emerged to date is that there are 'no specific contraindications noted for any socket design' (Schuch, 1988). As a practical matter, so long as the amputee is comfortable and the gait is satisfactory, any particular socket design may be considered acceptable.

An additional controversy exists over the advisability of flexible versus rigid socket walls. The decreased durability and tendency for the thermoplastic socket to shrink dimensionally in relation to the cast on which it is formed must be weighed against the claims of improved proprioception, heat dissipation and comfort that flexible sockets are said to offer. As materials and techniques improve, the percentage of flexible thermoplastic and ischial containment sockets seems to be increasing worldwide.

Although space limitations preclude a detailed discussion, it should be noted that every portion of the transfemoral stump makes some contribution to weight bearing; the forces on the pelvis are simply more pronounced. Authors such as Hanger (1964) and Foort (1979) have thoroughly discussed the biomechanical functions of each surface of the transfemoral socket.

Medio-lateral and antero-posterior stability

Comfortable weight bearing is necessary but not sufficient for successful ambulation. The stump must also be able to stabilize the torso over the prosthesis. Figures 15.2a and b, respectively, depict graphically the forces that are necessary for control and sagittal plane stabilization.

In essence, for trunk stability there must be an intimate coupling between the stump and the prosthetic socket and the ability comfortably to apply load over the major area and minimize pressure. This implies that the surgery should contrive to produce a stump of stable volume and shape. This in turn requires the elimination of oedema, usually in the post-operative period, but more importantly stabilization of the divided muscles. Inadequate surgical techniques resulting in weak, retracted muscles or skin that cannot tolerate the necessary pressures will obviously compromise stability. The forces necessary to

provide dynamic stability and propulsion during ambulation far exceed those necessary simply for weight bearing. This fact explains the common observation that a prosthesis may be quite comfortable when standing and yet painful when walking is attempted. In order to ensure stable shape and ensure maximum available power the major divided muscles must be attached in some way to the end of the stump. Myoplasty is successful but there is an inherent risk in later displacement of the mucle loops. Myodesis via drill holes in the femoral end produces a more stable situation. In order to achieve correct tension in the intact abductor muscles at the hip the adductor magnus assumes great importance and when attached to the end of the femur should ensure a hip position of small but significant secure adduction, thus simulating the position at stance in the intact limb and easier alignment of the prosthesis.

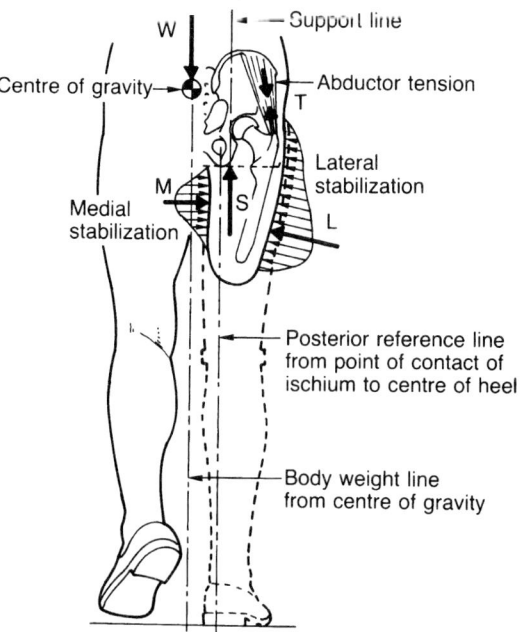

Fig. 15.2. (a) The tendency for the body to drop medially away from the prosthesis must be counteracted by the hip abductors. The distribution of force on the stump that results is depicted by the length of the thin vector arrows. (Reprinted with permission from Radcliffe, 1977, p. 159, Fig. 17.)

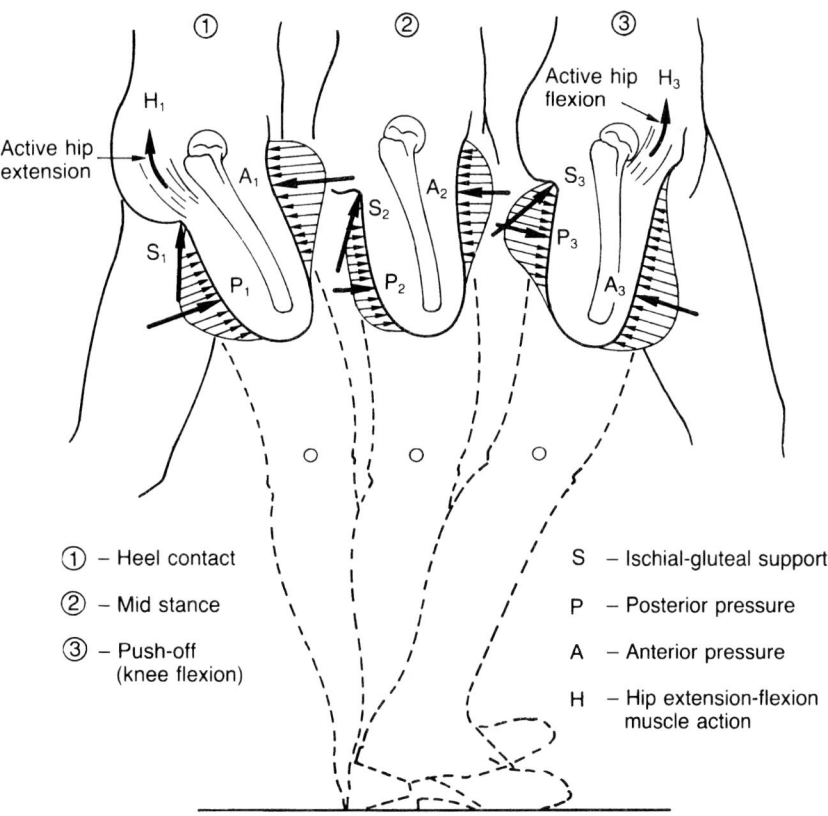

Fig. 15.2. (b) Forces generated to stabilize the trunk and control the prosthesis during stance phase are summarized. Note that virtually every portion of the transfemoral stump must tolerate significant pressure during ambulation. (Reprinted with permission from Radcliffe, 1977, p. 158, Fig. 16.)

Knee control

Propulsive control requires even more of the amputation stump. In addition to tolerating the pressures necessary to provide weight bearing and trunk stability, the amputee must also be able to generate and apply sufficient force to control the passive prosthetic knee mechanism. Due to the absence of functioning muscles below the amputation site, the amputee must create a knee hyperextension moment throughout stance phase to prevent the prosthetic knee from collapsing.

The prosthetist can influence the stability of the prosthetic knee by selection of components offering enhanced stance control (friction braking, polycentric mechanisms or fluid control) and by meticulous dynamic alignment (by placement of knee centre) during repeated walking trials. Both of these strategies tend to reduce the moment required of the amputee at the hip.

Too much inherent stability, however, creates another problem. Initiation of swing phase flexion becomes difficult or impossible. As Fig. 15.3 demonstrates, the prosthetist must create a design which is sufficiently stable from heel strike through midstance yet permits the initiation of swing phase flexion. Radcliffe has described this as optimizing the 'zone of voluntary stability'.

For the amputee who wishes to engage in heavy labour or competitive athletics, the biomechanical requirements are even more demanding. Throughout the world, more and more physically active amputees feel their rehabilitation has not been successful so long as vigorous activities remain difficult or impossible to achieve (Michael *et al.*,

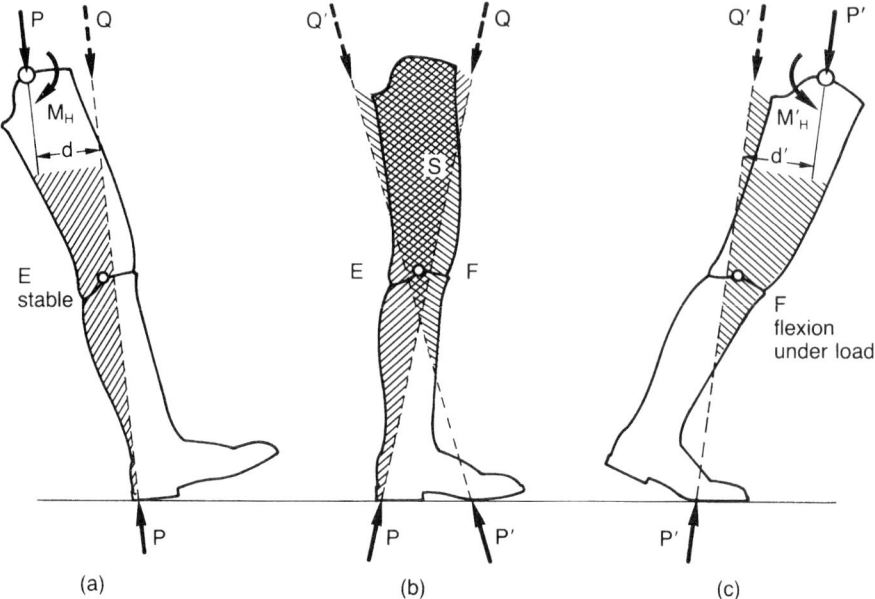

Fig. 15.3. Dual cross-hatched area is referred to as the *zone of voluntary stability.* Through iterative dynamic walking trials, the prosthetist must determine the unique zone for each individual transfemoral amputee and align the prosthetic knee centre within its confines. (Reprinted with permission from Radcliffe, 1977, p. 152, Fig. 7.)

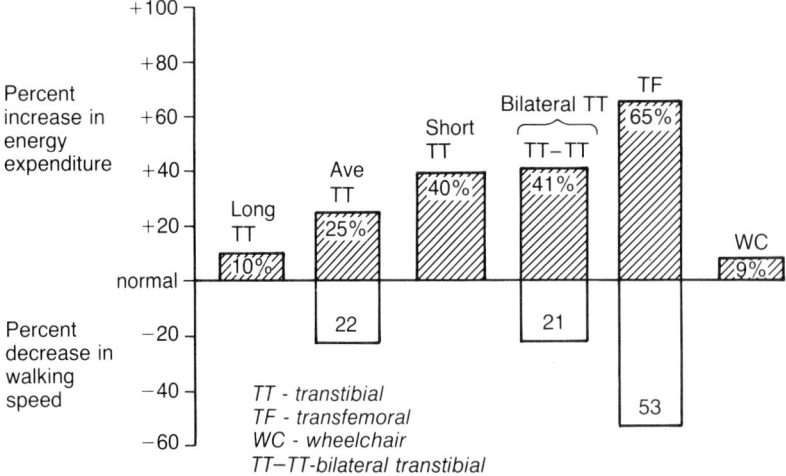

Fig. 15.4. Summary of energy cost and walking speed for various levels of amputation and for wheelchair use. Note that higher amputation significantly increases the energy consumption while slowing the velocity of gait. The higher the amputation level, the more critical optimal surgical technique becomes. (Reprinted with permission from Gonzalez et al., 1974, p. 118, Fig. 8.)

1990). Unfortunately, all too often a mediocre surgical result prevents transfemoral amputees from attaining their full functional potential.

Conclusion

Numerous studies have documented that significant functional losses are inevitable when higher levels of amputations are performed. Figure 15.4 summarizes the decrease in velocity and the increase in energy expenditure that transfemoral amputation creates (Gonzalez *et al.*, 1974). Dysvascularity has been shown to amplify these negative effects (Waters, 1992). Such data underscore the importance of maximizing the effectiveness of residual function through meticulous surgical technique. Because modern prosthetic care has advanced considerably over the past decades (Staats, 1985; Sabolich, 1991), the quality of the surgical result is now often the limiting factor in rehabilitation of the transfemoral amputee.

References

Chodera, J. D. (1970) Czechoslovakian above-knee socket design. In: *Prosthetic and Orthotic Practice* (ed. G. Murdoch), Edward Arnold, London, pp. 219–223.

Foort, J. (1979) Socket design for the above-knee amputee. *Prosthet. Orthot. Int.*, **3**, 73–81.

Gonzalez, E.G., Corcoran, P.J. and Reyes, R.L. (1974) Energy expenditure in below-knee amputees. *Arch. Phys. Med. Rehabil.*, **55**, 111–119.

Hanger, H. B. (1964) *Above-Knee Socket Shape and Clinical Considerations*. National Academy of Sciences, Washington, DC, 11 pp.

Kristinsson, O. (1983) Flexible above-knee socket made from low density polyethylene suspended by a weight-transmitting frame. *Orthot. Prosthet.*, **37**, 25–27.

Michael, J. W. (1990) Current concepts in above-knee socket design. In: *Instructional Course Lectures*, Volume XXXIX. American Academy of Orthopaedic Surgeons, Rosemont, III., pp. 373–378.

Michael, J.W., Gailey, R.S., and Bowker, J.H. (1990) New developments in recreational prostheses and adaptive devices for the amputee. *Clin. Orthop.*, **256**, 64–75.

Pritham, C.H., Fillauer, C. and Fillauer, K. (1985) Experience with the Scandinavian flexible socket. *Orthot. Prosthet.*, **39**, 17–32.

Radcliffe, C.W. (1977) The Knud Jansen Lecture – above knee prosthetics. *Prosthet. Orthot. Int.*, **1**, 146–160.

Sabolich, J. (1991) Prosthetic advances in lower extremity amputation. *Phys. Med. Rehabil. Clin. Am.*, **2**, 415–422.

Schuch, C.M. (1988) Modern above-knee fitting practice. *Prosthet. Orthot. Int.*, **12**, 89.

Staats, T.B. (1985) Advanced prosthetic techniques for below knee amputations. *Orthopaedics*, **8**, 249–258.

Waters, R.L. (1992) The energy expenditure of amputee gait. In: *Atlas of Limb Prosthetics: Surgical, Prosthetic and Rehabilitation Principles* (eds J.H. Bowker and J.W. Michael), Mosby, St Louis, pp. 381–387.

Section 6

The knee

16

Surgical techniques of knee disarticulation and femoral transcondylar amputations

J. S. Jensen

Goals

The primary goal with amputation surgery is to preserve life in patients with severe toxic gangrene of a dysvascular limb. The most frequent indication for amputation in the industrial world is peripheral vascular disease with or without diabetic complications, constituting more than 85% of the patients.

The second goal is to strive for independence and an acceptable quality of life for the amputees, who are usually elderly people averaging 70 years of age. In practical terms this means rehabilitating the patient so that walking on a prosthesis is the highest aim and as a minimum goal to enable the patient to transfer from chair to bed or toilet, independent of assistance from other people. It should be remembered that the remaining lifespan for dysvascular amputees is less than 3 months for some 20% of these elderly patients, and that the risk to survival increases dramatically if the patients have concomitant medical diseases.

The first goal applies to all life-threatening diseases or injuries, whereas the second goal applies to all categories of amputees and should be seriously considered in the selection process of the appropriate level of amputation.

Advantages of knee disarticulation

The classical knee disarticulation without any associated bone surgery is functionally characterized as a long, strong lever arm with undisturbed muscle attachments and excellent end-bearing capacity, with anatomical features providing good rotational control and the opportunity for very efficient suspension of the prosthesis. These advantages are fully exploited by modern prosthetists, who learn to fit the stump using contemporary principles based on a better understanding of the biomechanics and the benefits of modern technology. The unpopularity of the knee disarticulation amputation has most likely been caused by poor limb-fitting skills and a bad reputation for primary wound healing (Middleton and Webster, 1962; Newcombe and Marcusson 1972).

The energy cost of locomotion increases considerably following amputation (Waters *et al.*, 1976). Wheelchair mobility requires 10% more energy than normal level walking, a unilateral transtibial amputee 10–40% above normal and 40% above for a bilateral amputee. In spite of a halved walking speed, a transfemoral amputee requires 65% more energy in the case of a unilateral amputation and 110% for a bilateral transfemoral amputee. No physiological assessments relating to knee disarticulation have been identified, but there is good reason to believe that the energy costs will correspond most closely with the transtibial amputations because of the characteristics of the stump and the prosthetic suspension.

Patient selection

The obvious selection criterion for knee disarticulation is a high probability of bilateral amputation, not only as an alternative to a transfemoral amputation, but also in relation to the doubtful survival of a transtibial stump. Moreover, in hemiparetic patients and in particular in patients with a risk of ipsilateral knee contracture, the knee disarticulation should be considered as an alternative to a transtibial procedure.

The mortality in patients with knee disarticulation is half that of transfemoral amputees, but double that of transtibial amputees. The differences are essentially related to the toxicity which in turn is related to the magnitude of the gangrene, which is more extensive at the higher levels of amputation. Patient survival is, however, also related to the surgical trauma, which is considerably reduced by knee disarticulation. This procedure requires only transection of skin and tendinous structures, with no bleeding from muscle bellies or bone cuts. Consequently, the knee disarticulation should be considered the first choice in the feeble patient with a low chance of survival.

As an adjuvant to clinical examination in level assessment we have used the skin perfusion test and successfully performed knee disarticulation in the vast majority of patients with pressures of 20–40 mmHg as measured 10 cm distal to the knee joint (Thyregod *et al.*, 1983).

Repetitive surgery increases the risk of death, and our aim in amputation surgery is to achieve primary wound healing at the first amputation procedure. However, some re-amputations must be accepted and ablative surgery requires us to aim for that level which provides the best possible chance of rehabilitation to ambulation and independence.

For non-ambulators the long, strong stump and good muscle function of the knee disarticulation facilitates balance in the sitting position and transfers from bed to chair and prevents contractures. The pressure on the posterior aspect of the stump is distributed over a large area during sitting, thus increasing comfort.

Surgical technique

Although simple to describe, the surgical technique demands training and experience before being fully mastered and junior surgeons in training should be carefully supervised.

Skin

A circular skin incision is recommended some 8–10 cm distal to the knee joint, corresponding to 3–5 cm below the tibial tuberosity (Fig. 16.1). A short sagittal skin incision over the tibial tuberosity

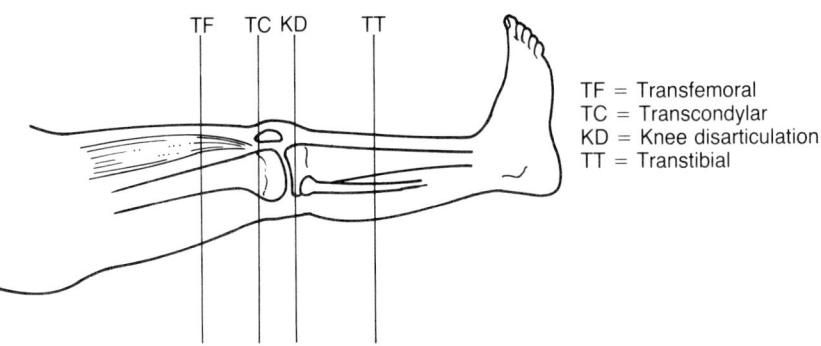

TF = Transfemoral
TC = Transcondylar
KD = Knee disarticulation
TT = Transtibial

Fig. 16.1. Levels of skin incision for amputations of lower limb.

is of help for the next steps in the procedure but the final skin trimming to provide side-flaps should be left until wound closure.

The circular skin incision is currently advocated, but skew skin flaps might be applied in traumatic cases or tumours, or where obvious differences in skin temperatures exist on the two sides of the leg. The circular incision makes it possible to delay the final design of the flaps as long as possible and facilitates the avoidance of skin tension over the medial condyle. Alternatively, sagittal or fish-mouth incisions with equally long anterior and posterior flaps have been used with equivalent results. Long anterior flaps have been disastrous with reamputation rates amounting to 30–50%.

With transcondylar procedures the skin incision should be placed level with the tibial tuberosity or slightly above.

Soft tissues

The succeeding steps of the knee disarticulation are detachment of the collateral ligaments, the hamstring tendons and the patellar tendon from their attachments on the tibia (Fig. 16.2). A broad transverse capsulotomy, below or above the semi-lunar cartilages (menisci), exposes the joint widely. The cruciate ligaments and the posterior joint capsule are divided at their distal attachments. In all of these procedures the knife dissects the soft tissue directly from the bone, leaving no soft tissue *in situ*.

Then the only incision through structures other than ligaments, tendons, vessels and nerves, is the division of the gastrocnemius bellies 2 cm below the joint level in order to preserve the blood supply to the posterior capsule from the superior genicular artery, thus minimizing the risk of deep necrosis.

The popliteal vessels are ligated and divided at joint level. The tibial and peroneal nerves are isolated, pulled down gently, ligated and divided. There is no evidence to support synovectomy.

The menisci are removed as they represent poorly vascularized tissue. Baumgartner (1979) recommended only removal of the menisci in vascular cases, but there is no evidence to show that they contribute in any way to the quality of the stump.

One of the major advantages of the knee disarticulation are the excellent possibilities of suspension and end-bearing as obtained by the bulbous, triangular stump profile (Fig. 16.3). This

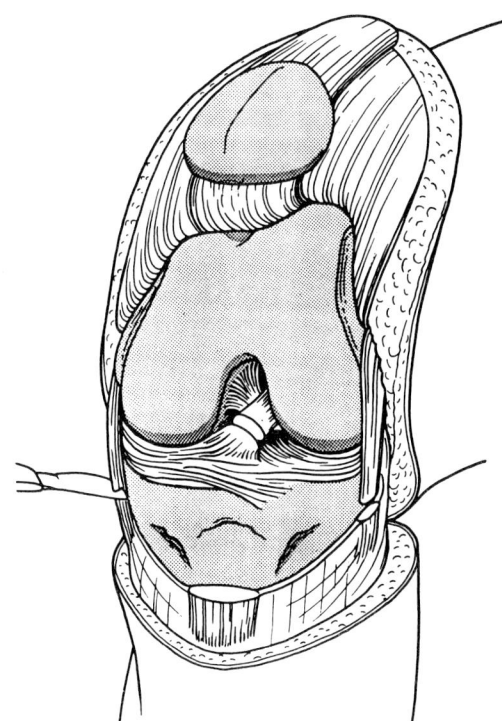

Fig. 16.2. Knee disarticulation. Patellar tendon transected and pulled upwards, side ligaments and cruciates transected at insertions on shank.

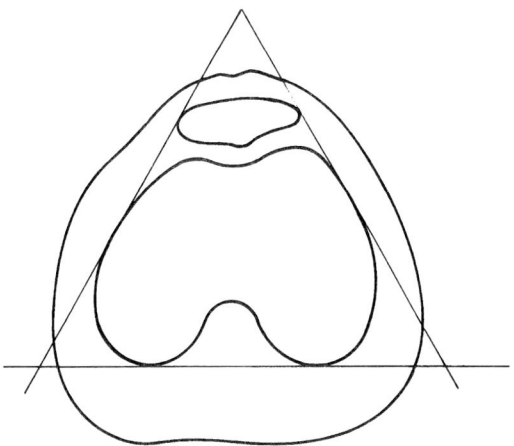

Fig. 16.3. Weight bearing area of knee disarticulation stump with patellar tendon sutured to cruciates.

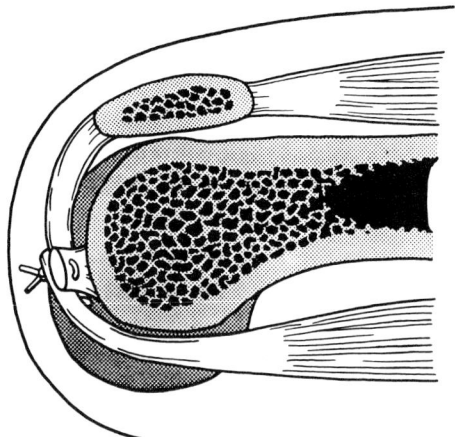

Fig. 16.4. Patellar tendon sutured to cruciates and the apex of the patella pulled to level of condyles.

is established by pulling the apex of the patella down to the level of the condyles and stitching the patellar tendon to the remnants of the cruciate ligaments, preferably with resorbable sutures (Fig. 16.4). The placement of the knee cap is crucial. Correctly placed, the stump end consists of three bone areas, the condyles and the apex of the patella, increasing the weightbearing area. If the patella is pulled down too far high peak pressures will be located over the prominent patella with a high risk of skin breakdown and discomfort. Secure fixation of the patellar tendon to the rump of the cruciates prevents retraction of the patella with the attendant weakening of the quadriceps

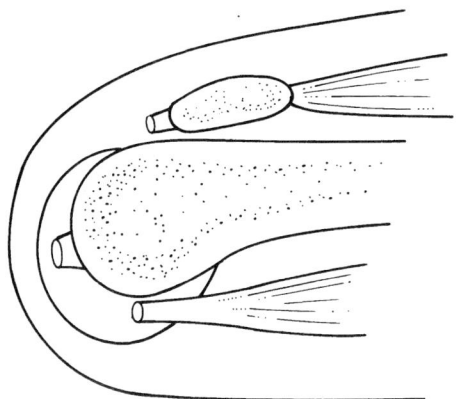

Fig. 16.5. Patellar tendon and ligaments cut at condylar level resulting in modest patella alta.

function. Baumgartner (1979) has advocated resecting the patellar tendon at the apex of the patella and similarly the ligaments at the same level (Green 1975), with the objective of avoiding deep suture materials left *in situ* and the potential risk of fistula formation. While the strong retinaculum around the knee prevents the patella from sliding markedly upwards (Fig. 16.5), the weight bearing area over the stump end is reduced (Fig. 16.6).

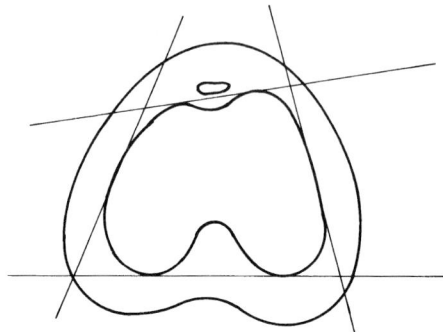

Fig. 16.6. Weight bearing area is reduced by patellar retraction.

Finally, both medial and lateral ligaments and the hamstring tendons can be sutured to the cruciates and the gastrocnemius bellies trimmed to cover the posterior condyles. These latter procedures are not mandatory and can be omitted.

The scar is placed sagittally between the femoral condyles, and a subcutaneous suction drain employed for 1–2 days. It is of utmost importance to ensure that in suturing the skin it remains slack with no tension whatsoever and especially so over the prominent medial condyle, where the highest risk of skin necrosis is located. Subcutaneously resorbable sutures are used and interrupted non-resorbable suture materials selected for the skin.

Cartilage and bone

There is no need to trim either the cartilage or the bone of the condyles. Only in cases with skin tension over the medial condyle should slices of the medial condyle be removed with the purpose of rescuing the knee disarticulation level.

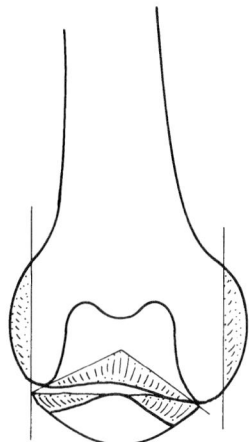

Fig. 16.7. Patella fused to wedge excision at intercondylar notch.

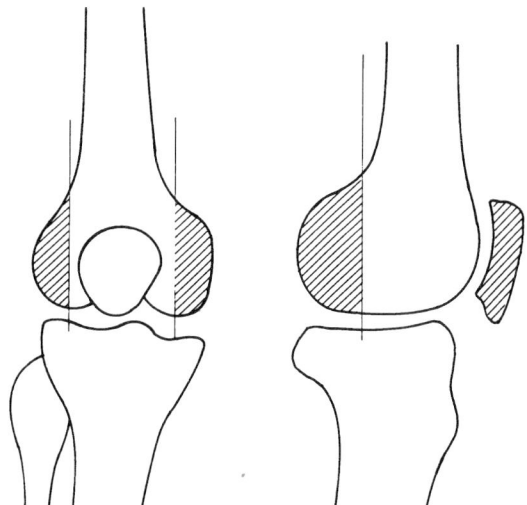

Fig. 16.9. Mazet procedure with bone resection medially, laterally and posteriorly of condyles along with removal of the patella.

Any attempt to improve the cosmesis of the bulky shape of the condyles inevitably leads to reduction of the weight bearing area and thus to increased local pressure with discomfort and even risk of stump breakdown. Even limited trimming of the condyles (Vaucher and Blanc, 1982) leads to oozing of blood from raw cancellous bone surfaces with potential risk of haematoma, infection and necrosis and should consequently be avoided.

Recently (Duerksen *et al.*, 1990) it has been suggested to shape the stump end conically by a wedge resection of the intercondylar notch with fusion of the patella into the area with the purpose of improving prosthetic fitting (Fig. 16.7). However, the weightbearing area is considerably

reduced and a high peak stress area created over the patella (Fig. 16.8).

Transcondylar amputation procedures have been described (Mazet and Hennessy, 1966) with sagittal resection of the condyles designed to slim them in all directions and associated removal of the patella (Fig. 16.9); fusion of the patella placed in a trough at the anterior aspect of the condyle has also been proposed. In all these techniques large areas of bleeding bone are exposed and the weight bearing capacity of the stump end significantly reduced (Fig. 16.10).

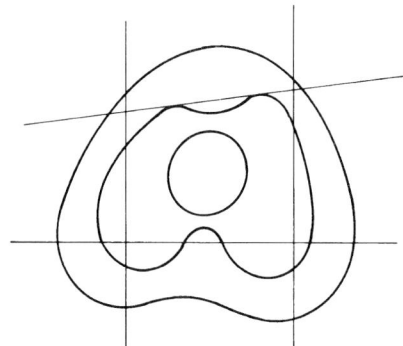

Fig. 16.8. Reduced weight bearing area by bone resection. High peak stress on patella, which constitutes main area of load.

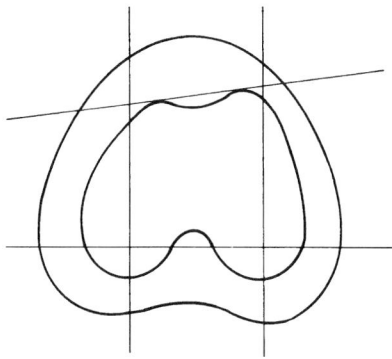

Fig. 16.10. Weight bearing area following Mazet amputation procedure is considerably reduced.

The Gritti–Stokes procedure (Gritti, 1857; Stokes 1870) includes oblique resection through cancellous bone at the level of the adductor tubercle and fusion of the sliced patella to the distal condylar bone (Fig. 16.11). A modification, the Callander procedure, includes removal of the patella. Both these techniques serve to shorten the length of the stump to facilitate prosthetic fitting and to avoid marked protrusion of the prosthetic knee mechanism in the sitting posture, but the stumps are not end-bearing and pose serious problems when the remaining portion of the patella fails to fuse.

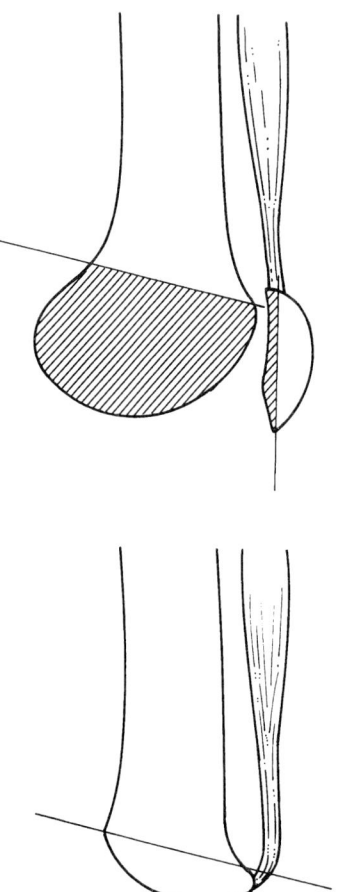

Fig. 16.11. Transcondylar amputations. The Gritti–Stokes procedure with fusion of patella to oblique transection line at level of adductor tubercle. In the Callander procedure the patella is removed. In both procedures weight bearing area is considerably reduced and end-bearing capacity doubtful.

Aftercare and complications

Post-operatively a soft dressing provides sufficient wound protection. The sutures can be removed after 3 weeks and an elastic stocking applied during the last week to reduce possible oedema. At the time of removal of the stitches the stump is ready for prosthetic fitting as further stump maturation is not required.

The major complications are skin necrosis and/or infection normally due to poor technique and frequently lead to stump breakdown and re-amputation at the transfemoral level (Jensen et al., 1982).

The formation of a synovial fistula in the middle of the wound in the intercondylar area is encountered in about 10% of cases. This is not a crucial complication as it can be successfully treated conservatively with patience and dry dressings until the fistula dries out spontaneously usually within months. Prosthetic fitting should not be delayed in the presence of a fistula and surgical intervention is rarely required.

Results of knee disarticulations and transcondylar femoral amputations

In scrutinizing the literature in connection with the ISPO Consensus Conference 1990 (Table 16.1) primary wound healing was found superior in transcondylar procedures (Gritti–Stokes, Callander) in comparison with both knee disarticulation and transtibial procedures. However, the differences were limited to a gain of in the order of 6–12%. Re-amputation at transfemoral levels were required in only 9% of transcondylar amputations, but found necessary in 21% and 18% of knee disarticulation or transtibial amputations. In the case of the transtibial procedure the use of the 'wedge resection' the requirement to amputate at transfemoral level can be reduced to 8% (Hadden et al., 1987)

The striking feature, however, was that the ability to obtain prosthetic fitting and ambulation was only 55% of transcondylar amputees as compared to 76% of knee disarticulations and 80% of transtibial amputees.

Table 16.1 Results of knee disarticulation and transcondylar femoral amputations

	No. Publications	No. Patients	No. Knee Disartic.	% Success	No. Gritti–Stokes	% Success	No. Transtibial	% Success	No. Transfemoral	% Success
Primary										
Wound Healing	18	1857	413	295 67%	760	556 73%	401	243 61%	227	186 82%
Re-amputations	30	4626	587	124 21%	634	58 9%	1805	333 18%	1456	87 6%
Prosthetic										
Ambulation	33	3885	577	441 76%	812	450 35%	1153	924 83%	912	372 41%

With the total aim of patient care it seems obvious that knee disarticulation is preferable to transcondylar procedures which appeals primarily to wound healing fanatics (Murdoch, 1977). A high degree of independence can be obtained by knee disarticulation. Modern prosthetic technology has improved the cosmesis and the results of rehabilitation even further for this group of patients. Transcondylar amputations are effectively no better than long transfemoral stumps and have no place in the amputation procedures recommended today.

References

Baumgartner, R.F. (1979) Knee disarticulation versus above-knee amputation. *Prosthet. Orthot. Int.*, **3**, 15–10.

Duerksen, F., Rogalsky, R.J. and Cochrane, I.W. (1990) Knee disarticulation with intercondylar patellofemoral arthrodesis. *Clin. Orthop.*, **256**, 50–56.

Greene, H.G. (1975) Amputation through the knee for the non-ambulatory patient. *Surg. Gynecol. Obstet.*, **140**, 771–773.

Gritti, R. (1857) Dell' amputazione del femore al terzo inferiore e della disarticolazione del ginocchio. *Annales Universitatis Milano*, **161**, 5–32.

Hadden, W., Marks, R., Murdoch, G. and Stewart, C.P. (1987) Wedge resection of amputation stumps: a valuable salvage procedure. *J. Bone Joint Surg.* B, **69(2)**, 306–308.

Jensen, J.S., Poulsen, T.M. and Trasnik, M. (1982) Through knee amputations. *Acta Orthop. Scand.*, **53**, 463–466.

Mazet, R. and Hennessy, C.A. (1966) Knee disarticulation: a new technique and a new knee joint mechanism. *J. Bone Joint Surg.*, **48A**, 126–139.

Middleton, M.D. and Webster, C.U. (1962) Clinical review of the Gritti–Stokes amputation. *BMJ*, **1 Sept**, 574–576.

Murdoch, G. (1977) Amputation surgery in the lower extremity. *Prosthet. Orthot. Int.*, **1**, 72–83.

Newcombe, J.F. and Marcuson, R.W. (1972) Through knee amputation. *Br. J. Surg.*, **59**, 260–266.

Stokes, W. (1870) On supra-condyloid amputation of the thigh. *Med. Chirurg. Transact. Lond.*, **53**, 175–186.

Thyregod, H.C., Holstein, P., and Jensen, J.S. (1983) The healing of through knee amputations in relation to skin perfusion pressure. *Prosthet. Orthot. Int.*, **7**, 61–62.

Vaucher, J. and Blanc, Y. (1982) Les desarticulation du genou. Technique operatoire–appareillage (Disarticulation of the knee. Surgical and prosthetic techniques.) *Rev. Chir. Orthop.*, **68**, 385–406.

Waters, R.L *et al.* (1976) Energy cost of walking of amputees: influence of level of amputation. *J. Bone Joint Surg.*, **58A**, 42–46.

Further recommended reading

Baumgartner, R. (1984) Die Exartikulation im Kniegelenk bei geriatrische Patienien: indikation, operative Technik und Nachbehandlung. *Med. Orthop. Tech.*, **104**, 5–7.

Hagberg, K., Berlin, O.K. and Renstrom, P. (1992) Function after through-knee compared with below-knee and above-knee amputation. *Prosthet. Orthot. Int.*, **16**, 168–73.

Howard, R.R.S. and Chamberlain, J. (1969) Through-knee amputation in peripheral vascular disease. *Lancet*, **2nd Aug**, 240–241.

Jansen, K. and Jensen, J.S. (1983) Operative technique in knee disarticulation. *Prosthet. Orthot. Int.*, **7**, 72–4.

Jensen, J.S. and Poulsen, T.M. (1983) Success rate of prosthetic fitting after major amputations of the lower limb. *Prosthet. Orthot. Int.*, **7**, 119–21.

Krause, U., Schmidt, G. and Littmann, K. (1984) Die Kniegelenksexartikulation – Alternative zur distalen Oberschenkelamputation bei arterieller Verschlusskrankhiet Stadium 3 and 4. *Zentralbl Chir.*, **109**, 436–40.

Luccia, N.D., Pinti, Guedes, J.P.B. and Albers, M.T.V. (1992) Rehabilitation after amputation for vascular disease. *Prosthet. Orthot. Int.*, **16**, 124–128.

Murdoch, G. (1968) Knee disarticulation amputation. *Bull. Prosthet. Res.*, **10**, 14–8.

Osterman, H.M. and Pinzur, M.S. (1987) Amputation: last resort or new beginning? *Geriatric Nursing*, **8**, 246–248.

Pinzur, M.S., Smith, D.G., Daluga, D.J. *et al.* (1988) Selection of patients for through-the-knee amputations *J. Bone Joint Surg.*, **70A**, 746–50.

Sethia, K.K., Berry, A.R., Morrison, J.D. *et al.* (1986) Changing pattern of lower limb amputation for vascular disease. *Br. J. Surg.*, **73**, 701–703.

Stirnemann, P., Mlinaric, Z., Desch, A. *et al.* (1987) Major lower extremity amputation in patients with peripheral arterial insufficiency with special reference to the trans-genicular amputation. *J. Cardiovasc. Surg.*, **28**, 152–158.

17

Biomechanics of the knee disarticulation prosthesis

E. Lyquist

Introduction

The biomechanics of the disarticulation prosthesis has previously been described in detail and the similarities to the biomechanics of the transfemoral prosthesis pointed out (Hughes, 1982).

The following is intended to highlight the difference between knee disarticulation and transfemoral biomechanics and to indicate any influence on socket design.

Knee disarticulation stump characteristics

The knee disarticulation stump presents the following characteristics:

(1) It is end-bearing, i.e. the body weight can be transferred to the socket through the distal part of the stump. Proximal weight bearing as employed in the transfemoral prosthesis is not required.
(2) The distal part of the stump is bulbous and triangular in shape: this provides an excellent means of suspension as well as preventing rotation of the stump within the socket.
(3) The long stump provides effective medio-lateral stability.
(4) The stump musculature has a good functional status and provides better control than the transfemoral level because the thigh muscles are largely intact.
(5) Proprioception is good.

Medio-lateral stability

As the socket does not provide ischial-gluteal support and the vertical load is transmitted through the very distal part of the socket during stance phase, the body weight causes a rotational effect about the distal point of support (Fig. 17.1). However, the stabilizing forces on the stump will act in the same general areas as in the transfemoral socket, i.e. on the lateral aspect distally and on the medial aspect proximally.

Due to the stump length, pressures will be less in the knee disarticulation socket compared to the transfemoral socket.

However, it must be emphasized that inadequate provision for application of medio-lateral stabilization forces on the lateral aspect distally will cause a distal shear force with serious stump problems as an unavoidable consequence.

Figure 17.2 shows the difference between the tendency of the stump to rotate within the socket both in the transfemoral and the knee disarticulation prosthesis. If stabilization is insufficient, the resulting motion may cause a roll of flesh to develop on the medial aspect proximally. The medial socket wall must consequently be kept high – normally about 2–3 cm below the level of the ischial tuberosity.

Antero-posterior stability

Those forces acting to provide stability and control of the prosthetic knee during the stance phase are

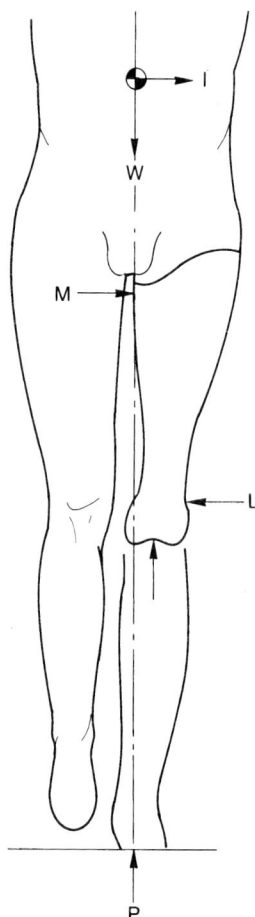

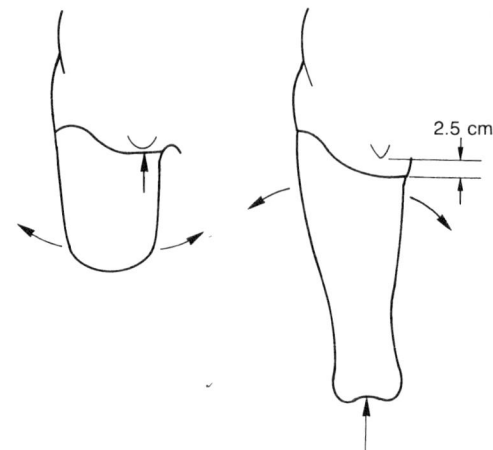

Fig. 17.2. Illustrates tendency to rotate in both transfemoral and knee disarticulation prostheses. Medial socket wall must be kept high to prevent a roll of flesh developing.

Fig. 17.1. Stabilization forces: on the medial aspect proximally and on lateral aspect distally. Precise fitting above and around the lateral condyle is essential to avoid shear problems.

similar to those in the transfemoral prosthesis. However, because of the long and powerful stump, it is normally possible to align the prosthesis with less 'built-in' or alignment stability.

Suspension

During the swing phase, forces due to gravity and inertia will tend to pull the socket off the stump. These forces may be balanced utilizing the flare of the condyles as shown in Fig. 17.3. In order that the stump passes the 'narrow' area proximal to the

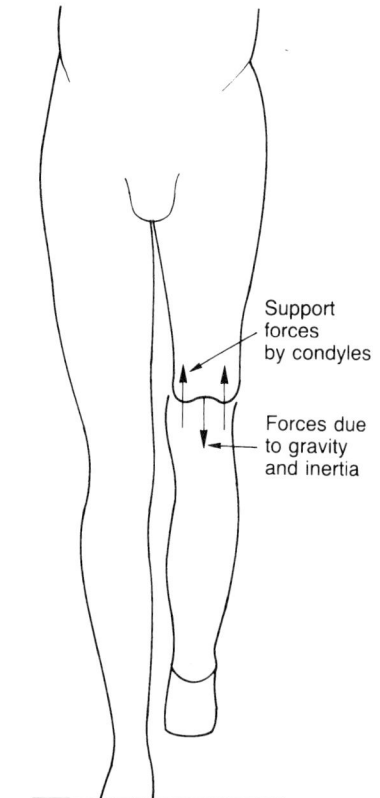

Fig. 17.3. Fitting around flare of condyles ensures balancing forces to combat forces of gravity and inertia tending to pull the socket off the stump.

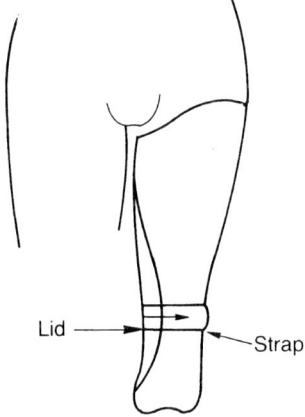

Fig. 17.4. Socket requires 'window' to permit bulbous stump end to pass the 'narrow neck' proximal to condyles. Window closed by a lid retained by a strap.

condyles, the socket must have a 'window', which is closed with a plate, lid (Fig. 17.4) or a soft insert made of a suitable foam.

Summary

In conclusion, the biomechanics pertaining to the knee disarticulation prosthesis are very similar to that of the transfemoral prosthesis.

Due to the use of end-bearing and the larger area available for stabilization, the knee disarticulation socket is more comfortable than the transfemoral socket.

It must be emphasized again that a socket which is too short may cause serious problems on the medial aspect proximally.

References

Hughes, J. (1983) Biomechanics of the through-knee prosthesis. *Prosthet. Orthot. Int.*, **7** (12.),

Section 7

The hip

18

Hip disarticulation, transiliac and sacroiliac amputations

B. M. Persson

History

Surgical disarticulation of the hip was first performed in 1816 by Kerr in Northampton, England on an 11-year-old girl with a tumour of the thigh but the patient died. Sir Astley Cooper published his first case in 1824. Larrey (1812) performed seven such operations on war victims. From 1839 Velpeau developed indications for civilian surgeons in cases of fracture, tumour and gangrene of an advanced stage.

Compared to hip disarticulation the transpelvic amputation is much more complicated, and some 70–80 years (Fig. 18.1) were to pass before reports of its use were available. Indeed, until the breakthrough of organ transplantations in about 1970, amputation of the whole leg and part of the pelvis was regarded as the most dramatic surgical intervention of all. When Pringle in Glasgow (1909) published three cases of his own it was the first performed in one stage with a surviving patient. Billroth in Vienna had performed the first case in 1891 in two stages but with early death. Girard in Bern is credited with the first surviving patient in 1895 but the operation was again performed in two stages. She was a 17-year-old girl first disarticulated at the hip in 1894 and then a hemipelvectomy was performed the following year for recurrent osteosarcoma of the ilium. The sacroiliac joint was disarticulated but the patient died 7 months later of a further recurrence of the tumour in the scar. When Pringle reviewed the literature in 1916 he uncovered 19 cases of sarcoma reported with an early mortality of 68%. Another five cases

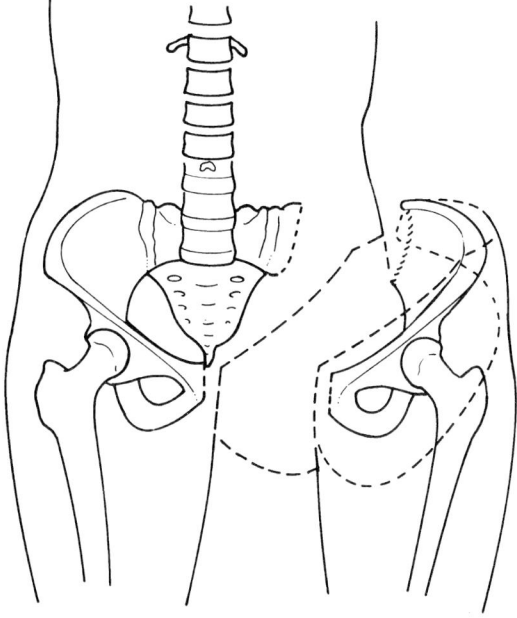

Fig. 18.1. Principal incisions for a left sacroiliac amputation with disarticulation of the sacroiliac joint and the symphysis pubis. A long posterior myocutaneous flap should be used when possible and the anterior part developed first to evaluate extension of pathology on the abdominal side of the pelvis before the point of no return is passed. After biopsy the incisions may need modification to ensure the inclusion of the biopsy site distal to the level of amputation.

had the same operation performed for tuberculosis, with three early deaths. Thus for long the operation did not offer very promising results, and, indeed, the first case of iliosacral amputation at the Mayo Clinic was not performed until 1943. With the advent of antibiotics and blood transfusion the procedure turned from being disastrous to successful. Mortality dropped from 50% to 5% within a few years. Pack (1946) published 25 hip disarticulations without any operative death as an example of this progress.

In 1945 Turnbull from Australia could report the first surviving traumatic hemipelvectomy. The patient was a 20-year-old sailor with his leg entangled in a wire hawser between two ships. With blood transfusion, suprapubic cystostomy, colostomy and penicillin he was resuscitated and survived. The number of similar cases had increased to 23 surviving patients as reviewed by Klasen in 1989. The youngest patient amputated through the hip was only 8 days old, the indication being fibrosarcoma (Svets *et al.*, 1976).

Terminology

The term amputation is derived from the Latin preposition 'ambi' which means around and 'putatio', which means trimming or pruning. It implies the separation by cutting of a terminal part of the body. Accordingly, the use of such terms like 'hemipelvectomy' and 'disarticulation' for certain types of amputations may be misleading. Those words do not express precisely that removal of a terminal part of the body has been performed. When Speed in 1938 introduced the term 'hemipelvectomy' for sacroiliac amputation he admitted that the word may be misleading. He used it because he believed interilioabdominal amputation was too cumbersome. The term is no longer appropriate to describe the procedure today, as a hemipelvectomy proper is often performed internally for tumour without amputation (Burri, 1979). As long ago as 1914 Napalkow collected 38 instances of pelvic resection without amputation, the first being performed in 1885. They produced good results, albeit with outward rotation of the leg, but no secondary amputation of the extremity was required. Therefore internal hemipelvectomy as a surgical procedure is even older than sacroiliac amputation or transiliac

amputation. Accordingly, the term 'hemipelvectomy' as applied to amputation should be discarded completely. Equally cumbersome descriptions that have been used are not appropriate, like the Spanish interabdomino-sacro-pubic disarticulation or interinomino-abdominal amputation (Robinson, 1950). The term 'hindquarter' amputation introduced in 1949 by Brittain and used by Werne in 1953 and McPherson in 1960 is rather popular but hardly exact. Therefore it is suggested that 'sacroiliac' amputation is used for the disarticulation through the sacroiliac joint with amputation of the hemipelvis and leg and 'transiliac' amputation is used for a less radical transpelvic amputation. Both the terms sacroiliac amputation and transiliac amputation are needed to specify the different extents of the procedures. The fact that the symphysis pubis is also divided during both these procedures is self evident but does not need specific mentioning because it complicates terminology unnecessarily.

Indication

Apart from the 23 published cases of traumatic sacroiliac amputation almost all cases today are performed for tumour but there are a few instances when the procedure might be indicated for congenital malformation or infection.

Obstructive vascular disease accounts for about 90% of all amputations in the industrial countries but is very seldom the cause of amputations through the hip or above the hip. The largest series of hip amputations for dysvascular cases (Laszlo and Kullman, 1987) shows only one of 27 being a diabetic and 17 of 27 surviving. Only two of the 27 had primary healing and only six of the 17 survivors received a prosthesis. The mean age was 50. The limited use of prosthesis for these proximal amputations is explained by the size of the prosthesis and the weakness of the patient (Waters, 1976). Nowroozi *et al.* (1983) also showed that more energy is needed during walking with than without the prosthesis in such cases. Only those with great motivation will use it daily, as in the case of a young woman amputated for tumour.

The choice between hip disarticulation and transpelvic amputation is usually simple. The compartmental principles of resection of soft tissue tumours around the hip often make a transpelvic amputation necessary. For tumours proximally in the femoral shaft a hip disarticulation amputation is sufficient as long as the soft tissue component is not so advanced and allows at least one uninvolved anatomical structure between the tumour and the resection (Stener and Stener, 1958).

Technique of the hip disarticulation amputation

The technique of hip disarticulation from an anatomical point of view is straightforward and any modified techniques described are usually related to individual circumstances in cases of tumour, trauma or ischaemia rather than any theoretical advantages (Fig. 18.2). The most elegant description is that by Harold Boyd from Memphis, Tennessee (1947), who developed a technique by which the dissection proceeds along fascial planes and all muscles about the hip are divided either at their origins from the pelvis or at their insertions on the femur. In this way bleeding is minimized as a minimum of capillary beds are transected. Both in tumour and vascular cases this technique also creates containment of involved spaces when excised. In the standard case a large gluteal flap forms an excellent myocutaneous stump coverage (Fig. 18.3). Boyd used an Esmarque tourniquet 2 inches below the level of the incision to exsanguinate the leg and reduce the shock but this detail has now merely historical interest and has been deleted from the description in *Campbell's Operative Orthopaedics* (1987). With the advent of blood transfusion techniques, which are comparatively cheap and safe, the use of a tourniquet should be avoided because it can expel tumour cells, necrotic material or venous thrombosis into the circulation; otherwise Boyd's description is highly recommended. The racquet shaped incision, beginning over the neurovascular bundle anteriorly, allows an inspection of the border of amputation before the point of no return has been reached. Boyd in his description also recommends that the nerves

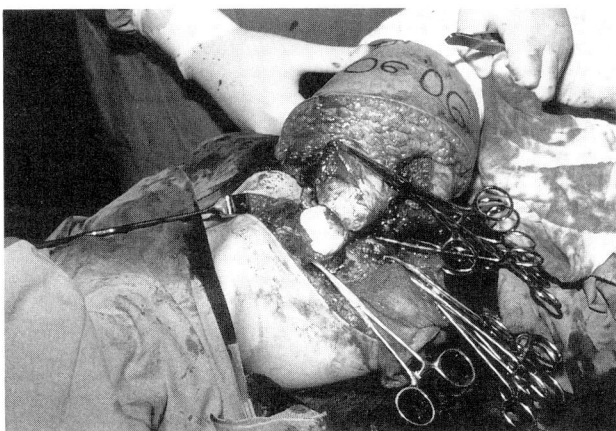

Fig. 18.2. Right hip disarticulation in a 12-year-old boy with a midshaft femoral osteosarcoma. Patient in supine or lateral position. Incisions marked with blue ink. Anterior incision just below the inguinal ligament proceeding into the fossa ovalis for wide exposure of femoral artery vein and nerve. Division of sartorius, pectinius, gracilis and adductors at their origins and the iliopsoas and obturators at their insertion. Obturator vessels and nerves usually bifurcate around the adductor brevis muscle, and these branches need to be secured to prevent retraction into the pelvis. Semimembranosus, semitendinosus and the long head of biceps are divided at their origin on the ischial tuberosity. The joint capsule is now incised and the hip disarticulated with completion of posterior skin incision and transection of the glutei muscles at the greater trochanter. After inspection the sciatic nerve is allowed to retract beneath the piriformis muscle. The pelvic bone is covered with muscular flaps by quadratus femoris to the iliopsoas, the obturator externus to the gluteus medius and the gluteal fascia to the inguinal ligament.

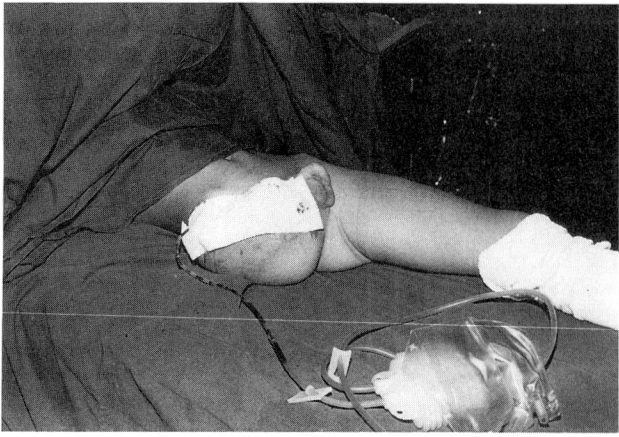

Fig. 18.3. The wound is closed with suction drain and urether catheter. Perioperative antibiotics are employed. A light elastic bandage is used to compress cavities and reduce pain. Plaster of Paris and direct prosthesis on table seldom indicated. It delays change of body image.

be injected with Novocain before division, but this recommendation is no longer supported as there is no scientific proof that shock is limited by such injections.

In older textbooks it is recommended that the femoral head, neck and trochanter should be preserved when possible because it could create an increased volume of the stump when sitting. However, since the introduction of the modern Canadian prosthesis described by McLaurin (1957) this notion is obsolete. Indeed, a short mobile bony stump within the socket increases the problem for those patients who do want to use a prosthesis. Furthermore, in tumour cases it destroys the compartmental approach to remove the whole of the femur and prevent 'skip metastases' (Brodie, 1970).

Flaps other than conventional may be indicated after biopsy for malignant tumours about the hip, e.g. to cover a large posterior defect a long rectangular anterior myocutaneous flap containing the femoral vessels and quadriceps can be created or a medial flap based on the obturator artery can be developed. Even in special cases a free vascularized graft can be deployed using the rectus abdominus muscle and the inferior epigastric artery with overlying skin. If necessary, the graft pedicled on the inferior epigastric artery above can be passed subcutaneously underneath the narrow strip of skin left between the flap site and the amputation (Mnaymneh, 1980).

Transiliac and sacroiliac amputation

During the last 40 years there have been many publications on the outcome of pelvic amputations. The largest is provided by Pack and Miller reporting on 101 cases (1964) 55% of which were soft tissue tumours; 20% melanoma; 7% metastatic cancer; 9% chondrosarcoma and 8% osteosarcoma. Ten survived without a recurrence for more than 5 years and a further 28 patients for more than 5 years, albeit with residual cancer. There were only two operative deaths among the 101 transpelvic amputations.

Hogshead (1971) analysed the management of 49 patients amputated for tumour. Blood loss, operative time and hospital stay proved to be greater for transpelvic than for disarticulation of hip amputation, with wound complications in 50% of the former and 33% of the latter. No correlation was found between the flap technique and wound complications whether the flaps were based anteriorly, posteriorly or medially depending on the pathological process and the site of biopsy. Stump problems at follow-up were minimal and troublesome amputation neuromata were not encountered. Troup and Bickel (1960) in an analysis of 264 cases of malignant tumours found a 5-year survival of 41.5% in transpelvic amputations. Only one hospital death was recorded after transpelvic amputation and none after 87 hip disarticulations. Capenna *et al.* (1990) reported the results of 76 transpelvic amputations performed from 1978 to 1988 at Instituto Rizzoli in Bologna for tumours. The subcutaneous gluteal flap technique gave healing complications in 50% while the classic King and Steelquist method (1943) with a gluteal myocutaneous flap proved to be the most reliable.

Technique of transpelvic amputation

There are three descriptions commonly referred to in modern textbooks and publications, i.e. King and Steelquist (1943) using a lateral position with the sound side downwards and the procedure divided into three parts, anterior, medial and posterior, whereas Gordon-Taylor divides the operation into one anterior and one posterior stage (Gordon-Taylor and Monroe, 1952), as do Sorrondo and Ferre (1948). A fourth elegant description is that of Ravitch and Wilson (1964) starting with the patient in a supine position. All four methods initially use an incision along Poupart's ligament with division of the rectus abdominis for inspection of the vessels and lymph nodes before the point of no return. The symphysis should be divided with knife or Gigli saw and the hemipelvis swung laterally with the sacroiliac joint as a hinge to simplify the posterior division. All use a Foley catheter and prophylactic antibiotics and operate with the leg draped free. All state biopsy should be performed at an earlier separate stage to confirm diagnosis and to discuss treatment alternatives with the patient (Figs. 18.4–5).

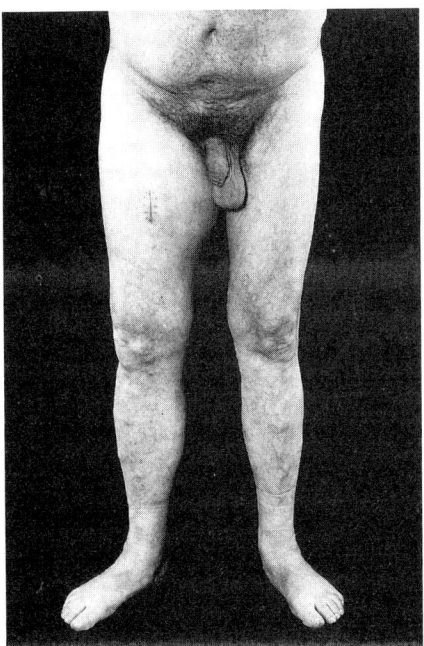

Fig. 18.4. Massive leiomyosarcoma of proximal medial thigh in a 60-year-old man with increasing pain, high sedimentation rate (ESR) and serum calcium. Cat scan and MRI show insufficient margins with hip disarticulation. The scar of biopsy is over the vertex of tumour anteromedially.

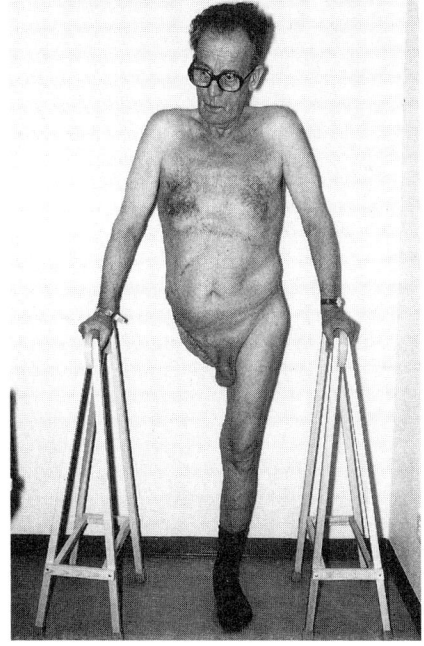

Fig. 18.5. One week after radical trans-pelvic amputation with preservation of upper iliac wing. Pain, was relieved and ESR and high serum calcium are now normal and a Canadian hip prosthesis could be used after 3 weeks supplemented with a separate filler for level sitting.

Special questions

Does pregnancy present a problem?

Transpelvic or sacroiliac amputation does not create any significant contraindication to pregnancy. In one series one of 13 patients had two children and in another series four patients out of 100 bore children without mishap (Schmidt-Peter, 1979). In a series of 60 patients, four had uncomplicated deliveries (Higginbotham *et al.*, 1966). A special corset can be made for support if abdominal muscles are deemed to be overloaded.

Division of common iliac artery or external iliac artery only?

This question is raised in many reports on pelvic amputations. Douglas (1975) analysing the case histories of 50 patients amputated for tumour observed skin necrosis in 80% requiring revision in 52% and skin transplantation in 20%. There was an infection rate of 28% and a 5-year survival of 32%. There was a highly significant correlation between surgical time above 4 h and necrosis. If surgery lasted longer than 4 h skin necrosis was recorded in 85%. There was no correlation associated with whether the common or external iliac artery had been divided. Pack (1964) in his review of 101 cases also arrived at the conclusion that the common iliac artery can be divided without any significant increase in skin problems. Pack believes that it is the subcutaneous tissue that has its nutrition from the internal iliac artery only and that the skin necrosis is secondary to fat necrosis. Gordon-Taylor and Monroe (1952) reporting 50 cases also supported the idea that the common iliac artery should be divided to reduce operative bleeding as does Koskinen (1967), reporting three cases without skin necrosis. Nilsonne *et al.* (1968) reported 13 cases and found equally good healing irrespective of how the iliac artery had been divided. Reporting 36 operations with 10 major wound problems, Ravitch and Wilson (1964) found them equally divided between patients who had the common iliac artery divided and patients who had only the external iliac artery divided. Only Coley believes division of the common iliac artery is contraindicated. He reports 43 cases between 1946 and 1958 with only three operative deaths and 35% 5-year survival (Coley and Higginbotham, 1958).

Urorectal problems

After pelvic amputation Nielsen (1977) examined 11 cases by urodynamic measurements without finding any insufficiency in spite of the one-sided denervation. Pack found no case of rectal incontinence.

Use of prostheses

Before McLaurin's construction of the Canadian hip disarticulation prosthesis successful prosthetic fitting was seldom achieved according to the publications available. Even if a prosthesis was made it was soon rejected. Sneppen *et al.* (1978) analysed 41 patients with 11 early deaths. Thirty received a prosthesis but only 15 tried to use it. One year later 13 still used it. Jones (1974) reported 17 personal cases with 10 survivors and only one of the 17 patients used the prosthesis 1 year later. Douglas *et al.* (1975) reviewed 50 cases, with 14 of them surviving 5 years. They state that 'independent crutch walking should be accomplished by the majority within a week but limb prostheses are cosmetic rather than functional and have generally not been well received except by patients highly motivated to retain the normal cosmetic appearance. Elderly patients may develop scoliosis after hemipelvectomy when sitting and a prosthetic socket for balance is of great benefit'.

Discussion and summary

In spite of an abundant literature of some 100 publications scrutinized there is, understandably, not one prospective randomized study. All three levels, whether it be through the hip, pelvis or sacroiliac joints, are rarely performed and together account for less than 1% of all amputations. Even when done in orthopaedic oncology centres the patients are so different concerning type, grade and location of tumour that randomized surgery can hardly be considered. In spite of this, all important questions that have been raised can be reasonably well answered from the many consecutive series in the literature analysed. Blood transfusion, proper anaesthesia and prophylactic antibiotics are necessary to create the low mortality of a few per cent that has now been achieved. The surgical team must be well prepared and the procedure concerning the dissection planned in detail beforehand. Magnetic resonance imaging and computerized tomography have replaced angiography except when embolization of vascular areas is needed to reduce peri-operative bleeding in selected cases. A freely draped leg and a lateral position is recommended by most authors and seems to be well supported. A short remnant of femoral head, neck and trochanter should not be preserved in hip amputation with the modern Canadian type of prosthesis for those patients who are planned to be walkers. The division of the iliac artery in transpelvic amputations may well be performed above the division into internal and external iliac artery to reduce the bleeding, but when so done the branches of the internal iliac artery have to be separately clamped, ligated and divided in any case to prevent retrograde bleeding at the separation of the hemipelvis from the internal organs and those of the other side. The operative time should be less than 4 h to reduce the number of flap necroses post-operatively (Douglas *et al.*, 1975). A Foley catheter should always be used but a ureteric catheter is indicated only in selected cases where the dissection of the urethra is expected to be troublesome. Both urinary and rectal incontinence are very unusual. The mortality from the procedure is negligible but the mortality from the malignant disease, which is normally the causal condition, is still very significant and therefore the operative procedure should be discussed in depth with the patient and agreement reached during the planning discussions.

The prosthesis functions well with three mobile joints according to the basic principles of McLaurin (1957) but a high proportion of patients still reject the prosthesis for daily use (Sneppen, 1978). After transpelvic amputation the loss of the ischial tuberosity creates a need for a cushion or a pelvic filler socket to ensure comfortable sitting in most patients. Problems with visceral herniation or post-operative neuroma are usually minimal in spite of the lack of muscular support after iliosacral amputation. At least 10 cases of pregnancy have been reported with no difficulties after transpelvic amputations.

In selected cases hip disarticulation amputations, transiliac and sacroiliac amputations are safe and predictable procedures in the hands of a well prepared team (Wagman and Terz, 1989).

References

Boyd, H.B. (1947) Anatomic disarticulation of the hip. *Surg. Gynaecol. Obstet.*, **84**, 346–349.

Brittain, H.A. (1949) Hindquarter amputation. *J. Bone Joint Surg.*, **31B**, 404–409.

Brodie, I.A.O.D. (1970) Lower limb amputation. *Br. J. Hosp. Med.*, **Nov**, 596–603.

Burri, C. (1979) Total internal hemipelvectomy. *Arch. Orthop. Trauma Surg.*, **94**, 219–226.

Capanna, R., Manfrini, M., Pignatti, G., Martelli, C., Gamberini, G. and Campanacci, M. (1990) Hemipelvectomy in malignant neoplasms of the hip region. *Ital. J. Orthop. Traumatol.*, **16**, 425–437.

Coley, B.L. and Higginbotham, N.L. (1958) Indications of hemipelvectomy. *Surgery*, **44**, 766.

Douglas, H.O., Razack, M. and Holyoke, E.D. (1975) Hemipelvectomy. *Arch. Surg.*, **110**, 82–85.

Gordon-Taylor, G. and Wiles, P. (1934) Interinnomino-abdominal amputation. *Br. J. Surg.*, **22**, 671.

Gordon-Taylor, G. and Monroe, R.S. (1952) Technique and management of hindquarter amputation. *Br. J. Surg.*, **39**, 536.

Gordon-Taylor, G., Wiles, P. and Warrwick, W.T. (1952) The interinnomino-abdominal operation. Observation on a series of fifty cases. *J. Bone Joint Surg.*, **34B**, 14–21.

Higginbottam, N.L., Marcove, R. and Casson, P. (1966) Hemipelvectomy. A clinical study of 100 cases with 5 yrs follow-up. *Surgery*, **59**, 706–708.

Hogshead, H.P. (1971) Experience with hip disarticulation and hemipelvectomy procedures. *J. Bone Joint Surg.*, **53A**, 1031.

Jones, P. (1974) Hindquarter amputations: review of a personal series. *Aust. N.Z. J. Surg.*, **44**, 1–11.

King, D. and Steelquist, J. (1943) Trans-iliac amputation. *J. Bone Joint Surg.*, **25**, 351–367.

Klasen, H.J. and Ten Duis, H.J. (1989) Traumatic hemipelvectomy. *J. Bone Joint Surg.*, **71B**, 291–295.

Koskinen, E.V.S. (1967) Hemipelvectomy for malignant tumours of bone: a study with preoperative arteriographic examination of the growth. *Ann. Chig. Gynaecol.*, **56**, 9–17.

Laszlo, G. and Kullman, L. (1987) Hip disarticulation in peripheral vascular disease. *Arch. Orthop. Trauma Surg.*, **106**, 126–128.

McPherson, J.H.T. (1960) Traumatic hindquarter amputation. *J. Med. Assoc. Georgia*, **49**, 494–495.

Mnaymneh, W. and Temple, W. (1981) Hemipelvectomy: new techniques of difficult skin coverage. *Orthop. Trans.*, **5**, 334.

Nilsonne, U., Hjelmstedt, A. and Hakelius, A. (1968) Surgical problems in hemipelvectomy. *Acta Orthop. Scand.*, **39**, 161–170.

Nielsen, M.L. (1977) Bladder function after hemipelvectomy. *Acta Orthop. Scand.*, **48**, 181–185.

Nowroozi, F., Salvanelli, M.L. and Gerber, L.H. (1983) Energy expenditure in hip disarticulation and hemipelvectomy amputees. *Arch. Phys. Med. Rehabil.*, **64**, 300–303.

Pack, G.T. (1956) Major exarticulations for malignant neoplasms of the extremities: interscapulothoracic amputation, hip-joint disarticulation and interilio-abdominal amputation: a report of 228 cases. *J. Bone Joint Surg.*, **38A**, 249–262.

Pack, G.T. and Miller, T.R. (1964) Exarticulation of the innominate bone and corresponding lower extremity for primary and metastatic cancer. A report of one hundred and one cases with end results. *J. Bone Joint Surg.*, **46A**, 91–95.

Pringle, J.H. (1909) The interpelvi-abdominal amputation. *Lancet*, **1**, 530–531.

Pringle, J.H. (1916/1917) The interpelvi-abdominal amputation, with notes on two cases. *Br. J. Surg.*, **4**, 283–296.

Ravitch, M.M. and Wilson, T.C. (1964) Long-term results of hemipelvectomy. *Ann. Surg.*, **159**, 667–682.

Robinson, R.A. (1950) Interinnomino-abdominal amputation. *J. Bone Joint Surg.*, **32A**, 446–448.

Schmidt-Peter, P. (1979) Hemipelvectomie. *Beitr Orthop. Traumatol.*, **26**, 233–238.

Sneppen, O., Johansen, T., Heerfordt, I., Dissing, I. and Petersen, O. (1978) Hemipelvectomy: a postoperative rehabilitation of 42 cases. *Acta Orthop. Scand.*, **49**, 175–179.

Sorrondo, J.P. and Ferre, R.L. (1948) Amputasion interilioabdominal. *Ann. Orthop. Traumatol.*, **1**, 143.

Stener, B. and Stener, I. (1958) Malignant tumours of the soft tissues of the thigh. *Acta Chir. Scand.*, **115**, 457–475.

Svets, W.R., Wagner, C. and Clark, M.W. (1976) A young hemipelvectomy patient. ICIB. **15**, 9–13.

Troup, J.B. and Bickel, W.H. (1960) Malignant disease of the extremities treated by exarticulation: analysis of 264 consecutive cases with survival rates. *J. Bone Joint Surg.*, **42A**, 1041–1050.

Turnbull, H.A. (1978) A case of traumatic hindquarter amputation. *Br. J. Surg.*, **65**, 390–392.

Unruh, T., Fischer, D.F. Jr and Unruh, T.A. (1990) Hip disarticulation. An 11 year experience. *Arch. Sur.*, **125**, 791–793.

Wagman, L.D. and Terz, J.J. (1989) Hemipelvectomy and translumbar amputation. In: *Lower Extremity Amputation* eds W.S. Moore, and J.M. Malone, Saunders, Philadelphia, pp. 157–176.

Waters, R.L. (1976) Energy cost of walking of amputees. The influence of level. *J. Bone Joint Surg.*, **58A**, 42–46.

Werne, S. (1953) Two cases of hindquarter amputation. *Acta Orthop. Scand.*, **22**, 90.

19

Biomechanics of the hip disarticulation prosthesis

E. Lyquist

Introduction

The biomechanics of the hip disarticulation prosthesis has been described in detail by McLaurin (1957, 1970).

The following highlights the most important features.

Stump characteristics

Disarticulation at the hip joint leaves the pelvis intact and provides an excellent means for vertical support through the ischial tuberosity.

However, when a transpelvic amputation is performed, support via any bony structure is not obtainable and the fitting of the socket or corset is significantly more complicated.

Medio-lateral stability

During stance phase the pelvis will tend to tilt medially around the vertical support point. This tendency is controlled by distal pressure on the amputated side and proximal pressure in the area of the iliac crest on the sound side (Fig. 19.1). Suspension is achieved by a precise fit over the iliac crest(s). The area over which the vertical support is distributed is almost horizontal in contrast to the situation in the transpelvic amputation socket where it slopes at about 35–45° to the horizontal.

In the transpelvic amputation in order to reduce the vertical movement of the pelvis within the

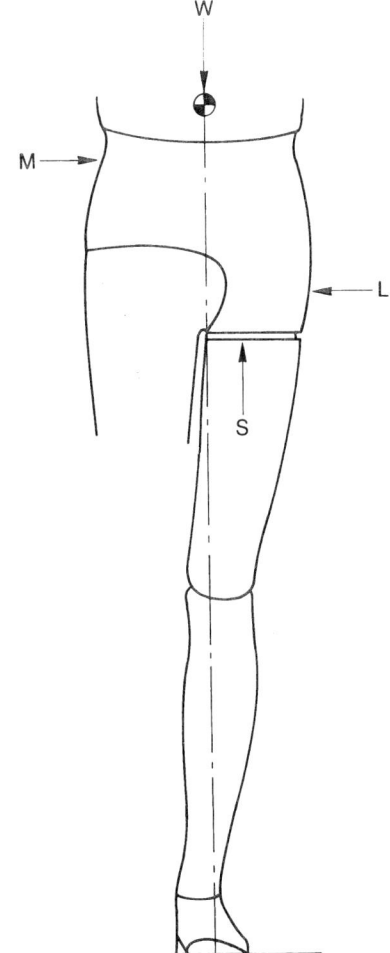

Fig. 19.1. Tendency for 'pelvis' to tilt medially around vertical support point (S) is controlled by distal pressure on amputated side (L) and proximal pressure in the area of the iliac crest on the sound side (M).

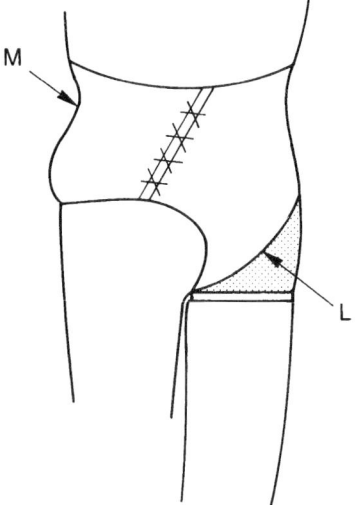

Fig. 19.2. Lack of bony support in transpelvic amputation requires force application in a more oblique direction via indentation of the socket under the iliac crest on the sound side (M) and at right angles to the 40–45° slope on the amputation side (L).

socket during stance phase and because of lack of bony support, the forces must be applied in a different manner as shown in Fig. 19.2.

Attempts to reduce the movement of the pelvis within the socket by utilizing support on the ischial tuberosity of the sound leg are successful in some cases (McQuirk, 1988).

Antero-posterior stability

Stability of the hip is achieved by positioning the joint on the anterior part of the socket and some distance anterior to the direction of the floor reaction force in mid-stance as shown in Fig. 19.3.

Again, to ensure stability, the knee joint in turn is placed posterior to the direction of the floor reaction force. As a result of the relative position of the hip and knee joints, the knee is in fact hyperextended. Inevitably there is relative elongation of the prosthesis during push-off.

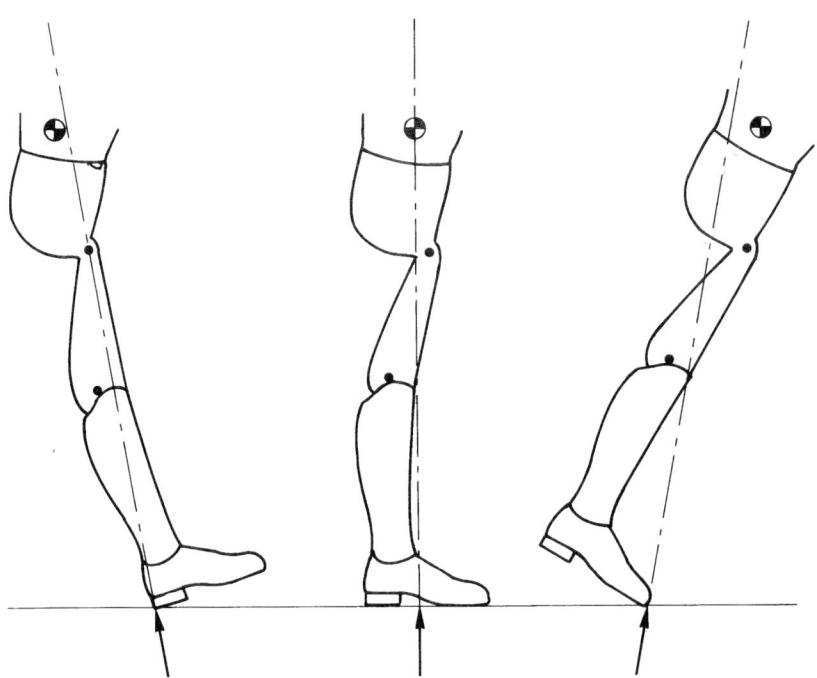

Fig. 19.3. Placement of hip joint on anterior part of socket anterior to floor reaction force at stance and the knee joints located posterior to it ensures antero-postero stability of hip and knee.

Summary

It is possible to achieve good results in fitting the hip disarticulation prosthesis to a true hip disarticulation amputee. The same quality of result is difficult, if not impossible, to achieve when fitting a patient with a transpelvic amputation.

References

McLaurin, C.A. (1957) The Canadian hip disarticulation prosthesis. *Artif. Limbs*, **Autumn**, 22–28.

McLaurin, C.A. (1970) The Canadian hip disarticulation prosthesis. In: *Prosthetic and Orthotic Practice* (ed. G. Murdoch), Edward Arnold, London, pp. 285–304.

McQuirk, A.W. (1988) Prosthesis for hip disarticulation and hemipelvectomy. In: *Amputation Surgery and Lower Limb Prosthetics* (ed. G. Murdoch), Blackwell Scientific Publications, London, pp. 236–244.

Section 8

Amputation in vascular disease

20

Amputation level selection

P. T. McCollum and D.K. Harrison

Introduction

In western civilisation, the vast majority of major limb amputations are performed for end-stage peripheral vascular disease, diabetes or a combination of these two disease entities (McCollum and Walker, 1992). While major limb amputations are performed for trauma in such places as Northern Ireland, Bosnia, Africa and other areas of serious civil unrest, relatively few are performed because of trauma in a normal civilian population. In such cases, the principal policy is generally one of preserving as much viable tissue as possible in limbs which have not generally had previous underlying arterial disease. The decision to amputate at any particular level is one which is also already effectively made by the level of injury and remaining viable tissue. No further specific attention will therefore be given to this group in terms of amputation level selection, as each will require to be dealt with on its own individual merits.

A second group which demands similar individual consideration is made up of limbs which have overwhelming sepsis present and which require major amputation as a life saving measure. Again, the over-riding consideration here is the elimination of any toxic and necrotic tissue so that the patient may be stabilized and more definitive treatment of the limb instituted later. In such cases where amputation is to be undertaken, the exact level of resection will depend upon the gravity of the situation coupled with the degree of tissue involvement in the infective process. In many cases, a guillotine amputation may be made as a temporary expedient with a view to more definitive amputation at a later date once the infection is settled and the viability of remaining tissue can be more accurately ascertained. These problems are addressed elsewhere and will not be discussed further in this chapter. Thus, it is really only in those patients who have end-stage peripheral vascular disease that there is a need for specific assessment of the optimum level of amputation.

In patients with end-stage peripheral vascular disease, it is now generally accepted that the more distal an amputation that can be achieved, the better the quality and chances of rehabilitation will be, within certain operational constraints (Castronuovo et al., 1980). The patient population is generally increasingly elderly, often having significant vascular disease present in the heart and other areas. These patients have a very limited lifespan and it is essential to return them back to a reasonable quality of life as soon as is practicable. It is therefore important not only to choose the most distal level of amputation possible but also to be as sure about achieving primary healing as possible; hence it is not reasonable to adopt a policy of 'amputate distally and see if healing occurs'. In particular, it is abundantly clear that patients with significant gangrene of the toes or foot, who do not have pure diabetic microangiopathy, are extremely unlikely to heal an attempted local amputation, be it a digital or part-foot procedure, in the absence of prior revascularization. Similarly, a Syme ankle disarticulation amputation is also a poor choice for peripheral vascular disease patients because the healing of the stump will depend upon collateral circulation around the calcaneum which will be inevitably already seriously compromised in these patients prior to any amputation.

It is therefore valid to suggest that the only decision to be made regarding amputation level selection in end-stage large-vessel vascular disease (with or without diabetes) relates to the healing potential or otherwise of the transtibial amputation as compared with the transfemoral level. In patients in whom a Syme ankle disarticulation or part-foot procedure is being contemplated, revascularization must be considered first as in these cases it is almost assured that a distal bypass will be possible and this will virtually guarantee primary healing of the amputation stump.

General principles

There are several difficulties encountered when comparing transfemoral and transtibial amputation rates between separate units and even countries. These frequently include a lack of knowledge about the underlying pathology, inconsistent indications for amputations, a failure to agree upon a reliable definition for critical ischaemia (Dormandy and Stock, 1990), and widely differing attempted revascularization rates combined with a great variation in available vascular services between areas.

Not only are there widely different rates of vascular intervention for end-stage vascular disease but there are also very different amputation rates between regions (Department of Health and Social Security, 1990). This disturbing fact underlines the point that reliable data on amputation rates is notoriously hard to collect and that great care must be taken before these data are interpreted and meaningful conclusions reached. For example, there is an increasing willingness on the part of vascular surgeons to attempt distal tibial bypasses to salvage the limb. Despite this, the amputation rate in many areas has not reduced and in general the ratio of transtibial to transfemoral amputations has not altered. Does this mean limb salvage surgery is no good? In fact, one might expect the number of transtibial procedures to decrease as it is in these patients that there is the greatest likelihood of a successful outcome with a leg bypass; by inference, an optimist might suggest that there has

therefore been an overall improvement in both net limb salvage rates and amputation services! Is the available data reliable enough to come to these sort of conclusions? Has any consideration been given to the increasing incidence of critical ischaemia being seen in western countries, largely as a result of an increasingly elderly population? Clearly, data in isolation can prove to be useless and even very misleading on occasions.

The prejudices of physicians involved in amputation surgery vary substantially between regions and even within regions and the consequences of these differences must not be underestimated. Such prejudices have perhaps the most major impact upon the final level of amputation decided upon in individual cases and thus, secondarily, on amputation levels in different series. It must also be remembered that, in most areas, major amputations are still carried out by general or vascular surgeons who may have little interest in the outcome beyond getting the patient out of 'their' surgical ward. These factors have a more clear relevance when it is recognized that the transtibial to transfemoral amputation ratio can vary from 3:1 or greater to 1:3 or even worse (Harris et al., 1988). Such a variation in clinical amputation practice is not simply secondary to differing case-mixes found in various regions but it is unfortunately readily explicable on the basis of personal beliefs and practices on the part of the clinicians in charge. This appalling lack of consistency emphasises that the more rigorous discipline of the team approach to the management of the amputee should be adopted universally if there is to be a substantial improvement in the management of end-stage, non-reconstructible peripheral vascular disease (Murdoch, 1977). Nowhere is this more starkly seen than in the approach to selecting the level of amputation in these patients.

Clinical evaluation

There is no doubt that over the years great faith has been placed upon the ability of the clinician to decide upon the optimum level of amputation by the simple 'laying on' of hands (Kendrick, 1956). This approach has been lent credence by several notable and distinguished amputation surgeons who learnt to assess patients carefully and had an enlightened approach towards performing transtibial amputations. Whilst it is possibly true that individuals may have been able to develop their own skills to a remarkable predictive degree, there is equally no doubt that most clinicians do not possess the clinical ability to assess the level of probable healing for an amputation very accurately. In particular, entities such as skin edge bleeding and palpation of temperature are very poor predictors of subsequent healing (Perlow, 1962; Stahlgren and Otteman, 1965; Barnes *et al.*, 1981; Nicholas *et al.*, 1982).

Certain clinical factors are, however, very relevant to consider when trying to reach a rational decision about where best to place the amputation level. Patients with severe fixed flexion contractures of the knee or hip are generally unsuitable for prosthetic fitting and may be better off with a transfemoral amputation where healing is more assured. Similarly, patients with a dense hemiplegia of the contralateral limb are almost impossible to mobilize and careful consideration should be given to doing a transfemoral amputation rather than risking non-healing in a transtibial operation. If, however, the hemiplegia involves the limb to be amputated, there may be a good case for doing a transtibial amputation if control of the knee joint is reasonable.

The use of palpable pulses to determine the optimum level of amputation is generally to be discouraged (Dwyer and Edwards, 1985). However, there are several circumstances where this very subjective assessment may be of value. If amputation must be done in patients without a palpable femoral pulse (rarely the case because of the likely success of aorto-femoral or axillo-femoral bypass to effect limb salvage), then a transfemoral amputation will usually be required. In contrast, in patients with a palpable popliteal pulse who require amputation (usually diabetics), a transtibial amputation should always heal. Most importantly, the absence of ankle or popliteal pulses is extremely common in patients with end-stage vascular disease and does *not* mean that a transtibial amputation will fail to heal – indeed most will heal very successfully.

Angiograms

The role of angiography in amputation surgery is not designed to decide upon amputation level, but rather to attempt to screen out those patients who should be offered alternative vascular reconstruction. Despite the fact that angiography does not always demonstrate distal vessels even when they are present, very few patients with end-stage vascular disease should ever undergo amputation without the benefit of angiography and an expert vascular opinion. It is therefore likely that in most cases an angiogram will be available in these patients even if reconstruction is not feasible. Given the availability of an angiogram, are there any findings on the angiogram which may help in deciding upon healing potential? Several workers have found arteriography to be of very little value (Lim *et al.*, 1967; Persson, 1974; Robbs and Ray, 1982). Roon, Moore and Goldstone (1977) stressed the importance of finding a patent profunda femoris in the ischaemic leg before attempting a transtibial amputation. This is a logical conclusion in keeping with the need to palpate a femoral pulse in the leg. Even chronic occlusion of both profunda and superficial femoral vessels is very likely to lead to non-healing of a poor transtibial amputation.

Doppler pressure measurements

Doppler derived segmental pressure measurements are now commonly available in every significant centre which purports to practice vascular surgery. Although there are difficulties with the technique, they have been utilized in the assessment of peripheral vascular disease for many years and have proved to be very valuable. However, there has been serious difficulty in setting out strict guidelines for the definition of critical ischaemia based upon ankle pressure criteria (Dormandy and Stock, 1990). While this is a desirable and much needed parameter designed to identify patients with critical ischaemia and so allow meaningful comparison between series, the criteria currently proposed are of little or no help when considering the level at which to amputate.

There was considerable initial enthusiasm for ankle Doppler pressure measurements to predict major amputation stump healing (Dean *et al.*, 1975; Barnes *et al.*, 1976; Pollock and Ernst, 1980; Lepantalo *et al.*, 1982; Nicholas *et al.*, 1982), but this was modified by the failure of later studies to substantiate early reports (Mehta *et al.*, 1980; Schwartz *et al.*, 1982; Van Soest *et al.*, 1985; Lepantalo *et al.*, 1987; Welch *et al.*, 1988). Ankle pressure measurements have been especially disappointing in that they are of no value in part-foot amputations or in diabetics (Gibbons *et al.*, 1973). This is because of the difficulty in compressing calcified vessels leading to an artificially high 'measured' ankle systolic pressures and perhaps also because of arterio-venous shunting in the foot giving rise to non-nutritional blood flow. In contrast, lower thigh pressure measurements do appear to be relatively predictive of success in transtibial amputation surgery. Several workers have demonstrated that transtibial amputation healing is likely to occur at thigh pressures greater than 50 mmHg and will almost always occur at pressures above 70 mmHg (Dean *et al.*, 1975; Holstein and Lassen, 1985; McCollum *et al.*, 1988). The data reflect flow into the profunda femoral artery to a large extent. Universal adoption of this simple recommenda-tion alone would be likely to reduce the number of transfemoral amputations currently being performed.

Plethysmography

Strain gauge plethysmography, which is a tech-nique relying upon a measurable change in tissue (usually toe or finger) volume with each heart beat, has been used to predict healing in trans-tibial amputations (Lee *et al.*, 1979). The main disadvantage with this technique is its insensitiv-ity in critically ischaemic limbs because of the very small volume changes encountered as a consequence of poor perfusion and consequent unreliable waveforms (Burgess and Matsen, 1981). More recently, there have been encourag-ing results using photoplethysmography to derive skin perfusion pressures (Stockel *et al.*, 1982; Ovesen and Stockel, 1984; Christensen *et al.*, 1988; Van der Broek *et al.*, 1988). While this is more of a local measurement rather than a regional assessment, it is nevertheless a fairly reproducible and efficient methodology. At pres-ent the data on this method still require further validation but it may provide useful additional quantifiable information, unaffected by the prob-lem of mesial arterial stiffness.

Infrared thermography

Infrared thermography is an exciting concept in which the radiated heat from a surface (the skin) is picked up by a special camera and translated into a colour image (Plate 4). It then requires inter-pretation and quantification if this thermal image is to be considered a reflection of skin blood flow (Windsor, 1971; Henderson and Hackett, 1978). Whereas in most patients with critical ischaemia there is a good correlation between infrared thermography and areas of non-viability, this is not universal and while the principle of trying to identify specific skin flaps for lower limb amputa-tions is not new (Spence *et al.*, 1981; Spence and Walker, 1984), there are many interpretative diffi-culties. Although there is a good relationship between skin blood flow and thermography, recent work by Wilson and Spence (1989) suggests that much of the 'heat' seen on the thermogram derives from two main sources: convective heat where the transfer to the skin surface is primarily from arterioles of >50 um and heat conducted to the skin surface where there is often a temperature gradient from the deeper structures to the skin. Excellent results have been obtained from specialized units using this technique (McCollum *et al.*, 1985a; Luk *et al.*, 1986), but the cost and great difficulty of interpretation of such images combined with an inability to accurately quantitate the methodology other than by additional skin blood flow or perfusion measurements probably renders it useful only as a development tool in specialist centres. This is especially disappointing as it is one of the few tests that may provide an indication of specific skin flap viability (Plate 5).

Skin fluorescence

The principle of outlining viability of specific skin flaps by the presence or absence of uptake of fluorescein when injected intravenously has been pioneered by a few groups in relation to healing potential in amputations (McFarland and Law-rence, 1982; Tanzer and Horne, 1982). More recently, the technique of quantitative fluorometry has been developed (Silverman and Wagner, 1983; Silverman *et al.*, 1987) in an effort to provide more objective criteria for the method. Significantly, there is still some difficulty in interpretation of results especially where there is associated inflammation. The initial difficulties with anaphylaxis secondary to sodium fluorescene have been largely eliminated and it appears to be a safe methodology.

Unfortunately, Burnham *et al.* (1990) were not able to show a good correlation with healing using this method and clearly further work is required before it can be recommended as a completely reliable test. Given that it is both cheap and relatively easy to perform, it is to be hoped that the earlier promise in this methodology found by Silverman and his colleagues (1983) will eventually be further developed and the current problems eradicated.

Skin blood flow measurements

There is little doubt that healing of wounds is dependent upon an adequate blood flow to the tissues involved. The early use of measurements to assess tissue viability inevitably encompassed attempts to evaluate blood flow (Yao, 1988). The principle of the isotope blood flow technique is to measure the washout of an injected radioisotope tracer and from this to derive a result which is a function of capillary blood flow (Fig. 20.1). Using a formula devised by Kety (1949), who looked at the myocardium, an absolute value for blood flow can be calculated. When applied to the skin, a quantifiable measurement of skin blood flow (SBF) can be obtained. Particular advantages of this technique are that it appears to measure nutritional blood flow and it provides a reproducible value. The objections to it are that it is essentially invasive and uses a radioactive substance. It is also relatively slow to use and requires multiple measurements in order to avoid the pitfall of finding an isolated high or low result. A number of tracers have been used including xenon-133 (Kostuik *et al.*, 1976), technicium-99 (Ristkari *et al.*, 1988), 1–131 and 1–125 iodoantipyrine (Spence and McCollum, 1986). Xenon is difficult to use because of its biphasic clearance, gaseous properties and affinity to adipose tissue (Daly and Henry, 1980); nevertheless, marvellous, reproducible results were achieved, principally by Moore's group in San Francisco (Malone *et al.*, 1981; Moore *et al.*,

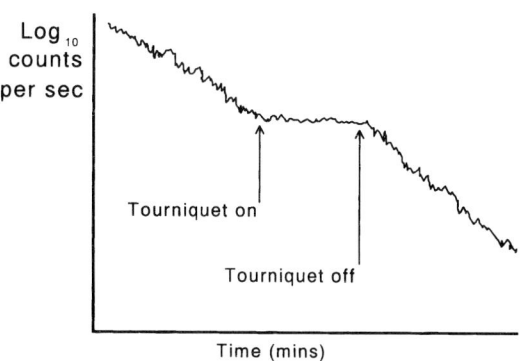

Fig. 20.1. The washout curve from an intradermal injection of I^{125} 4-iodoantipyrine. The logged trace shows a clear linear relationship with time and the blood flow can be calculated using Kety's equation: $F = K \times \lambda \times 100$ in ml/100g/min, where λ is the tissue/blood partition coefficient – approximately unity for $[^{125}I]$ 4-iodoantipyrine – and K is the washout rate constant where:

$$K = \frac{\ln 2}{t1/2} = \frac{0.693}{t1/2}$$

and t1/2 is the half-time of the isotope in the tissue derived from the washout curve. The abolition of blood flow in the skin with a temporary blood pressure cuff tourniquet can easily be seen.

1981). These results have been emulated using 1–125 iodoantipyrine more recently (McCollum *et al.*, 1985c) and again, a good correlation between healing and mean SBF levels greater than 2.5 ml/100 g/min was achieved. While providing good and reliable data, especially in the lower limb (Spence *et al.*, 1984). SBF measurements in the foot are not at all constant and a high skin blood flow does not predict a successful outcome to amputation. The probable explanation for this is that a local point measurement (such as SBF) is not representative of a region as a whole, and especially so in the foot (McCollum *et al.*, 1985b). However, it is at the transtibial level that it is most reliable and it is healing at this level that in practice requires the most accurate prediction. Regardless of these drawbacks, SBF measurements are still the most accurate single measurements to assess skin viability currently available, particularly around the knee joint level.

Skin perfusion pressure

The main pioneers of this technique came from Denmark where it has become popular as an adjunctive method for selecting amputation level (Neilson *et al.*, 1973; Holstein *et al.*, 1977). In principle, the method depends upon observing the blood pressure at which the capillary blood flow to the skin is abolished. The method used for detection of flow may be either the clearance of a radioisotope or the use of a photospectrometer (Mylrea *et al.*, 1986). The method has the advantage of ignoring absolute values of SBF but has the disadvantage of depending upon blood pressure measurements using a double sphygmomanometer cuff system. In effect, it is an indirect measurement of diastolic blood pressure. Further difficulties are seen in deriving absolute values at the very low clearance rates found in critically ischaemic skin and it is of course in these patients that there is the most need for such an assessment method. Several workers have noticed quite a degree of variation in definition as to what absolute skin perfusion pressure should be used to predict healing and, while this may reflect the methodology involved, it makes comparisons between units very difficult and factual advice about interpretation very suspect (Holstein, 1973; Hammersgaard and Baadsgaard, 1977; Thyregod *et al.*, 1983).

Trans-cutaneous oxygen measurements

Logic would suggest that the main reason for non-healing of amputation wounds is likely to be inadequate oxygen delivery at the cellular level. It therefore seems sensible to try to use a method which appears to assess the availability of oxygen in the skin. Perhaps because of this, since transcutaneous oxygen pressure (tcpO$_2$) measurements were first used to aid in the assessment of tissue viability by Keller *et al.* (1978), there has been an epidemic of publications on its value (or otherwise) in peripheral vascular disease (Franzeck *et al.*, 1982; White *et al.*, 1982; Dowd *et al.*, 1983; Mustapha *et al.*, 1983; Cina *et al.*, 1984; Matsen *et al.*, 1984; Spence *et al.*, 1985; Feenstra *et al.*, 1988) and especially in relation to amputation level (Burgess *et al.*, 1982; Ito *et al.*, 1984; Ratcliff *et al.*, 1984; Harward *et al.*, 1985; McCollum *et al.*,

1986; Bongard and Krahenbuhl, 1988; Oishi *et al.*, 1988; Falstie-Jensen *et al.*, 1989). The transcutaneous oxygen electrode was originally designed to reflect arterial pO$_2$ in the neonate and since then its application has been extended to include the evaluation of ischaemic tissue. Unfortunately, the electrode design has not been modified to take account of the great differences in skin characteristics between the newborn and the adult (Spence *et al.*, 1986) and, not surprisingly, there has been a great variation in results between different workers using different machines. In normal skin, the variables affecting tcpO$_2$ measurements such as PaO$_2$, skin thickness, electrode response, skin blood flow, local oxygen availability and others are overshadowed by a maximal vasodilatory response to the integral heater. Ischaemic skin is quite different in its physiological response and is much more affected by these variables, especially by the oxygen consumption at the electrode and by the underlying perfusion pressure.

Certainly, the great appeal of the transcutaneous method is its simplicity and relative ease of use. However, the difficulties of calibration and interpretation are always underestimated and result in

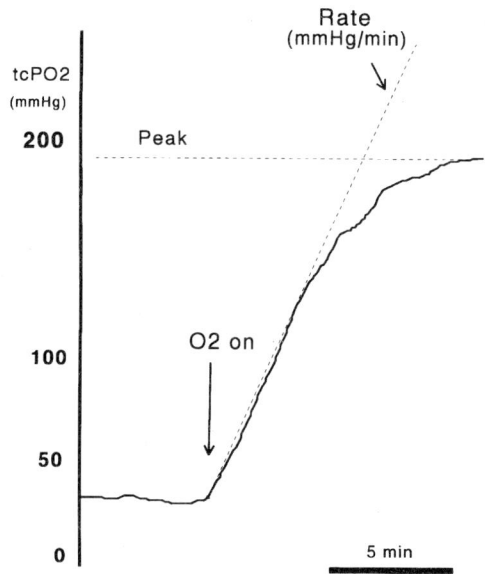

Fig. 20.2. The effect of breathing 100% oxygen on transcutaneous oxygen level at calf. The dramatic increase can be seen clearly, and an estimate of the rate of change of tcpO$_2$ can be made from the slope of the response curve.

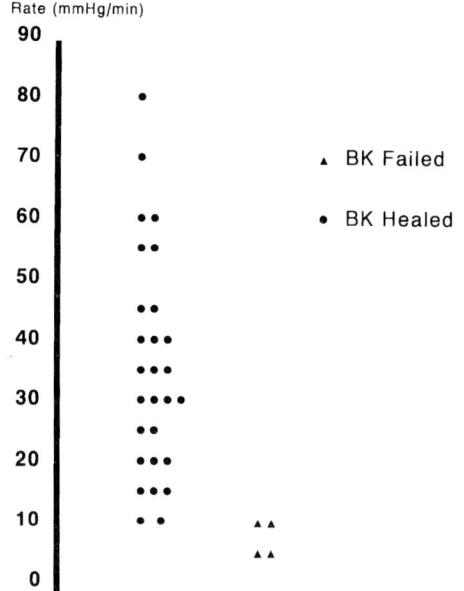

Fig. 20.3. The improved sensitivity obtained when using the parameter of 'rate of change of tcpO$_2$' in response to 100% O$_2$ inhalation. Of 32 transtibial amputations, four failed and all were predicted by a rate of change of less than 9 mmHg/min with only one amputation healing below this level.

Other measurements

Several other techniques have also been tried to aid in amputation level selection. These include laser Doppler studies (Holloway and Burgess, 1983; Matsen *et al.*, 1984), pulse volume recording (Holstein and Lassen, 1985), radioactive microspheres (Fee *et al.*, 1977), skin pH studies (Harrison and Walker, 1981), muscle pH studies (Young and Couch, 1978), and others. Of these, the laser Doppler has the most potential, offering the ability to interrogate the microcirculation in an non-invasive mode but, despite much work in the area, efforts to quantitate the method have been disappointing.

Summary

There are many difficulties which are encountered when attempting to analyse the available methods of ancillary amputation level assessment. The single most important factor is the enormous variation in clinical approach on the part of individual surgeons and centres. Unfortunately, for the most part, this reflects personal prejudice which is extremely difficult to overcome. If we are to examine the benefits of a certain test to predict the viability of, say, transtibial amputations, then we must be sure that the optimum environment exists in which that amputation is to be performed. As suggested, this is clearly not the case in much of the published literature and this makes the value of using any technique to aid in amputation level selection in such units very doubtful. While there is no question that such tests could dramatically improve the transtibial amputation rate, they are of no value to the surgeon who chooses to ignore them anyway. Put another way, if there is a preference for transfemoral amputations in a unit, the value of laboratory methods to predict amputation outcome is considerably diminished. Presumably, most of those amputations, clinically considered to be marginal transtibials, would have transfemoral amputations and the remainder ought to heal at the transtibial level without problem. This is a particularly important point because transfemoral/transtibial amputation ratios vary enormously and it is likely that it is only in those centres which offer a 2:1 or greater transtibial to transfemoral

data which at best is insensitive and poorly evaluated. In early tcpO$_2$ reports, there were wide variations in deductions as to what tcpO$_2$ value reliably predicted amputation healing at the below knee level. This was compounded by the fact that several workers found that healing took place in a number of amputations where the tcpO$_2$ was zero. The addition of 100% oxygen inhalation undoubtedly improved the sensitivity of the technique considerably (Harward *et al.*, 1985; McCollum *et al.*, 1986; Oishi *et al.*, 1988) but it is inconvenient, difficult to perform and time consuming (Figs 20.2 and 20.3).

There are enough data on tcpO$_2$ assessments to be sure of two points. One is that the method is not sufficiently sensitive to predict healing in many cases. Secondly, if a tcpO$_2$ level of 35 mmHg is used as a cut-off point for transtibial amputation healing, then nearly all such amputations will heal primarily. If oxygen inhalation is added to the test, then discrimination between healers and non-healers is improved dramatically. TcpO$_2$ values in the foot are of little predictive value.

amputation ratio that these measures have a significant role to play in level selection.

The second major problem relates to the attitude of the clinician towards amputation. Obviously, if there is a preference for primary amputation rather than for attempted reconstructive surgery, better transtibial amputation healing rates should occur because many of these patients will not have such strictly 'end-stage' vascular disease. In such instances, any attempt to perform non-invasive vascular tests would be largely irrelevant as, again, most transtibial amputations ought to heal. It is only when we are performing surgery at the edge of tissue viability that these sophisticated techniques have a valuable role to play.

If there is to be a significant impact upon amputation level by increasing the transtibial amputation rate, the primary goal must be to dispel the notion that the transtibial amputation will not heal. In the best scenario, this would be accomplished by utilizing the team approach where each member serves as a peer audit for the unit and there is a pride in keeping the amputation rate as low as possible but, where necessary, to perform as many as feasible at the transtibial level.

What technique then, if any, should the present-day amputation surgeon employ to help him in his selection of the optimum amputation level? To achieve the highest transtibial amputation rate the obvious method is to perform all major amputations at this level. This is clearly unwise, although such a policy is not without some intrinsic merit. Unfortunately, there is no one single test which can be said to be infallible. Many of the techniques are expensive and difficult to perform correctly and while they may provide useful data to a particular specialist unit, the methods often prove quite difficult to reproduce effectively elsewhere. In particular, no method has proved particularly useful for the purposes of predicting part-foot amputation healing, although skin perfusion pressure techniques are probably most predictive. Much more work requires to be carried out on the critically ischaemic foot to ascertain why current techniques are insensitive as predictors of foot healing.

The simplest readily available test is undoubtedly Doppler derived systolic pressure measurement. Although relatively insensitive, there is considerable evidence to support the use of a cut-off range of 70 mmHg at the lower thigh for transtibial amputations. Using this standard, many transfemoral amputations could be avoided with very few transtibial failures as a consequence. It is probably only in those limbs with marginal skin flap viability that other techniques of evaluation have a major role to play. Of these, skin blood flow measurements offer the most precise objective criteria. Although not necessarily representative of an entire region, multiple blood flow measurements appear to confer an increased reliability on the method and correlate well with outcome. A mean skin blood flow of less than 2.5 ml/100 g/min at the transtibial level should be regarded as likely to result in flap failure. It is worth corroborating these data with some other, regional, methodology to look for specific well vascularized flaps. Flourescein angiography or thermography offer the best methods for this approach.

New developments are under way. The main drawback to skin blood flow assessment is the use of radioactive substances. Assessment of blood flow by alternative non-invasive, non-radioactive techniques such as trans-cutaneous hydrogen clearance is now possible, although further work is required (McCollum et al., 1993). Assessment of tissue oxygenation is possible with near-infrared tissue spectrophotometry (Harrison et al., 1992) and, although this has yet to be used in amputation level selection, may provide an excellent index of tissue viability.

It is obvious that there is still considerable controversy with regard to many of the methods discussed in this chapter. It is equally clear that significant refinement and more experience is needed with all of the techniques before any can be recommended for routine use as a reproducible objective assessment of tissue viability. The results of assessment of foot viability have been especially disappointing in this regard and much research and development is required in this area. Without question, the main way of delivering an amputation service in which less transfemoral amputations are being performed and more attention given to preserving the knee joint, is by changing the prevalent attitude of surgeons who feel that the transfemoral amputation is the 'gold standard'. Until this is done – ideally by developing specific amputation/vascular units and fostering the team approach – all the methods described in this chapter are largely irrelevant!

References

Barnes, R.W., Shanik, G.D. and Slaymaker, E.E. (1976) An index of healing in below knee amputation: leg blood pressure by Doppler ultrasound. *Surgery*, **79**, 13–20.

Barnes, R.W., Thornhill, B., Nix, L., Rittgers, E. and Turley, G. (1981) Prediction of amputation wound healing. *Arch. Surg.*, **116**, 80–83.

Bongard, O. and Krahenbuhl, B. (1988) Predicting amputation in severe ischaemia. The value of transcutaneous pO_2 measurement. *J. Bone Joint Surg.*, **70(B)**, 465–467.

Burgess, E.M. and Matsen, F.A. (1981) Determining amputation levels in peripheral vascular disease. *J. Bone Joint Surg.*, **63–A**, 1493–1497.

Burgess, E.M., Matsen, F.A., Wyss, C.R. and Simmons, C.W. (1982) Segmental transcutaneous measurements of pO_2 in patients requiring below-the-knee amputation for peripheral vascular insufficiency. *J. Bone Joint Surg.*, **64(A)M**, 378–382.

Burnham, S.J., Wagner, W.H., Keagy, B.A. and Johnson, G. (1990) Objective measurement of limb perfusion by dermal fluorometry. A criterion for healing of below-knee amputation. *Arch. Surg.*, **125**, 104–106.

Castronuovo, J.J. Jr, Deane, L.M., Deterling, R.A. Jr, O'Don-nell, T.F. Jr, O'Toole, D.M. and Callow, A.D. (1980) Below knee amputation: is the effort to preserve the knee joint justified? *Arch. Surg.*, **115**, 1184–1187.

Christensen, K.S., Falstie-Jensen, N., Christensen, E.S. and Brochner-Mortensen, J. (1988) Results of amputation for gangrene in diabetic and non-diabetic patients. Selection of amputation level using photoelectric measurements of skin perfusion pressure. *J. Bone Joint Surg.*, **70–A**, 1514–1519.

Cina, C., Katsamouris, A., Mergerman, J., Brewster, D.C., Strayhorn, E.C., Robison, J.G. and Abbott, W.M. (1984) Utility of transcutaneous oxygen tension measurements in peripheral arterial occlusive disease. *J. Vasc. Surg.*, **1**, 362–369.

Daly, M.J. and Henry, R.E. (1980) Quantitative measurement of skin perfusion with Xenon-133. *J. Nucl. Med.*, **21**, 156–160.

Dean, R.H., Yao, J.S.T., Thompson, R.G. and Bergan, J.J. (1975) Predictive value of ultrasonographically derived arterial systolic pressure in determination of amputation level. *Am. Surg.*, **41**, 731–737.

Department of Health and Social Security and Office of Population Censuses and Surveys (1990) *Hospital Inpatient Enquiry.* HMSO, London.

Dormandy, J.A. and Stock, G. (1990) Definition and epidemiology of chronic limb ischaemia. In: *Critical Leg Ischaemia; Its Pathophysiology and Management* (eds J.A. Dormandy and G. Stock), Springer, Berlin.

Dowd, G.S.E., Linge, K. and Bently, G. (1983) Measurement of transcutaneous oxygen pressure in normal and ischaemic skin. *J. Bone Joint Surg.*, **65(B)**, 79–83.

Dwyer, K.J. and Edwards, M. (1985) The association between lowest palpable pulse and wound healing in below knee amputations. *Ann. R. Coll. Surg. Engl.*, **67**, 232.

Falstie-Jensen, N., Christensen, K.S. and Brochner-Mortensen, J. (1989) Selection of amputation level is not aided by transcutaneous pO_2 measurements. *Acta Orthop. Scand.*, **60**, 483–485.

Fee, H.J., Friedman, B.H. and Siegel, M.E. (1977) The selection of amputation level with radioactive microspheres. *Surg. Gynecol. Obstet.*, **144**, 89–90.

Feenstra, B.W., Meiss, L., Montauban van Swijndregt, A.D., Stigter, H. and Van Urk, H. (1988) Assessment of peripheral vascular obliterative disease by transcutaneous oxygen tension tests. *Eur. J. Vasc. Surg.*, **2**, 19–26.

Franzeck, U.K., Talke, P., Bernstein, E.F., Golbranson, F.L. and Fronek, A. (1982) Transcutaneous pO_2 measurements in health and peripheral arterial disease. *Surgery*, **91**, 156–163.

Gibbons, G.W., Wheelock, F.C., Siembieda, C., Hoar, C.S., Rothbotham, J.L. and Persson, A.B. (1979) Non-invasive prediction of amputation level in diabetic patients. *Arch. Surg.*, **114**, 1253–1257.

Hammersgaard, E. and Baadsgaard, K. (1977) Healing of below knee amputations in relation to perfusion pressure of skin. *Acta Orthop. Scand.*, **48**, 335.

Harris, J.P., Page, S., Englund, R. and May, J. (1988) Is the outlook for the vascular amputee improved by striving to preserve the knee joint? *J. Cardiovasc. Surg.*, **29**, 741–745.

Harrison, D.K. and Walker, W.F. (1981) Microelectrode measurement of skin pH in post-operative intensive care patients. In: *Progress in Enzyme and Ion Selective Electrodes* (eds D.W. Lubbers, H. Acker, R.P. Buck, G. Eisenman, M. Kessler and W. Simon), Springer, Berlin/Heidelberg/New York.

Harrison, D.K., Evans, S.D., Swanson-Beck, J. and McCollum, P.T. (1992) Spectrophotometric measurements of haemoglobin saturation and concentration in skin during the tuberculin reaction in normal human subjects. *Clin. Phy. Physiol. Measure.*, **13**, 349–363.

Harward, T.R.S., Volny, J., Golbranson, F., Bernstein, E.F. and Fronek, A. (1985) Oxygen inhalation-induced transcutaneous pO_2 changes as a predictor of amputation level. *J. Vasc. Surg.*, **2**, 220–225.

Henderson, H.P. and Hackett, M.E.J. (1978) The value of thermography in peripheral vascular disease. *Angiology*, **29**, 65–75.

Holloway, G.A. and Burgess, E.M. (1983) Preliminary experience with laser Doppler velocimetry for the determination of amputation levels. *Prosthet. Orthot. Int.*, **7**, 63–66.

Holstein, P. (1973) Distal blood pressure as guidance in choice of amputation level. *Scand. J. Clin. Lab. Invest.*, **31**, 245–248.

Holstein, P. and Lassen, N.A. (1985) Methods for prediction of amputation wound healing based on regional blood flow or blood pressure measurements. In: *Diagnostic Techniques for Vascular Surgeons* (ed. R.M. Greenhalgh), Grune & Stratton, London, pp. 343–350.

Holstein, P., Lund, P., Larsen, B. and Schomacker, T. (1977) Skin perfusion pressure measured as the external pressure required to stop isotope washout. *Scand. J. Clin. Lab. Invest.*, **37**, 649–659.

Ito, K., Ohgi, S., Mori, T., Urbanyi, B. and Schlosser, V. (1984) Determination of amputation level in ischaemic legs by means of transcutaneous oxygen pressure measurements. *Int. Surg.*, **69**, 59–61.

Keller, H.P., Klaue, P. and Lubbers, D.W. (1978) Transcutaneous pO2 measurement for evaluating the oxygen supply of skin allo-and autografts. *Eur. Surg. Res.*, **10**, 272–282.

Kendrick, R.R. (1956) Below-knee amputation in arteriosclerotic gangrene. *Br. J. Surg.*, **44**, 13–17.

Kety, S.S. (1949) Measurement of regional circulation by the local clearance of radioactive Sodium. *Am. Heart J.*, **38**, 321–328.

Kostuik, J.P., Wood, D., Hornby, R., Feingold, S. and Mathews, V. (1976) Measurement of skin blood flow in peripheral vascular disease by epicutaneous application of Xenon-133. *J. Bone Joint Surg.*, **58–A,** 833–837.

Lee, B.Y., Trainor, F.S., Kavner, D., McCann, W.J. and Madden, J.L. (1979) Non-invasive hemodynamic evaluation in selection of amputation level. *Surg. Obstet. Gynecol.*, **149**, 241–244.

Lepantalo, M.J., Haajanen, J., Lindfors, O., Paavolainen, P. and Scheinin, T.M. (1982) Predictive value of pre-operative segmental blood pressure measurements in below-knee amputations. *Acta Chir. Scand.*, **148**, 581–584.

Lepantalo, M., Isoniemi, H. and Kyllonen, L. (1987) Can the failure of a below knee amputation be predicted? *Ann. Chir. Gynaecol.*, **76**, 119–123.

Lim, R.C., Blaisdell, F.W., Hall, A.D., Moore, W.S. and Thomas, A.N. (1967) Below knee amputation for ischaemic gangrene. *Surg. Gynecol. Obstet*, **125**, 153–158.

Luk, K.D.K., Yeung, P.S. and Leong, J.C.Y. (1986) Thermography in the determination of amputation levels in ischaemic limbs. *Int. Orthop.*, **10**, 79–82.

Malone, J.M., Leal, J.M., Moore, W.S. Henry, R.E., Daly, M.J., Patton, D.D. and Childers, S.J. (1981) The 'gold standard' for amputation level selection: xenon 133 clearance. *J. Surg. Res.*, **30**, 449–455.

Matsen, F.A., Wyss, C.R., Robertson, C.L., Oberg, P.A. and Holloway, G.A. (1984) The relationship of transcutaneous pO2 and laser Doppler measurements in a model of local arterial insufficiency. *Surg. Gynecol. Obstet*, **159**, 418–422.

McCollum, P.T. and Walker, W.F. (1992) Major limb amputation for end-stage peripheral vascular disease: level selection and alternative options. In: *Atlas of Limb Prosthetics*, (eds. J.H. Bowker and J.W. Michael), Mosby Year Book, St. Louis.

McCollum, P.T., Spence, V.A., Walker, W.F. and Murdoch, G. (1985a) A rationale for skew flaps in amputation surgery. *Prosthet. Orthot. Int.*, **9**, 100–104.

McCollum, P.T., Spence, V.A. and Walker, W.F. (1985b) Circumferential skin blood flow measurements in the ischaemic lower limb. *Br. J. Surg.*, **72**, 310–312.

McCollum, P.T., Spence, V.A., Walker, W.F., Murdoch, G., Swanson, A.J. and Turner, M. (1985c) Antipyrine clearance from the skin of the foot and the lower leg in critical ischaemia: clinical implications. In: *Practical Aspects of Skin Blood Flow Measurement* (eds. V.A. Spence and C.D. Sheldon), Biological Engineering Society, London.

McCollum, P.T., Spence, V.A. and Walker, W.F. (1986) Oxygen induced changes in the skin as measured by transcutaneous oxymetry. *Br. J. Surg.* **73**, 882–885.

McCollum, P.T., Walker, W.F. and Spence, V.A. (1988) Amputation for peripheral vascular disease: the case for level selection. *Br. J. Surg*, **75**, 1193–1195.

McCollum, P.T., Harrison, D.K., Abi-Raad, R., Newton, D., Holdsworth, R.J. (1993) H2 Clearance as a non-invasive technique for measuring skin blood flow in critical ischaemia. *Br. J. Surg.*, **80**, 145.

McFarland, D.C. and Lawrence, P.F. (1982) Skin fluorescence, a method to predict amputation site healing. *J. Surg. Res.*, **32**, 410–415.

Mehta, K., Hobson, R.W., Jamil, A., Hart, L. and O'Donnell, J.A. (1980) Fallibility of Doppler ankle pressure in predicting healing of transmetetarsal amputation. *J. Surg. Res.*, **28**, 466–470.

Moore, W.S., Henry, R.E., Malone, J.M., Daly, M.J., Patton, D. and Childers, S.J. (1981) Prospective use of Xenon Xe–133 clearance for amputation level selection. *Arch. Surg.*, **116**, 86–88.

Murdoch, G. (1977) Amputation surgery in the lower extremity. *Prosthet. Orthot. Int.*, **1**, 72–83.

Mustapha, N.M., Redhead, R.G., Jain, S.K. and Wielogorski, J.W. (1983) Transcutaneous partial oxygen pressure assessment of the ischaemic lower limb. *Surg. Gynecol. Obstet.*, **156**, 582–584.

Mylrea, K.C., McVeigh, L.J., Bean, B.A. and Malone, J.M. (1986) A noninvasive method for assessing the adequacy of surface tissue circulation. *J. Clin. Eng.*, **11**, 135–143.

Neilson, P.E., Poulsen, H.L. and Gyntelberg, F. (1973) Arterial blood pressure in the skin measured by a photoelectric probe and external counterpressure. *Vasa*, **2**, 65–74.

Nicholas, G.G., Myers, J.L. and DeMuth, W.E. (1982) The role of vascular laboratory criteria in the selection of patients for lower extremity amputations. *Ann. Surg.*, **195**, 469–473.

Oishi, C.S., Fronek, A. and Golbranson, F.L. (1988) The role of non-invasive vascular studies in determining levels of amputation. *J. Bone Joint Surg.*, **70–A**, 1520–1530.

Ovensen, J. and Stockel, M. (1984) Measurement of skin perfusion pressure by photoelectric technique – an aid to amputation level selection in arteriosclerotic disease. *Prosthet. Orthot. Int.*, **8**, 39–42.

Perlow, S. (1962) Amputation for gangrene because of occlusive arterial disease. *Am. J. Surg.*, **103**, 569–574.

Persson, B.M. (1974) Sagittal incision for below-knee amputation in ischaemic gangrene. *J. Bone Joint. Surg.*, **56B**, 110–114.

Pollock, S.B. and Ernst, C.B. (1980) Use of Doppler pressure measurements in predicting success in amputation of leg, *Am. J. Surg.*, **139**, 303–306.

Ratcliff, D.A., Clyne, C.A.C., Chant, A.D.B. and Webster, J.H.H. (1984) Prediction of amputation wound healing: the role of transcutaneous p02 assessment. *Br. J. Surg.*, **71**, 219–222.

Ristkari, S.K., Vorne, M. and Mokka, R.E. (1988) Early assessment of amputation level in frostbite by 99mTc-pertechnetate scan. Case Report. *Acta Chir. Scand.*, **154**, 403–405.

Robbs, J.V. and Ray, R. (1982) Clinical predictors of below knee stump healing following amputation for ischaemia. *S. Afri. J. Surg.*, **20**, 305–310.

Roon, A.J., Moore, W.S. and Goldstone, J. (1977) Below knee amputation: a modern approach. *Am. J. Surg.*, **134**, 153–158.

Schwartz, J.A. Schuler, J.J., O'Connor, R.J.A. and Flannigan, D.P. (1982) Predictive value of distal perfusion pressure in the healing of amputation of the digits and the forefoot. *Surg. Gynecol. Obstet.* **154**, 865–869.

Silverman, D. and Wagner, F.W. (1983) Prediction of leg viability and amputation level by fluorescein uptake. *Prosthet. Orthot. Int.*, **7**, 69–71.

Silverman, D.G., Roberts, A. and Reilly, C.A. (1987) Fluorometric quantification of low-dose fluorescein delivery to predict amputation site healing. *Surgery*, **101**, 335–341.

Spence, V.A. and McCollum, P.T. (1986) Quantitative assessment of cutaneous blood flow using radioisotope tracers. In: *Clinical Investigation of the Microcirculation* (eds. J.E. Tooke and L.H. Smage), Martinus and Nighoff, London.

Spence, V.A. and Walker, W.F. (1984) The relationship between temperature isotherms and skin blood flow in the ischaemic limb. *J. Surg. Res.*, **36**, 278–281.

Spence, V.A., Walker, W.F., Troup, I.M. and Murdoch, G. (1981) Amputation of the ischaemic limb: selection of the optimum site by thermography. *Angiology*, **32**, 155–169.

Spence, V.A., McCollum, P.T., Walker, W.F. and Murdoch, G. (1984) Assessment of tissue viability in relation to selection of amputation level. *Prosthet. Orthot. Int.*, **8**, 67–75.

Spence, V.A., McCollum, P.T., McGregor, I.W., Sherwin, S.J. and Walker, W.F. (1985) The effect of the transcutaneous electrode on the variability of dermal oxygen skin tensions. *Clin. Phys. Physiol. Measure.*, **6**, 139–145.

Stahlgren, L.H. and Otteman, M. (1965) Review of criteria for selection of the level for lower extremity amputation for atherosclerosis. *Ann. Surg.*, **162**, 886–892.

Stockel, M., Ovesen, J., Brochner-Mortensen, J. and Emneus, H. (1982) Standardised photoelectric technique as routine method for selection of amputation level. *Acta Ortho. Scand.*, **53**, 875–878.

Tanzer, T.L. and Horne, J.G. (1982) The assessment of skin viability using fluorescein angiography prior to amputation. *J. Bone Joint Surg.* **64–A**, 880–882.

Thyregod, H.C., Holstein, P. and Steen-Jensen, J. (1983) The healing of through-knee amputations in relation to skin perfusion pressure. *Prosthet. Orthot. Int.*, **7**, 61–62.

Van der Broek, T.A.A., Dwars, B.J., Rauwerda, J.A. and Hakker, F.C. (1988) Photoplethysmographic selection of amputation level in peripheral vascular disease. *J. Vasc. Surg.*, **8**, 10–13.

Van Soest, M.G.A., Breslau, P.J., Jorning, P.J.G. and Greep, J.M. (1985) The clinical value of preoperative indirect systolic pressure measurements after amputation of the lower extremity. *Neth. J. Surg.*, **37**, 75–78.

Welch, G.H., Leiberman, D.P., Pollock, J.G. and Angerson, W. (1988) Failure of Doppler ankle pressure to predict healing of conservative forefoot amputations. *Br. J. Surg.*, **72**, 888–891.

White, R.A., Nolan, L., Harley, D., Long, J., Klein, S., Tremper, K., Nelson, R., Tabrisky, J. and Shoemaker, W. (1982) Noninvasive evaluation of peripheral vascular disease using transcutaneous oxygen tension. *Am. J. Surg.*, **144**, 68–75.

Windsor, T. Vascular aspects of thermography. (1971) *J. Cardiovasc. Surg.*, **12**, 379–388.

Wilson, S.B. and Spence, V.A. (1989) Dynamic thermographic imaging method for quantifying dermal perfusion: potential and limitations. *Med. Biol. Eng. Comput.*, **27**, 496–501.

Yao, J.S.T. (1988) Choice of amputation level. *J. Vasc. Surg.*, **8**, 544–545.

Young, A.E. and Couch, N.P. (1978) Muscle perfusion and the healing of below knee amputations. *Surg. Gynecol. Obstet.*, **146**, 533–534.

21

The dilemma: amputation or vascular reconstruction?

K.P. Robinson

John Woodall, in his treatise, the Surgeon's Mate (1639), wrote as follows:

> Amputation, or dismembering, is the most lamentable part of surgery, it were therefore the honour of a surgeon never to use dismembering at all if it were possible for him to heal all that he undertaketh.

This sentiment is particularly applicable to the dilemma of primary amputation or vascular surgical reconstruction. The clinical situation is changing with the passage of each year. Over the past decade, vascular surgery has made enormous progress so that concepts and attitudes evident 10 years ago are no longer applicable and it is hoped that, with continuing progress, the situation will remain one of change with the emphasis on fewer primary amputations.

The involvement of the vascular surgeon when this dilemma must be faced occurs, first, in the case of the traumatized limb where arterial and venous injury is part of the situation and a decision must be made as to whether restorative surgery is pursued or to proceed to a primary or delayed amputation. The second situation is where a patient develops spontaneously an acutely ischaemic limb due to an arterial embolus or thrombosis due to disease. Thirdly, and the most commonly encountered situation is where a chronically ischaemic limb deteriorates to the extent that rest pain, ischaemic ulceration, threatened or actual gangrene, bring the patient to the point where surgical treatment has to be considered and either primary amputation or vascular reconstructive surgery performed without delay (Ebskov, 1980).

The dilemma is summarized in Table 21.1.

Table 21.1.

Primary amputation	Revascularization
For	
Usually single operation	Single operation may succeed
Decisive solution of problems	Prolonged but atraumatic surgery
Good function with prosthesis	Limb may be fully restored
Limited inpatient care	Good function if other limb affected
Against	
Short but traumatic operation	Multiple procedures may be needed
Destructive procedure	Uncertain result
Disability inevitable	Uncertain duration of improvement
Poor function if other limb affected	Appreciable rehabilitation needed
Complex rehabilitation	Needs vascular surgeon
Prosthetic supply	Vascular services helpful
Wheelchair supply	Eventual amputation when fails
Often loss of independence	Prolonged hospitalization
High cost	High cost

Trauma

The surgical management of the severed limb includes the possibility of replantation highlighting the need for an extremely critical and accurate assessment. In the case of a mangled extremity with multiple and complicated damage to arteries, veins, nerves, muscles, bones, joints and soft tissues, even if revascularized, it would be functionally useless. Amputation is clearly the best alternative.

To generalize in a highly specialized field, it appears that severance of a lower limb beyond the growth period is always an indication for an amputation and prosthetic replacement. The distance the nerves require to regenerate leaves the virtual certainty of an unstable, deformed and possibly painful extremity far inferior to a sound prosthesis and satisfactory stump. In the young child, the place for replantation of a severed lower limb must be an individual and highly specialized decision. In the upper limb the situation is more in favour of a replantation. The more distal the injury, be it digital, palmar, wrist or forearm, the more likely the probability of a functional extremity that will exceed the capabilities of the most sophisticated arm prosthesis. At higher levels in the arm in older patients, reimplantation becomes more controversial and, again, decisions depend on the individual patient and a careful multi-disciplinary assessment.

In injuries short of complete severance a number of grading systems (Howe *et al.*, 1987; Johansen *et al.*, 1990; Russell *et al.*, 1991) have been devised in order to aid prediction as to the likely outcome of limb salvage procedures and alternatively indicators for primary early amputation. These systems are examined and discussed by Hunter elsewhere in this volume.

Whatever system is used the need for accurate assessment is emphasized by Pozo *et al.* (1990) in a review of 35 patients who came to amputation late after failed initial treatment, of which seven were required for vascular insufficiency 1 month after the injury, 13 for late infection and 15 for ununited fractures. It was emphasized that these 35 patients underwent an average of 12 operations following initial treatment and spent some 8 months in hospital.

The accurate assessment of the degree of vascular injury may require imaging prior to the initial surgical exploration, ideally, by arterio-graphy; the best image is obtained by the DSA technique but duplex doppler imaging may have a place, when available.

In the final analysis, surgical exploration of the vessels, if necessary with angiography on the operating table, will determine the likelihood of success of the arterial reconstruction. A decision may have to be made during the course of the initial debridement and restorative procedure.

Acute arterial ischaemia

The management of the acutely ischaemic limb has undergone marked changes in the last 5 years. Firstly, the treatment of acute arterial embolus has changed considerably. In the past, rheumatic heart disease resulted in a high incidence of atrial thrombus and peripheral embolus often at an age before degenerative arterial disease was a major problem and arterial embolectomy was usually successful and uncomplicated. It is now apparent that embolectomy performed with the Fogarty embolectomy catheter does not always give good results (Ljungman *et al.*, 1991). The poor results are identified with an embolus impacted in a vessel already diseased with atheroma.

To improve the results of the treatment of arterial embolism, the patient must be assessed and treated in a specialist vascular unit. This allows pre-operative imaging of both the occluded vessels in the affected limb and the vessels on the contralateral limb to assess the likelihood of atheroma complicating the embolic phenomenon and thus going some way to resolving the diagnostic difficulty between arterial embolus and spontaneous thrombosis. The preliminary angiogram will allow placement of a catheter at or in the upper limit of the occluded artery and thus permit the effective administration of a thrombolytic agent; this has now emerged as the first line treatment for acute arterial ischaemia. There remains a place for Fogarty balloon catheter embolectomy in those patients where the occlusion affects the proximal vessels and is of short duration (Wagner, 1992).

Otherwise, initial treatment by systemic urokinase, streptokinase or tissue plasminogen activator with the catheter placed in the thrombus material has emerged as the most effective treatment, tissue plasminogen activator being preferred for its lack of patient sensitization and consistent activity. In addition, the use of the arterial catheter to break up the clot mechanically and aspirate the dissolving debris has significantly increased the success of lytic treatment (Wagner and Stark 1992) and greatly reduced the need for amputation due to failed embolectomy.

Where an embolus has caused the occlusion, it is necessary then to proceed to a full cardiac assessment and conventional anticoagulant therapy. Where the occlusion has been due to an acute arterial thrombus, usually supervening on existing disease, the atheroma must be treated once the thrombus is lysed. Often a balloon angioplasty is successful in preventing a further thrombotic episode but bypass surgery may well be required.

However, in a few cases the arterial occlusion will have been present for more than 48 h and already complicated by compartment syndrome and muscle necrosis in the lower leg or lower arm with the possibility of overt gangrene before lytic treatment can be applied (Gregson, 1991). In these patients, there is still the possibility that one or more of the distal vessels have not been involved in the occlusive process and is available for an arterial bypass graft. The opportunity should not be lost to treat these patients by distal bypass before irreversible changes are established with decompression of the affected compartments and debridement of any necrotic toes or ulcers.

The dilemma emerges in evaluating the long-term benefit of these procedures for the patient in this context. The situation is similar to those patients with progressive chronic peripheral vascular disease.

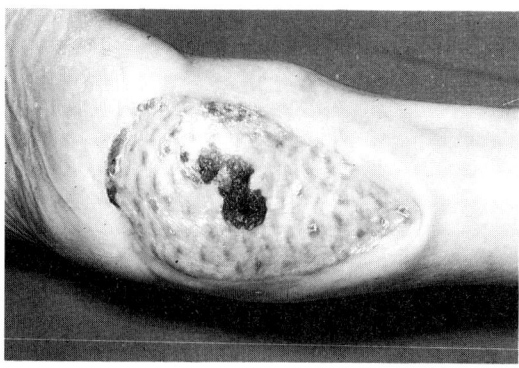

Fig. 21.1. A meshed skin graft applied to a previously gangrenous heel following resection of part of the calcaneum. A further small débridement was required before complete healing was obtained.

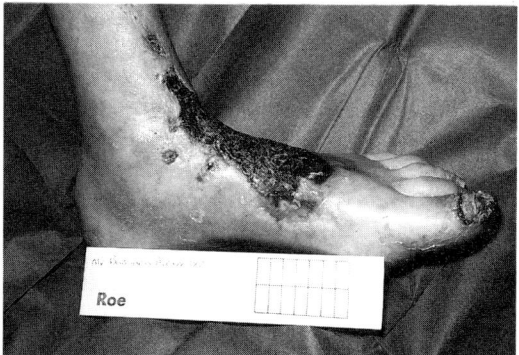

Fig. 21.2. Illustration of extensive gangrene and skin loss involving the dorsum of the foot which, following a distal bypass graft, débridement and skin grafting, eventually healed completely.

Progressive chronic peripheral vascular disease (Figs 21.1, 21.2)

In patients with progressive peripheral vascular disease, whether due to thromboangiitis obliterans, which is extremely rare, diabetes with peripheral vessel calcification combined with premature atheroma or in patients with progressive atheromatous disease, the development of the disease process involves more and longer lengths of the vessels in the lower limb. Eventually it results in the blood flow being reduced to the point where the distal part of the extremity, usually the foot,

develops rest pain, ischaemic ulcers and ultimately ischaemic gangrene, progressing proximally from the toes or heel to the forefoot, sole of the foot and lower leg (Fig. 21.3). Once there is established gangrene of the forefoot, the sole of the foot or a large volume of the heel, the indication is for primary amputation without consideration of restorative surgery. Ill-considered vascular surgery would leave a useless extremity with function worse than that which could be achieved with an amputation and prosthesis (Powell, *et al.*, 1984; Robbs *et al.*, 1984; Ouriel *et al.*, 1988).

Occasionally, the limb affected by vascular disease is also the site of other disease or

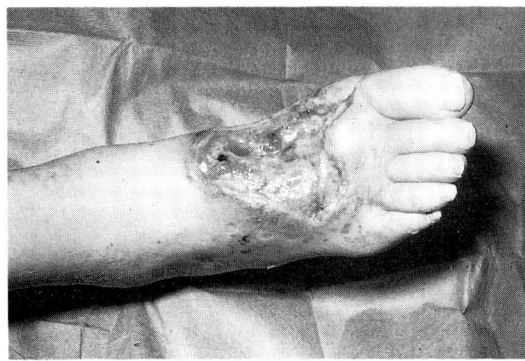

Fig. 21.3. Ulceration of the dorsum of the right foot in a patient with peripheral vascular disease which healed completely following débridement, skin graft and a femoro-distal bypass graft.

deformity, making primary amputation the best option. This may be the case in hemiplegia after a stroke. If the patient is not ambulant and already bed-bound with intercurrent disease, any intervention must be critically assessed against the option of pain relief and purely palliative medication. In the case of an alert well-motivated bedridden patient with a useless and painful, often ulcerated, extremity, an amputation might be regarded as a kindness. However, very often, patients in this situation will refuse surgery and opt for medical control of their pain and symptomatic treatment only.

Whether arterial reconstruction should ever be considered to improve the level of amputation is a highly controversial matter, be it angioplasty or bypass graft (Johansen *et al.*, 1981). If amputation surgery is performed and later fails, the patient is left with the need for a further amputation. Alternatively, it is possible to postulate a situation in which an easily remediable proximal occlusion could be treated simultaneously with the amputation procedure to obtain the benefit of a lower level of amputation.

Where the patient has severe rest pain, threatened gangrene of the fore foot, limited ulceration of the heel, instep or over the malleoli then arterial reconstruction should be the first option (Tunis *et al.*, 1991).

Whether the operative procedures necessary for restorative vascular treatment and the physical demands entailed would determine a decision in favour of amputation remains a highly con-troversial point. Despite taking several hours, it is contended that the operative risk of a complex vascular procedure in an attempt at limb salvage is less than most major amputations (Bunt *et al.*, 1990).

The results of *in situ* vein bypass procedures to circumvent the occluded vessels indicate that they present the best options, whether it be from the femoral artery in the groin, the upper popliteal or even the lower popliteal artery. The reversed saphenous vein, if necessary harvested from the opposite leg or the cephalic vein from the arm, is an acceptable substitute. As an alternative, the use of a human umbilical vein bypass graft (Dardik *et al.*, 1982) or a reinforced polytetrafluoroethylene graft combined with a Miller cuff, Taylor patch, autogenous vein or a composite segment with autogenous vein will allow revascularization of the otherwise 'lost' ischaemic extremity, albeit with an inferior patency rate.

Profundoplasty and lumbar sympathetic interruption procedures remain of uncertain status and are becoming increasingly superseded by bypass grafts. A not uncommon situation is for a proximal balloon angioplasty to be accompanied by a distal *in situ* vein graft constituting a relatively atraumatic procedure well tolerated by the frail and weak patient.

The effect of a failed graft on the ultimate amputation level is not yet resolved. The suspicion that, after a failed graft, a higher level of amputation is needed has been made but the results of a number of studies have produced conflicting results (Fig. 21.4). It is suggested that a failed distal anastomosis is unlikely to result in a higher

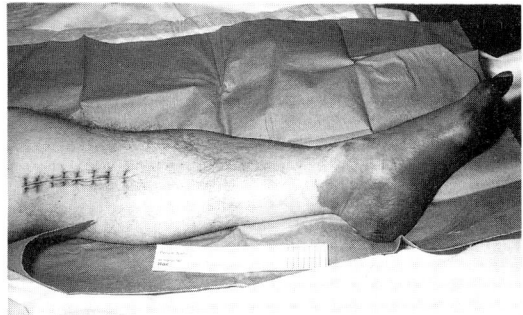

Fig. 21.4. A failed limb revascularization with a femoro-distal popliteal bypass graft necessitating a transtibial amputation: in this case, without prejudice to the amputation level from the vascular surgery.

amputation level (Hunter, 1978; Kazmers *et al.*, 1980; Dardik *et al.*, 1982; Samson *et al.*, 1982; Haimovici, 1985; Tsang *et al.*, 1991).

However, in the final analysis, it is the duration of the 'success' of the revascularization procedure which is significant in competition with a primary amputation that is most cogent. Gregg (1985) stated that to be cost effective and clinically effective, a revascularization procedure must obtain at least 1 year of successful, asymptomatic function for the limb and, ideally, should obtain asymptomatic function for the rest of the patient's lifespan. All the revascularization procedures have a limited patency rate which brings an end to the improvement at some stage and, while improved techniques and materials are increasing this duration year by year, so also is the progress in medical treatment of disease allowing the patient's lifespan also to be extended. It may be that for this reason the overall rate of amputations remains unchanged as observed in a community setting by Tunis *et al.* (1991), who noted that in Maryland the use of percutaneous angioplasty resulted in an increase in the number of patients having arterial imaging, an increase in the amount of peripheral vascular surgery but no change in the overall rate of amputation surgery. However, the carefully compiled statistics of the Danish Amputation Register (Chapter 4) demonstrate a decrease in major amputations of 27% in the decade 1980–1990 and that the improvements in the medical treatment, vascular surgery (Chapter 22) and a publicly supported podiatric service is considered to be of importance. Clearly, the actuarial considerations and quality of survival following revascularisation surgery are all important. The subject of cost has been extensively considered by Gupta *et al.* (1988), who demonstrated that the overall charges and costs accumulated by a patient having a primary amputation with supply of a prosthesis, rehabilitation and on-going care were compared with the costs of patients following a revascularization procedure and rehabilitation, with the conclusion that the cost is approximately equal. In either case, limb threatening atherosclerosis and its treatment by vascular reconstruction or by primary transtibial amputation are extremely expensive and they conclude that decisions regarding the economics of treatment should be based on patient factors and quality of life and not on cost alone.

The scales become progressively weighted in favour of revascularization as improved techniques bring better results with long-term relief of rest pain, cessation of ischaemic tissue loss and normal function. The problem which is as yet unresolved is in the group of patients treated by revascularization surgery with regard to the management of the limb and the restitution of a functional extremity. Gangrenous toes, if mummified and uninfected, can be allowed to proceed to auto-amputation and indeed all five toes, if necessary, can be treated in this way. With special footwear, even part of the forefoot can be allowed to separate in the same way. This is, of course, a socially unpleasant treatment in that the gangrenous tissue is offensive both in smell and appearance and episodes of cellulitis may result and require treatment with antibiotics and occasionally even hospitalization. If the separation involves the forefoot, the remaining wound takes an appreciable period to heal. If only one digit is affected and one associated metatarsal then community care may allow the defect to be dressed and treated until healing is complete, but even with a perfect blood supply this may take several months to achieve. Slightly better results can be achieved by an amputation performed with meticulous debridement at the time of the vascular reconstruction, which is possible provided there is no established infection or non-biological graft material at the site. A further point in favour of an *in situ* or an autogenous vein bypass graft is of particular importance in diabetic patients, where an unhealed defect in the foot remains a portal of entry for a possibly serious systemic infection. A well cleaned amputation of multiple toes or metatarsal heads, or even a Lisfranc or ray amputation may heal uneventfully within weeks of the primary vascular procedure and is an ideal to be sought. To achieve this, more of the foot may have to be resected than would be necessary with auto-amputation and its associated long-term healing. The problem of the time taken for the foot to heal and become functional increases as more of the volume of the foot is lost by gangrene established prior to the revascularization procedure.

The tissue loss may be precipitated by what would otherwise be seen as minor trauma, such as pressure over a malleolus, or the heel, a burn on the instep, a diabetic perforating ulcer, an ulcerated bunion, or badly advised chiropody producing infective gangrene in a critically ischaemic foot

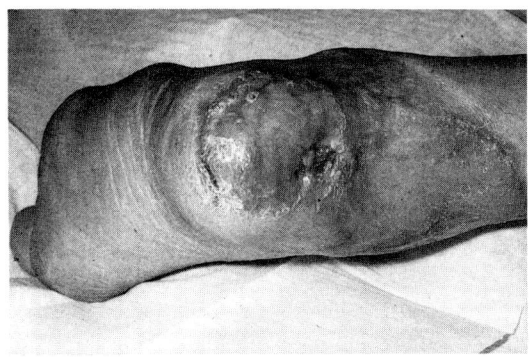

Fig. 21.5. A successful reconstruction of a gangrenous heel following a femoro-popliteal bypass graft in a diabetic patient with excision of half the calcaneum and skin grafting with limb salvage achieved.

(Fig 21.5). Hospital records contain cases where part of a calcaneum has been removed, dead tissue débrided, granulation tissue allowed to develop with subsequent skin grafting and function successfully restored; again, where a burned instep has been debrided and grafted; where a Lisfranc amputation or a ray amputation has been performed following revascularization. Unfortunately the process of healing with occasionally successive débridements and multiple skin grafts has required up to as many as 12 or more weeks in hospital, bringing the virtues of a primary amputation and prosthetic fitting into closer focus (Mikulin *et al.*, 1986).

The diabetic patient remains a problem, particularly with regard to healing of the extremity after revascularization (Fig 21.6). Where there is proximal atheroma that has been treated by a bypass graft, the foot can remain affected by the other manifestations of diabetes. The microangiopathy remains, as does the problem of hyperglycaemia. Peripheral neuritis may produce trophic changes and the diminished resistance to infection may make it extremely difficult to eliminate continued sepsis and tissue destruction. In some patients, revascularization alone, debridement, control of the diabetes and antibiotic treatment has not always achieved a satisfactory resolution of the problem or the avoidance of major amputation (Desai *et al.*, 1980). Unfortunately, it is impossible to predict which patients will fail to respond to revascularization. Certainly there is a poor prognosis in patients who continue to smoke (Myers *et al.*, 1978) and they would clearly be best treated by

primary amputation, usually at the transtibial level. However, for the present, revascularization (Cossart *et al.*, 1983) should be attempted if it is likely to be technically satisfactory but the patient must be warned that the way in which diabetes behaves may determine whether limb salvage will be successful. Meticulous débridement, appropriate and adequate antibiotic medication, careful control of the diabetes with an insulin perfusion, continued for a considerable period if necessary, may provide the revascularized diabetic foot the best chance of healing.

Clearly, no general rules can be laid down about the choice between revascularization and primary amputation but it is as well to be cognisant of the limitations of revascularization when a prolonged

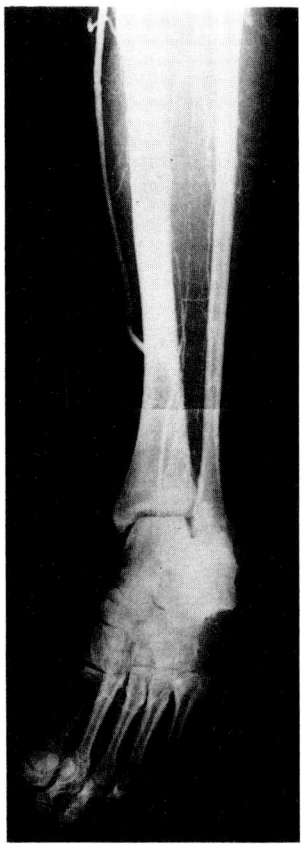

Fig. 21.6. A postoperative arteriogram of a femoro-distal bypass graft into the posterior tibial artery revascularizing a diabetic foot with ischaemic gangrene of the hallux which has been partially amputated at the time of the bypass.

healing programme is necessary before full function is restored. The emphasis is on more accurate assessment of the candidates for peripheral vascular reconstruction.

Even when technically successful, a distal bypass graft may not necessarily successfully revascularize the extremity and we have experience of one patient with an entirely satisfactory femoro-tibial bypass graft in whom the fore foot, against all predictions, remained poorly perfused with unacceptable rest pain and an unhealed forefoot following digital amputation and regrettably a transtibial amputation had to be performed despite a functioning graft. This example represents poor selection of the vascular reconstruction procedure reflecting over-optimistic interpretation of the angiograms obtained at the time of the bypass procedure.

In conclusion, all agree that in severe injury where vascular damage has occurred, a vascular surgeon should be consulted to ensure the best chance of revascularization. Indeed every patient with severe ischaemia should be referred to a centre where vascular surgical expertise is available with all facilities for vascular imaging, thrombolytic therapy, angioplasty and bypass surgery. Similarly, with chronic ischaemia progressing to severe and critical levels, the patient should be assessed by a vascular surgeon for accurate imaging, not necessarily invasive, with a view to revascularization but who is well aware that primary amputation may yet prove to be the more realistic overall solution. There is no doubt at all that every patient should have the option of the revascularization procedure and the more frail, the more elderly and disabled the patient is, the more likely there will be benefit from a well-judged and effective revascularization.

References

Bunt, T.J., Manship, L.L., Bynde, R.P.H. and Haynes, J.L. (1984) Lower extremity amputation for peripheral vascular diseases, a low risk operation. *Ann. Surg.* **50(11)**, 581–584.

Cossart, L. de., Randall, P., Turner, P., and Marcuson, R.W. (1983) The fate of the below knee amputee. *Ann. R. Coll. Surg.*, **65**, 230–232.

Dardik, H., Kahn, M., Dardik, I., Sussman, B. and Ibrahim, I.M. (1982) Influence of failed vascular bypass procedures on conversion of below knee to above knee amputations levels. *Surgery*, **91(1)**, 64–69.

Desai, Y., Robbs, J.V. and Keenan, J.P. (1986) Staged below knee amputations for septic peripheral ischaemia. *Br. J. Surg.*, **73(5)**, 392–394.

Ebskov, B. (1980) Incidence of re-amputation and death after gangrene of the lower extremity. *Prosthet. Orthot. Int.*, **4(2)**, 77–80.

Ebskov, B. and Ebskov, L. (199) The Epidemiology of Lower Limb Amputation.

Gregg, R.O. (1985) Bypass or amputation, concomitant review of bypass arterial grafting and major amputation. *Ann. J. Surg.*, **149(3)**, 397–402.

Gregson, R.H.S. (1991) (Editorial) Thrombolysis for peripheral arterial occlusion, *Br. J. Hosp. Med.*, **46**, 79.

Gupta, S.K., Veith, F.J., Ascer, E., Flores, S.A. and Gleidman, M. (1988) Cost factors in limb threatening ischaemia due to infra inguinal arteriosclerosis. *Eur. J. Vasc. Surg.* **Jan 2(8)**, 151–154

Gwynn, B.R., Shearman, G.P. and Sims, M.H. (1989) Amputation for peripheral vascular disease (letter). *Br. J. Surg.*, **76(6)**, 654.

Haimovici, H. (1985) Failed grafts and level of amputation. *J. Vasc. Surg.*, **2(3)**, 371–374.

Howe, H.R., Poole, G.V. Jr. Hansen, K.J. *et al.* (1987) Salvage of lower extremities following combined orthopaedic and vascular trauma. A predictive salvage index. *Am. Surg.*, **53**, 205–208.

Hunter, G. (1978) Major amputation following reconstructive procedures. *Can. J. Surg.*, **78 21(5)**, 456–458.

Johansen, K., Burgess, E.M., Zorn, R., Holloway, G.A., Dee, M. and Bach, A. (1990) Improvement of amputation level by lower extremity revascularisation. *Surg. Gynaecol. Obstet*, **153(5)**, 707–709.

Kazmers, M., Satiana, B. and Evans, W.E. (1980) Amputation level following unsuccessful distal limb salvage operations. *Surgery*, **87(6)**, 683–687.

Kikta, M.A. (1987) Mortality and limb loss associated with infected infrainguinal bypass grafts. *J. Vasc. Surg.*, **5(4)**, 566–571.

Ljungman, C., Adami, H.O., Bergquist, D., Sparen, P. and Bergstrom, R. (1991) Risk factors for early lower limb loss after embolectomy for acute arterial occlusion: a population based case control study. *Bri. J. Surg.*, **78**, 1482–1485.

Mikulin, T., M Hopkinson, B.R. and Makin, G.S. (1986) Major amputations compared with graft occlusion as the end point for assessing results of bypass surgery in lower limb ischaemia. *Br. J. Surg.*, **73(3)**, 200–203.

Myers, K.A., King, R.B., Scott, D.F., Johnson, N. and Morris, P.J. (1978) The effect of smoking on the late potency of arterial reconstruction in the legs. *Br. J. Surg.*, **65**, 267–271.

Odland, M.D., Gisbert, V.L., Gustilo, R.B., Ney, A.L., Blake, D.P. and Bubrick, M.P. (1990) Combined orthopaedic and vascular injury in the lower extremities; Indications for amputation. *Surgery*, **108(4)**, 660–666.

Ouriel, K., Fiore, M.D. and Geary, J.E. (1988) Limb threatening ischaemia in vascularly compromised patients – amputation or revascularisation. *Surgery*, **104(4)**, 667–672.

Powell, T.W., Burnham, S.J. and Johnson, G. (1984) Second leg ischaemia – lower extremity bypass versus amputation in patients with contra-lateral lower extremity amputation. *Ann. Surg.*, **50(11)**, 577–580.

Pozo, J.L., Powell, B., Andrews, B.G., Hutton, P.A.N. and Clarke, J. (1990) The timing of amputation for lower limb trauma. *J. Bone Joint Surg.*, **72(2)**, 288–292.

Robbs, J.V., Human, R.R. and Rajaruthnam, P. (1984) Bypass versus primary major amputation in patients with femoro-popliteal distal disease and ulcerated limbs. *S. Afr. Med. J.*, **66(21)**, 809–12.

Russell, W.L., Sailors, D.M., Whittle, T.B., Fisher, D.F. and Burns, R.P. (1991) Limb salvage versus traumatic amputation. *Ann. Surg.*, **213(5)**, 473–481.

Samson, R.H., Gupta, S.K., Scher, L.A. and Veith, F. (1982) Level of amputation after failure of limb salvage procedures. *Surg. Gynaecol. Obstet.*, **154(1)**, 56–58.

Tsang, G.M.K., Crowson, M.C., Hickey, W.C. and Simms, M.H. (1991) Failed femorocrural reconstruction does not prejudice amputation level. *Br. J. Surg.*, **78**, 1479–1481.

Tunis, S.R., Bass, E.B. and Steinberg, E.P. (1991) The use of angioplasty, bypass surgery and amputation in the management of peripheral vascular disease. *N. Eng. J. Med.*, **325(8)**, 556–562.

Wagner, H.J. and Stark, E.E. (1992) Acute embolic occlusion of infra-inguinal arteries: Percutaneous aspiration embolectomy in 102 patients. *Radiology*, **182**, 403–407.

Woodall, J. (1639) *The Surgeon's Mate.*

22

Distal bypass: the influence on major amputations

P. E. Holstein

Introduction

Nationwide surveillance by the Danish Amputation Register (DAR) has shown that the number of major amputations due to vascular disease has decreased in recent years (Chapter 4; see also Ebskov, 1991). This paper intends to discuss possible reasons for this favourable development and, further, as to whether arterial reconstructions actually save limbs or does such activity simply postpone inevitable amputations. Finally, the risk of converting transtibial amputation levels to transfemoral levels when arterial reconstructions fail will be addressed.

Pattern of amputations in a city community hospital

In the late 1970s the *in-situ* saphenous vein bypass was reintroduced (Leather *et al.*, 1979) as a durable arterial conduit to tibial and pedal arteries albeit with improved techniques. The vascular unit in Bispebjerg Hospital in Copenhagen began to use this method in 1988 and it was noted that the number of patients referred for amputation was reduced significantly. For this reason the files on all patients subjected to leg amputation in the hospital during the period 1981 through 1990 were scrutinized (Petersen *et al.*, 1994). This survey demonstrated that there were 1383 amputations including re-amputations resulting in loss of the leg in 1167 cases, i.e. final level amputations. Of

these 665 were in men, median age 71 years (range 38–97 years) and 502 were in women, median age 76 (37–97 years). Out of the 1167 final level amputations 1001 patients were admitted from

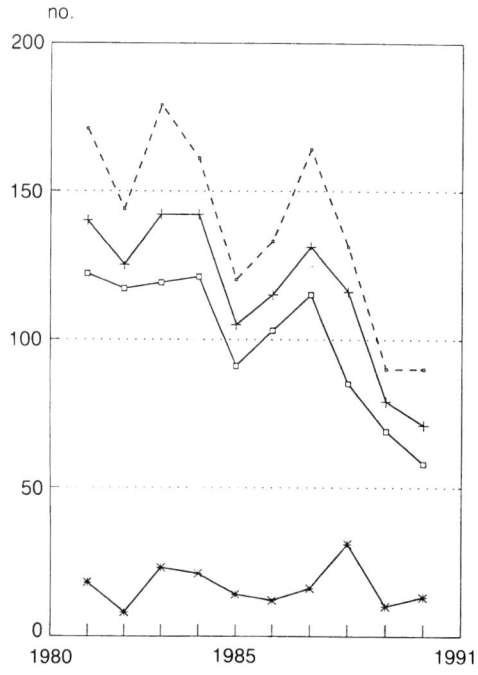

Fig. 22.1. Major amputations during a 10-year period at Bispebjerg Hospital. From Petersen *et al.* (1994) with permission.

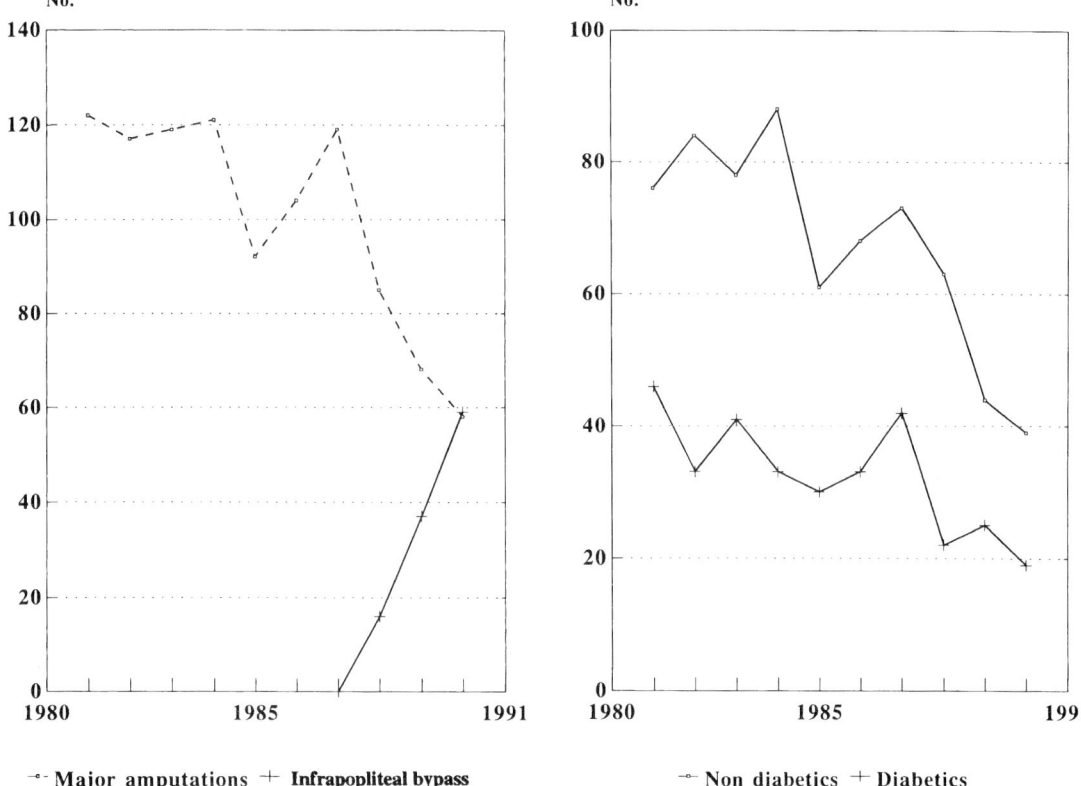

-°- **Major amputations** -+- **Infrapopliteal bypass**

Fig. 22.2. Decrease in major amputations synchronously with increase in infra-popliteal bypass operations. From Petersen *et al.* (1994) with permission.

-°- **Non diabetics** -+- **Diabetics**

Fig. 22.3. Decrease in major amputations in diabetic patients as well as non-diabetics.

their own home and 166 were transferred from other wards, e.g. dealing with the chronic sick. Of those admitted from home there were 482 transtibial (TT) and 476 transfemoral (TF) amputations, eight ankle disarticulations, 33 knee disarticulations and two hip disarticulations.

Figure 22.1 demonstrates that the number of major amputations decreased by about 50% during the period studied. The total number decreased from 191 to 90 per year, re-amputations from 35 to 21 and the final amputations in those patients admitted from home from 122 to 58. The number of amputations in those patients, transferred from other wards, however, remained fairly constant over the years. These last figures are shown because chronic ward patients in the centre of Copenhagen are rarely considered for limb salvage surgery due to chronic mental organic deficiency.

Figure 22.2 shows the number of major amputations in those patients admitted from home as compared with the number of infra-popliteal bypass operations, which were initiated in 1988. The reduction in amputation numbers becomes more marked throughout the same period.

Figure 22.3 show that major amputations in diabetic patients as well as in non-diabetic patients decreased by about 50%.

Figure 22.4 shows the number of transfemoral (TF) and transtibial (TT) amputations during the period studied. In the last 3 years, i.e. the period where infra-popliteal bypass surgery was employed, the ratio of TT/TF amputations was 80/118 = 0.67 as compared to 402/358 = 1.12 for the years 1981 through 1987 ($P<0.005$). It can be calculated that there were some 24 TF amputations more in the last period than would have been

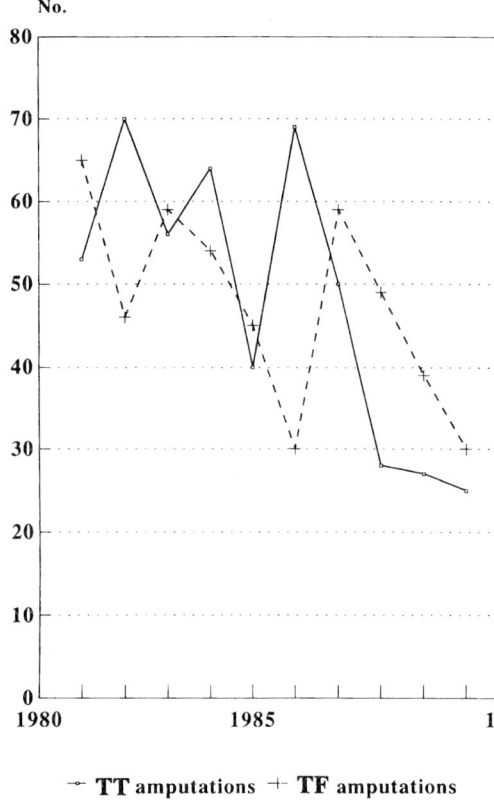

Fig. 22.4. The number of TT and TF amputations per year. From Petersen *et al.* (1994) with permission.

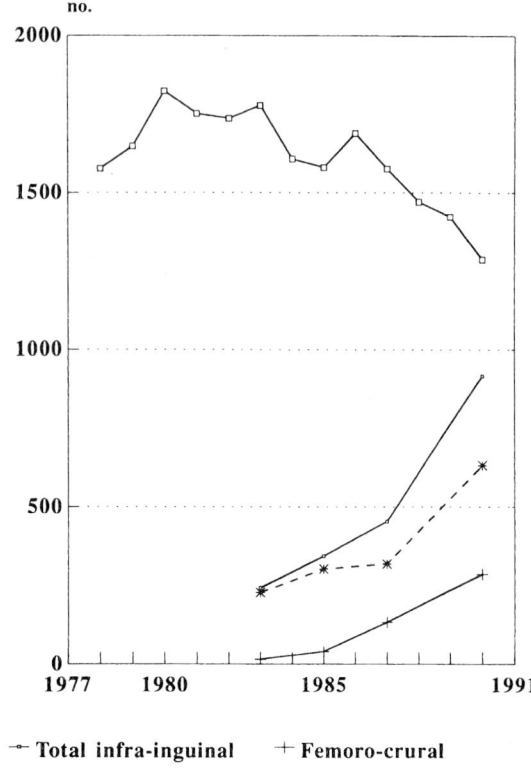

Fig. 22.5. Decrease in amputations in Denmark synchronously with the increase in distal bypass operations. From Ebskov *et al.* (1994) with permission.

expected from the ratio of the previous years. This should be seen against the background of the decrease in the number of legs amputated which in the same years amount to about 140.

Pattern of amputations in Denmark

The Danish Amputation Register has supplied evidence for decreasing numbers of amputations in Denmark from 1777 in 1983 (34.5/100 000) to 1288 in 1990 (25/100 000), i.e. a decrease of 27.5% (Fig. 22.5). During the same period the vascular surgical activity increased significantly from 34 per 100 000 to 74 per 100 000 inhabitants (Danish National Health Board Report (Vascular Surgery), 1991). In particular, it is of interest to

note that infra-inguinal arterial reconstructions increased from 243 to 916, comprising femoro-popliteal reconstructions as well as bypass operations to crural and pedal arteries (Fig. 22.5). This latter group increased from 15 to 286 per year (Ebskov, Schroeder and Holstein, 1994).

Discussion

In spite of evidence of the efficacy of various arterial reconstruction procedures in critical limb ischaemia, those reports on large surveys have hitherto failed to demonstrate a significant reduction in major amputations. On the contrary, reports from Sweden (Liedberg and Persson, 1983), the United States (Ernst *et al.*, 1987), Finland, (Pohjolainen and Alaranta, 1988), Denmark, (Ebskov,

1988), the Netherlands (Dwars, 1990) and the United Kingdom (Harris and Moody, 1990) have shown that the numbers of amputations have been increasing during the 1970s with a tendency to level off in the 1980s. In direct contrast the DAR has shown a significant decrease of 27.5% during a period of 8 years (Ebskov *et al.*, 1994).

Whether or not this decrease is caused by vascular surgical interventions should be discussed in detail. The possibility exists that improved risk factor modifications such as cessation of smoking, exercise, less intake of fat and treatment of hypertension might be important. However, there is no evidence of any decrease in the prevalence of gangrene and the elderly part of the population continues to increase. Several drugs have been introduced in the medical treatment of peripheral vascular disease, but no substantial benefit has been demonstrated.

A report from Birmingham (Hickey *et al.*, 1991) showed that of a consecutive series of 329 limbs with critical ischaemia undergoing revascularization as many as 73% required an infra-popliteal bypass and 20% a femoro-popliteal bypass. The message is clear: in patients with 'end stage' ischaemia the majority require arterial reconstruction not previously available due to poor patency of the grafts. The patency of these grafts is now, fortunately, satisfactory and typically the chance of successfully retaining the limb 2 years after an infra-popliteal bypass is about 90%. This is the result of the reappraisal of the *in situ* saphenous vein technique, although it is likely that several technical improvements integral to the operative procedure, such as vastly improved lighting, magnifying glasses and more meticulous techniques, are important contributory factors.

This chapter serves to demonstrate that the decrease in amputations performed in a city community hospital and reflected in the national statistics is synchronous with the increasing use of pertinent types of revascularization procedures, and suggests a direct relationship. Moreover, it is demonstrated that the decrease in amputation rates in the various communities in Denmark is statistically correlated to the increase in vascular surgery activity (Ebskov, Schroeder and Holstein, 1994).

Furthermore, in diabetic patients the decrease in amputation rates in the city community hospital took place in conjunction with the introduction of bypass surgery to crural and pedal arteries. The large majority of diabetic patients were previously considered unsuitable for vascular reconstruction due to occlusions in crural and pedal arteries. Today we can see that this 'diabetic' pattern of occlusive arterial disease is almost the same as that found in the old non-diabetic population (Mantoni and Holstein, 1990). However, the occlusive arterial process in both groups of patients typically leaves one or more distal arterial segments intact and perfectly suitable as outflow channels for arterial grafts. This lesson has been learned following improved arteriograms, which today can display crural and pedal arteries in such detail in all those patients considered for distal arterial grafting that terms such as 'small artery disease' and 'microangiopathy', inherently tinged with notions like 'no treatment by surgical means', need no longer exist in the considerations of gangrene in diabetics.

It should, however, be emphasized that amputations in diabetic patients began to decrease in Denmark some years before the observed increase in the number of vascular reconstructions (Ebskov, 1991). This points to factors other than vascular surgery and there is probably an increasing understanding of the special risk factors in diabetic cases. The National Health Board in Denmark in 1981 recommended centralization of treatment of foot problems. A subsidy available to the public of 60% of the actual costs for regular foot care by authorized chiropodists has been introduced and the number of chiropodists has correspondingly increased.

In considering those factors influencing amputation rates, other reports attempting to relate a reduction with increasing arterial reconstructions can be identified. The most significant is that of Veith *et al.* (1990) who have reported that amputations in first-time threatened limbs but not including those limbs treated solely by thrombo-embolectomy decreased from 59% to 29% during a 14-year period. This was achieved by aggressively attempting revascularization in most patients whose limbs were threatened by ischaemia. Our results confirm this experience, although our reports include not only 'first-time' ischaemia amputations but all amputations in dysvascular patients.

A 25% reduction in amputations has been reported in two other series. One centre ascribed the reduction observed over a 5-year period to increasing use of femoro-popliteal bypass procedures (Jeans *et al.*, 1986). In the other the reduction took place over a 3-year period after the establishment of a new vascular unit in an area formerly geographically remote to vascular centres (Lindholt *et al.*, 1994). Although difficult to quantitate in relation to limb salvage it has to be said that during the 1980s vascular intervention by angioplasty and by fibrinolytic substances, sometimes in combination (Jorgensen *et al.*, 1991), has expanded the armamentarium of vascular units.

The classical objections to vascular surgery in critical limb ischaemia are that vascular intervention simply postpones inevitable amputation and that failed vascular surgery jeopardizes subsequent major amputation in that TT levels are converted to TF levels. As to the first objection, the decrease in absolute amputation rates is now substantial over several years. Since the risk of graft failure is maximum during the first month after reconstruction, it can be stressed that amputations are avoided and not just postponed.

As to the second objection, most authors dealing with this problem find that there is a risk of converting TT amputations to TF amputations in graft occlusion (Kazmers *et al.*, 1980; Sethia *et al.*, 1986; Stern, 1988), whereas other investigators do not confirm this finding (Larsson and Risberg, 1988; Cook *et al.*, 1992). In the literature the TT to TF ratio in primary amputated patients is often compared with the same ratio in patients with failed reconstructions. The possibility exists, however, that the ratio in both groups is unfavourably influenced by vascular surgery activity, i.e. TT levels are lost in the group with graft occlusions and the primary amputated group consists of patients with a very poor health and deemed suitable only for TF amputation. This is turn highlights the need to consider the number of saved limbs, a subject rarely addressed.

In the present series recording the experience of a city community hospital there were 24 TT level amputations more than one would expect following the introduction of the infra-popliteal bypass. The number of avoided amputations in this period was about 140. These figures are necessarily incomplete, because the number of saved limbs from other vascular interventions is not included (and cannot be defined) and because patients with revascularization might become part of that group of patients ultimately deemed suitable for a TT amputation.

In summary, the evidence suggests that decreasing amputation rates are a consequence of improved and more frequently used distal bypass operations, in particular those performed on crural and pedal arteries. Since these techniques are not yet employed as often as is appropriate, it is likely that amputation rates will decrease even further. Nevertheless, those patients undergoing amputations will probably present more difficult problems regarding surgery as well as rehabilitation.

References

Cook, T.A., Davies, A.H., Horrocks, M. and Baird, R.N. (1992) Amputation level is not adversely affected by previous femoro-distal bypass surgery. *Eur. J. Vasc. Surg.*, **6**, 599–601.

Dwars, B.J. (1990) *Amputation Level Selection*. Thesis, VU University Press, Amsterdam, p. 12.

Ebskov, B. (1988) Trends in lower extremity amputation (Denmark, 1978–1983). In *Amputation Surgery and Lower Limb Prosthetics*. (eds G. Murdoch and R. Donovan) Blackwell Scientific: Oxford, 3–8.

Ebskov, L.B. (1991) Epidemiology of lower limb amputation in diabetics in Denmark 1980 to 1989. *Int. Orthop.* (SICOT) **15**, 285–288.

Ebskov L.B., Schroeder T.V., Holstein P. (1994) Epidemiology of major amputations. The influence of infra-popliteal bypass. *B. J. Surg.*, **81**, 1596–1599.

Ernst, C.B., Rutkow, I.M., Cleveland, R.J., Folse, J.R., Johnson, G. and Stanley, J.C. (1989) Vascular surgery in United States. *J. Vasc. Surg.*, **6**, 611–621.

Harris, P. and Moody, P. (1990) Amputations. In *Critical Leg Ischaemia. Its Pathophysiology and Management* (eds J.A. Dormandy and G. Stock), Springer: Berlin, pp. 87–95.

Hickey, N.C., Thomason, I.A., Shearman, C.P. and Simms, M.H. (1991) Aggressive arterial reconstruction for critical lower limb ischaemia. *Br. J. Surg.*, **78**, 1476–78.

Jeans, W.D., Danton, R.M., Baird, R.N. and Horrocks, M. (1986) The effects of introducing balloon dilatation into vascular surgical practice. *Br. J. Radiol.*, **59**, 457–459.

Jorgensen, B., Tonnesen, K.H., Nielsen, J.D., Holstein, P., Bulow, J., Jorgensen, M. and Andersen, E. (1991) Segmentally enclosed thrombolysis in percutaneous transluminal angioplasty for femoropopliteal occlusions: a report from a pilot study. *Cardiovasc. Intervent. Radiol.* **14**, 293–298.

Kazmers, M., Satiani, B. and Evans, W.E. (1980) Amputation level following unsuccessful distal limb salvage operations. *Surgery*, **87**, 683–687.

Larsson, P.A. and Risberg, B. (1988) Amputations due to lower-limb ischaemia. *Acta Chir. Scand.*, **154**, 267–270.

Leather, R.P., Powers, S.R. and Karmody, A.M. (1979) Reappraisal of the in situ saphenous vein arterial bypass: Its use in limb salvage. *Surgery*, **86**, 453–461.

Liedberg, E. and Persson, B.M. (1983) Increased incidence of lower limb amputation for arterial occlusive disease. *Acta*

Orthop Scand. **54,** 230–234.

Lindholt, J.S., Bovling, S., Fasting, H. and Winther Henneberg, E. (1994) Vascular surgery reduces the frequency of lower limb major amputations. *Eur. J. Vasc. Surg.*, **8,** 31–35.

Mantoni, M.Y. and Holstein, P. (1990) Aorto-femoral digital subtraction angiography in old patients with symptoms of peripheral arterial disease. *Danish Med. Bull.*, **37,** 192–193.

Petersen, A.E., Olsen, B.B., Krasnik, M., Ebskov, L.B., Leicht, B.P., Sager, P., Helgstrand, U. and Holstein, P. (1994) Halving the number of leg amputations. The influence of infra-popliteal bypass. *Eur. J. Vas. Surg.* **8,** 26–30.

Pohjolainen, T. and Alaranta, H. (1988) Lower limb amputation in southern Finland 1985–86. *Prosthet. Orthot. Int.*, **12,** 9–18.

Report from the Danish National Health Boards group of experts on vascular surgery (1991) Danish National Health Board 1–49.

Sethia, K.K., Berry, A.R., Morrison, J.D., Collin, J., Murie, J.A. and Morris, P.J. (1986) Changing pattern of lower limb amputation for vascular disease. *Br. J. Surg.*, **73,** 701–703.

Stern, P.H. (1988) Occlusive vascular disease of lower limbs: diagnosis, amputation surgery and rehabilitation. *Am. J. Phys. Med. Rehab.*, **67,** 145–154.

Veith, F.J., Gupta, S.K., Wengerter, K.R., Goldsmith, J., Rivers, S.P., Bakal, C.W., Dietzek, A.M., Cynamon, J., Sprayregen, S. and Gliedman, M.L. (1990) Changing arteriosclerotic disease patterns and management strategies in lower limb threatening ischaemia. *Ann. Surg.*, **212,** 402–413.

23

Amputation following vascular surgery

A. S. Jain

The basic principle of amputation surgery is that whatever the causal condition, if a functional, painless limb can be salvaged by any other means then that procedure should be preferred to amputation.

It is becoming increasingly clear that more and more distal bypasses are undertaken in an attempt to save limbs. Any procedure which helps to salvage the limb is an important one and should be encouraged. Taylor, Hamre, Dalman *et al.* (1991) from the Oregon Health Sciences University reviewed 498 patients with 627 critically ischaemic legs. They highlighted that patent revascularization had a significant influence on limb salvage but that renal failure was one of the factors which put these limbs at risk. Miller, Dardik, Woolridger *et al.* (1991) from the Vascular Surgical Service, Engelwood Hospital, New Jersey reviewed 160 trans-metatarsal amputations over a 12-year period. Three groups were identified: (1) those with non-reconstructable vessel disease; (2) trans-metatarsal amputation in conjunction with distal revascularization and (3) reconstructable vessel disease but trans-metatarsal amputation performed without simultaneous revascularization. This well conducted comparative study demonstrated that a distal bypass performed at the same time can save a trans-metatarsal amputation provided vascularization is patent. However, there remain patients who will require primary amputation and who should be assessed thoroughly prior to any projected distal bypass. A successful distal bypass

procedure, however, does improve the patient's quality of life and preserves the limb.

It is important to realize that injudicious attempts at arterial reconstruction when amputation appears inevitable may adversely affect the subsequent level of amputation thus jeopardizing rehabilitation. In Dundee we surveyed 192 consecutive amputations which were carried out during 1990 and 1991. There were 54 transfemoral, 133 transtibial and five at other levels. Of these 72 patients had previous vascular surgery prior to amputation (37.5%) and 120 had no previous vascular surgery. The causal condition in both groups was similar, constituting a high percentage of vascular disease and diabetes. The two groups were compared with respect to (1) site of amputation; (2) smoking habits; (3) final level of amputation; (4) wound healing and (5) revision surgery (Table 23.1).

The study clearly shows that smokers are more likely to undergo vascular surgery prior to amputation. More transfemoral amputations are carried out in the vascular surgery group. There was very little difference in primary wound healing but there was a higher rate of delayed wound healing in the vascular surgery group and also a slightly higher rate of revision.

These findings support the previously published results from Sethia, Berry, Morrison *et al.* (1986), who reported that 58% of their patients required reamputation following previous vascular surgery. In that series it was also concluded that they had to perform Gritti–Stokes and transfemoral amputa-

Table 23.1   Vascular review 1990–1991

Vascular surgery Total number in this group – 72		No vascular surgery Total number in this group – 120	
Diagnosis			
PVD	50	PVD	58
Diabetes	19	Diabetes	38
Others	3	Others	24
Site of amputation			
Left	45	Left	72
Right	27	Right	48
Smoking habits			
Smokers	53	Smokers	67
Non-smokers	19	Non-smokers	53
Level			
TF	25	TF	29
TT	43	TT	90
Others	4	Others	1
Ratio TF:TT	25–43	Ratio TF:TT	29–90
	1:1.72		1:3.1
Wound healing			
Primary healing	46	Primary healing	84
Delayed healing	18	Delayed healing	25
Revision surgery local	3	Revision surgery local	8
Higher revision	5	Higher revision	3

tion rather than transtibial, but they fail to support these findings with reasons. Moreover, they fail to mention the methods used for level selection. Our experience over the years has shown that the acute vascular incident and renal failure probably lead to high levels of amputation, whereas chronic vascular disease provides the opportunity to save the knee joint.

It is important to highlight that even in this group of patients meticulous care in level selection cannot be over-emphasized.

References

Miller, N., Dardik, H., Woolridger, F. *et al.* (1991) Trans-metatarsal amputation; the role of adjunctive revascularization. *J. Vasc. Surg.*, **13** (5), 705–711.

Sethia, K.K., Berry, A.R., Morrison, J.D. *et al.* (1986) Changing pattern of lower limb amputation for vascular disease. Br. J. Surg., **73** (September), 701–703.

Taylor, L.M. J, Hamre, D., Dalman, R.L. *et al.* (1991) Limb salvage vs amputation for critical ischamia. The role of vascular surgery. *Arch.* Surg., **126** (10), 1257–1258.

Section 9

Trauma

24

Decision making in limb salvage and/or amputation of the lower limb following trauma

G. A. Hunter

'One hundred and fifty years ago, an open fracture was virtually synonymous with death, and generally necessitated amputation. Amputation itself carried with it a very high mortality, usually from hemorrhage and sepsis' (Colton, 1992). As recently as 1985, Lange *et al.* reported on 23 patients with open tibial fractures complicated by limb threatening vascular injuries; 14 (61%) limbs were ultimately amputated.

In the 1990s, why are we even considering amputation of a severely injured lower limb when there have been significant advances in the care of patients with multiple trauma? Surely such advances allow the surgeon to salvage a limb which would previously have been amputated (Fig. 24.1). Rapid air transportation and efficient resuscitation of the patient at the scene of the accident and in the trauma room undoubtedly save many lives.

On admission to any general hospital, the surgeon is now able to provide the patient improved care by:

(1) early wound irrigation and debridement;
(2) sensible use of a broad range of antibiotics;
(3) adequate blood and fluid replacement;
(4) a better understanding of the classification of fracture (Gustilo *et al.*, 1984; Tscherne, 1984) and the metabolic effects of trauma.

In specialized units, the patient will benefit from new techniques in stabilizing fractures combined with repair and reconstruction of the injured soft tissues. For up-to-date information, the reader is referred to recent texts:

(1) Internal and/or external fixation of fractures with early bone grafting (Behrens, 1992; Green, 1992)
(2) Wound coverage by skin grafting, myofascial, local and muscle flaps (Sherman and Ecker, 1992)
(3) Evaluation and treatment of vascular injuries (Felliciano, 1992)
(4) Treatment of nerve injuries and reconstruction by neurolysis, nerve repair or grafting and tendon transfer (Mackinnon and Dellon, 1988)
(5) Treatment of nonunion and malunion (Rosen, 1992)

Now that early rigid stabilization of open fractures is acceptable treatment, and with methods to treat vascular and soft tissue injuries, we can now salvage a limb; but at what cost to the patient, the family and society? Preserving a limb does not by itself guarantee a useful limb. In spite of repeated, lengthy and elaborate reconstructive operations, there may be severe limitation in function, pain, anaesthesia of the sole of the foot, chronic osteomyelitis, shortening, stiff joints and deformity. When these complications are combined, the patient will request an amputation after two or three years of useless treatment (Fig. 24.2). Hansen (1987) noted that protracted attempts at limb salvage may destroy a person physically, psychologically, socially and financially with adverse consequences for the entire family.

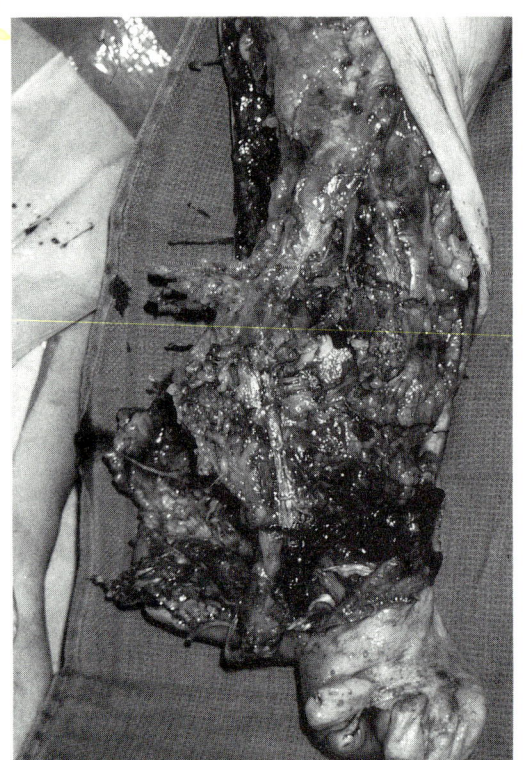

Fig. 24.1. A severe injury of the lower limb.

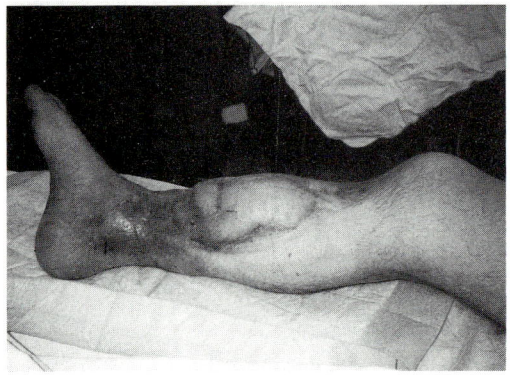

Fig. 24.2. Patient requests amputation after 14 operations.

Early amputation, when indicated, is safer for the patient with polytrauma, and is a rapid and less costly method of restoring function to the young adult than prolonged attempts at limb salvage. What are the results of early primary amputation for lower limb trauma? Millstein, Hunter and Kellan (1990) reported that 35/72 (49%) patients had a good result following early amputation of lower limbs with vascular injuries. Purry and Hannon (1989) studied 25 young adults who had amputations for severe lower limb trauma. They concluded that most healthy young adults who undergo transtibial amputation will rehabilitate well. Giorgiadis et al (1991) examined 18 primary amputees following trauma, and noted that 72% had pain and 39% had problems with skin breakdown. These figures are not surprising in that many of such patients have pre-existing social problems, and emergency amputations are often carried out by surgeons inexperienced in the care of the amputee.

Are there any methods we can use to predict whether the patient would be best treated by limb salvage or primary early amputation? Gregory *et al.* (1985) devised a grading system, 'Mangled Extremity Syndrome Index' (MESI) related to the injury severity score, shock, age and pre-existing disease of the patient, injury to skin, nerve, blood vessels and bone. They stressed that a lag time of up to 6 hours after injury was critical. Retrospective analysis suggested an MESI score of 20 was the dividing line below which functional limb salvage could be expected, and above which limb salvage was improbable. Howe *et al.* (1987) described a 'Predictive Salvage Index' (PSI) based on the level of arterial injury, the degree of bone and muscle injury, and the time interval from injury to arrival in the operating room. If the total score was greater than 8 points, amputation was preferred to salvage. Pozo *et al.* (1990) suggested a limb injury score based on wound contamination, and damage to skin, bone, muscle, arteries and nerves. The patient with an injury score of eight or greater was unlikely to finish up with a useful lower-limb. Johansen *et al.* (1990) used a 'Mangled Extremity Severity Score' (MESS) to predict limb salvage or primary amputation in lower limb trauma requiring revascularization. The MESS score used skeletal and soft tissue injury, the age of the patient, the degree of limb ischaemia and shock. All patients with a score of seven or more were best treated by amputation. Russell *et al.* (1991) formulated a limb salvage index (LSI) based on the following factors: Injury to veins and arteries, nerves, bone, skin and muscle. They also stressed the period of ischaemia between injury and treatment. All patients with an LSI score of six or greater required amputation as opposed to limb salvage.

A number of criticisms may be levelled at most of these scoring systems:

(1) There may be considerable 'observer error' in scoring shock and the extent of early bone and soft tissue damage.
(2) The system should include life threatening injuries in polytrauma patients, and the site of the injury, i.e. above or below the knee joint.
(3) The scoring systems give no indication of long-term function in the salvaged limb.
(4) The patient, the family or the ever-present lawyers will not be impressed by a scoring system.

This author's preference is to deal with each particular patient by referring to the 'personality of the injury', i.e. the patient, the injured limb and the health care available.

The patient

(1) Age.
(2) Underlying or previous disease.
(3) Multisystem injuries including the extent of shock, disseminated intravascular coagulation (DIC) and adult respiratory distress syndrome (ARDS).
(4) Work and recreational activities of the patient.
(5) Desires and realistic expectations of the patient and family.
(6) Pre-existing psychosocial problems which are common in this group of patients.

The injured limb

(1) Mechanism of injury.
(2) Location of the injury, i.e. above or below the knee joint.
(3) Degree of damage to bone and soft tissues including skin, muscle, nerve and vascular injuries.
(4) Duration of the ischaemic period before successful revascularization.

Health care

(1) Available facilities for surgery and rehabilitation.

(2) Socioeconomic costs to the patient, the family and society.

As a further guide, it is suggested that, in addition to the scoring system, there are a number of absolute and relative indications for primary amputation of the lower limb following severe trauma.

Absolute indications for primary lower limb amputation following severe trauma

(1) Complete amputation at the time of the accident (Fig. 24.3).
(2) Irreparable sciatic or posterior tibial nerve injury in association with a Type III C fracture of the tibia.
(3) Ischaemic period of more than 6–8 h.
(4) Associated life threatening injuries with prolonged shock, DIC or ARDS.
(5) A cadaveric foot at initial examination (Fig. 24.4).

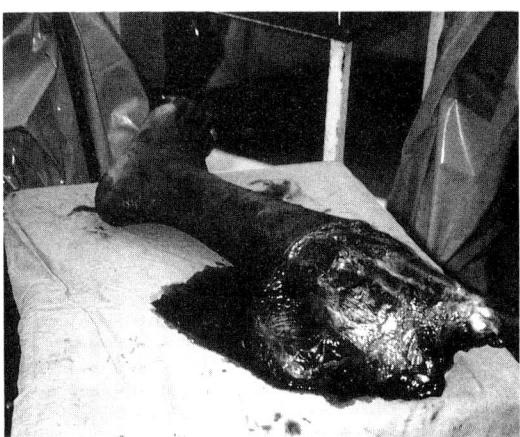

Fig. 24.3. Complete amputation of the lower limb following injury.

Relative indications for primary amputation of the lower limb following severe trauma

(1) Type III C injuries of the tibia/fibula.
(2) Crush injuries of the lower limb and ipsilateral foot.

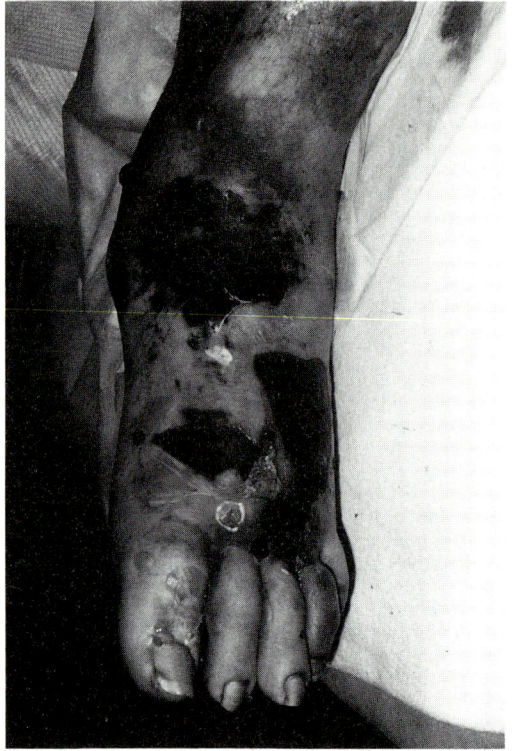

Fig. 24.4. A cadaveric foot on admission.

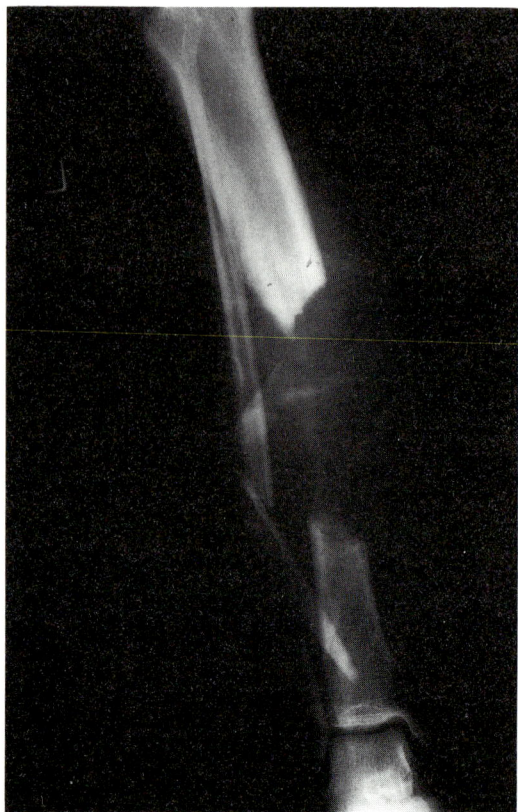

Fig. 24.5. Significant tibial bone loss suggesting a relative indication for transtibial amputation.

(3) Significant tibial bone loss or associated severe damage to knee/ankle joints (Fig. 24.5).
(4) Age more than 50 years.
(5) Isolated sciatic or posterior tibial nerve injury.
(6) Inadequate health care facilities available.

The ultimate decision lies in the hands of the surgeon, who should be prepared to make a clear recommendation to the patient and the family.

Technique and timing of amputation

The patient and family will usually refuse to sign a consent form for lower limb amputation, because they have unrealistic expectations of the medical profession. Early stabilization of the fractures, usually by external fixation, will allow the patient to reconsider the situation a few days after the injury, but it is important to stress that delay in amputation may increase the patient's chances of requiring a more proximal amputation because continued inappropriate attempts at limb salvage increase both morbidity and mortality following severe lower limb trauma. The surgeon should forget the rule of 'saving all length possible' when dealing with severe foot trauma. Too often, in an attempt to preserve as much of the foot as possible, the patient is left with a scarred painful partial foot amputation with insensitive skin that ulcerates on minimal pressure (Fig. 24.6). Associated hindfoot pathology resulting from the injury may be due to fractures or incongruity of the joint surface leading to malunion and non-union; these will progress to early arthritis of the ankle and subtalar joints.

The good reputation of distal foot amputation, and the poor reputation of midfoot amputations are

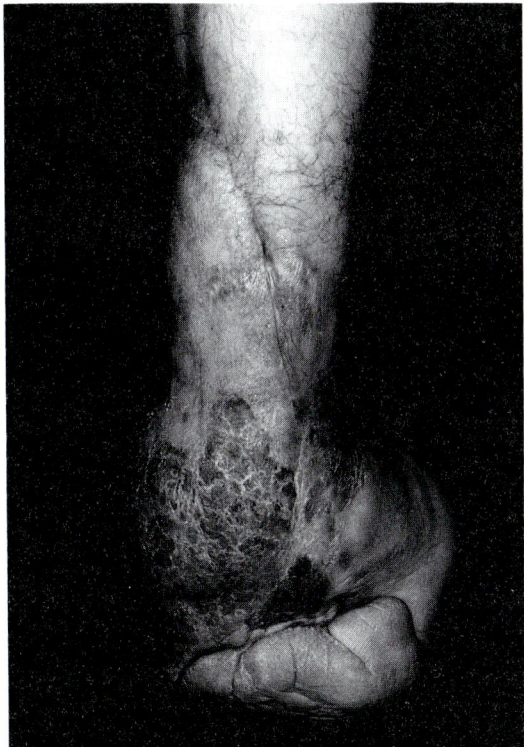

Fig. 24.6. A scarred, painful partial foot amputation requiring revision to transtibial amputation.

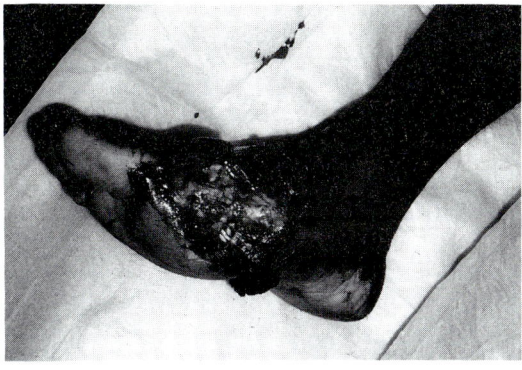

(a)

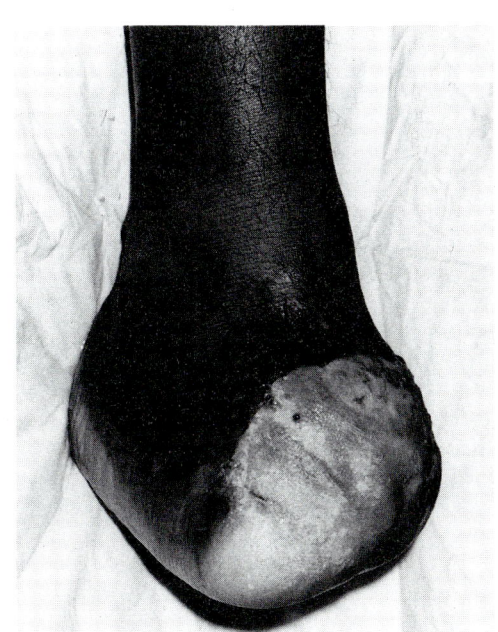

(b)

Fig. 24.7 (a) Severe bone and soft tissue crushing injury of the foot. (b) Partial foot amputation at the Lisfranc level.

not borne out by critical review (Millstein *et al.*, 1988). Multiple procedures to save the forefoot will not relieve pain or improve function and will delay the patient's rehabilitation. Early in the post-traumatic period, consideration should be given to an elective transverse amputation at the metatarsal or mid-tarsal level preserving the extrinsic foot muscles and using a healthy plantar skin flap when possible (Fig. 24.7a, b). Avoid skin grafts, especially on the terminal or plantar areas of the partial foot amputation. Treat associated hindfoot bone and joint injuries by early open reduction and rigid internal fixation (Wright *et al.*, 1987).

Syme's ankle disarticulation following trauma is successful because there is usually a good blood flow in the posterior tibial vessels; there must be viable skin available in the heel pad to ensure a satisfactory stump and subsequent prosthetic fitting (McElwain *et al.*, 1985).

Transtibial amputation is the usual method of dealing with the severely injured lower limb in the hope of preserving the knee joint. All contaminated and necrotic muscle and soft tissues should be removed at the initial amputation or at an obligatory 'second look' procedure at 48–72 h; all amputations following trauma should be left open for a few days before delayed skin closure or skin grafting is attempted. Split skin grafts may be used over soft tissues, but should be avoided over the exposed tibial bone surfaces.

Use of tissue expanders at a later date may remove the skin grafted areas, while still preserving the knee joint (Wood *et al.*, 1987). Rarely, weight bearing skin may be transferred from the foot as a neurovascular flap to cover exposed bone; such expertise is rarely available in the emergency trauma situation. The standard long posterior myofasciocutaneous flap is often not available, and consideration should then be given to the 'skew flap' technique or sagittal skin flaps. The skin flaps should permit delayed skin closure, without tension, rather than using a transverse guillotine operation, which will certainly require revision to a more proximal amputation. At least one muscle compartment should be preserved to cover the exposed tibia. The standard tibia length preserved should be at least 12–15 cm below the knee joint. If in order to preserve the knee joint a very short transtibial stump is considered feasible, it should be recognized that there is a limit as to how short the transtibial stump can be for successful prosthetic fitting. Given that the prosthetic skills are available, a useful guide is to divide the tibia no more proximal than two fingers breadth distal to the insertions of the hamstring tendons. If a short tibial stump is the only alternative, it is probably better to excise the entire fibula to enhance prosthetic fitting.

Knee disarticulation may be used in the trauma situation if the upper tibial segment is smashed beyond repair. One must remember that generous sagittal or antero-posterior skin flaps are necessary to cover the exposed femoral condyles. Resection of the prominent femoral condyles may avoid excessive tension (and inevitable failure) on the skin flaps.

If a transtibial amputation or knee disarticulation is impossible, or has failed to heal, it is necessary to perform a transfemoral amputation 10–12 cm above the knee joint using a myodesis technique. Adequate skin healing will allow early prosthetic fitting and enable the patient to return to work and a reasonable lifestyle following amputation of the lower limb.

References

Behrens, F. (1992) Fractures with soft tissue injuries. In: *Skeletal Trauma* (Vol. I) (eds B.D. Browner *et al.*), W.B. Saunders, Philadelphia, pp. 311–336.

Colton, C.L. (1992) The history of fracture treatment. In: *Skeletal Trauma* (Vol. I) (eds B.D. Browner *et al.*), W.B. Saunders, Philadelphia, pp. 1–30.

Feliciano, D.V. (1992) Evaluation and treatment of vascular injuries. In: *Skeletal Trauma* (Vol. I) (eds B.D. Browner *et al.*), W.B. Saunders, Philadelphia, pp. 269–284.

Giorgiadis, G.M., Beyrens, F., Joyce, M. *et al.* (1991) End results after severe open tibial fractures treated by limb salvage or early below knee amputation. Paper presented at the *Seventh Annual Meeting of the Orthopaedic Trauma Association, Seattle, USA.*

Green, S. (1992) The Ilizarov method. In: *Skeletal Trauma* (Vol. I) (eds B.D. Browner *et al.*), W.B. Saunders, Philadelphia, pp. 543–570.

Gregory, R.T., Gould, R.J., Peclet, M. *et al.* (1985) The mangled extremity syndrome (MES). A severity grading system for multisystem injury of the extremity. *J.Trauma*, **25**, 1147–1150.

Gustilo, R.B. Mendoza, R.M. and Williams, D.N. (1984) Problems in the Management of Type III (Severe) open fractures: a new classification of Type III open fractures. *J. Trauma*, **24**, 742–746.

Hansen, S.T. Jr (1987) The type III C tibial fracture. Salvage or amputation? *J. Bone Joint Surg.*, **69A**, 799–800.

Howe, H.R., Poole, G.V. Jr, Hansen, K.J. *et al.* (1987) Salvage of lower extremities following combined orthopaedic and vascular trauma. A predictive salvage index. *Am. Surg.*, **53**, 205–208.

Johansen, K., Daines, M., Howey, T. *et al.*, (1990) Objective criteria accurately predict amputation following lower extremity trauma. *J. Trauma*, 568–572.

Lange, R.H., Bach, A.W., Hansen, S.T.Jr *et al.* (1985) Open tibial fractures with associated vascular injuries. Prognosis for limb salvage. *J. Trauma*, **25**, 203–207.

Mackinnon, S.E. and Dellon, A.L. (1988) *Surgery of the Peripheral Nerve*. Thieme Medical Publishers, New York, pp. 108–110.

McElwain, J.P., Hunter, G.A. and English, E. (1985) Syme's amputation in adults: a long term review. *Can. J. Surg.*, **28**, 203–205.

Millstein, S.G., McCowan, S.A. and Hunter, G.A. (1988) Traumatic partial foot amputations in adults: a long term review. *J.Bone Joint Surg.*, **70B**, 251–254.

Millstein, S.G., Hunter, G.A. and Kellam, J.F. (1990) Injuries of the lower limb leading to revascularization and/or amputation in polytrauma. Paper presented at the *Sixth Annual Meeting of the Orthopaedic Trauma Association, Toronto, Canada.*

Pozo, J.L., Powell, B., Andrews. B.G. *et al.* (1990) The timing of amputation for lower limb trauma. *J. Bone Joint Surg.*, **72B**, 288–292.

Purry, N.A. and Hannon, M.A. (1989) How successful is below knee amputation for injury? *Injury*, **20**, 32–36.

Rosen, H. (1992) Nonunion and malunion. In: *Skeletal Trauma* (Vol. I) (eds B.D. Browner, *et al.*), W.B. Saunders, Philadelphia, pp. 501–541.

Russell, W.L., Sailors, D.M., Whittle, T.B. *et al.* (1991) Limb salvage versus traumatic amputation. A decision based on a seven-part predictive index. *Am. Surg.*, **213,** 473–480.

Sherman, R., Ecker, J. (1992) Soft tissue coverage. In: *Skeletal Trauma* (Vol. I) (eds B.D. Browner, *et al.*), W.B. Saunders, Philadelphia, pp. 337–366.

Tscherne, H. (1984) *Fractures with Soft Tissue Injuries.* Springer, Berlin.

Wood, M.R., Hunter, G.A. and Millstein, S.G. (1987) The value of stump split skin grafting following amputation for trauma in adult upper and lower limb amputees. *Prosthet Orthot Int.*, **11,** 71–74.

Wright, J., Worlock, P., Hunter, G. *et al.* (1987) The management of injuries of the midfoot and forefoot in the patient with multiple injuries. *Techniques in Orthopaedics*, **2,** 71–79.

Section 10

Tumours

25

Amputations for tumours

B. M. Persson

Amputation of a limb for a tumour is a challenge complicated because of the location, size and type of tumour, thus making every patient a very special problem. Is amputation at all necessary or can a local radical resection be performed?

There are several alternatives, ranging from a rotationplasty according to Borggreve and Van Nes, in which there is a segmental resection with preservation of the distal limb shortened and rotated through 180° to make the ankle function as a knee, or a radical local resection be it replaced by a custom-made megaprosthesis or with joint fusion and bone transplantation instead? If the latter, is a microvascular autograft the best or rather a massive homograft with its slower healing?

Tumours of the locomotor system may be localized skeletally or in the soft tissues. They might be benign or malignant, the malignant either primary or secondary. In time they often cause increasing threat to function, often with pain, fractures and ultimately possible loss of life. Early detection is of utmost importance for all the treatment options to be available.

Bone tumours usually present as a mass when benign, but most often this is preceded by pain when malignant. Pain at night usually comes before pain on weight bearing, which is a later but important symptom of impending pathological fracture. In advanced stages amputation may be the only treatment left to suggest for a possible cure but it should be remembered during counselling that the sacrifice of a limb is no guarantee of cure. Certainly in grade IV malignancies like most osteosarcomas and Ewing sarcomas and a majority

of malignant fibrous histiocytomas, the most common type of soft tissue sarcomas, amputation was for long most commonly followed by recurrence in spite of radical surgery. The introduction of adjuvant chemotherapy about 1980 for both high grade osteosarcoma and Ewing sarcoma has drastically improved the disease-free 5-year survival after radical surgery from 20% to 80%. Radical local resection over the last 10 years has also replaced amputation in the majority of such cases. For soft tissue sarcoma, on the other hand, adjuvant chemotherapy has not improved the survival. Early and precisely performed surgery here is the most important improvement (Enneking *et al.*, 1980; Kreicbergs and Rydholm, 1992).

Biopsy

In all cases of limb tumours except exostotic osteochondroma and non-ossifying fibroma, a biopsy is necessary to define the type of tumour. It should never be forgotten that some cases of osteomyelitis may present a radiographic appearance similar to a bone tumour like Ewing sarcoma making a well planned early biopsy even more mandatory. Comprehensive assessment with sedimentation rate, C-reactive protein, serum electrophoresis, alkaline and acid phosphates, calcium in serum, blood sugar and haemoglobin is essential.

The next stage is subject to controversy. Should the biopsy be open or through a cannulated drill for core material? Is conventional pathological analysis enough or are histochemistry and DNA analysis needed? Can a fine needle aspiration cytology ever be sufficient in combination with X-ray examination and clinical findings to present a decision of amputation? Is magnetic resonance imaging (MRI) now a mandatory step in the preoperative analysis? (Szendroi *et al.*, 1993).

All these questions and alternatives regarding both biopsy and definitive treatment have made it necessary to create special orthopaedic oncology centres and cooperative national protocols to follow and analyse the results of different pro-grammes. The incidence of osteosarcoma is only three, of chondrosarcoma two and of Ewing's sarcoma one per million people. The correspond-ing incidence of soft tissue sarcomas of the locomotor system is 14. Of these, 75% are of high grade malignancy and 50% are above the age of 60 years (Rydholm *et al.*, 1983).

This rarity is a further reason for the organiza-tion of special centres. In the USA such centres are to be found in Boston, Chicago, Gainesville, Houston, Los Angeles, New York, Rochester; in Europe in London, Glasgow, Paris, Vienna, Bologna, Stockholm, Gothenburg, Lund and Oslo, to mention some of the most famous. A special musculoskeletal tumour society has been formed both in the USA and in Europe and in Scandinavia a special interdisciplinary sarcoma group co-ordinates the efforts to improve the results.

Highly interesting outcome studies are now running including a grouping of more than 250 patients for comparison of the results in terms of both function and survival after transfemoral amputation for osteosarcoma of distal femur com-pared to local resection with mega-prosthesis including knee and distal femur. The cases have been matched as regards grade and stage of tumours, etc. and still it seems worthwhile to preserve the limb when possible. It can be expected, however, that prosthetic interface loos-ening will create an increasing problem later (Simon, 1988; Harris *et al.*, 1990).

Biopsy in some centres is performed in one stage with definitive surgery by use of frozen section examination of the pathology, closure of the wound and fresh prepping and draping. This is, however, a difficult routine as it gives a short time for the pathologist and can even be hazardous. Many centres therefore use a two stage biopsy/ surgery routine. It has been proven by randomized

studies many years ago that this delay of a few days does not increase the risk of spread.

With regard to the use of fine needle aspiration biopsy the traditions are different and special competence has to be created for safe use. DNA flow cytometry to check diploidy or anaploidy has increased the precision in diagnosis as with fine needle biopsy (Akerman *et al.*, 1985, 1987; White *et al.*, 1988).

Thus in making all the points above it is clear that a proper biopsy is both difficult and influential in respect of the outcome.

First, it should be done after clinical examina-tion with functional palpation to determine size, depth, tenderness, borders, mobility and after sufficient X-ray, scintigraphy, MRI and blood chemistry mentioned above.

Second, the pathway should be chosen so that further surgery is not jeopardized regarding the goal of local radical excision without unnecessary loss of function or cosmesis. Contrary to what some believe, it is better to approach bone for biopsy through covering muscle rather than between two muscles that will cause spread along fascial planes. The first way creates an enclosure of the biopsy tract which is safer, provided that the muscle in question can be removed later en-bloc with the involved bone. The biopsy scar and tract always have to be removed en-bloc with the tumour in orthopaedic oncology programmes. Sarcoma spread locally and by veins to the lungs but very seldom through lymph channels to regional nodes.

Third, it should be remembered that the biopsy must include a typical and representative part of the lesion. In bone TV-amplified X-ray should be used to guide the chisel and curette or the core-drill.

Chondrosarcomas and dedifferentiated lipomas, for instance, may have areas of a different nature in the same mass and advanced necrosis in some high grade sarcomas may make diagnosis impossible on material from the centre only. Usually lid, margin and centre should be taken separately.

Fourth, after a bone biopsy, if bleeding is difficult to stop, it should never be solved by close suturing of soft tissues above, as the post-operative bleed then may cause distant spread in the surgical area making further local surgery unpredictable (Fig. 25.1). Diathermy, spongostan packings or bone cement must be used as needed to prevent this. Bone cement also marks the site of biopsy

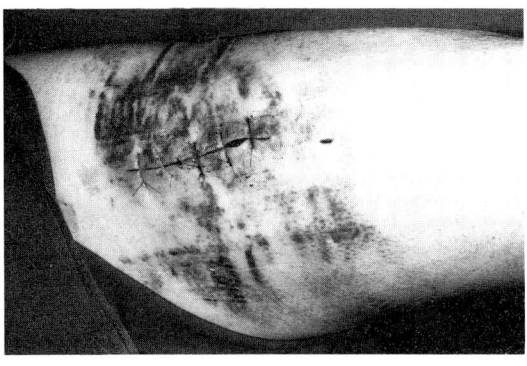

Fig. 25.1. Post-operative appearance 1 week after open biopsy for bone lesion in the trochanteric area in a 65-year-old man. In spite of suction drain post-operative bleeding from bone caused unacceptable soft tissue haematoma with uncontrolled capacity to spread tumour cells. Bone defect from open biopsy should be cemented if bleeding is difficult to control otherwise.

well and does not create any new harmful effects even if the pathology should be benign.

Fifth, take material for both fixed and unfixed preparation, imprint cytology and bacterial culture and preferably involve the pathologist in selecting material.

Amputation

When finally a decision to amputate has been taken after assessment and discussions with the patient and if young, with his family, it is important to consider the compartmental rules. An amputation in itself looks radical and safe to the patient but is so only when correctly planned (Table 25.1). If it passes across a corner of the tumour it is not much

better than an incisional biopsy debulking. If it passes within millimetres of the tumour it is about as safe as a marginal excision, certainly not enough for malignant or even an aggressive benign tumour like a desmoid, a nodular fasceitis or an intramuscular lipoma. The amputation has to be wide, passing several centimetres away from the tumour or preferably on the other side of a fascia covering the involved compartment. These rules have to be respected, but still it is possible sometimes to preserve stump length in cases of soft tissue sarcomas. If it is located in the anterior thigh (quadriceps) compartment, for instance, a long posterior flap can be created and turned around anteriorly to cover the defect, thus allowing a moderate transfemoral amputation rather than a hip disarticulation or similarly in the medio-lateral direction (Fig. 25.2).

When a bone tumour is located in the femur, sometimes much more bone than soft tissue has to be sacrificed. The redundant soft tissues are difficult to use because of the lack of stability in a socket. In such cases an elongation prosthesis has been used (Plenk *et al.*, 1978) and also a Moore prosthesis (Marcove *et al.*, 1979). This may also be combined with a tissue expander, or callus distraction reconstructions to increase the functional stump length (Persson and Broome, 1993).

'Skip' metastases can occur in osteosarcoma, making at least a 5 cm free margin necessary. It was long supposed that the whole involved bone had to be removed but that has not stood the test of time (Marcove *et al.*, 1979). The whole concept of surgery for osteosarcoma and Ewing sarcoma has reversed and now there is proof that 2 months or longer pre-operative chemotherapy after biopsy for typing and grading gives a superior survival and makes more conservative surgery possible without loss of limbs or lives.

Malignant melanoma and metastatic cancer with pathological fractures and advanced soft

Table 25.1.   Alternative margins at surgery of tumours

Margin	Local resection	Amputation
Intralesional	Curettage or in pieces	Through tumour
Marginal	Exposing surface of tumour	Exposing surface
Wide	Centimetres away en-bloc	Centimetres above
Compartmental	Outside involved compartment	Above involved compartment

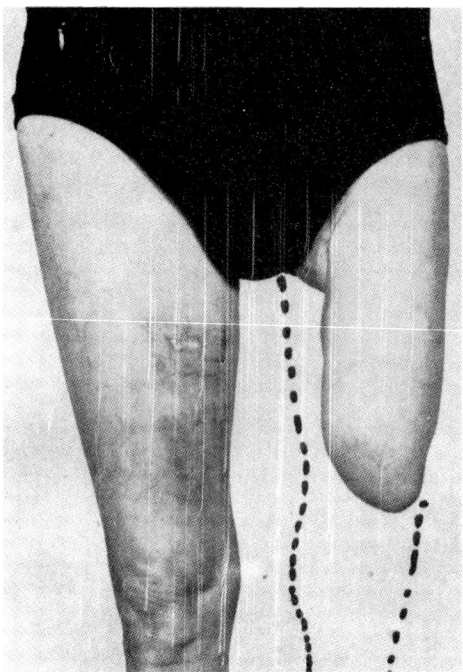

Fig. 25.2. Example of stepshaped transfemoral amputation for intramuscular soft tissue sarcoma of the medial thigh allowing compartmental radical resection of soft tissues with preservation of functional stump length.

tissue involvement may sometimes require an amputation, provided the lungs are without metastases. Otherwise the stage is usually too advanced to allow any rehabilitation post-operatively. In general, metastatic cancer can often be treated with radiotherapy and possibly internal fixation which reduces the need of amputation. For soft tissue sarcomas pre-operative radiotherapy has been helpful in cases with large tumours to make later local resection possible instead of amputation.

The most mutilating of all amputations, lower hemicorporectomy or translumbar amputation, has occasionally successfully been undertaken for tumour cases usually in advanced tumours of low malignancy and when the patient demanded it. It necessitates colostomy, cystostomy and sexual hormone supplementation. Several long-time surviving cases have been published (Miller *et al.*, 1966; Wagman and Terz, 1989).

In general with tumour amputations the patients are young with a good healing and rehabilitation potential, sometimes temporarily set back by the chemotherapy programme. They will become demanding prosthetic users and give highly rewarding experiences both to the surgical and the rehabilitation teams.

References

Akerman, M., Rydholm, A. and Persson, B.M. (1985) Aspiration cytology of soft tissue tumours. 10 years experience at an orthopaedic oncology group. *Acta Orthop. Scand.*, **56**, 407–412.

Akerman, M., Killander, D., Rydholm, A. and Rooser, B. (1987) Aspiration of musculoskeletal tumours for cytodiagnosis and DNA analysis. *Acta Orthop. Scand.*, **58**, 523–528.

Enneking, W.F., Spanier, S.S. and Malawer, M. (1980) A system for the surgical staging of musculoskeletal sarcoma. *Clin. Orthop.*, **153**, 106–120.

Harris, I., Leff, A., Gitelis, S., Simon, M. (1990) Function after amputation arthrodesis, or arthroplasty for tumours about the knee. *J. Bone Joint Surg.*, **72A**, 1477–1485.

Kreicbergs, A. and Rydholm, A. (1992) Soft tissue sarcoma. *Curr. Opin. Orthop.*, **3**, 777–783.

Marcove, R.C., McMillan, R.D. and Nasr, E. (1979) Preservation of the functional above-knee stump following hip disarticulation by means of an Austin-Moore prosthesis. *Clin. Orthop.*, **141**, 217–222.

Miller, T.R., McKenzie, A.R. and Randall, H.T. (1966) Transulmar amputation for advanced cancer. *Ann. Surg.*, **164**, 514.

Plenk, H., Salzer, H., Locke, N., Stark, N., Punzet, G. and Zweymuller, K. (1978) Extracortical attachment of biomechanic prosthesis to long bones. *Clin. Orthop.*, **32**, 260.

Rydholm, A., Berg, N.O., Persson, B.M. and Akerman, M. (1983) Treatment of soft tissue sarcoma should be centralized. *Acta Orthop. Scand.*, **54**, 333–339.

Simon, M.A. (1988) Current concepts review: limb salvage for osteosarcoma. *J. Bone Joint Surg.*, **70A**, 307–320.

Szendroi, M., Antal, J., Liszka, G., Monostori, Z. (1993) Examination and evaluation of surgical margins in bone tumours. Comparative pre- and postoperative CT and MRI. *Int. Orthop.*, **17**, 93–97.

Wagman, L.D. and Terz, J.J. (1989) Hemipelvectomy and translumbar amputation. In: *Lower Extremity Amputation* (eds W.S. Moore and J.M. Malone), W.B. Saunders, Philadelphia, pp. 157–176.

White, V.A., Fanning, C.M., Ayala, A.G., Raymond, K., Carasco, C.M. and Murray, J.A. (1988) Osteosarcoma and the role of fine-needle aspiration. A study of 51 cases. *Cancer*, **2**, 1238–1246.

Section 11

Rotationplasty

26

The role of rotationplasty

I. P. Torode

This contribution deals with patients who fall into two major groups. First are those patients who present with a gross deficiency of the femur. Their deficiencies are usually discovered at birth or less commonly are due to a disease process causing extensive bone loss. Secondly, there are patients who present with a malignant lesion of the femur where treatment necessitates an extensive resection of the femur and the surrounding soft tissues.

Rationale of rotationplasty

In situations where a patient presents with a major deficiency of the femur, the foot rests at approximately the level of the contralateral knee. Apart from the fact that the foot is rendered useless by the distance from the ground, the ankles' range of motion is of little benefit to the patient and thus the presence of the foot serves only to complicate prosthetic fitting.

The options for the patient include fitting a prosthesis around the foot, removal of the foot and fitting the patient as an 'above knee amputee' or reconstruction of the limb by means of a rotationplasty. The concept of rotationplasty is to place the foot and ankle at an appropriate level and orientation so that the patient has the ability to function as a 'below knee amputee' with the ankle providing the new 'knee' joint and the ankle dorsiflexors and plantarflexors providing the control and power to the modified 'below knee' prosthesis.

In patients where treatment of a malignant tumour requires major femoral resection, the situation described above pertains, if the remnant of the limb is retained or alternatively the patient is left with either a very short transfemoral amputation or with a hip disarticulation. In those patients the rationale of rotationplasty is to make use of the healthy remnant of the limb, distal to the resection margin, to allow the patient the function of a transtibial amputee.

Benefits to the patient

The functional benefits to the patient are seen by examining the function of an 'above knee' amputation with that of a 'below knee' amputee. It can be demonstrated that in terms of energy consumption, oxygen utilization and gait speed a Syme ankle disarticulation amputee is more efficient than a transtibial amputee who in turn is more efficient than a transfemoral amputee (English, 1981).

Furthermore, the control by the patient of an innervated anatomical joint is far superior to that of even the most sophisticated prosthetic joint. Patients with a rotationplasty retain the ability to walk up and down steps or stairs leading with the deficient limb. They are able to kick with the prosthetic foot or stand on the prosthetic foot and kick with the sound limb. Gait studies suggest that following rotationplasty the patients develop reflexes that protect them from stumbling and falling. Such reflex patterns cannot be duplicated in a prosthetic joint (Glynn, 1983).

Negative aspects of a rotationplasty

The most commonly voiced concern of rotationplasty is the appearance of a foot that is back-to-front and at the level of the opposite knee (Fixsen, 1983). However, this concern must be seen in the context that without any intervention the foot will be at the level of the contralateral knee and not in the context of a foot being in a normal position. As 'normality' is not an option when discussing surgical options where the choices are of a foot in an abnormal position, a foot at the appropriate level but back-to-front or no foot at all.

With regard to prosthetic fitting there is no doubt that a greater degree of skill and patience is required to fabricate and fit a prosthesis for a patient who has undergone a rotationplasty than for a patient who has had a straightforward transfemoral amputation.

The history of rotationplasty

The actual use of a rotationplasty was described in print by Borggreve (1930). Following the publication of a further three cases by Van Nes (1950) the operation became recognized as the Van Nes procedure. The three cases described were all congenital in origin.

A number of authors have published the results of rotationplasties performed by a variety of techniques. In the series published by Kostuik *et al.* (1975) the rotation was performed through a tibial osteotomy which was in most cases secured by a plate and screws. Kritter (1977) used a fibular strut as a means of fixation with rotation through a tibial osteotomy. The concern in most series has been the tendency of the distal portion of the limb to derotate with time towards the original orientation, particularly if the initial rotationplasty has been performed through the tibia.

The results in some of these patients have been disappointing, requiring either a revision rotationplasty or a Symes amputation. In some

the combined motion of the ankle and subtalar joint has allowed enough free movement for the rotationplasty to provide adequate function despite some derotation. This derotation may be in part due to the spiral effect of the anterior and posterior tibial muscles if the tibia is rotated between the origin and insertion of these muscles, particularly in the child with open epiphyseal plates. Many patients have undergone staged surgical procedures to effect the rotationplasty.

The technique of combined knee fusion and rotationplasty for deficiencies of the femur discovered at birth and here described addresses the problems of loss of rotation and multiple operations. This technique can be used in the young child with open epiphyseal plates and allows earlier definitive prosthetic fitting; the technique follows the original description of Torode and Gillespie (1983).

Rotationplasty has also been utilized in the management of malignant neoplasms of the femur. The technique is varied according to the anatomical level of the primary tumour. The description has thus been subdivided into two sections following the original descriptions of Winkelman (1986) for the proximal lesions and of Kotz and Salzer (1982) for lesions of the distal two-thirds of the femur.

Deficiencies of the femur at birth

The prerequisites for this procedure concern a child with a femoral deficiency discovered at birth but with a largely complete foot and a reasonably stable ankle with good motor power. As there is a frequent association of fibular and foot deficiencies with femoral deficiencies, the feet and ankle joints in these children are usually not wholly normal. However, as the ankle will be encompassed by hinges between the foot and leg sections of the prosthesis a degree of instability is acceptable. Furthermore, a foot that lacks one or two of its lateral rays may still have good dorsiflexors and plantarflexors and thus still be able to provide good control to the 'below knee' section of the prosthesis.

The best results are found in those children whose femur is hypoplastic but intact in so far as the musculature around the hip is more normal in this situation than in those children whose deficiency is a true proximal femoral focal deficiency (PFFD) (Gillespie and Torode, 1983). However, improvements in leg lengthening techniques have meant that more children with a very short but otherwise intact femur are candidates for femoral lengthening.

The surgical management of these patients usually involves both rotationplasty and knee fusion. This is not difficult to accommodate as the ideal method of performing the rotationplasty is to rotate the tibia through 180° at the level of the knee joint at the time of knee fusion.

Surgical technique

At the time of surgery the patient must be positioned and draped so that the surgeon can ascertain the appropriate length of the limb and also determine the appropriate orientation of the foot with respect to the trunk.

The incision (Fig. 26.1)

The incision for the rotationplasty is sited over the anatomical knee joint extending proximally to allow exposure of the femoral artery proximal to the adductor magnus insertion and distally to allow exposure of the peroneal nerve as it enters the anterior compartment of the leg. The incision can be extended to allow exposure of the tibial diaphysis if required.

The femoral artery and popliteal artery

The femoral artery must be exposed as it enters the adductor hiatus at the insertion of adductor magnus. That muscle in these children usually approaches the femur at a more obtuse angle than in the normal adult, which may confuse the surgeon. The insertion must be divided in order to allow the popliteal artery to slide in an anterior direction around the medial aspect of the femur as the leg is externally rotated (Fig. 26.2).

The peroneal nerve

The peroneal nerve must be isolated and protected throughout the rotationplasty. In patients with an intact fibula the nerve will lie in its normal site

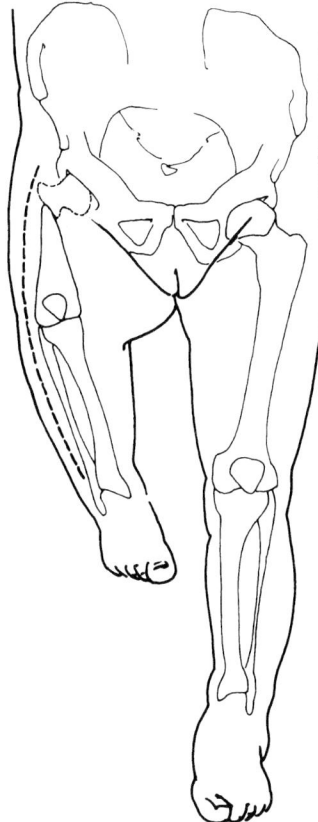

Fig. 26.1. Diagrammatic representation of proximal focal femoral deficiency. The cartilaginous proximal femur is shown in broken outline and the proposed incision is shown as a line of dashes.

adjacent to the proximal fibula. In patients without a normal fibula great care must be taken, as the nerve may lie adjacent to the proximal tibia. It is imperative that the nerve must be free to rotate with the leg during the external rotation of the rotationplasty and inadvertent medial rotation must be avoided as this will put traction on the nerve.

The knee joint

The capsule of the knee joint must be defined with great care, particularly on the lateral aspect as the peroneal nerve may be in close proximity. The capsule and collateral ligaments are divided circumferentially. The knee is flexed to allow the popliteal vessels to fall away posteriorly as dissection proceeds. The deficiencies of the cruciate

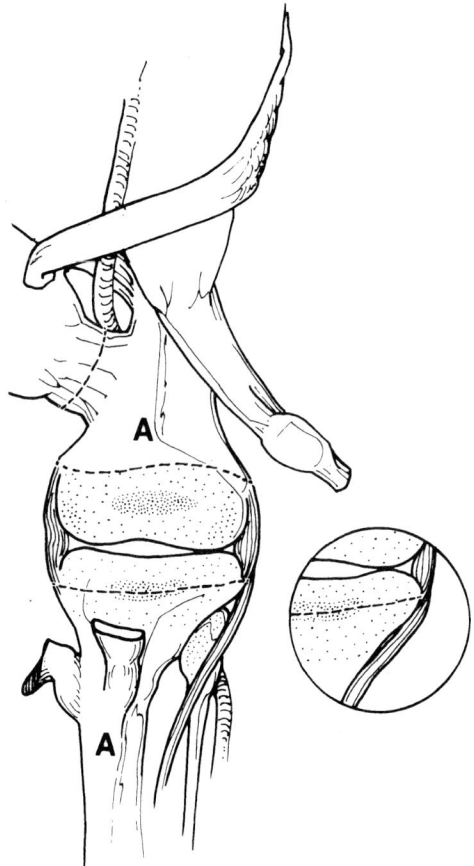

Fig. 26.2. Sartorius and quadriceps have been retracted to expose the adductor hiatus and femoral artery. The peroneal nerve is displayed in its usual location and with fibular hemimelia (inset). The proposed bone cuts are shown as dashes, with the ossific centres shaded (A = anterior).

ligaments, the lack of development of the lateral condyle, the intercondylar notch and the tibial spines are readily apparent.

The adjacent musculature

The patellar tendon is divided anteriorly and the two heads of gastrocnemius are divided posteriorly. The insertion of adductor magnus and the insertion of the medial hamstrings are divided medially. When the leg is externally rotated through 180° the two heads of gastrocnemius will come to lie anteriorly and can be sutured to the

quadriceps complex. The adductor magnus tendon can then be sutured to the medial aspect of the femur or to what was formerly the lateral aspect of the leg.

The growth plates

It is possible to perform the rotationplasty and retain either the distal femoral or proximal tibial growth plates. In some patients neither will be retained. This decision is made by a careful comparison of the length of the contralateral femur with the combined length of the deficient femur and ipsilateral tibia, and taking into consideration

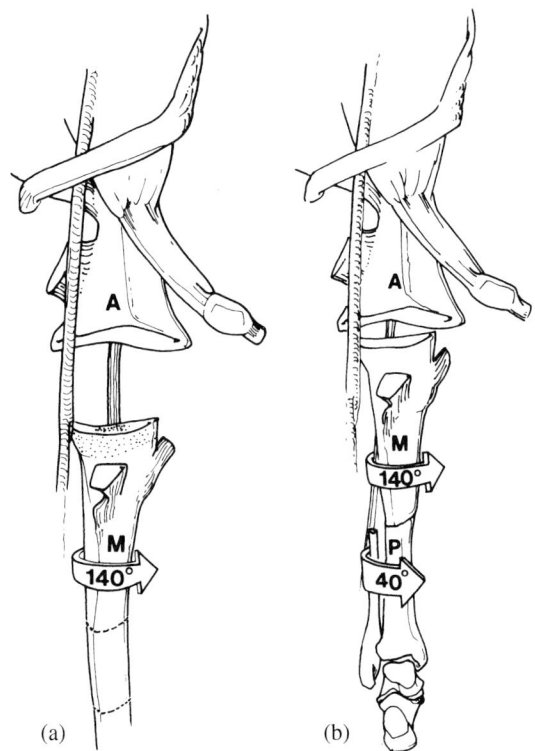

Fig. 26.3. (a) Portions of the distal femur and proximal tibia have been removed allowing rotation of the proximal tibia. The femoral and popliteal arteries can rotate freely with the distal segment freed from the tether of the adductor magnus insertion; (b) the proximal tibial segment is rotated as much as possible and the remaining rotation required obtained through the tibial shaft osteotomy. (A = anterior, M = medial, P = posterior).

the remaining growth (Anderson *et al.*, 1963) in the relevant epiphysial plates. Ideally, at maturity the original ankle joint will approximate to the level of the contralateral knee joint (Fig. 26.3).

Knee fusion and fixation

The knee fusion is performed by dividing the distal femur and proximal tibia at the appropriate levels with regard to both length and growth plates. Usually the proximal tibial ossific nucleus is exposed but care is taken that the sawcut does not violate the growth plate. The distal femoral epiphysis is removed. An intramedullary rod is then passed proximally up the femur and out through the buttock. The rod is then drilled down the tibia, transfixing the knee fusion and the tibial osteotomy.

Tibial osteotomy (Fig. 26.4)

The tibial osteotomy can be utilized to provide rotation where there is difficulty in obtaining the required complete rotation at the level of the knee joint. However, its greater value is in allowing secondary adjustment in the post-operative period if the rotational alignment is thought to be inappropriate. A fibular osteotomy is also performed to allow this rotary component. Fasciotomies of the adjacent compartments are performed at this stage.

The rotationplasty

Having completed the appropriate osteotomies and made sure the vessels and nerves are not impeded, the leg is rotated at the level of the knee joint

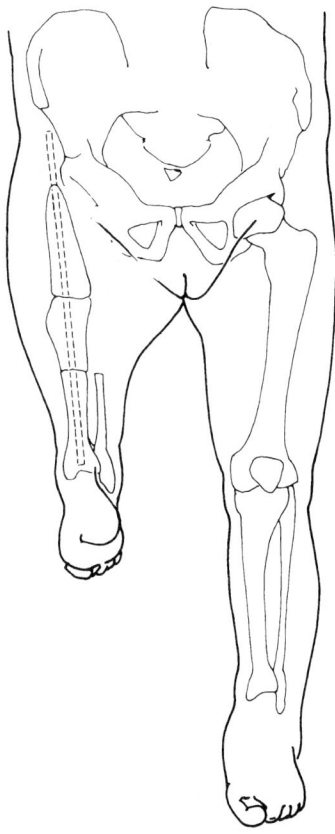

(a)

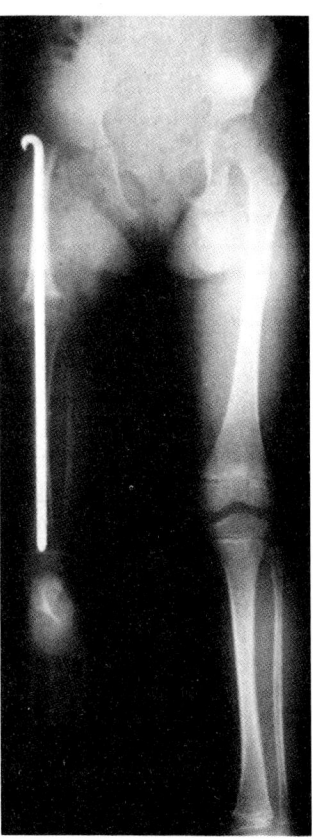

(b)

Fig. 26.4 (a and b) The intramedullary rod fixes both the fusion and the osteotomy allowing good stability and simple instrumentation.

initially. It is entirely possible to obtain the full 180° of rotation at that site. The vessels must be palpated at the completion of the rotation. If the desired rotation cannot safely be attained, then the tibia is fixed to the femur with a crossed wire and rotation is completed at a tibial osteotomy.

Post-operative care

In the younger child a hip spica is the most satisfactory method of protection and used for the first 6–8 weeks. In the older patient where a larger intramedullary rod can be used without concern for the growth plates a spica may not be necessary. As soon as there is union across the osteotomy sites the patient is mobilized and encouraged to crawl and kneel using the recently rotated foot in weight bearing. Standing against a chair or stool is also encouraged, so that the patient can use the foot for balance and impor-tantly for practice in using the muscles of the leg in a manner the reverse of the pre-operative state.

The prosthesis

The prosthesis (Fig. 26.5) can be fitted as soon as knee fusion is assured. Although the ankle joint in these children is less precise with regard to the axis of motion when compared to children without a limb deficiency, care must be taken to align the hinges of the new 'knee' joint in concert with the axis of rotation of the talus in the ankle mortise. If the patient is not using the limb after surgery and the prosthesis fitted to the extent that physical examination would suggest is possible, then the patient should be subjected to further radiological examination with an image intensifier to see if the axis of rotation of the prosthesis does indeed match the axis of rotation of the talus in the ankle joint.

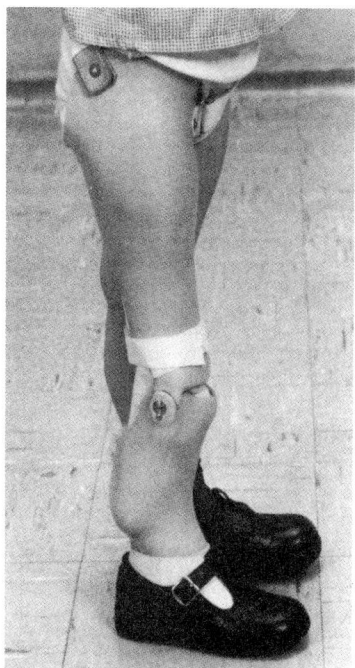
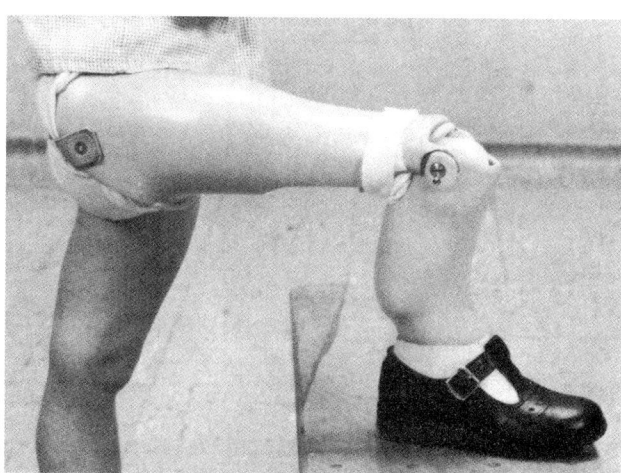

Fig. 26.5. (a and b) Three-year-old child with right proximal focal femoral deficiency fitted with a modified below-knee prosthesis showing the flexion and extension of her new 'knee' joint.

Tumour of the femur

Lesions of the femur which on radiological examination are suspicious of malignancy should be investigated as such. These investigations should include computed tomography and magnetic resonance imaging accurately to assess the involvement of soft tissues and in particular the involvement of the neuromuscular bundles. Tumour involvement of the sciatic or peroneal nerves is a contraindication to surgery. Furthermore, the biopsy which is part of the investigative process must be performed in the expectation that the lesion is malignant and will require further surgical resection and thus should be placed in a site that can be excised at the time of any 'en bloc' resection.

Many primary malignant lesions of the femur can be resected adequately and reconstructed with a prosthetic femoral component, allograft or autograft or a total hip or knee arthroplasty with success. However, where an adequate resection will render the limb useless even if the femur is reconstructed or where the lesion's proximal extent would necessitate an amputation so high as to result in a functional hip disarticulation, then consideration should be given to retention of the tumour-free components of the limb and utilising a rotationplasty. In this way function equivalent to a transtibial amputation can be achieved with a modified 'below-knee' prosthesis.

Lesions evident at birth should not necessarily require the services of a vascular surgeon. However, in the case of tumour resection, as the procedure does involve division of both femoral and popliteal vessels and reanastomosis of the femoral to the popliteal vessels at the completion of the tibial rotationplasty, a vascular surgeon should be involved. In tumour surgery the resection of the lesion is paramount and where doubt exists it is safer to err on the side of caution and accept that removal of all of the thigh except the sciatic nerve in performing a rotationplasty effectively amounts to an intercalary amputation.

The surgical technique will vary depending on the location of the tumour within the femur and the extraosseus extension of the lesion. To attempt to clarify the description the topic has been subdivided into tumours of the proximal third of femur, tumours of the diaphysis and those in the distal aspect of the femur. This surgery has greater application in young patients of an age when arthroplasty surgery is otherwise contraindicated.

Tumours of the proximal third of the femur

This small group of patients includes those in whom an 'en bloc' resection of the proximal femur will destroy hip joint function. In this situation the ability to sit and stand can be preserved following resection by fusing the proximal end of the residual femoral segment to the ilium after rotating the limb through 180°.

The rotated knee joint provides a 'hip' that can flex and extend although obviously not rotate. The rotated ankle becomes the patient's 'knee' joint to provide the patient with flexion and extension and the power to control a 'below knee' prosthesis.

Incision

The incision for these patients is shown in Fig. 26.6. This incision allows resection of the primary lesion without actually encountering the tumour. The plane of the proximal skin incision is between the anterior superior iliac spine and the gluteal fold. The distal incision passes around the distal one third of the thigh. The flap created must match the defect over the buttock and hip. Thus the medial margin of the flap will be significantly more proximal than the lateral margin. The posterior connecting incision follows the line of the sciatic nerve.

Excision of the tumour

The technique of the resection is described by Winkelmann (1986). The proximal line of resection depends on whether the hip joint is not involved (Type 1) or involved (Type 2) in which case the line of resection must include the ilium. Anteriorly, the femoral nerve is isolated and divided, the saphenous vein is ligated and the femoral artery and vein are ligated sufficiently distal to provide access for the revascularization. The anterior thigh muscles are removed from their origins.

Medially, the obturator nerves and vessels are dissected and ligated. The obturator externus is divided and the psoas tendon detached from its insertion. Posteriorly and laterally the glutei and external rotators of the hip are detached from the femur but leaving the insertions on the bone. The capsule of the hip joint is divided in the Type 1

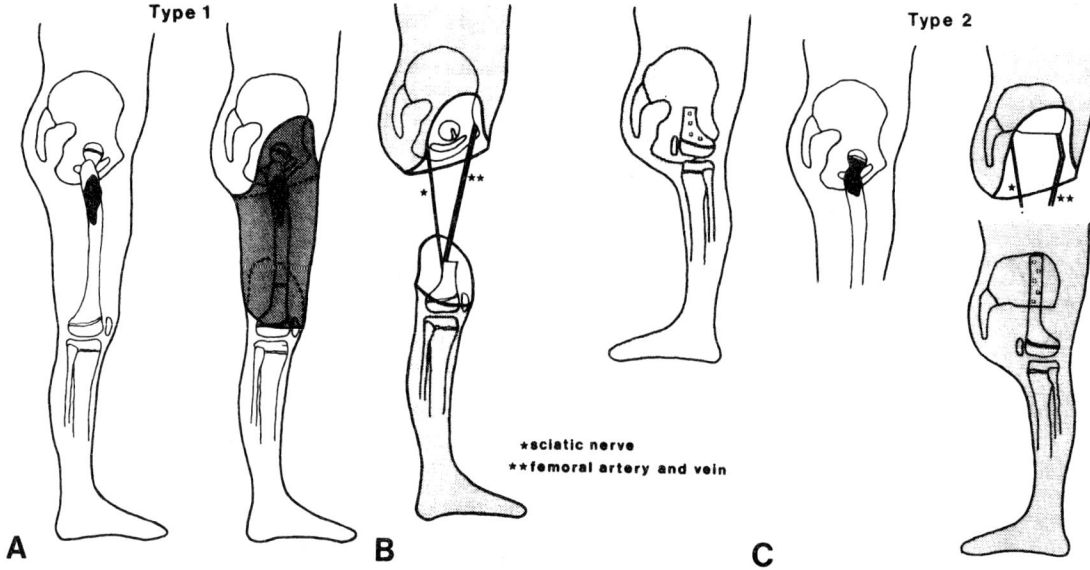

Fig. 26.6. Schematic diagrams of the surgical techniques.

patients, allowing disarticulation of the hip and removal of the tumour. In the Type 2 patients the proximal line of resection is through the ilium proximal to the acetabulum. Clearly a greater degree of dissection is required and then the tumour specimen is removed 'en-bloc' with the hip joint. In some patients it may be necessary to use a bone graft joining the ilium to the pubis to stabilize the pelvis.

Distally, the quadriceps, adductors and hamstrings are divided. The femoral artery and veins are isolated and divided for re-anastomosis. The femur is osteotomized at a predetermined level to provide the appropriate overall length with the knee joint at the level of the ischium. The age of the patient must also be taken into account to allow for future growth of the adjacent growth plates.

Posteriorly, the sciatic nerve is isolated and protected. This will allow removal of the thigh including all malignant tissue.

The reconstruction

After the resection, reconstruction proceeds with the limb carefully rotated laterally through 180°. The proximal femur is bevelled to fit the lateral aspect of the wing of the ilium. In both Types 1 and 2 the femoral fragment is fixed to the pelvis by screws. The residual femoral segment is aligned so that the limb will move in the sagittal plane through the reversed knee joint and fixed to lie in the coronal plane. This fixation is supplemented by a spica cast until union is assured.

Posteriorly, the quadriceps are sutured to the glutei and anteriorly the psoas is sutured to the hamstrings. The femoral artery is re-anastomosed to either the distal aspect of the femoral artery or to the popliteal artery. The sciatic nerve is looped and placed under the gluteal flap.

Post-operative care

During the post-operative period the patient will require assistance and training in order to use the reconstructed limb. When the union of the femur to the pelvis is sound the patient is fitted with a rotationplasty prosthesis and further gait training instituted.

Tumours of the distal two-thirds of the femur

The patients in this group are those where an adequate margin of tumour-free bone of the femur can be preserved, allowing fusion to the proximal tibia and maintenance of the function of the hip joint without compromising resection of the tumour. This usually means that the proximal tumour margin is at least 5 cm distal to the lesser trochanter.

The incision

A rhombus shaped skin incision is made with the long axis on the anterior aspect of the thigh. The distal margin of the incision is at the level of the proximal tibia. The proximal margin of the incision can be adjusted depending on the planned length of the limb and the proximal margin of the tumour.

Because of difficulties encountered with the rhombus incision, Gebhart et al. (1987) devised a modified incision. This incision incorporates a circumferential incision around the thigh proximally and a fish-mouth shaped incision around the leg distally.

Resection of the tumour

The femoral artery and vein are exposed in the proximal aspect of the wound and ligated and divided. The femoral nerve is also divided at this level. The muscles of the anterior thigh are divided at the proposed level of femoral transection, thereby exposing the femur. The level of femoral transection should be at least 5 cm proximal to the proximal margin of the tumour as defined by the pre-operative scans.

Posteriorly, the sciatic nerve is exposed and traced distally. The common peroneal nerve is then traced down the fibula. The tibial nerve is also traced distally. Involvement of these nerves in the tumour mass is a contraindication to the continuation of the procedure.

The remaining muscles of the thigh are then divided proximally and the profunda vessels ligated at the level of the femoral transection.

At the level of the proximal tibia the popliteal vessels are isolated, ligated and divided. The peroneal nerve must be protected throughout the dissection. The proximal aspect of the tibia is then exposed just distal to the capsule of the knee joint and sectioned usually distal to the proximal epiphysis. The specimen is then removed 'en bloc' and can be sectioned longitudinally macroscop-ically to ascertain adequate margins of resection. Microscopic examination of the tissues at the ends of the specimen should also be carried out.

The reconstruction

The limb is then externally rotated through 180° and the distal end of the femur and the proximal end of the tibia joined by either an intramedullary nail or by a plate and screws. The advantage of an intramedullary nail is that it does allow for further rotational adjustment in the post-operative period. Prior to performing the osteosynthesis, the length of the residual limb is determined. Allowance is made for future growth discrepancies between the ipsilateral tibia and the contralateral femur. Initially, the axis of the rotated ankle joint should lie approximately 4–6 cm distal to the contralateral knee joint.

The femoral vessels are then anastomosed to the popliteal vessels. Once circulation is restored the residual quadriceps muscle is sutured to the calf muscles and the adductors and hamstrings are sutured to the periosteum and tissues around the proximal tibia. The redundant sciatic nerve is tucked into a space deep to the proximal ham-strings. The skin is closed over a drain in a routine manner.

The patient is then protected in a hip spica cast. The time in the cast depends primarily on the stability of the fixation but it is usually possible to begin mobilization of the limb within 3 weeks. When stability is assured and bone healing observed a prosthesis can be fabricated.

Prosthetic requirements

The prosthesis for patients who have had a rotationplasty for either congenital or surgical deficiencies may look outwardly similar. However, following surgical resection in tumour manage-ment the hip is invariably stable. Such is not always the case in PFFD patients and it may be necessary to use a pelvic band to control rotation, particularly in the younger patients (Bochmann, 1980).

At the ankle joint or new 'knee' joint it is necessary to take considerable care in positioning the hinges. In PFFD patients there is commonly some degree of ankle instability due to congenital anomalies, so that the axis of rotation of the ankle is not as precisely defined as in a normal ankle and thus the joint is more forgiving of minor malpositioning of the hinges. In tumour cases, however, the ankles have normal stability, the patients are usually older and positioning of the hinges is more critical.

If a patient is forced to wear a prosthesis where the ankle's axis of motion is not the same as the prosthetic hinges then either the range of motion will be restricted or the ankle joint will deteriorate and become painful, or both.

Summary

The principles of rotationplasty and the surgical techniques applicable for patients with deficiency of the femur or malignant tumours of the femur have been described. This procedure remains a valuable component of the armamentarium of the surgeon involved with these patients. The procedure can provide the patient with a level of function that is superior to that of an 'above knee' amputation and yet allows conversion to a stump that is of a more desirable length should either the function or cosmesis fail to meet with the patient's needs.

In the case of deficiency of the femur of the severe type, i.e. PFFD, some form of prosthetic assistance is obligatory and the goal is to provide the optimal function for the patient. In the case of malignant tumours of the femur the goal must be to resect the primary lesion adequately and then to make the best use of the residual healthy tissue. In the author's experience in both groups of patients the level of function attained in the 'below knee' prosthesis has surpassed that of comparable 'above knee' amputees and the concern regarding cosmesis has not proved to be a problem. To date, not one patient has requested conversion to an above-knee stump.

References

Anderson, M., Green, W.T., and Messner, M.B. (1963) Growth and predictions of growth in the lower extremities. *J. Bone Joint Surg.*, **49-A**, 1–14.

Bochmann D. (1980) Prosthetic devices for the management of proximal femoral focal deficiency. *Ortho-pros.* (Can.), 4–29.

Borggreve J. (1930) Kniegelenksersatz durch das in der Beinlangsachsce um 180 gedrehte Fusgelenk. *Arch. Orthop. Chir.*, **28**, 175–8.

English, E. (1981) The energy costs of walking for the lower-extremity amputee. In *Amputation Surgery and Rehabilitation* (ed. J.P. Kostuik) Churchill Livingstone, Edinburgh, pp. 311–314.

Fixsen, J.A. (1983) Rotationplasty. Editorial. *J. Bone Joint Surg.*, **65B**, 529–530.

Gebhart, M.J., McCormack, R.R.,Jr., Healey, J.H., Otis, J.C., and Lane, J.M. (1987) Modification of the skin incision for the Van Nes limb rotationplasty. *Clin. Orthop.*, **216**, 179–182.

Gillespie, R. and Torode, I.P. (1983) Classification and management of congenital abnormalities of the femur. *J. Bone Joint Surg.*, **65-B**. 557–568.

Glynn, M.K. (1983) *Biomechanics of Gait following Van Nes Rotationplasty in Children with Proximal Femoral Focal Deficiency.* Thesis for M.Ch. National University of Ireland.

Kostuik, J.P., Gillespie, R., Hall, J.E. and Hubbard, S. (1975) Van Nes rotational osteotomy for treatment of proximal femoral focal deficiency and congenital short femur. *J. Bone Joint Surg.*, **57-A**, 1039–46.

Kotz, R. and Saltzer, M., (1982) Rotationplasty for childhood osteosarcoma of the distal part of the femur. *J. Bone Joint Surg.*, **64-A**, 959–969.

Kritter, A.E., (1977) Tibial rotation-plasty for proximal femoral focal deficiency. *J. Bone Joint Surg.*, **59A**, 927–934.

Torode, I.P., and Gillespie, R. (1983) Rotationplasty of the lower limb for congenital defects of the femur. *J. Bone Joint Surg.*, **65-B**, 569–573.

Van Nes, C.P. (1950) Rotationplasty for congenital defects of the femur: making use of the ankle of the shortened limb to control the knee joint of the prosthesis. *J. Bone Joint Surg.*, **32-B**, 12–16.

Winkelmann, W.W., (1986) Hip rotationplasty for malignant tumours of the proximal part of the femur. *J. Bone Joint Surg.*, **68-A**, 362–369.

Section 12

Amputation and growth

27

Amputations in the growth period including deficiencies present at birth

G. Neff

Introduction

Whenever amputation is considered in a child or a juvenile it is important to be aware that a child is not simply a 'small adult'. Frantz and Aitken (1959) explained rather clearly, that a child is psychologically, emotionally and physically 'in flux'. Furthermore, they point out that there exists economical dependence on the family which is an influence on the child, in contrast to the independent situation of the adult.

In addition, essential differences in the physiology and development of the growing body have to be taken into consideration in decision making before, during and after amputation and especially so with respect to discussing the possible conversion of a longitudinal limb deficiency birth into an amputation stump.

In general the skeleton is growing both in a longitudinal and a circumferential manner which sometimes necessitates additional corrective surgery throughout the growth period unless precautions have been taken. In contrast to the adult, the tissue of a child and especially with regard to the skin has a much greater tolerance with respect to both tension and pressure. This attribute may permit the saving of more length of a severed limb as skin closure can heal despite the tension and quite large areas of skin grafts can contribute to a longer stump.

Origin and aetiology of amputation in childhood and adolescence

Amputations can be classified as acquired or present at birth (still better known as 'congenital'). Acquired amputations in children and juveniles are predominantly due to trauma, mainly caused by machinery, traffic (Fig. 27.1, a, b), explosive and thermal injuries.

Malignant and in selected cases benign tumours (Fig. 27.2, a, b), infection and post-traumatic as well as haematogenous osteomyelitis may result in amputation as well as extensive leg-length discrepancy, sometimes combined with serious deformation and joint instability; furthermore, paralysis and neurotrophic disorders with markedly impaired function may be an indication for amputation (Murdoch, 1977).

Transverse deficiencies present at birth represent a stump-like deficit of the limb (Fig. 27.3), whereas longitudinal deficiencies may be a case for conversion to an amputation stump depending on function and sometimes viability, e.g. constriction band syndrome (Fig. 27.4, a, b, c) or appearance. Critical judgement is required in considering the options – amputation, corrective surgery or simply refraining from any kind of surgery in the limb deficient child. Thoughtless

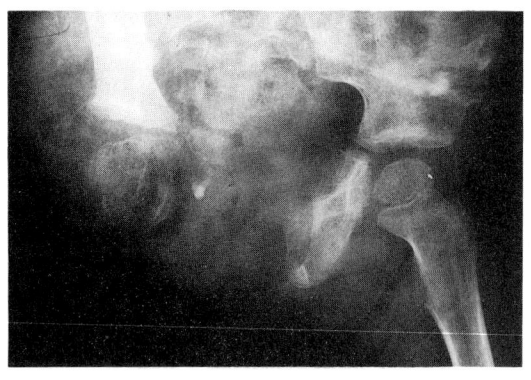

(a)

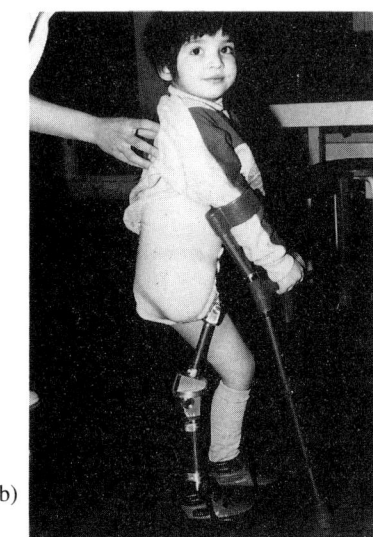

(b)

Fig. 27.1(a,b) Acquired amputation due to trauma; the little girl was run over by a bus, resulting in a partial through pelvic amputation (hemipelvectomy) besides other serious abdominal and thoracic injuries.

and hasty surgery (Fig. 27.5) may interfere seriously with the overall function of such a child and may create more dependence than prior to surgery.

Though longitudinal deficiencies are more common in the upper extremity (Gibson, 1980) with virtually no indication for amputation, there is a much higher incidence of amputation in lower limb deficiencies to allow for better stance and gait with an appropriate prosthesis (Fig. 27.6). In the multiple limb deficient child, however, apparently useless looking feet in defective position may be

(a)

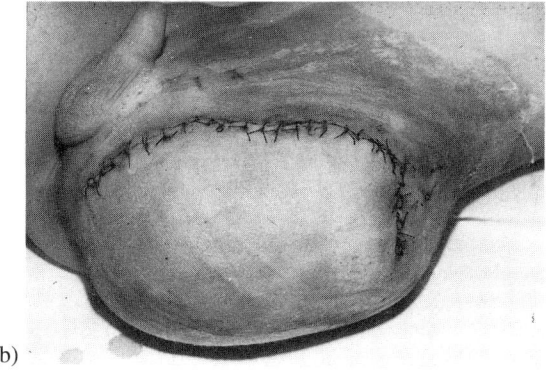

(b)

Fig. 27.2(a,b) Typical osteosarcoma of the distal femur with 'skip' metastases requiring hip disarticulation.

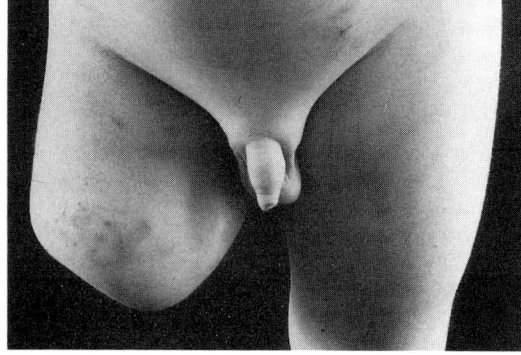

Fig. 27.3. Transverse deficiency – above-knee 'stump'.

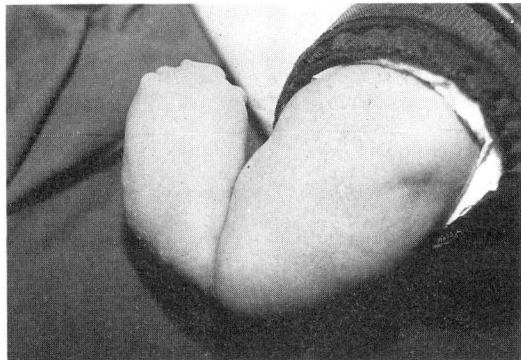

(a)

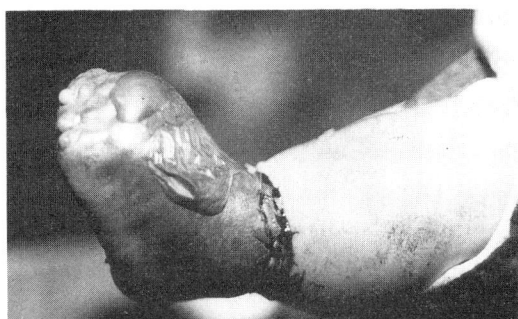

(b)

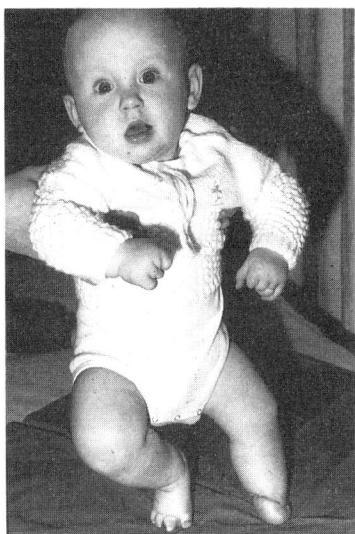

(c)

Fig. 27.4.(a–c) Constriction band syndrome: (a) with subtotal amputation present at birth; (b) after excision of the constriction band with incomplete blood circulation and (c) after definitive surgery creating a weightbearing hind-foot stump.

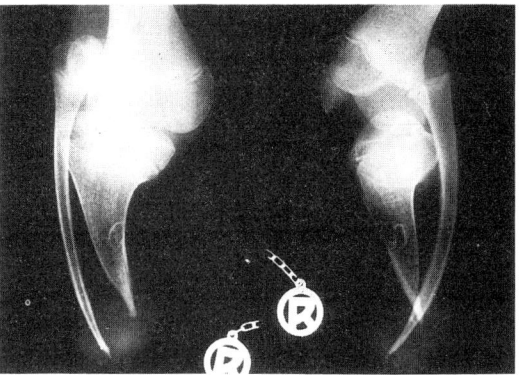

Fig. 27.5. Transtibial stump after amputation due to longitudinal limb deficiency with bowing tibia and fibula and imminent osseous outgrowth of the bone spikes.

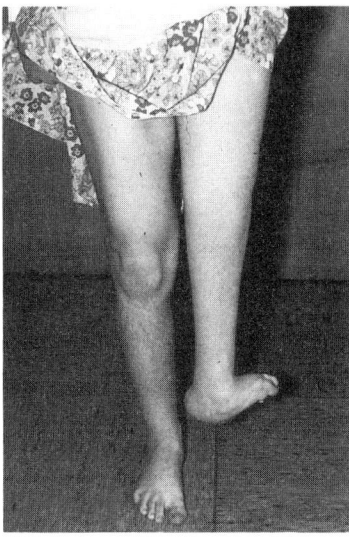

Fig. 27.6. Excessive leg length discrepancy with stiff knee joint, dislocated hip joint and deformed rotated foot prior to Pirogoff amputation and prosthetic fitting.

indispensable to compensate for a serious loss of function in both upper extremities (Kruger, 1980; Marquardt, 1981 a, b, c). Aitken and Frantz (1964) and later on Lambert *et al.* (1976), respectively, reported on 10% and 6.2% of conversions in upper limb longitudinal deficiencies, whereas in the case of the lower limb, the incidence of amputation was 50% and 28.8%.

The stump of a child amputee

For children and in the adolescent the loss of a limb by knee disarticulation, Syme's ankle disarticulation or a modified Pirigoff amputation such as Boyd's results in total weight bearing as in the adult. In contrast, in the growing skeleton an amputation through the shaft of a long bone, especially below the knee joint and above the

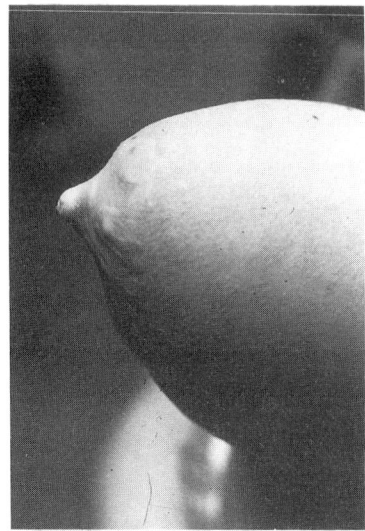

(a)

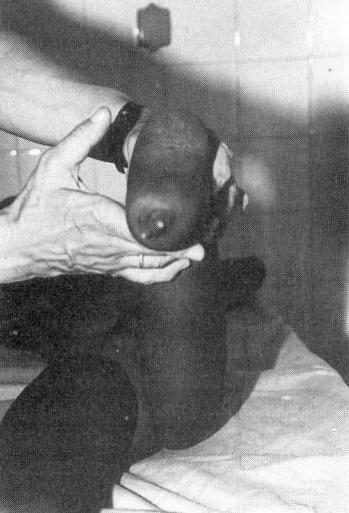

(b)

Fig. 27.7 Clinical appearance of bony outgrowth in (a) transhumeral stump and (b) in a transtibial amputation.

elbow joint, is very likely to develop terminal osseous overgrowth with the spiky bone end penetrating through the soft tissue, thus making stump revision inevitable (Abraham *et al.*, 1968) (Fig. 27.7). Aitken and his collaborators in a series of contributions (1953, 1959a, b, 1962, 1963, 1964, 1968, 1981) saw the reason for this development only in terms of appositional growth and at the same time in the rarification at the bone end of the stump. We now know that the different growth potential of the growth plates (Jeannopoulos, 1969) immediately proximal to the stump end has an important input: the higher the contribution to length, the higher is the risk of developing terminal osseous overgrowth. This correlates very well with clinical experience and Aitken's (1963) statement that overgrowth is to be expected 'in stumps of the humerus, fibula, tibia, femur in that order of frequency'.

Instead of continuous re-amputations with an increasing shorter stump, an angulation osteotomy of the humerus or stump capping with an osseous-cartilage graft, both methods developed by Marquardt (Marquardt and Neff 1974; Marquardt *et al.*, 1983; Warquardt 1976, 1981 a, b), prevent further occurrence of osseous overgrowth and contribute to a growing length of the remaining limb (Blasius *et al.*, 1990; Bernd *et al.*, 1991). Silastic plugs (Swanson, 1972) used for the same purpose proved unsuccessful.

Friedmann and Friedmann (1985) successfully used continuous skin and soft tissue traction as practised by Marquardt (1981 a) in seven amputees to overcome the problem of osseous overgrowth. This method, however, is very demanding for the patient as well as the family and requires absolute compliance for a long period.

If an amputation can be planned without urgency and risk of infection a proper osteoplasty with closure of the open bone end by a periosteal flap or in the transtibial amputation by a periosteal bridge (Fig. 27.8) between the tibial and fibular ends may prevent osseous overgrowth later on (Ertl, 1949; Dederich, 1970).

The connection between tibia and fibula is blamed for causing varus deformity due to a faster growth of the fibula; nevertheless, we prefer this procedure where applicable because both valgus and varus deformities in children's transtibial stumps occur after ordinary amputation without any osteo-myoplastic procedure. The same argument applies to our advocacy of stump capping

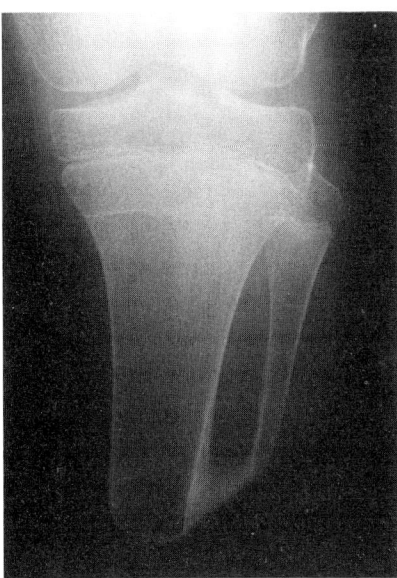

Fig. 27.8. Radiograph of a bone bridge following an osteomyoplasty amputation.

which is perhaps even better than the periosteal flap or bridge because it provides a 'quasi-disarticulation' with full weight bearing on the stump end.

Phantom limb, phantom pain and neuroma

According to the literature (Aitken, 1963; Tooms 1981, 1986), revision for neuroma was necessary in a very small percentage of the child amputee population.

It was also revealed that phantom limb and phantom pain did not appear if the amputation was performed in very young children, e.g. under 5 years of age; up to 10 years of age while phantom limb may be present, phantom pain was seldom experienced. The frequency of phantom limb and phantom pain increases with ongoing age but appears to be of lower severity in comparison to adults (Lambert, 1972). It is worthwhile to note that Jorring (1971) did not find phantom problems in adults who had been amputated during early infancy. In this respect, given that there is adequate shortening of the nerves during amputation addi-

tional procedures, such as injection of alcohol, cortisone and other substances into the nerve, should be avoided.

Level of amputation, specific surgical procedures and rehabilitation

In the upper extremity unilateral amputation, especially so at high transhumeral levels there is a risk of developing a secondary spinal deformity in a growing individual. Furthermore, a unilateral arm amputation has as a possible consequence the establishment of one-handed prehensile patterns in the remaining contralateral hand and arm, thus excluding the entire affected arm from both the body function and image. This can be avoided to a certain extent by fitting an appropriate functional prosthesis as soon as possible after amputation, preferably by early fitting immediately after surgery, maintained during the healing period with an early start to prosthetic training and daily use. The higher the level of amputation and loss of function, the more difficult is the restoration of significant function with a prosthetic system. Even the best available prosthesis has a fundamental lack; namely, sensation and sensory feedback. In addition, overprotective care may lead to an increase of functional neglect with a reduction in physiological functional stress necessary for a normal development of the skeleton as well as of soft tissues.

In bilateral acquired high level arm amputation it is crucial that these children are given every opportunity to use their feet and legs instead of hands and arms for every kind of daily activity possible (Fig. 27.9). Sophisticated high tech arm prostheses in some instances have proved to be at least a contribution to a society which is disgusted at the appearance of an armless individual. Beyond these social interactive aspects this kind of prosthetic expression often turned out to be more an additional handicap or at least an embarrassment for the children than of a real advantage in daily life.

This applies even more in upper limb deficiencies present at birth where missing arms are spontaneously 'replaced' by the prehensile function of the feet and the range of motion of the lower limb without any training or education;

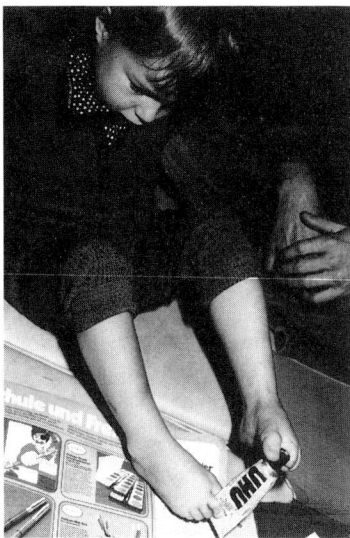

Fig. 27.9. Deficiency present at birth; both arms missing, showing use of both feet to replace missing hands.

however, it is important to leave the feet free of socks and trousers to allow for grasping, touching and in gaining skills and sensory feedback qualities in a natural and spontaneous manner.

Sometimes simple devices like a bandage around the stump to hold a spoon or a pencil proves to be more helpful than a prosthesis, especially so in triple or quadruple limb deficiencies or acquired amputations. Nevertheless, these child amputees develop considerable skills in using their mouth, chin and other parts of the body to grasp, catch, push or to to develop special manoeuvres where we would normally use our hands. It is surprising and fascinating to realize the innovative potential of many of these children.

It is obvious that the complex task of serving those child amputees to the optimal level can scarcely be the job of a single person; the team approach and the dedication of every team member based on professional qualification and skills will create the necessary overall input to reach the goals of rehabilitation.

The following paragraphs draw attention at different amputation levels to related specific details which differ from the adult situation.

Transhumeral amputation

Transhumeral stumps are very likely to develop osseous overgrowth as already mentioned above. Repeated re-amputation and loss of length should be avoided where possible and the stump capping procedure developed by Marquardt (1981a, b, c, 1985; Marquardt *et al.*, 1983) employed. This can be performed best at the occasion of primary amputation by using an autogenous cartilage and bone graft from the amputated limb or bone bank (Fig. 27.10). The stump will eventually provide

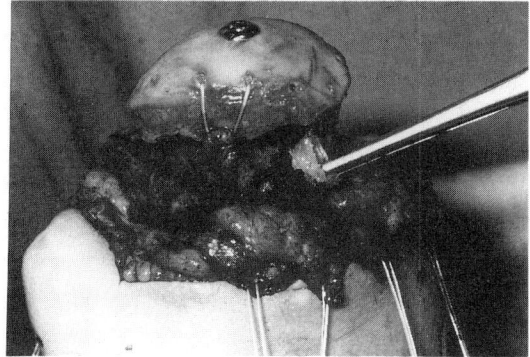

(a)

(b)

Fig. 27.10.(a,b) Stumpcapping procedure as developed by Marquardt in a boy with an acquired transhumeral stump and imminent osseous overgrowth (a) by using a cartilage–bone graft from the bone bank and (b) resulting in full weight bearing of the stump end.

excellent weight bearing capacity but the most important gain is further growth of the remaining part of the humerus rather than repeated stump revisions for osseous overgrowth which should belong to the past.

In long transhumeral stumps a primary or secondary angulation osteotomy of the distal stump end (Marquardt, 1974, 1976, 1981a, b, c) prevents osseous overgrowth in both children and juvenile amputees; in addition, the angulation

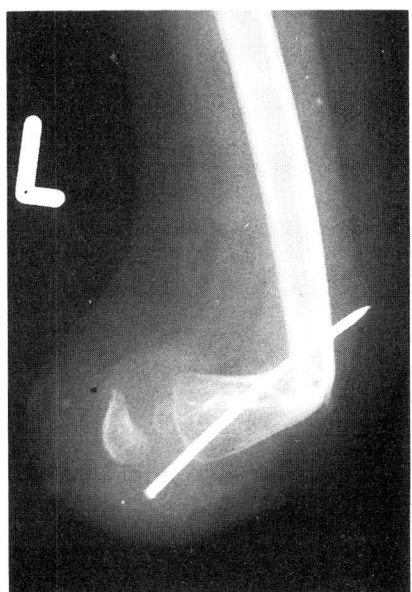

(a)

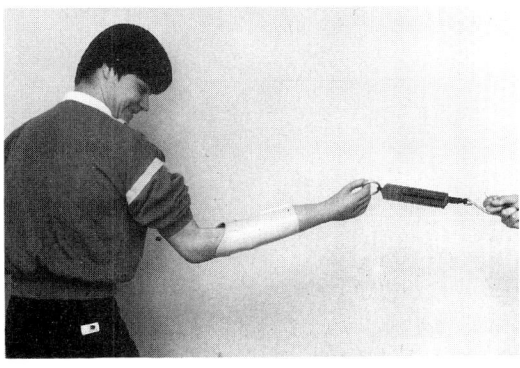

(b)

Fig. 27.11.(a) Angulation osteotomy according to Marquardt, (b) and after surgery providing excellent suspension and rotational stability for a functional prosthesis.

provides excellent suspension and rotational stability for a functional prosthesis (Fig. 27.11 a, b). Thus the full range of motion in the shoulder joint and shoulder girdle is maintained and can also be used for positioning of the terminal device (Neff, 1979; Gillespie, 1981).

Elbow disarticulation

Elbow disarticulation stumps can sustain considerable end-bearing, avoid osseous overgrowth and allow for sufficient self-contained suspension of a prosthesis fitted to the shape of the epicondyles and the supracondylar undercut. After acquired amputation in childhood the loss of length during the growth period is minimal.

In contrast, transverse deficiencies present at birth at this level may be subject to reduced growth because of hypoplastic growth plates. Due to the missing flexor and extensor muscles of the forearm and hand, the ulnar and radial epicondyles are less prominent. For proper suspension and direct transfer of forces and movements to a functional prosthesis an angulation osteotomy will be necessary in these cases.

Transradial amputation

In the upper limb transverse deficiencies of the proximal forearm are the most frequent ones present at birth.

In forearm amputation, too, as much length as possible should be preserved with respect to future function and direct use of the stump with and without a prosthesis; thus a long transradial stump can be fitted with an open end prosthetic socket (Kuhn, 1979) which allows for tactile sensation with the stump end (Fig. 27.12). This is important especially in bilateral amputations, both in acquired and those present at birth.

In bilateral amputation a conversion of one or better, both forearm stumps into a Krukenberg 'forceps' stump (Krukenberg, 1917; Jain, this volume) is advantageous and improves prehensile and tactile function for better independence in daily activities for child and juvenile amputees as well (Swanson, 1964; Swanson and Swanson,

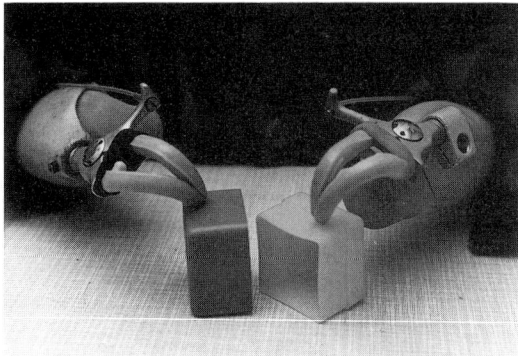

Fig. 27.12. Open-end prosthetic sockets in long congenital transverse deficiencies of both forearms allowing sensory feedback with the free stump ends.

1980). Marquardt (1981c) described specific surgical technique for the preservation of good sensory abilities in the tips of both 'Krukenberg' branches.

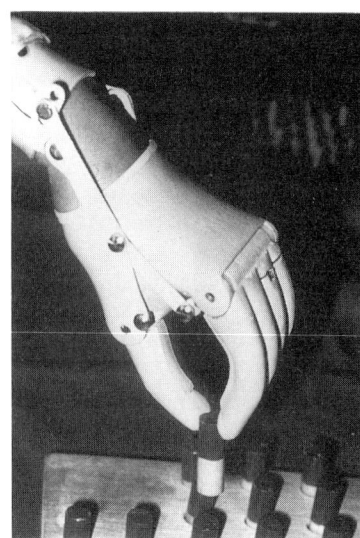

Fig. 27.13. 'Totalbad hand' for an adult amputee with an acquired through-carpal stump, which opens and closes the 'hand' by direct wrist movement transfer to the prosthesis.

Through-wrist disarticulation

Disarticulation stumps of the wrist offer a bulky stump end for self-contained prosthetic sockets, thus allowing for a full range of motion in the elbow joint and also for good pronation and supination of the forearm. In this respect the distal end of the radius and the ulnar should not be smoothed in order to reduce bulk; only a very prominent styloid of the radius or the ulnar need be removed.

Carpal disarticulation

Carpal stumps, be they present at birth or acquired, are normally used spontaneously without any prosthetic device for tactile and prehensile function and for hand-in-hand function with the contralateral hand. This obtains even more importance in bilateral amputation. The active movement can also be used to control a prosthetic hand, e.g. the 'Tobelbad-Hand' (Zrubecky, 1976), constructed according to the principle of 'wrist extension-finger flexion' of a flexor hinge orthosis (Fig. 27.13).

Hemipelvectomy and hip disarticulation

Amputations in these levels are predominantly tumour related; even so in recent years an increasing number of children have survived serious traffic accidents with this very high level of traumatic amputation. Every attempt should be made to preserve as much as possible of the pelvis, soft tissues, and local skin for future prosthetic

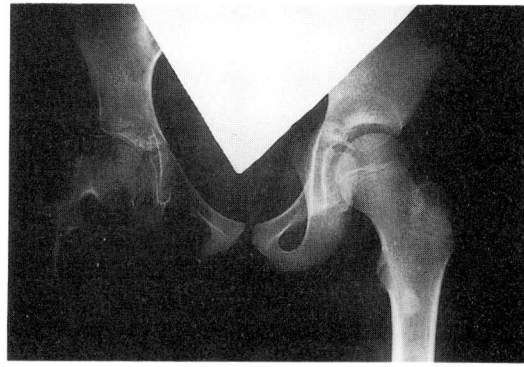

Fig. 27.14. Hypoplastic femoral stump after traumatic amputation in early childhood.

fitting. Indeed all the tissues, including the skeleton on the amputation side, will end at maturity more or less hypoplastic in comparison to the normally developed contralateral extremity (Fig. 27.14). This may cause problems for satisfactory fitting of a prosthesis and related proper stance and gait.

Transfemoral amputation

According to Baumgartner (1979), trauma is the major reason for amputation at transfemoral level in children with a majority of boys. The second most frequent cause is malignant tumour; change is likely with modern treatment concepts avoiding amputation and in favour of resection with limb salvage.

After transfemoral amputation in children an increasing shortening of the remaining femur is inevitable due to the lack of length contribution from the missing distal growth plate; the younger the child, the greater the loss of length by the end of growth (Neff, 1992).

In addition, reduced function despite regular prosthetic use has a negative impact on physiological stress related growth stimulation. Stump capping for weight bearing and perhaps also for additional growth provided by an active growth plate within the graft could be an answer to this defective development of transfemoral amputation stumps. The distal tibia has been used for stump capping while performing a transfemoral amputation in a boy for osteosarcoma; the distal growth plate was still open 2 years after amputation and the stump end had developed reasonable weight bearing properties (Marquardt and Correll, 1984; Marquardt, 1987).

Knee disarticulation

Because of these deficits in further growth of an amputated thigh a knee disarticulation stump should be performed not only to avoid osseous overgrowth, but by preserving the distal growth plate to ensure as much length as possible by the end of the growth period. Furthermore, the full weight bearing capacity of the stump permits the fitting of precise prostheses with physiological

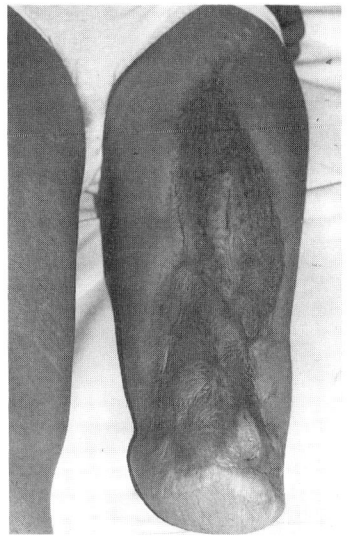

Fig. 27.15. Large area covered with split thickness skin grafts to preserve knee-disarticulation stump after severe trauma.

transfer of forces between the stump end and the prosthetic socket.

A medial and a lateral flap technique is preferred to a long anterior flap to cover the condyles. In children, however, one can use any and every part of the remaining skin after trauma in creating flaps to ensure skin closure as long as there is a minimum of blood circulation. Major damage and loss of skin may necessitate resection of part of the femoral diaphysis, saving the distal end of the femur with the growth plate (Marquardt, 1985, 1987). If necessary extensive split thickness skin grafts may save a knee disarticulation stump (Fig. 27.15). The obvious advantages compared to a transfemoral stump apply equally well to both children and juvenile amputees (Neff, 1987).

Transtibial amputation

In transtibial stumps terminal osseous overgrowth occurs next in frequency after transhumeral amputation. Stump capping with a cartilage and bone graft should be performed to prevent further stump revisions (Marquardt, 1976, 1981a, b, 1988; Blasius *et al.*, 1990; Bernd *et al.*, 1991).

Compared to the normally expected growth range, those stumps with cartilage and bone stump caps developed almost the expected length by the end of the growth period (Bernd *et al.*, 1991). In contrast, Christie *et al.* (1979), in an investigation of 18 children with 20 transtibial stumps, described a remarkable reduction of the stump growth. Only three stumps achieved approximately the expected length at maturity; an average of only 36% of the expected growth was observed in eight congenital related transtibial stumps; traumatic transtibial stumps achieved 53% only of the precalculated length. Seven patients in all underwent stump revisions for osseous overgrowth before reaching maturity.

Taking these facts into consideration, planning for a transtibial amputation should include the management of an osteoplastic procedure for the stump end, whether by stump capping, preferably with an autogenous cartilage and bone graft from the amputated part of the limb, or by bridging well shaped bony ends of the tibia and the fibula with a tube constructed from periosteal flaps to create a solid bone bridge with similar properties to a stump cap (Ertl, 1949; Neff, 1986a; Dederich, 1970). Sometimes a varus or less frequently a valgus deviation of the stump may occur, but this also occurs in transtibial stumps without any kind of osteoplasty. In the author's experience alignment of the prosthetic socket to accommodate the deviation has proved sufficient and additional corrective osteotomies or epiphysiodesis have not proved necessary.

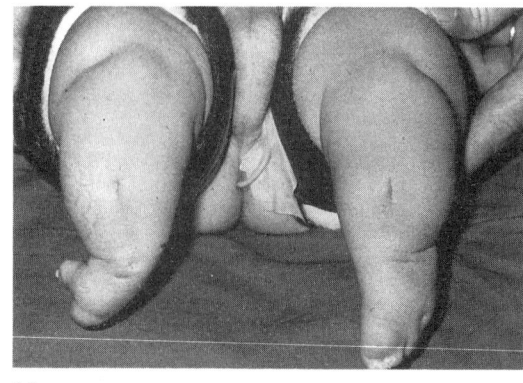

(a)

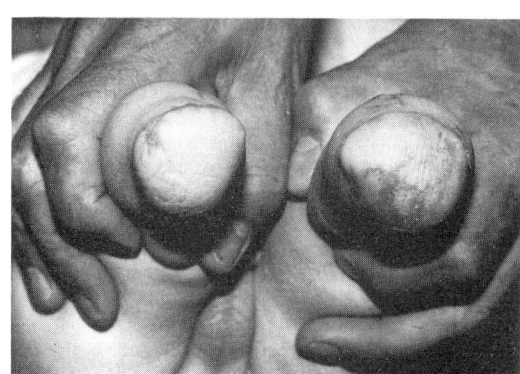

(b)

Ankle disarticulation/hindfoot amputation

In the English speaking world the Syme ankle disarticulation has enjoyed clear priority compared to Boyd or modified Pirigoff amputations (Wood *et al.*, 1968; Mazet, 1968; Greene and Cary, 1982; Anderson *et al.*, 1984; Ferguson *et al.*, 1987). Most of these follow-up papers presented generally satisfactory or excellent results, but specific problems of the Syme stump were also identified, predominantly heel pad migration and scar problems.

Frankovitch and Farrell (1984) recommended the Boyd amputation as the 'choice for children's amputations' with rare complications compared to

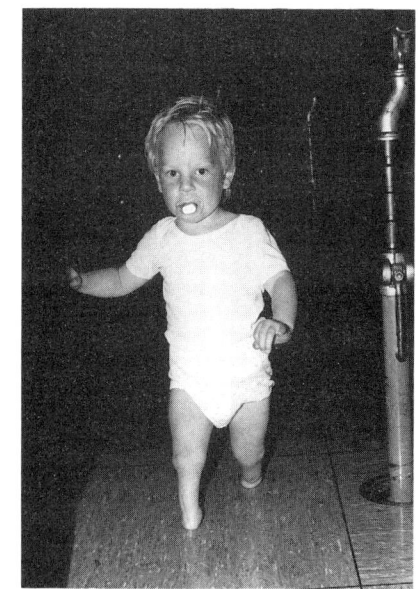

(c)

Fig. 27.16.(a) bilateral longitudinal fibular limb deficiencies **(b)** converted to Syme amputation and **(c)** demonstrating end bearing.

Syme stumps. The modified Pirigoff stump is advocated rather than either the Boyd or Syme procedures, as it retains the pressure durable hindfoot skin in the weight bearing area (Neff *et al.*, 1977, Neff, 1986c). If a Syme amputation has to be performed due to lack of sufficient calcaneal bone and/or normal weight bearing hindfoot skin a simple ankle disarticulation is indicated combined with a limited resection of the malleoli but leaving untouched the cartilage surface of the distal tibia. Fixation of the heelpad is obtained by a few well placed sutures from the capsular remnants of the ankle joint to the tendon of the tibialis anterior muscle or alternatively via drill holes in the bone. Finally, the employment of the temporary fixation of a well shaped plaster of Paris cast ensures uneventful healing and shrinking of the skin. These steps provide for a satisfactory result for the Syme amputation in children and juvenile amputees whatever the causal condition (Fig. 27.16, a, b, c).

Even better when possible will be a tenomyoplastic Chopart amputation developed by Marquardt (Marquardt, 1973; Neff *et al.*, 1977; Neff, 1986 b, c). In order to prevent a contracture as usually occurs in the original Chopart stump (Lisfranc, 1815), it is mandatory to suture the tendons of the tibialis anterior, extensor hallucis longus and extensor digitorum communis muscles to the plantar capsular and ligament structures. A second antagonistic force to the unimpaired triceps surae muscles is created by deploying a long musculo-cutaneous flap of the sole and suturing the short flexor muscles of the toes to the fascia and anterior capsule of the ankle joint. By re-establishing a powerful counterforce to the triceps surae the full range of active motion in the ankle joint is preserved and a degree of 'push-off' can be exerted while walking barefoot on the stump or with specifically designed orthopaedic footwear (Fig. 27.17).

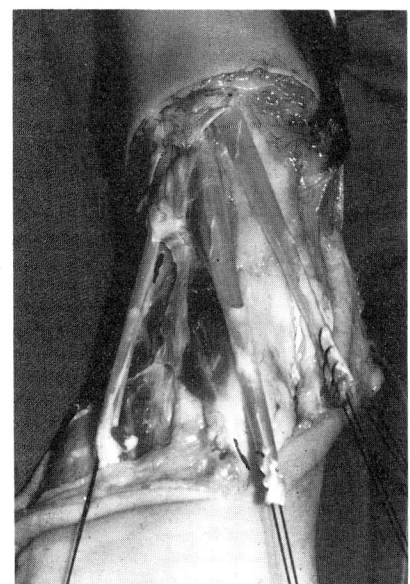

(a)

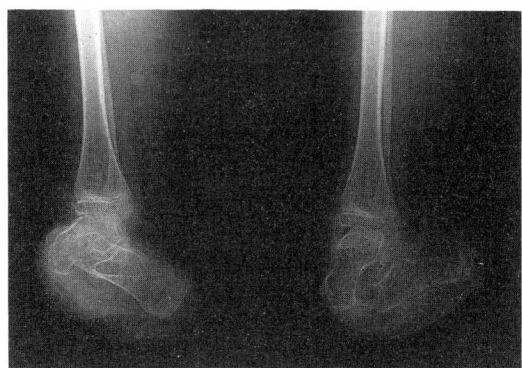

(b)

Fig. 27.17.(a,b) Tenomyoplastic Chopart amputation according to Marquardt. (a) The tendons of the tibialis anterior, the extensor hallucis longus and the extensor digitorum communis muscles are sutured to the capsular and associated ligaments. (b) A long musculo-cutaneous flap of the sole provides additional power once sutured to the anterior capsular structures of the ankle joint providing controlled planti-and dorsi-flexion.

Transtarsal disarticulation transmetatarsal and toe amputation

The longer the stump, the better the function provided there is sufficient pain-free motion and satisfactory soft tissue and skin cover of the stump (Neff, 1986 a, c). This is certainly true for forefoot stumps in children and juvenile amputees.

Lisfranc disarticulation or, even shorter, Bonna-Jager transtarsal stumps have an advantageous longer lever arm for the extension of the ankle joint as long as the tendons of the extensor muscles are properly inserted at the stump end (Lisfranc, 1815; Bonna-Jager).

Transmetatarsal amputations should be restricted to the spongy part of the bones in the proximal third, as stumps of the diaphyses may end up with spiky bone tips with growth and necessitate stump revision.

Beyond these considerations, availability of soft tissue with weight bearing skin from the sole of the foot will dictate the limits for a proper amputation level. Metatarsal stumps as well as disarticulation stumps in the metatarso-phalangeal (MTP), proximal interphalangeal (PIP) or distal phalangeal (DIP) level should be covered with soft tissue specifically designed for weight bearing, thus preventing future pressure sores at the stump end.

If one or more toes have to be amputated, usually a disarticulation should be performed with the exception of the big toe: a trans-phalangeal amputation will preserve the best 'push-off' effect as long as there is a full active range of motion in the MTP I and PIP I joints, respectively.

In contrast to more proximal amputation levels, tendons should not be used for tenoplastic sutures as this will compromise the range of motion in the neighbouring toes similar to the 'Quadriga-syndrome' in the hand (Verdan, 1972).

Sometimes it is necessary to amputate longitudinally one or more rays more or less entirely. In these cases the minimum of remaining functional rays to be preserved should be either the first and the second rays (complete or partial) or the third, fourth and fifth (complete or partial) to to comply with best function and the fitting with orthopaedic footwear.

Amputations in longitudinal limb deficiencies present at birth

Decision making with a regard to possible conversion of a longitudinal deficiency into an amputation stump is difficult and each case requires very serious discussion (Kruger, 1976; Marquardt, 1981 b) which can only be solved by a holistic approach to the individual child and the entire impairment, whatever the cause, and must include all disabilities including mental ones. In particular, the home environment and especially the attitude of the parents and the entire family, with its economic and overall social situation, have to be taken into

consideration. It is obvious that only a team of qualified specialists in their respective fields can cope with these requirements, but responsibility for the definitive decision, the team's performance and the final outcome of the entire process of diagnosis, treatment and rehabilitation is unquestionably that of the responsible orthopaedic surgeon. The duty imposed includes those critical considerations leading to a decision to amputate or undertake corrective surgery, including limb lengthening procedures. The dangers lie in neglecting the necessity for a clear understanding of the outcome when it is applied to limb deficiencies at birth which might be quite different from its usual application. As a guideline, it is generally accepted that ablative surgery can be performed up to the third year of life or alternatively to postpone amputation until a juvenile age which will allow the opinion of the boy or girl in question to be taken into consideration in the decision making process (Kruger, 1976; Marquardt, 1981 b). Last but not least, weight bearing stumps should be the preferred choice and amputation through the shaft of bones should only be used in combination with stump capping or similar procedures to gain more function and to avoid future stump problems, such as osseous overgrowth.

Conversion into an amputation stump is rarely indicated in upper limb deficiencies present at birth and for this reason no detailed information is provided beyond some of the specific procedures mentioned here.

Proximal femoral focal deficiencies (PFFD)

In short and subtotal absence of the femur, mainly in type C and D according to the classification of Aitken (1968) or type 2 and 3 according to Blauth (1967), but exhibiting a stable ankle joint with a well developed shank and foot, a resection rotationplasty first described by Borggreve (1930) and reintroduced by Van Nes (1950) for malformation in lower limbs discovered at birth uses the ankle joint as a 'knee-joint' and the foot, now facing backwards, as a 'below-knee' stump. Additionally, arthrodesis of the knee joint improves function and cosmetic appearance (Kostuik et al., 1975; Kritter, 1977; Lange et al., 1978; Kruger, 1980; Gillespie, 1981; Marquardt, 1981 b; Miller, 1983).

This kind of limb salvage procedure does not prove to be as successful in longitudinal deficiency of both the femur and fibula because of the instability of the ankle joint and the reduction of the foot and its skeletal structure. In these cases an ankle disarticulation or the modified Pirigoff amputation (Neff *et al.*, 1977; Neff, 1986 b, c) provides excellent weight bearing and is a pre-requisite for better prosthetic fitting (Fig. 27.18). The situation is further improved if combined with

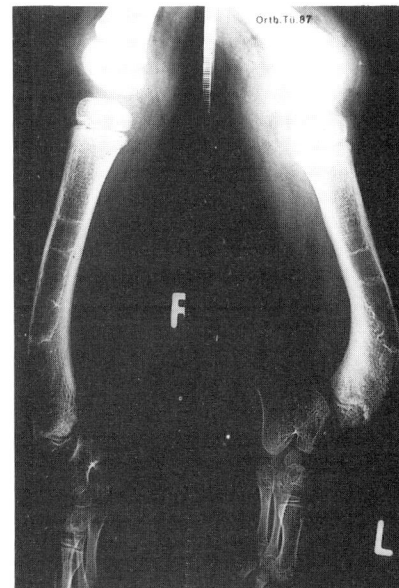

(a)

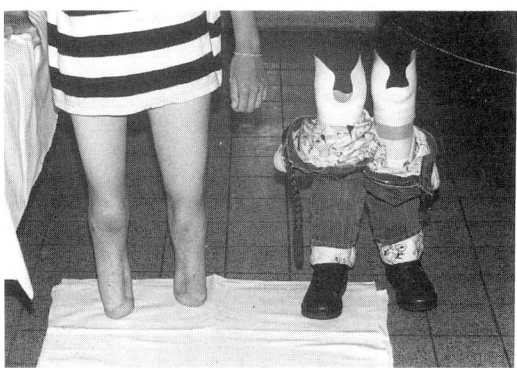

(b)

Fig. 27.18. Conversion of (a) a bilateral longitudinal fibular deficiency into (b) fully end bearing Pirogoff/Le Fort stumps and prosthetic fitting at the age of 13.

arthrodesis of the knee, and, if necessary, resection of one of the growth plates to avoid excessive length of the stump of the limb at the end of the growth period (Kritter, 1977; Gillespie *et al.*, 1981; Marquardt, 1981 b, c, 1988; Frankovitch and Farrell, 1984). These recommendations can only apply to unilateral PFFD. In the case of bilateral PFFD which is often combined with serious upper limb deficiencies, neither amputation or rotation-plasty is indicated (Blauth, 1967; Marquardt, 1981 b).

Kritter (1977) stressed the importance of a stable hip joint for gait performance in PFFD. Marquardt (1990) performed 'stump capping of the spiky proximal end of the femur' using one femoral condyle on the occasion of an arthrodesis to improve 'hip joint' function and weight bearing of the previous painful proximal end of the femur.

Fibular deficiencies

Throughout the English language literature the Syme ankle disarticulation is recommended for conversion of type II or type III fibular deficiencies according to the classification of Coventry and Johnson (1951) and supported by others (Kruger and Talbott, 1961; Wood *et al.*, 1968; Cary, 1977; Achterman and Kalamchi, 1979; Gillespie, 1981; Anderson *et al.*, 1984; Crandall, 1987). The author's view is that the modified Pirigoff stump is preferred to avoid heelpad migration and scar dislocation after early amputation (Marquardt, 1981 b, c; Frankovitch and Farrell, 1984; Neff, 1986 b, c). After the conversion into a weight bearing hindfoot stump, correction of any bowing of the tibia and if need be a supracondylar osteotomy to correct valgus deformity should be performed to align the limb for proper prosthetic use (Marquardt, 1990).

Tibial deficiencies

A completely absent tibia (Type I a and b according to the classification of Jones *et al.*, 1978) (Fig. 27.19) creates an unstable situation for the entire leg; despite the presence of the fibula there is no weight bearing capacity due to the upward

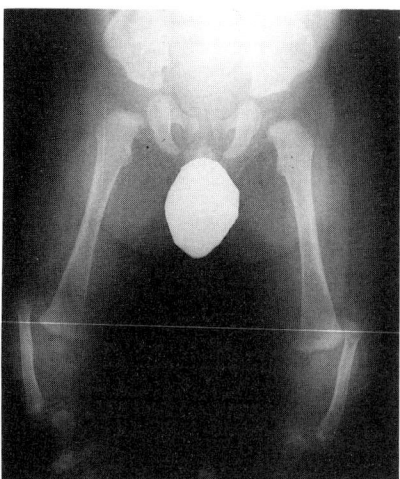

Fig. 27.19. Complete absence of tibia.

dislocation of the fibula, flexion contracture at knee joint level and the supinated attitude of the foot with the sole facing upwards.

To overcome this serious defect (Brown, 1965; Brown and Pohnert, 1972), a surgical solution was developed by transferring the proximal end of the fibula under the femoral condyles. Despite temporary fixation with Kirschner wires, the reconstruction of a ligament and tendon relocation dislocation occurs necessitating additional surgery. The main reason for knee disarticulation proved to be a persistent flexion contracture of the new 'knee joint'. The best results obtained were when the patients were fitted with a long leg brace to stabilize the 'knee joint' and to align the foot in a more or less normal position or to accommodate an ankle disarticulation stump after ablation of the foot.

In these circumstances the experience of those who used the Brown procedure have led them now to discourage any further application; the advice is to perform early knee disarticulation for both unilateral and bilateral absent tibia (Jayakumar and Eilert, 1979; Kalamchi and Dawe, 1985; Loder and Herring, 1987 a, b).

A Type II deficiency with a proximal tibial remnant allows the transfer and fusion of fibula to tibia and provides both sufficient knee joint function and stability of the shank; it may be combined with an ankle disarticulation (Schoenecker *et al.*, 1989). Alternatively, the foot may be fused to the distal end of the fibula as proposed by

Blauth (1965, 1967) and this may sometimes be followed by partial foot amputation (Marquardt 1981 c, 1988).

In a Type IV deficiency with diastases of the distal ends of tibia and fibula a hindfoot amputation may be required despite the normal appearance of a well developed foot, unless stability of the limb can be established by other surgical procedures (Jones *et al.*, 1978; Schoenecker *et al.*, 1989). Even a successful reconstruction of an ankle joint in Type IV deficiencies may still require an ankle disarticulation for a major leg length discrepancy (Schoenecker *et al.*, 1989; Kruger, 1980).

In combined longitudinal deficiencies of the femur and tibia an early fusion of the fibula and the short femur along with a weight bearing hindfoot stump will result in a 'knee disarticulation level amputation' by the end of the growth period and enable a satisfactory prosthetic supply and function (Marquardt, 1985 b).

Marquardt has stressed strongly the completely different situation in multiple deficiencies, where so called 'defective' lower limbs can at least partially compensate for the deficits in upper limbs with respect to sensory and prehensile function (Marquardt, 1981 b, c, 1985).

Other conditions for amputation in the growth period

Constriction band syndrome

Constriction band syndrome may lead to amputation already present at birth or within the first days and weeks of life because of the constriction of soft tissues, and especially vessels and nerves. Unless immediate surgery with excision of the constricting bands restores sufficient blood circulation, the guidelines of amputation should be followed as described.

Congenital pseudarthrosis of tibia and fibula

Jacobsen *et al.* (1983) and McCarthy (1982) performed an ankle disarticulation amputation in children with congenital pseudarthrosis of the tibia

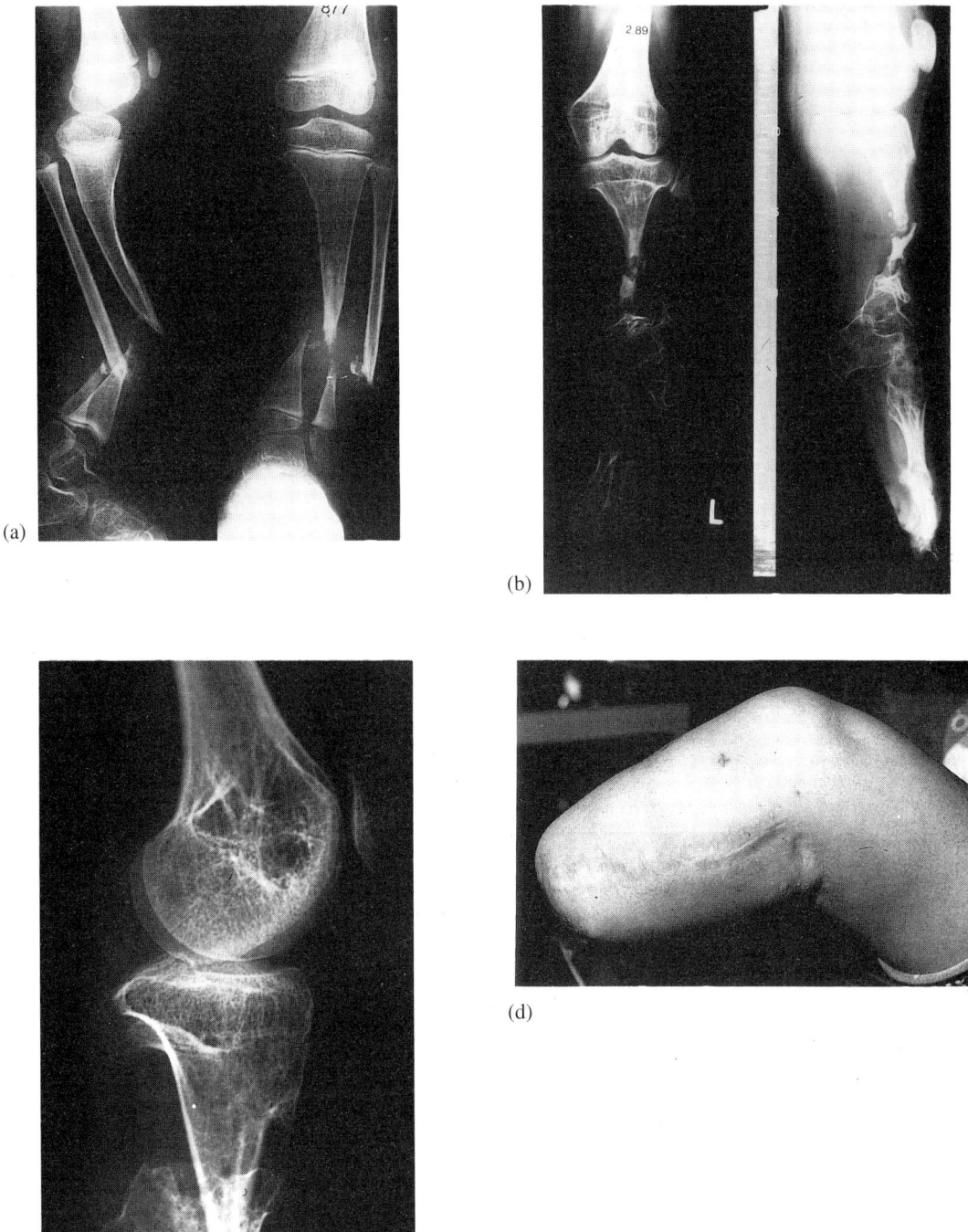

(a)

(b)

(c)

(d)

Fig. 27.20. Congenital pseudarthrosis (a) before and (b) after 19 surgical attempts to 'heal' the defect: (c) below knee amputation using the calcaneus as a stump cap and for stump lengthening; (d) final result.

to improve function and growth despite a persistent pseudarthrosis. They recommended this procedure rather than expose the child to repeated hospitalization for countless surgical attempts to heal the pseudarthrosis to be followed by an inevitable transtibial amputation (Fig. 27.20, a, b, c, d).

There is hope that microsurgical transfer of the contralateral fibula may prevent both ankle disarticulation and transtibial amputation.

Macrodactylism giantism

'Idiopathic hyperplasia' in the extremities leads to early overgrowth of the affected limb and may require resection and occasionally amputation (Fig. 27.21).

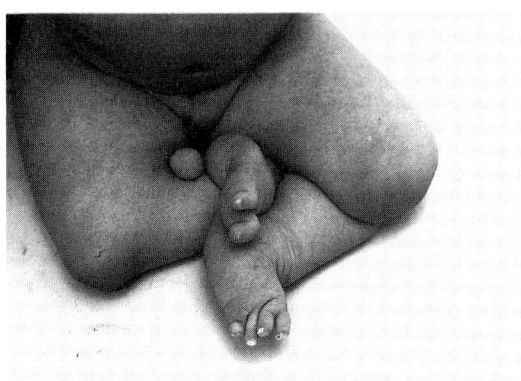

Fig. 27.21. Macrodactylism.

Morbus Klippel–Trenaunay

The Klippel–Trenaunay disease with enlarged vessels increases growth in length and circumference, leading to hyperplasia of a total extremity or parts of it (Fig. 27.22). Amputation may be the last resort to re-establish leg length equality. Sometimes amputation has to be performed due to ulceration or necrosis of the periphery (Fig. 27.23).

Conclusion

Taking into consideration the large variety of different conditions and the varying roles of

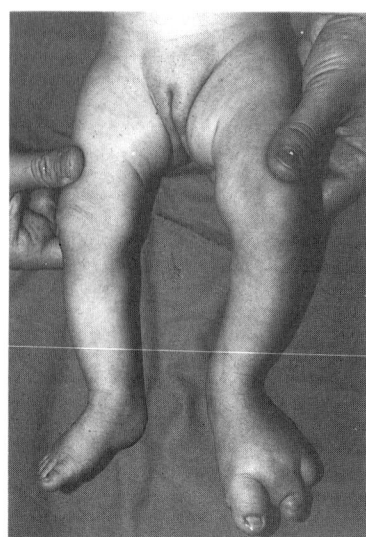

Fig. 27.22. Klippel–Trenaunay syndrome.

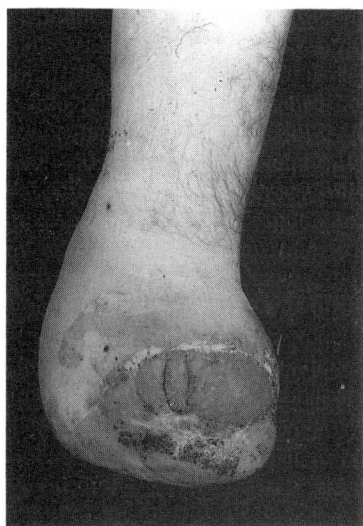

Fig. 27.23. Necrosis in periphery in case of Klippel–Trenaunay disease.

amputation in the growth period, it is clear that centres of excellence must be established for diagnosis and treatment in this field on the basis of a team approach with available qualified specialists including experienced orthopaedic surgeons, nurses, prosthetists, physiotherapists, occupational therapists, social workers and psychologists if we are to offer the best solution for each individual child requiring a unique solution.

References

Abraham, E., Pellicore, R.J., Hamilton, R.C., Hallmann, B.W. and Gosh, L. (1968) Stump overgrowth in juvenile amputees. *J. Pediatr Orthop* **6**, 66–71.

Achterman, C. and Kalamchi, A. (1979) Congenital Deficiency of the fibula. *J. Bone Joint Surg.*, **61B**, 133–137.

Aitken, G.T. (1959a). Amputation as a treatment for certain lower-extremity congenital abnormalities. *J. Bone Joint Surg.*, **41A**, 1267–1285.

Aitken, G.T. (1959b) Overgrowth of the amputation stump. *ICIB*, **I(11)**, 1–8.

Aitken, G.T. (1963) Surgical amputation in children. *J. Bone Joint Surg.*, **45A**, 1735–1741.

Aitken, G.T. (1968a) The children with an acquired amputation. *ICIB*, **7(8)**, 1–15.

Aitken, G.T. (1968b) Proximal femoral focal deficiency: definition, classification and management. In: *Proximal Femoral Focal Deficiency, Congenital Anomaly*: Symposium (ed. G.T. Aitken) National Academy of Sciences, Washington, DC.

Aitken, G.T. and Frantz, C.H. (1953) The juvenile amputee. *J. Bone Joint Surg.*, **35A**, 659–654.

Aitken, G.T. and Frantz, C.H. (1964) The juvenile amputee: a fourteen year follow up (abstract). *J. Bone Joint Surg.*, **46A**, 1376.

Aitken, G.T. and Pellicore, R.J. (1981) Introduction to the child amputee. In: *Atlas of Limb Prosthetics Surgical and Prosthetic Principles*. American Academy of Orthopaedic Surgeons, C.V. Mosby, St Louis, pp. 493–500.

Anderson, L., Westin, G.W. and Oppenheim, W.L. (1984) Syme amputation in children: indications, results and long term follow up. *J. Pediatr. Orthop.*, **4**, 550–54.

Baumgartner, R.F. (1979) Above-knee amputation in children. *Prosthet. Orthot. Int.*, **3**, 26–30.

Bernd, L., Blasius, K., Lukoscheck, M. and Lucke, R. (1991) The autologous stump plasty – treatment for bony overgrowth in juvenile amputees. *J. Bone Joint Surg.*, **73B**, 203–206.

Blasius, K., Berndt L., Gooden, Mvd. and Lucke, R. (1990) Erfahrungen mit der Stumpfkappenplastik nach E. Marquardt. *Med. Orthop. Tech.*, **110**, 223–230.

Blauth, W. (1965) Die operative FuBunterstellung bei angeborener Tibiaaplasie. *Arztl. Prax.*, **17**, 4.

Blauth, W. (1967) *Der kingenitale Femurdefekt*. Enke, Stuttgart.

Borggreve, (1930) Kniegelenkersatz durch das in her Beinlangsachse um 180° gedrehte FuBgelenk. *Arch. Orthop. Unfall. Chir.*, **28**, 175–178.

Brown, F.W. (1965) Construction of a knee joint in congenital total absence of the tibia (paraxial hemimelia tibia). A preliminary report. *J. Bone Joint Surg.*, **47A**, 695–704.

Brown, F.W. and Pohnert, W.H. (1972) Construction of a knee joint in meromelia tibia (congenital absence of the tibia). A fifteen-year follow-up study. (Abstract). *J. Bone Joint Surg.*, **54A**, 1333.

Cary, J.M. (1977) Fibular hemimelia with tibial angulation: the identification of a problem. *ICIB*, **16(7/8)**, 1–6, 10, 17.

Christie, J., Lamb, D.W., McDonald, J.M. and Britten, S. (1979) A study of stump growth for children with below-knee amputations. *J. Bone Joint Surg.*, **61B**, 464–465.

Coventry, M.B. and Johnson, E.W. (1951) Congenital absence of the fibula. *J. Bone Joint Surg.*, **34A**, 941–956.

Crandall, R.C. (1987) Complete longitudinal deficiency of the fibula: comparison of foot ablation to retention in long-term follow-up. *J. Assoc. Child Prosthet. Orthop. Clin.*, **22**, 73–78.

Dederich, R. (1970) *Amputationen der Unteren Extremitat*. Thieme, Stuttgart.

Donaldson, W.F. (1962). Knee disarticulation in childhood. *ICIB*, **1(7)**, 5–9.

Ertl, J. (1949) Uber Amputationsstumpf. *Chirurg.*, **20**, 218.

Fergusson, C.M., Morrison, J.D. and Kenwright, J. (1987) Leg-length inequality in children treated by Syme's amputations. *J. Bone Joint Surg.*, **69B**, 433–436.

Frankovitch, K.F. and Farrell, W.J. (1984) Syme and Boyd amputations in children (Abstract). *ICIB*, **19**, 61.

Frantz, C.H. and Aitken, G.T. (1959) Management of the juvenile amputee. *Clin. Orthop.*, **14**, 30–49.

Friedmann, L.W. and Friedmann, L. (1985) The conservative treatment of the bony overgrowth problem in the juvenile amputee. *ICIB*, **20(2)**, 17–23.

Gibson, D.A. (1980) Child and juvenile amputee. In: *Rehabilitation Management of Amputees*, (ed. S.J. Banerjee) Williams & Wilkins, Baltimore, pp. 391–414.

Gillespie, R. (1981) Congenital limb deformities and amputation surgery in children. In: *Amputation Surgery and Rehabilitation: the Toronto Experience* (eds J. Kostuik and R. Gillespie) Churchill Livingstone, New York, pp. 105–136.

Greene, W.B. and Cary, J.M. (1982) Partial foot amputations in children: a comparison of the several types with the Syme amputations. *J. Bone Joint Surg.*, **64A**, 438–443.

Jacobsen, S.T., Crawford, A.H., Millar, E.A. and Steel, H.H. (1983) The Syme amputation in patients with congenital pseudarthrosis of the tibia. *J. Bone Joint Surg.*, **65A**, 533–537.

Jayakumar, S.S. and Eilert, R.E. (1979) Fibular transfer for congenital absence of the tibia. *Clin. Orthop.*, **139**, 97–101.

Jeannopoulos, C.L. (1969) Postnatal growth of the human limbs: some classical implications. In: *Limb Development and Deformity: Problems of Evaluation and Rehabilitation* (ed. C.A. Swinyard) C.C. Thomas, Springfield.

Jones, D., Barnes, J. and Lloyd-Roberts, G.C. (1978) Congenital aplasia and dysplasia of the tibia with intact fibula. Classification and management. *J. Bone Joint Surg.*, **60B**, 31–39.

Jorring, K. (1971) Amputation in children: a follow up of 74 children whose lower extremities were amputated. *Acta Orthop. Scand.*, **42**, 178–186.

Kalamchi, A. and Dawe, R.V. (1985). Congenital deficiency of the tibia. *J. Bone Joint Surg.*, **67B**, 581–587.

Kostuik, J.P., Gillespie, R., Hall, J.E. and Hubbard, S. (1975) Van Nes rotational osteotomy for treatment of proximal femoral focal deficiency and congenital short femur. *J. Bone Joint Surg.*, **57A**, 1039–1046.

Kritter, A.E. (1977) Tibial rotation-pasty for proximal femoral focal deficiency. *J. Bone Joint Surg.*, **59A**, 927–934.

Kruger, L.M. (1976) Decision making for the child with multiple limb deficiencies. *ICIB*, **15(1/2)**, 13–19.

Kruger, L.M. (1980) Recent advances in surgery of lower limb deficiencies. *Clin. Orthop.*, **148**, 97–105.

Kruger, L.M., and Talbott, R.D. (1961) Amputation and prosthesis as definitive treatment in congenital absence of the fibula. *J. Bone Joint Surg.*, **43A**, 625–642.

Krukenberg, H. (1917) *Uber die plastische Umwertung von Amputationsstumpfen*. Enke, Stuttgart.

Kuhn, G.G. (1979) Eigenkraft-Armprosthesen. *Z. Orthop.*, **117**, 631–637.

Lambert, C.N. (1972) Amputation surgery in the child. *Orthop. Clin. N. Am.*, **3**, 473–482.

Lambert, C.N., Pellicore, R.J., Hamilton, R.C., Sciora, J. and Birch, A. (1976) Twenty-three years of clinical experience. *ICIB*, **15(3/4)**, 15–20, 25.

Lange, D.R., Schoenecker, P.L. and Baker, C.L. (1978) Proximal femoral focal deficiency: treatment and classification in forty-two cases. *Clin. Orthop.*, **135**, 15–25.

Lisfranc, J. (1815) *Nouvelle methode operatoire pour l'amputation partielle du pied dans son articulation tarsometatarsienne; methode precedes des nombreuse modifications qu'a subies celle de Chopart*. Gabon, Paris.

Loder, R.T. and Herring, J.A. (1987a) Fibular transfer for congenital absence of the tibia: a reassessment. *J. Pediatr. Orthop.*, **7**, 8–13.

Loder, R.T. and Herring, J.A. (1987b) Disarticulation of the knee in children. *J. Bone Joint Surg.*, **69A**, 1155–1160.

Marquardt, E. (1973) Die Chopart-Exarticulation mit Tenomyoplastik. *Z. Orthop.*, **11**, 584–586.

Marquardt, E. (1976) Plastische Operationen bei drohender KnochendurchspieBung am kindlichen Oberarmstumpf. Eine Vorlaufige. *Z. Orthop.*, **114**, 711–714.

Marquardt, E. (1981a) The Knud Jansen Lecture: the operative treatment of congenital limb malformation – part II: case study. *Prosthet. Orthot. Int.*, **5**, 2–6.

Marquardt, E. (1981b) The Knud Jansen Lecture: the operative treatment of congenital limb malformation – part III. *Prosthet. Orthot. Int.*, **5**, 61–67.

Marquardt, E. (1981c) The multiple limb-deficient child. In: *Atlas of Limb Prosthetics* (ed. American Academy of Orthopaedic Surgeons) C.V. Mosby, St Louis.

Marquardt, E. (1983) Amputation bei angeborenen Erkrankungen der unteren Extremitat (Amputation in congenital disability of the lower limb). *Therapiewoche*, **33**, 5010–5021.

Marquardt, E. (1985) Plastischer Eingriff an den unteren Extremitaten beim Tibiadefekt und ihre Bhedeutung fur Funktion und Asthetik der unteren Extremitat. In: *Die Asthetik von Form und Function in der plastischen und Wiederherstellungschirurgie* (ed. G. Pfeifer) Springer, Heidelberg.

Marquardt, E. (1987) Indikation und spezielle Amputation-stechnik am Aberschenkel. *Unfallheilk*, **189**, 809–819.

Marquardt, E. (1988) GliedmaBendefekte der unteren Extremitat, ihre operative behandlung in Kooperation mit Orthopadie-Technik. *Z. Orthop.*, **126**, 227–238.

Marquardt, E. (1990) Valgus deformities and fibula deficiencies. *Semi. Orthop.*, **5(1)**, 32–50.

Marquardt, E. and Neff, G. (1974) The angulation-osteotomy of above-elbow stumps. *Clin. Orthop.*, **104**, 232–238.

Marquardt, E. and Correll, J. (1984) Amputations and prostheses for the lower limb. *Int. Orthop.*, **8**, 139–145.

Marquardt, E., Martini, A.K. and Banniza von Bazan, U. (1983) Stumpfkappenplastik und Stumpfverlangerung bei traumatischen Amputationen. In: *Regionale plastische und rekonstructive Chirurgie im Kindesalter* (ed. W. Kley and C. Naumann) Springer, Berlin-Heidelberg.

Mazet, P. (1968) Syme's amputation: a follow-up study of fifty-one adults and thirty-two children. J. Bone Joint Surg. 50A, 1549–1563.

McCarthy, R.E. (1982) Amputation for congenital pseudarthrosis of the tibia: indications and techniques. *Clin. Orthop.*, **166**, 58–61.

Miller, E.A. (1983) Van Nes turnaround procedure (abstract). *Orthop. Trans.*, **7**, 564.

Murdoch, G. (1977) Amputation surgery in the lower limb – part 2. *Prosthet. Orthot. Int.*, **1**, 183–192.

Neff, G. (1979) Prothetische Versorgung nach Winkelosteotomie. *Orthop. Tech.*, **30**, 1–5.

Neff, G. (1986a) Allgemeine Amputationslehre. In: *Praxis der Orthopadie* (eds. M. Jager and C.J. Wirth) Thieme, Stuttgart.

Neff, G. (1986b) Spezielle Stumpfformen. In: *Praxis der Orthopadie* (eds M. Jager and C.J. Wirth) Thieme, Stuttgart.

Neff, G. (1986c) Amputationen am KinderfuB (foot amputations in children). *Orthopadie*, **15**, 264–272.

Neff, G. (1987) Knieexartikulation versus Oberschenkelamputation – Operationstechnik, prothetische versorgung und rehabilitation (Knee disarticulation versus above-knee amputation: operative technique, prosthetic management and rehabilitation). *Med. Orthop. Tech.*, **107**, 92–100.

Neff, G., Winkler, E. and Waigand, H. (1977) Die modifizierten stumpfformen nach Chopart, Pirigoff und Syme: Ihre indikation, operationstechnik und die orthopadie-technische Versorgung. *Orthop. Tech.*, **28**, 1–4.

Schoenecker, P.L., Capelli, A.M., Millar, E.A., Sheen, M.R., Hahner, T., Aiona, M.D. and Meyer, L.C. (1989) Congenital longitudinal deficiency of the tibia. *J. Bone Joint Surg.*, **71A**, 278–287.

Swanson, A.B. (1964) The Krukenberg procedure in the juvenile amputee. *J. Bone Joint Surg.*, **46A**, 1540–1548, 1610.

Swanson, A.B. (1972) Silicone-rubber implants to control the overgrowth phenomenon in the juvenile amputee. *ICIB*, **11(9)**, 58.

Swanson, A.B. and Swanson, G.D. (1980) The Krukenberg procedure in the juvenile amputee. *Clin. Orthop.*, **148**, 55–61.

Tooms, R.E. (1981) Acquired amputations in children. In: *Atlas of Limb Prosthetics: Surgical and Prosthetic Principles* (ed. American Academy of Orthopaedic Surgeons). C.V. Mosby, St Louis, pp. 553–559.

Tooms, R.E. (1986) The amputee. In: *Pediatric orthopaedics*, 2nd edn (eds W.N. Lovell and R.B. Winter) Lippincott, Philadelphia, pp. 979–1030.

Van Nes, C.P. (1950) Rotation-plasty for congenital defects of the femur. Making use of the ankle of the shortened limb to control the knee joint of a prosthesis. *J. Bone Joint Surg.* **32B,** 12–16.

Verdan, C. (1972) Die eingriffe an muskeln, sehnen und sehenscheiden. In: *Die Operationen an der Hand* (eds W. von Wachsmuth and A. Wilhelm) Springer, Berlin.

Wood, W.I.., Zlotsky, N. and Westin, G.W. (1968) Congenital absence of the fibula. Treatment by Syme amputation – indications and technique. *J. Bone Joint Surg.*, **47A,** 1159–1169.

Zrubecky, G. (1976) Eine neue Eigenkraft-Prosthese der Hand. *Orthop. Techn.*, **27,** 2–4.

28

Prosthetic considerations during the growth period

J. W. Michael

Introduction

The aphorism, 'The child is *not* a small adult' summarizes the philosophical foundation for prosthetic management during the growth period. Although the external appearance of the child's prosthesis may resemble a miniature of the adult version to the casual observer, in fact it is based on a set of unique criteria much different from its adult analogue (Michael, 1990). Due to the complexity of many malformations present at birth, the most advanced treatment is typically available from specialty clinics devoted to paediatric practice.

This contribution will highlight the basic parameters which guide prosthetic prescription and design for this population rather than focus on specific prostheses. The interested reader is referred to the growing body of specialized literature for more detailed discussions of specific devices and cases.

Overview

The child is in growth a dynamic organism, physiologically and psychologically different from the ageing, decelerating adult. It is *not* sufficient for the paediatric prosthesis to be technically well designed; it must also be provided precisely when appropriate for the abilities and maturity of the child (Aitken and Pellicore, 1981). Therefore, a keen appreciation of the normal development sequence of human maturation is the hallmark of effective prosthetic intervention during the growth period. Table 28.1 (Novotny and Swagman, 1992) summarizes typical cognitive and psychological milestones and suggests specific prosthetic interventions for each developmental stage.

In addition to a sometimes rapidly changing developmental readiness, the child is also characterized by significant skeletal and muscular growth. A variety of prosthetic strategies have evolved to accommodate the resulting increase in circumferential and longitudinal dimensions of the stump. The most common approach is to create a series of 'nesting' sockets, one inside the other, at the time of the initial fitting. As the child grows, the innermost socket can be removed as if it were one of the concentric segments of an onion, creating a larger cavity to accommodate volumetric growth (Sauter *et al.*, 1987). Since longitudinal growth is often more rapid than circumferential (Blakeslee, 1963), a removable elastomer end pad in the socket is frequently encountered.

The surgical implications of skeletal immaturity are discussed in this chapter. As a result there is a preponderance of disarticulations as opposed to trans-diaphyseal amputations. While disarticulation offers such widely acknowledged advantages as distal end bearing, inherent suspension, rotary stability and normally functioning musculature (Jensen and Lyquist, 1992), it also restricts the choice of prosthetic components and compromises the cosmetic appearance by its bulk (Michael, 1988). Fortunately, the negative aspects are largely eliminated as the child grows (if the disarticulation is performed early enough), since the slower growth on the involved side usually results in a relative

Table 28.1 Although the specific timing varies among individuals, each stage of paediatric development may be characterized by typical psychosocial, cognitive, adaptive, and coping issues. This table summarizes this information and suggests typical intervention strategies and prosthetic issues for each stage. (Adapted, with permission, from Michael (1990), p. 192, Table 1).

Stages	Coping Processes				Suggested Interventions for Professionals	Prosthetic Issues and Approaches
	Erikson Psychosocial	**Piaget Cognitive**	**Adaptive Tasks**	**Coping Skills**		
Infancy *0–1 year old*	*trust/mistrust* emotional stability eye-hand coordination sensory discrimination motor skills	*N/A*	integrating environmental experiences	motor activity (goal gratification or tension reduction) crying kicking	● take time with new families to answer questions, assess needs and establish parental trust ● recognize the families' level of acceptance provide reassurance acknowledge concerns ● encourage contact with other families ● review rationale for treatment along with demonstration of device application	Upper Limb: *Using prosthesis for sitting balance to allow other hand to explore* ★ Cannot operate body powered hook but may use electric hand ★ Many centres fit passive terminal device initially Lower Limb: *Use prosthesis for standing balance* ★ Fit to allow pull-to-stand (9 months old is average) ★ Accomodate infants flexed, abducted stance ★ Use solid or locked prosthetic knee
Toddlerhood *1–3 years old*	*autonomy/shame and doubt* self-control language/ functional communication fantasy and play elaboration of locomotion assertion of independence interpersonal control	egocentrism questioning through play transductive reasoning	mastery through fantasy and play separation from parent exploring the environment improving communication coping with frustration	repetitive play activities for mastery and anxiety reduction protest fantasy comfort time out experimentation restricting environment	● value parents' input about their child ● involve parents in all aspects of care for support and education ● willingly answer questions and offer educational materials as needed ● recognize unpredictable situations and allow for flexibility in scheduling	Upper Limb: *Use prosthesis for bimanual activities and to assist other hand* ★ Activate body powered terminal device when child has sufficient strength ★ Parents can encourage 2-handed play activities Lower Limb: *Use prosthesis to improve strength, balance* ★ Follow frequently to accommodate growth by lengthening prosthesis ★ Modify alignment, socket fit as development dictates

Table 28.1 (continued)

Stages	Coping Processes				Suggested Interventions for Professionals	Prosthetic Issues and Approaches
Early Childhood *3–5 years old*	*initiative/guilt* exploration and mastery independent activity	rudimentary conscience understands right and wrong magical thinking	independence role and identity in family maintain and protect body integrity	routines ritualized activity	● provide a variety of safe toys in waiting and fitting areas ● support established routines of family whenever possible ● encourage parents to be present to decrease anxiety ● provide age-appropriate information for child and parent	Upper Limb: Use prosthesis for more complex tasks to develop fine control ★ Encourage Activities of Daily Living skills with prosthesis ★ Encourage grasp and release play activities Lower Limb: *Refine balance and gait* ★ Unlock prosthetic knee when balance permits (later for bilaterals) ★ Reduce flexion, abduction alignment as gait matures
Middle Childhood *6–12 years old*	*industry/ inferiority* skill mastery contending with inferior feelings concern with peers and societal expectations	perceives past and future questions other points of view questions beliefs trial and error problem solving	peer relationships acceptance	problem solving humor withdrawal aggression control of behaviour	● prepare child for responses of peers in school setting ● promote child's participation/ self-direction in procedures ● encourage interaction with others of similar circumstances ● use humor in relationships	Upper Limb: *Explore capabilities of prosthesis* ★ Encourage use of prosthesis in school and sports activities ★ Consider more complex componentry or special sports designs to expand options Lower Limb: *Explore capabilities of prosthesis* ★ Encourage use of prosthesis in school and sports activities ★ Consider more complex componentry or special sports designs to expand options

| Adolescence
13–16 years old | *identity*
self-image in relation to peers
future orientation
testing autonomy
emotional independence | formal operations
strategic interventions
interpretation of earlier experience | self-conscious
competitive
transition to adult
integrating disability with identity
emotional stability | denial
intellectualization
conformity
emotional control
motor activity
withdrawal | ● respect confidentiality and privacy needs
● provide opportunities for discussion of individual needs and concerns
● support independence
● encourage choices and participation in decisions
● acknowledge individual priorities related to functions and cosmesis | Upper Limb:
Incorporate prosthesis into adult self-concept
★ Interest in cosmesis is often high
★ Conversion to myoelectric control often successful
★ Consider more complex componentry or special sports designs to expand options
Lower Limb:
★ Interest in cosmesis is often high
★ Conversion to endoskeletal design often successful
★ Consider more complex componentry or special sports designs to expand options |

'shortening' as compared to the contralateral side (Cummings and Kapp, 1992). Figure 28.1 illustrates this feature. Prior to performing ablative surgery, it is prudent to discuss the inevitably bulky appearance of the child's first few prostheses so that the parents will not be disappointed.

A broad variety of components, socket designs and suspension alternatives is available for the growing child. Since there are virtually no scientific data to guide the choices among the options, specific recommendations are generally made by the clinic team and often reflect local custom as much as any other factor. Children are very adaptable and will successfully utilize almost any of the alternatives presented to them.

Even though the specific choice of components is subjective, it need not be irrational. Over many decades of experience, a general consensus has emerged among the specialty clinics providing for the needs of the growing child. In general, prosthetic design should reflect a desire to:

(1) maximize durability;
(2) minimize weight;
(3) maximize performance;
(4) protect from injury (Cummings and Kapp, 1992).

Maximize durability

During the 'rough and tumble' years of childhood, most authorities agree that the primary goal of prosthetic design is durability. Particularly during the early years of infancy and 'toddlerhood', this is usually accomplished by keeping the design as

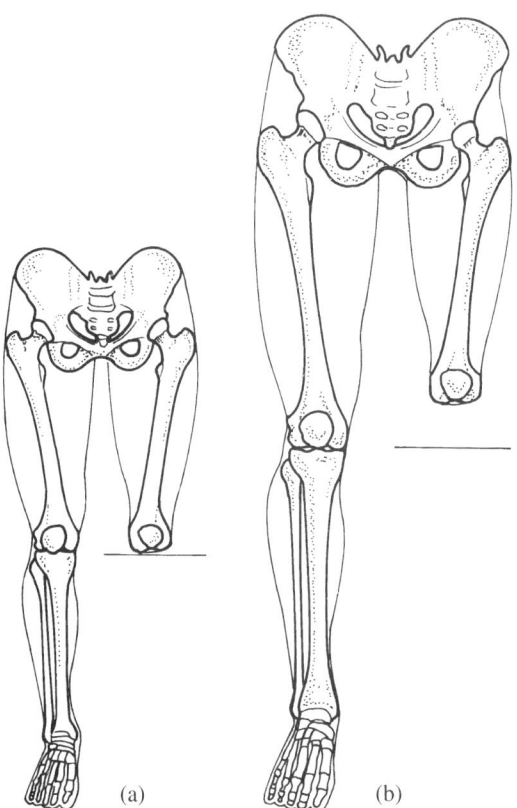

Fig. 28.1.(a) When disarticulation is first performed, the stump is bulky and protrudes beyond the contralateral knee axis. It is impossible to create a prosthesis with a symmetric, cosmetic appearance at this time. (b) After several years of differential growth, the amputated side is now smaller than the contralateral side and the cosmetic appearance is markedly improved.

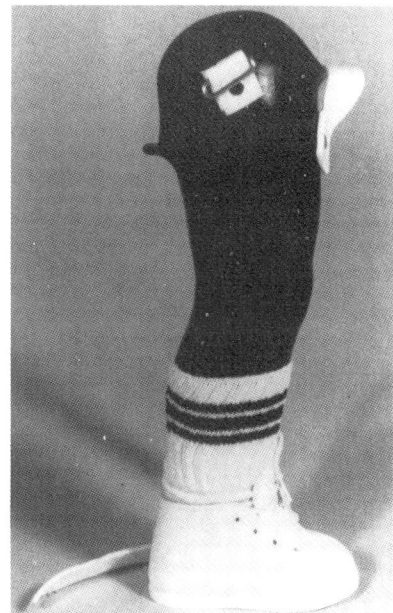

Fig. 28.2. Transfemoral prosthesis for an infant first pulling up to stand is non-articulated and reflects the child's early posture of hip flexion, knee flexion and ankle dorsiflexion. It is usually constructed from lightweight balsa wood or polyurethane foam, covered with a thin plastic shell. Suspension is typically by a soft fabric (Silesian) belt. (Reprinted with permission from Cummings and Kapp (1992), p. 200 Fig. 4).

simple as possible (Oglesby and Tablada, 1992). Figure 28.2 illustrates a typical infant's transfemoral prosthesis. Note the typical bent-knee stance of this age group. Since the very small child cannot control a free knee, the structural elements of such prostheses are often sculpted from a single lightweight piece of balsa wood or polyurethane foam, covered with a very thin lamination of plastic (Sabolich, 1991).

When the child appears ready for more active knee control, it is often useful to provide a manually locking knee (which can be unlocked when desired) to ease the transition to a free knee. Because of its rigid outer shell, the exoskeletal type survives the rigors of sandboxes and playground activities well and is therefore commonly used. Endoskeletal designs, with a soft covering, offer identical function but are not as durable.

Minimize weight

To some degree, the dictum to minimize weight conflicts with the goal of maximizing durability. Particularly for juveniles who participate in competitive sports activities, it is frequently necessary to provide additional reinforcement at points of high stress in the prosthesis. Children will tolerate without complaining a proportionally heavier prosthesis in contrast to the adult.

Excess weight translates into increased energy cost and often into less active use of the limb. The use of modern materials that have a favourable strength-to-weight ratio such as carbon composites and titanium provides a practical means of combining both light weight and high strength. Figure 28.3 demonstrates the substantial weight savings that modern materials offer.

Maximize performance

Maximizing performance is a relatively new concept in paediatric prosthetics. Until the past decade, the accepted viewpoint was that children had unlimited energy reserves and therefore virtually any component would suffice. Traditionalists discourage or delay the use of more complex, modern components such as dynamic response feet and electrically powered hands.

Fig. 28.3. Functionally identical endoskeletal knee mechanisms fabricated from steel (top) and titanium (bottom). Weight savings of 30–40% are common when modern alloys of metal and plastic are used.

At the same time, numerous authors have reported consistent, long-term success with advanced technology (Sorbye, 1980). More aggressive clinicians will argue that it is wrong to withhold such developments (Hutnick, 1990). One of the most significant unresolved debates in paediatric prosthetics is the degree to which high performance components can be justified (Fisk, 1992).

The controversy is even more pronounced when considering upper limb prostheses for absences present at birth. Unlike traumatic amputees, the child born with an absence has not 'lost' any previous function; there is simply a difference from the norm. The ability of many who lack one or both arms to master activities of daily living without prostheses is well documented (Edelstein, 1992), as is the near universal rejection of upper limb prostheses (regardless of complexity), by high level bilateral amputees (Marquardt, 1992). Can the provision of an artificial arm for this population be criticized as creating an unnatural dependency (Baughn, 1992)? Can this concern be reconciled with the vociferous cry from parents of limb deficient children who quote tremendous satisfaction with and functional use of complex myoelectric limbs by children with unilateral hand absence, even from early infancy (Leingang, 1992)? Although such debates will undoubtedly rage on for years, they probably represent a maturation in thinking on paediatric treatment. Many clinicians now agree that the 'rules' of previous decades are no longer defensible given modern technology, but they are intellectually honest enough to acknowledge that the next generation of guidelines is yet to emerge (Dillon, 1992).

Protect from injuries

Protection of the knee joint from injuries is also a controversial subject. Many malformations, epitomized by fibular absence, are associated with anomalous or lax ligaments proximal to the deficit. Varus and valgus deformities are also common. Particularly for children who participate in vigorous activities, consideration should be given to extending the socket trimlines as proximally as feasible to buttress the knee. Careful observation at regular intervals is always indicated, and additional protection can be provided if progression of the instability or deformity is observed. Marked laxity may require the addition of hinged joints at the knee to brace it against further deforming forces.

The basic requirement of paediatric practice is regular follow-up and frequent changes to the device to reflect accurately the growing maturity of the child. Evaluation at least quarterly is strongly recommended, as linear growth can be surprisingly rapid and must be accommodated regularly or the

prosthesis becomes progressively too 'short'. It is common practice to add 1 cm or more of length to the lower limb device every 4 months or so during a growth spurt. Volumetric adjustments must also be accommodated. It is equally important to prescribe a new prosthesis with more advanced components or controls whenever the child is developmentally ready, and to anticipate impending maturity in the prescription lest the child receive a new device that is inappropriately simple in function.

Non-standard devices

As previously mentioned, many anomalies found at birth present a very complicated clinical picture. Although surgical conversion to an amputation is common, it is not always the best course, particularly for the child with multiple abnormalities.

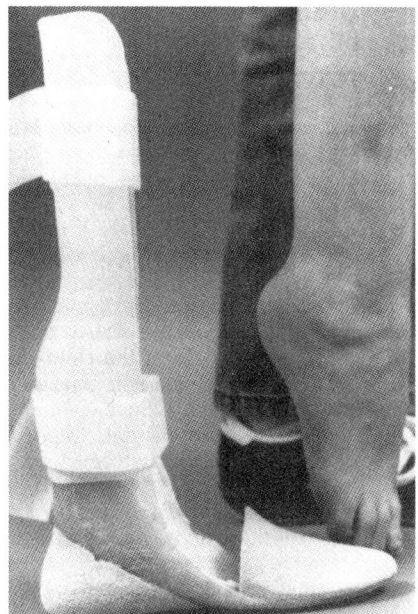

Fig. 28.4. Orthotic/prosthetic hybrids such as the lightweight thermoplastic device pictured here may be used in lieu of surgical corrections, either temporarily or definitively. This design accommodates a fused ankle plus leg length discrepancy and allows ambulation with conventional footwear. (Reprinted with permission from Cummings and Kapp (1992), p. 198, Fig. 1).

Parents understandably are often reluctant to permit surgery when the child is very young, and prefer a trial with a device that combines aspects of both prosthetics and orthotics.

Figure 28.4 presents one example of such a device, used to accommodate a leg length discrepancy secondary to proximal focal femoral deficiency. If this approach proves satisfactory, it can be continued as a definitive treatment. Otherwise, it does allow a relatively normal progression of ambulatory development and serves as an interim technique prior to definitive surgical measures. Hundreds of such devices have been described in the literature, with new variants reported almost every month. Creative solutions to one-of-a-kind problems are often necessary in paediatric prosthetic practice.

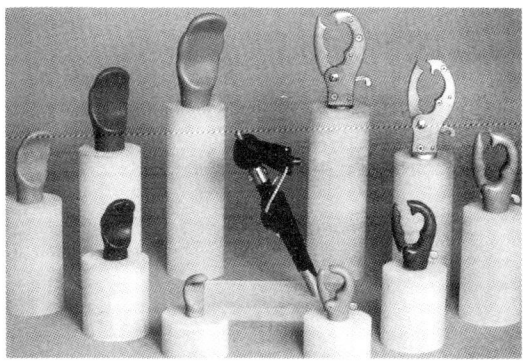

Fig. 28.5. These specialized terminal devices facilitate sports activities for both child and adult amputees. Note the flexible elastomer mitts for ball sports such as soccer and the ski pole attachment. (Reprinted by courtesy of TRS, Inc.; Boulder, Col.).

Sports and recreation

Children with a limb absence, whether present at birth or acquired, almost always desire to participate in vigorous sports and recreational activities as do many of their adult counterparts (Epps and Bryant, 1991). It has been estimated that 15–20 000 amputees participate in modern sports in the United States alone (Riley, 1984). Modern prosthetic technology can often help make this goal a realistic one (Michael et al., 1990).

In some circumstances, specialized components are necessary to facilitate participation in particular sports, especially for the upper limb amputee (Radocy, 1987). Figure 28.5 illustrates just a few of the specialized adaptations available for this group. More commonly, readily available components that are suitable for normal walking, such as dynamic response feet or fluid controlled knees, may be utilized in combination with a well-fitted socket (Michael, 1992). It should be noted that a prosthesis is not necessary for many recreational activities, particularly those performed in water or in snow-covered areas (Michael, 1989).

Numerous organizations have been formed to assist the amputee to participate in these recreational activities. A recent publication listed over 200 sources in the United States alone to help amputees master such diverse activities as sailing, golfing, water and snow skiing, skating, scuba diving, horseback riding, wheelchair sports and mountain climbing (Kegel et al., 1978). Since running is a key facet of many competitive sports, much effort has been focused on developing technology that allows amputees to run actively (Kegel et al., 1978; Kegel et al., 1980; Mensch and Ellis, 1986; Sabolich, 1987). Selected individuals with all types of unilateral and bilateral amputation including the transfemoral level, given that they were properly fitted, the appropriate components were made available and relevant training provided, have been able to learn to run in a near-normal 'leg over leg' style, at least for short distances (Gailey and Clark, 1992). Paediatric clinic teams, in particular, now acknowledge that rehabilitation has not been completed until regular participation in sports and similar vigorous activities is feasible for the child amputee (Page and Messner, 1992).

Conclusion

Due to the complexities presented by normal human development and the myriad malformations seen at birth, individualized prosthetic treatment specific to the individual is the hallmark of the best paediatric care (Cummings and Kapp, 1992). As a general rule, complexity in design must be clearly justified by functional advantage: the best device is the least device necessary to accomplish the desired result (Michael, 1989). Prosthetic success is not solely dependent on the selected components but rather upon the harmony between the interventions offered and the developmental needs and abilities of the particular child. Clinicians throughout the world are eager for valid scientific studies of the many options currently available to help choose wisely for the growing child (Supan, 1992).

References

Aitken, G.T. and Pellicore, R.J. (1981) Introduction to the child amputee. In: *Atlas of Limb Prosthetics* (eds J.H. Bowker and J.W. Michael), American Academy of Orthopaedic Surgeons, C.V. Mosby, St. Louis, pp. 493–500.

Baughn, B. (1992) Caveat Emptor? *J. Prosthet. Orthot.*, **4(4)**, 180.

Blakeslee, B. (editor) (1963) *The Limb Deficient Child*, University of California Press, Los Angeles, p. 137.

Cummings, D.R. and Kapp, S.L. (1992) Lower limb pediatric prosthetics: general considerations. *J. Prosthet. Orthot.*, **4(4)**, 196–206.

Dillon, S.S. (1992) New technology and the status quo. *J. Prosthet. Orthot.*, **4(4)**, 178–179.

Edelstein, J. (1992) Rehabilitation without prostheses: functional skills training. In: *Atlas of Limb Prosthetics: Surgical, Prosthetic and Rehabilitation Principles* (eds J.H. Bowker and J.W. Michael), C.V. Mosby, St Louis, pp. 721–728.

Epps, C.H. and Bryant, D.D. (1991) Sports participation for children with limb deficiencies. *J. Assoc. Child. Prosthet. Orthot. Clin.*, **20**, 12.

Fisk, J.R. (1992) Introduction to the child amputee. In: *Atlas of Limb Prosthetics: Surgical, Prosthetic and Rehabilitation Principles* (eds J.H. Bowker and J.W. Michael), C.V. Mosby, St Louis, pp. 731–734.

Gailey, R.S. and Clark, C.R. (1992) Physical therapy management of adult lower limb amputees. In: *Atlas of Limb Prosthetics: Surgical, Prosthetic and Rehabilitation Principles* (eds J.H. Bowker and J.W. Michael), C.V. Mosby, St Louis, pp. 569–597.

Hutnick, G.F. (1990) Advanced technology. *Orthot. Prosthet. Today*, **1(2)**, 1–2.

Jensen, J.S. and Lyquist, E. (1992) Through-knee amputations: In: International Society for Prosthetics & Orthotics *Report of Consensus Conference on Amputation Surgery*. ISPO, Copenhagen, pp. 69–80.

Kegel, B. (1992) Adaptations for sports and recreation. In: *Atlas of Limb Prosthetics: Surgical, Prosthetic, and Rehabilitation Principles* (eds J.H. Bowker and J.W. Michael), C.V. Mosby, St Louis, pp. 623–654.

Kegel, B. *et al.* (1978) Functional capabilities of lower extremity amputees. *Arch. Phys. Med. Rehabil.*, **59**, 109–120.

Kegel, B. *et al.* (1980) Recreational activities of lower extremity amputees: a survey. *Arch. of Phys. Med. Rehabil.*, **61**, 258–264.

Leingang, A.F. (1992) Myoelectrics for infants: sine qua non? *J. Prosthet. Orthot.*, **4(4)**, 181–183.

Marquardt, E. (1992) The multiple-limb-deficient child. In: *Atlas of Limb Prosthetics: Surgical, Prosthetic and Rehabilitation Principles* (eds J.H. Bowker and J.W. Michael), C.V. Mosby, St. Louis, pp. 839–884.

Mensch, G. and Ellis, P.E. (1986) Running patterns of transfemoral amputees: a clinical analysis. *Prosthet. Orthot. Int.*, **10**, 129–134.

Michael, J.W. (1988) Component selection criteria: lower limb disarticulations. *Clin. Prosthet. Orthot.* **12(3)**, 99–108.

Michael, J.W. (1989) New developments in prosthetic feet for sports and recreation. *Palaestra*, **5(2)**, 21–35.

Michael, J.W. (1990) Pediatric prosthetics and orthotics. *Phys. Occup. Ther. Pediatr.* **10(2)**, 123–146.

Michael, J.W. (1992) Prosthetics. In: *McGraw-Hill Encyclopedia of Science & Technology*, McGraw-Hill, Inc., New York, pp. 437–438.

Michael, J.W., Gailey, R.S. and Bowker J.H. (1990) New developments in recreational prostheses and adaptive devices for the amputee. *Clin. Orthop. Rel. Res.*, **256**, 64–75.

Novotny, M. and Swagman, A. (1992) Caring for children with orthotic prosthetic needs. *J. Prosthet. Orthot.*, **4(4)**, 191–195 (adapted, with permission)

Oglesby, D.G. and Tablada, C. (1992) In: *Atlas of Limb Prosthetics: Surgical, Prosthetic and Rehabilitation Principles* (eds J.H. Bowker and J.W. Michael), C.V. Mosby, St Louis, pp. 835–838.

Page, C. and Messner, D.G. (1992) Juvenile amputees: sports and recreation program development. In: *Atlas of Limb Prosthetics: Surgical, Prosthetic and Rehabilitation Principles* (eds J.H. Bowker and J.W. Michael), C.V. Mosby, St Louis, pp. 901–907.

Radocy, B. (1987) Upper-extremity prosthetics: considerations and designs for sports and recreation. *Clin. Prosthet. Orthot.*, **11(3)**, 131–153.

Riley, R. (1984) The amputee athlete. *Sports Med.*, **4**, 31–32.

Sabolich, J. (1987) The OKC above-knee running system. *Clin. Prosthet. Orthot.*, **11(3)**, 169–172.

Sabolich, J. (1991) Prosthetic advances in lower extremity amputation. *Phys. Med. Rehabil. Clin. N. Am.*, **2**, 415–422.

Sauter, W.F., Dakpa, R., Galway, R. *et al.* (1987) Development of layered "onionized" silicone sockets for juvenile below-elbow amputees. *J. Assoc. Child. Prosthet. Orthot. Clin.*, **22**, 57–59.

Sorbye, R. (1980) Myoelectric prosthesis fitting in young children. *Clin. Orthop. Rel. Res.*, **148**, 34–40.

Supan, T.J. (1992) Upper limb deficiencies: prosthetic and orthotic management. In: *Atlas of Limb Prosthetics: Surgical, Prosthetic and Rehabilitation Principles* (eds J.H. Bowker and J.W. Michael), C.V. Mosby, St Louis, pp. 761–765.

Section 13

The developing world

29

Amputation surgery: the African perspective

K.C. Rankin

Amputation is regarded as one of the oldest surgical operations and has been of the greatest importance during wars. The practice of surgery in Europe developed and amputation surgery was refined through two global conflicts because surgeons whose skill and expertise focused on this branch of surgery were available to make improvements. In contrast, the past 3 decades in Africa have seen wars varying from low level guerrilla campaigns to large scale offensives using powerful weaponry causing immense human mutilation. Additionally, there has been episodic civilian strife and increasing accidents, both of which have caused further loss of limb. The resulting surgical needs have not been matched by the required surgical skills. As the wars of independence reached their climax and change was at hand, medical professionals fled the new situation, causing a vacuum which has not yet been filled.

Successful rehabilitation of the amputee requires the dual approach of surgical skill for the amputation and prosthetic expertise to overcome the resulting disability. In many countries both sectors are deficient. The patient in need of amputation will be fortunate to find a surgeon to remove his damaged limb adequately and a prosthetist to provide a good appliance.

This situation must be seen against the background of great population growth and ever decreasing medical resources. The need for surgical services, although recognized as indispensable, must take its place in competition with the demand for preventive services and surgeons can find themselves trying to sustain an already existing service against the threat of reduced finance. Prosthetic services have fared somewhat better. Partly because of the wars, the amputee has a high international profile and may be supported outwith normal health budgets, at least until such time as those facilities established by international organizations are handed over to local governments. Unfortunately, the amputee from either industrial or road accidents has had to take second place until the need of the war amputees is fully satisfied. Throughout Africa there is the pervasive problem of the victims from road accidents. Roads which are good encourage speeding and new cars are fast and traffic discipline poor. Equally, those roads which are in bad condition present a different type of accident but sadly with the same effect.

The present situation with regard to amputation surgery and prosthetics can be stated as follows:

(1) Surgical services are overloaded and have to compete with other budget imperatives.
(2) Orthopaedic specialists are few in number in relation to overall populations; probably not more than 1 per 1 000 000 in Central Africa, e.g. Zambia, Mozambique and Malawi.
(3) Orthopaedic surgeons work within Departments of General Surgery and development has to compete with other surgical specialities.
(4) Specialist orthopaedic training does not exist in most countries and amputation surgery must be taught to all surgeons in training.
(5) Provincial hospitals may have few surgeons on staff and usually no orthopaedic surgeon.

(6) Both district and mission hospitals are staffed by young general medical officers with little or no formal surgical training. Those few who were trained in Europe may start work with little or no understanding of African conditions.

(7) Large areas of Africa are served solely by Rural Health Centres with no medical officer, having only nursing staff.

It will be realized from this that conditions leading to successful amputations and a resultant good stump will not be found in many African countries.

Logistic problems relating to amputation surgery

The size and complexity of the African Continent is difficult to grasp. Africa is a mosaic of individual countries ranging in size from tiny communities with populations of less than 500 000 to countries with individual populations in excess of 50 million and more. The population, ethnic diversity and economic circumstances are so varied that some countries occupying a large area may have a widespread but small population whereas others are densely populated in some parts and sparsely in others. Each country will have various population groupings and a diversity of language which makes generalizations difficult. Combined with this are the problems of scale, e.g. many African countries have a capital city with a population in the order of 1 million with several large provincial centres some 300–400 miles away. For example, in Zambia the capital Lusaka is situated 350 miles from the provincial centres of the Western and Eastern provinces, and 280 miles from the provincial centre of the Southern Province. Moreover, it is 180 miles from the industrialized copperbelt which in turn is 400 miles from the northern border with Tanzania.

A capital city normally has a medical school and several central hospitals of high standard all able to offer a wide range of medical facilities. The patient attending a teaching hospital in the capital is likely to get high quality medical care. This may also apply to some private hospitals. In contrast, most mission and district hospitals are located in remote rural areas of Africa where roads are poor and rail links non-existent.

In the provincial centres there are usually surgeons who are capable of carrying out most standard operations but they are not necessarily skilled in orthopaedic procedures. Medical treatment is extremely variable from capital city to provincial centre; even more variation is seen in the services provided by district and mission hospitals. In one you might find an excellently run tightly knit mission hospital and in the other a poorly equipped and poorly staffed district hospital. It is from these centres that many of the amputation problems arise.

The population densities also produce their own problems, e.g. in East and Central African countries the populations now range from 10 to over 25 million. The number of new amputees per year can therefore be considerable.

Surgical problems of an acute nature require urgent attention and must be dealt with straightaway; the question of transfer to other hospitals is usually uncertain and difficult.

Many amputations performed as a life saving measure are carried out by the inexperienced. The resulting stumps are a testimony to this problem. It also follows that access to prosthetic fitting may be very haphazard and those closest to the capital will benefit most.

The practice of amputation surgery and prosthetics has important differences on the African continent. These can be dealt with under the following headings:

(1) Indication for amputation and cause of limb loss.
(2) The decision to amputate and attitude to amputation.
(3) Amputation levels.
(4) Orthopaedic workshops. Their establishment and support.
(5) A plan for progress.

Indications for amputation

(1) *Trauma*, e.g. compound fractures resulting from road accidents, gunshot wounds, land mines, industrial and farming accidents.
(2) *Uncontrolled infection*
   (a) Osteomyelitis
   (b) Soft tissue infection, e.g. pyomyositis
   (c) Infected wounds leading to ulceration.

(3) *Snake bite.*
(4) *Diabetic infection with or without gangrene.*
(5) *Vascular disease.*
(6) *Neuropathic conditions.*
(7) *Tumours.*

Compound fractures

Many compound fractures which present to hospitals in Africa are in grade 3, whether it is as a result of a high speed road accident or consequent on civil strife in the shape of gunshot wounds or mine and grenade injury.

A poor understanding of the basic principles of compound fracture management means that many patients will arrive at the first hospital only to be given delayed or unsatisfactory treatment. This leads to unnecessarily high complications such as tetanus, gas gangrene and subsequent amputation. A patient will be taken to a rural hospital initially for first aid or basic surgical treatment and then be transferred to a district hospital only when transport becomes available or the patient's condition deteriorates. Thus at the ultimate referral hospital patients are seen in a state where limb salvage is impossible.

A compound fracture should be treated by wide wound excision, external fixation and if necessary by full thickness skin flap cover but the centres that can offer this form of treatment in Africa are few in number. Delay can lead to the additional complication of Volkmann's ischaemia.

Paramount in any campaign to preserve limbs must be training at district hospital level in the correct management of a compound fracture and impending ischaemia.

Uncontrolled infection

Osteomyelitis and pyomyositis are invariably caused by the penicillin resistant *Staphylococcus aureus* and late presentation often occurs. This leads to extensive damage to bone and soft tissue and the resulting limb is sometimes rendered functionless. Amputation is therefore the only recourse (Fig. 29.1)

Infected wounds arising from penetrating injuries, e.g. thorns, metal spikes and wooden splinters, are often neglected in the early stages. This can lead to severe infection, ulceration and

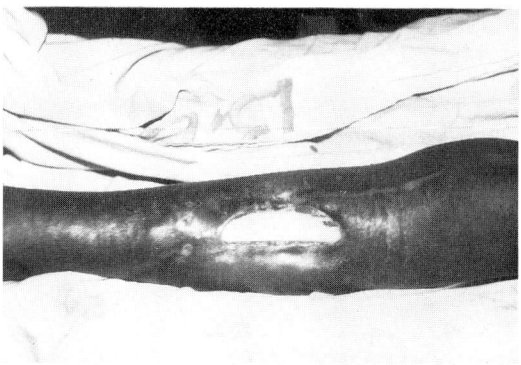

Fig. 29.1. Late presentation with resultant extensive damage to bone and soft tissue. Amputation the only solution.

destruction of skin and underlying tissue. The foot may be rendered useless and the only way to rehabilitate the patient is to amputate the leg.

Snake bite

This is most common in the rural areas where the collecting of firewood is a daily task. Depending on the type of snake and the degree of poison injected, the immediate reaction can vary from the moderately mild case with severe pain and local swelling to extensive tissue necrosis, systemic haemolysis and possible respiratory arrest. In the severe bites extensive tissue necrosis is inevitable and unless the patient can reach surgical help within 6 h to have decompressive fasciotomy the limb segment will certainly die with the inevitable need for amputation. Many patients are seen days and even weeks after the original bite with withered limbs from dry gangrene.

Diabetes

There is a high prevalence of diabetes in some parts of Africa. The presence of undiagnosed diabetes is also high and this comes to light only after the onset of infection in the foot or ischaemia of one or more toes. It is therefore important to investigate all patients with peripheral vascular disease coming to amputation in the expectation of discovering a certain percentage of undiagnosed diabetics.

Vascular disease

Peripheral vascular disease is becoming more common. In countries where tobacco is grown cigarettes are cheap and smoking is common. This adds to the likelihood of occlusive vascular disease. Most patients do not come for treatment in the early stages of claudication when reconstructive procedures would be possible. They present with early gangrene in the toes and a clear indication for a transtibial or even a transfemoral amputation. There are few centres where vascular reconstruction can be undertaken and therefore it must be accepted that most patients with distal ischaemia will require amputation.

Neuropathic conditions

Leprosy is still common in some countries although the incidence overall appears to be declining. Other neuropathic conditions also occur but are less likely to result in amputation.

There is also a group of patients who suffer neuropathic ulceration in the feet in whom leprosy cannot be conclusively diagnosed. These patients attend regularly with recurrent ulceration on the soles of the feet. They are improved by bed rest, elevation and antibiotics and are then discharged only to return later with the same or similar condition. Amputation may well become necessary.

Tumours

The commonest tumour leading to amputation is the squamous cell carcinoma secondary to an area of chronic skin damage (Fig. 29.2) A large squamous carcinoma arising from a burn scar, a chronic sinus from osteomyelitis or indeed any wound in which there is skin de-pigmentation is likely to appear after a period of 10 years, particularly in the lower extremity and when the depigmented scar is exposed to sunlight. Although local resection is sometimes feasible, when the tumour has started to invade bone, amputation is inevitable. Provided the tumour margin is excised adequately, a more distal amputation is possible in contrast to osteogenic sarcoma.

Osteogenic sarcomata occur regularly in Africa. Biopsy is required prior to amputation but because

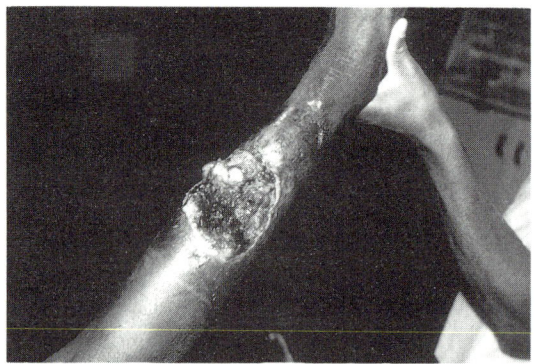

Fig. 29.2. Squamous cell carcinoma secondary to chronic skin damage.

of the reluctance to undergo limb ablation patients will often delay and the tumour can reach an enormous size before an amputation is undertaken.

Chondrosarcomata are encountered, as are all the soft tissue sarcomas. It has also been noted that due to the prevalence of HIV there is now a slowly increasing incidence of skeletal malignant tumours.

The emergency amputation

The medical personnel most likely to be confronted with this problem are as follows, in an increasing order of frequency:

(1) An orthopaedic surgeon (rarely).
(2) Surgeon (more commonly).
(3) General duties medical officer or district medical officer (commonly).
(4) Recently graduated doctor (commonly).
(5) A surgical technician or medical assistant (varying frequency).

The decision to amputate

This is the most difficult aspect of the problem for the inexperienced doctor. Unless the limb is literally hanging by some fragments of tissue, the decision of whether or not to amputate the limb will be difficult. Particularly in the early stages of severe trauma or a gunshot wound it may be difficult to decide that amputation will be the best solution in the long term rather than embark on a long and difficult attempt to salvage the limb. Where resources are scarce and conditions poor, limb salvage will be much less successful than in more favourable circumstances but medical staff may be afraid to perform amputation because of the consequences in terms of social and domestic strife. The attitude to amputation in most traditional African circles is one of abhorrence. Medical reasoning regarding the consequences of saving limbs carries no weight.

Patients have been seen with frankly necrotic limbs who would rather be taken home to die because both patient and family insist on this course of action rather than have a life saving amputation.

Attitudes to amputation surgery in Africa

Cultural practices are deeply ingrained and it is important to realize that in many situations faced with the prospect of limb ablation consent for amputation will not be automatic. The decision to allow an amputation may not be one taken by the patient alone. The family are deeply involved and it is only possible sometimes after the family has consulted all or many of its members that the decision can be reached. In this situation the male members of the family have a dominant role. Ultimately, a surgeon is faced with one or a combination of the following circumstances.

(1) The patient and the family agree to amputation at the level indicated necessary by the surgeon.
(2) The relative and patient agree to an amputation but at a lower level than indicated.
(3) The patient agrees to an amputation but the family do not. It is usually the older members of the family in discussion with others who decide that amputation should not be carried out.
(4) If the patient agrees and the family refuse, the patient may overrule the family decision, particularly if male.
(5) A husband has complete authority over his wife and may refuse to allow his wife to undergo amputation even if she agrees and is suffering from a malignant condition.
(6) Children under the age of 15 or 16 never have authority to authorize their own amputations.
(7) Males usually decide; females do not.
    It follows from the above that where the question of consent for amputation is concerned, full and frank discussions take place with the patient and all the family members.
(8) Regional variation.

Generally in an unbiased setting, permission is granted, particularly among the educated groups.

Once an amputation has been carried out and the patient has recovered from the trauma of surgery the rehabilitation phase can be extremely rewarding. Family and friends usually come to terms with the limb loss and the patient is then able to proceed with his rehabilitation.

It is probable at this stage that the patients in Africa are just as discerning as those elsewhere with regard to the type and quality of prosthesis with which they are supplied. One fact is quite certain, that is that many patients will go to great lengths to obtain an artificial limb when they realize that independent mobility is thereby possible. Patients will even make their own appliance to enable them to walk (Fig. 29.3).

It is a common experience also that patients having been fitted with a prosthesis then depart for

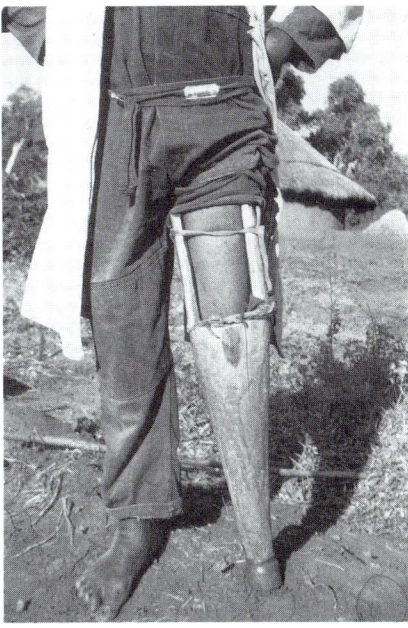

Fig. 29.3. Self designed and fabricated prosthesis.

Fig. 29.4. Self designed prosthesis subsequently repaired by 'designers'.

remote areas and carry out their own repairs in order to keep themselves mobile (Fig. 29.4).

Considerations of cosmesis, weight and cost are all relevant in the African patient. There is also a profound difference between patients who will return to live in rural areas and those who remain in towns. Those in rural areas may accept the Jaipur type of foot but those who live in towns will prefer a standard SACH foot and normal shoes.

Amputation levels

All levels of amputation carried out in Africa are dependent on the skill of the surgeon and the centre where major surgery is carried out.

Transpelvic amputation and hip disarticulation

Because of the late presentation of disease, patients will often come with tumours in a very late stage and major amputations around the pelvic girdle will be necessary principally as palliation but occasionally as definitive treatment.

Example: A 13-year-old boy presented to the Central Hospital in Maputo in 1980 with an extensive destructive lesion of the proximal femur involving the hip joint (Figs 29.5 and 29.6). Radiological appearance suggested that this was a Ewing's sarcoma and chest X-ray showed one specific small opacity in the right lower zone. A hip disarticulation was carried out principally to relieve pain and make the patient more comfortable. Initial biopsy had been unclear and subsequent biopsy material was sent to a bone tumour centre in Europe. The result from the centre showed that this tumour was a benign chondroblastoma and curable by amputation. The opacity in the right mid-zone remained unchanged over a year-and-a-half, indicating that it was an old calcified tuberculous focus. The decision to operate on these large and late tumours may be very worthwhile.

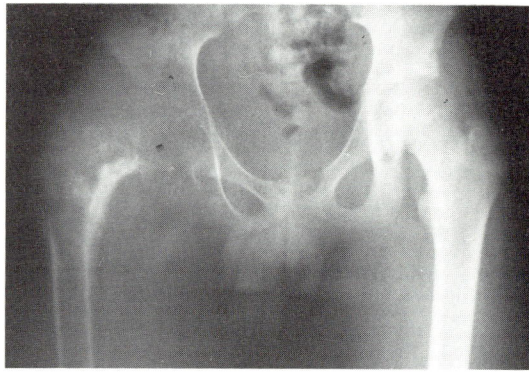

Fig. 29.5. Benign chondrosarcoma originally and pre-operatively diagnosed as a Ewing's Sarcoma.

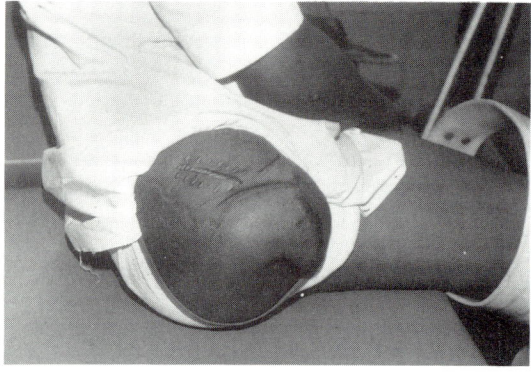

Fig. 29.6. 'Worthwhile' amputation in hip region.

Transfemoral amputations

These procedures are necessary usually because of trauma but also because of peripheral vascular disease with subsequent gangrene. Unfortunately, patients in the early stages of peripheral vascular occlusion will not often be suitable for reconstructive treatment or will present late with incipient gangrene and consequently transfemoral amputations are often necessary.

Prosthetic fitting for the transfemoral amputee can be difficult principally because of the lack of artificial knee joints. A locally made joint may have to suffice or the local workshop may supply a fixed knee prosthesis.

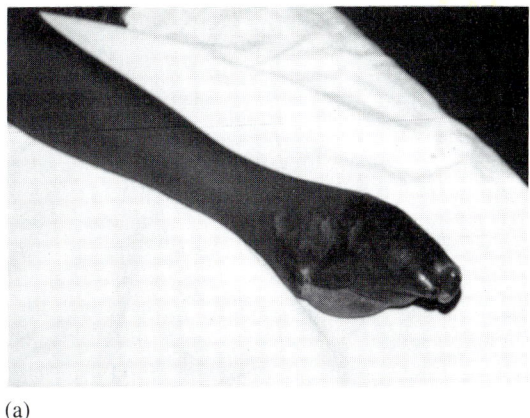

(a)

Knee disarticulation

These are almost always performed for trauma where there is insufficient tibial length to allow a transtibial amputation. The long anterior flap procedure should be avoided and instead a circular incision developing medial and lateral flaps employed. The resultant scar lies posteriorly in the intercondylar notch.

Transtibial amputations

Although this is the commonest amputation performed, it is nearly always badly performed. The medical staff who have no experience of limb fitting do not appreciate the importance of stump length and quality, soft tissue configuration and a mobile scar. Trauma is the commonest reason for this amputation, closely followed by diabetes.

Transtibial amputation is sometimes possible in peripheral vascular disease and the importance of the posterior myocutaneous flap cannot be overemphasized. The contrast between the need to keep all length in a transtibial amputation for trauma where possible and the ideal stump length in vascular disease is often not appreciated.

Ankle disarticulation (Syme)

This is possible in situations where trauma to the foot has resulted in a destroyed and fixed equinus foot. It is an ideal amputation for Africa in that the

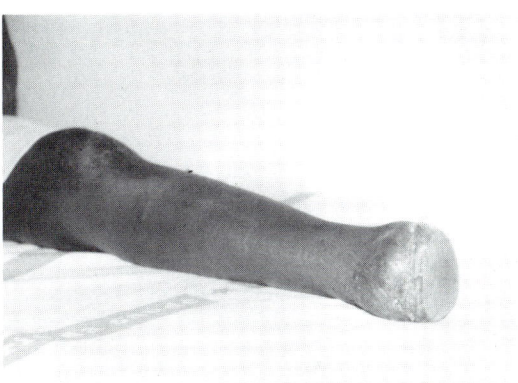

(b)

Fig. 29.7. (a) Destroyed forefoot. **(b)** Ideally treated by Syme ankle disarticulation.

end bearing stump can be used without a prosthesis in the domestic scene. Figure 29.7 shows a destroyed forefoot and is an ideal case for the Syme amputation.

Foot amputations

Chopart and Pirigoff amputations are useful procedures because of the ability to walk without an appliance. These amputations are also sometimes indicated in leprosy when there are penetrating ulcers in the forefoot but sensation is retained in the heel pad.

Orthopaedic workshops

In the making of prostheses in Africa many of the components will have to be made physically by hand. It is therefore necessary to have a large staff of artisans who are able to do simple metal work, woodwork and leather work in order to complete the fabrication. It is probable that a ratio of one prosthetist to five or six artisans is necessary. This requires a workshop of considerable size in terms of staff but is essential to perform the tasks.

An orthopaedic workshop requires the supervision of an orthopaedic surgeon and a prosthetist. The orthopaedic surgeon is needed for the overall clinical supervision including the assessment of stumps and the check-out of prostheses.

The artisan staff require recognition and this is best achieved by formal testing of their skills. A certificate of competence can then be issued giving the worker a feeling of belonging to the establishment.

There is usually a government workshop in the capital city. It is this centre which will dictate what type and quality of appliance the patient receives. Accordingly, these centres must have the correct orientation towards patient needs. In Zambia, during 1971–1976 the standard modern laminated plastic leg was produced. It was much appreciated by the patients and it was possible to fit all ranges of amputations from hip disarticulation to bilateral transtibial amputations in children (Figs 29.8 and 29.9)

By contrast, recent attempts in Zimbabwe to utilize what were considered to be local materials such as stainless steel, wood and leather resulted in an extremely heavy and unsightly prosthesis. Although patients were provided with these free of charge, after a short period of use they discarded them and usually managed to find alternatives. They sought the modern laminated prosthesis often at considerable financial expense rather than continue with the one made from local materials.

The prostheses must provide the function necessary for the patient to perform manual work and must also be sufficiently robust to reduce the need for repairs and renewal.

Also to be taken into account are the financial constraints and practical possibilities of limb fitting in Africa. Whereas it might be ideal to have all patients with a transfemoral amputation fitted with a knee hinge modular assembly, ankle joint and a specialized foot, in practice it may have to be

Fig. 29.8. Employment of laminated plastic prosthesis in fitting child with hip disarticulation.

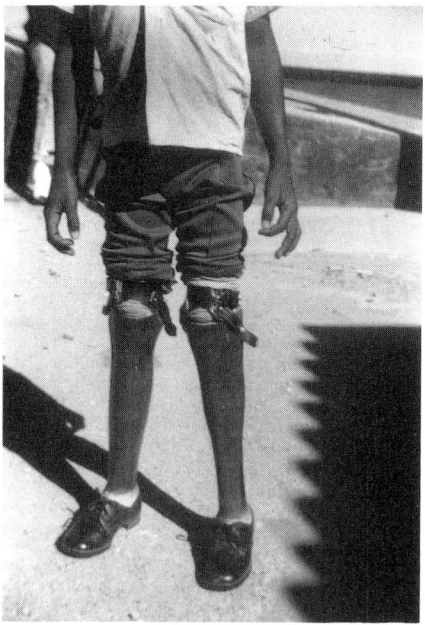

Fig. 29.9. Bilateral transtibial amputations fitted with laminated plastic prosthesis.

accepted that some patients will require to walk with a rigid knee and a local adaptation of the SACH foot.

Given the large number of very short transtibial amputations with flexion contractures of the knee, a flexed knee prosthesis may be necessary. Such decisions can only be made by each individual workshop, depending on the resources available. However meagre these resources may be, if there is no workshop nothing can be achieved.

Summary

In the African scene the management of compound fractures is all important. This should consist of:

(i)   Early adequate wound excision and irrigation.
(ii) Stabilization of the fracture by external fixation.
(iii) The wound(s) must be left open.
(iv) Exposed bone should be covered within 7 days with skin, ideally of full thickness by flap or graft.

Other conditions that may lead to amputation include:

(i)   Acute osteomyelitis; requires urgent adequate treatment by antibiotics and where necessary, surgery.
(ii) Penetrating wounds and tropical ulcers; again, the emphasis is on early treatment.
(iii) Snake bite; must be seen as an emergency requiring immediate fasciotomy.

When amputation is seen to be inevitable it is essential to explain in detail what is involved. Both patient and relatives should be fully informed and in depth discussion encouraged. Agreement to proceed with amputation should not be expected immediately. Patience and persistence must be the watchwords. The nursing staff know the patient best and can be of fundamental importance in reaching a satisfactory outcome.

If the advice to proceed to amputation is not acceptable, then maintain a caring attitude and after an interval try once again to obtain agreement. Whatever the outcome of discussion, the patient will still require support.

Once amputation is inevitable and agreed with the patient there are certain points that require emphasis.

(i)   Maintain all possible length in transtibial amputations.
(ii) Amputation in the case of immediate trauma requires retention of all possible length provided the operation is carried out through viable tissue leaving flaps for closure later. Bone division should be more proximal than muscle and skin.
(iii) In the case of gangrene the wound must be left open.
(iv) In late trauma with infection, the stump wound should be left open.
(v)   In amputation for tumour always amputate through healthy bone and when possible through the non-involved more proximal bone.
(vi) A rigid dressing is recommended. After 10 days a pylon can be added and supervised ambulation encouraged.

30

Amputation surgery: an Indian perspective

P. K. Sethi

Introduction

Poverty, scarcity of limb fitting services and inappropriate designs of prostheses for a barefoot walking, floor sitting population residing in the hot climate of a tropical country are responsible for the majority of Indian amputees still compelled to move around on crutches. The surgeon, when removing a limb, is primarily concerned with saving the life of a patient or getting rid of a diseased or badly injured part of a limb under adverse conditions when patients report so late that the surgical exercise is carried out in haste, with ablation rather than reconstruction governing the technique of amputation. The possibility of fashioning a stump which should become an effective motor and sensory end organ to activate an artificial limb painlessly and efficiently hardly occurs to him. This inevitably results in a disturbingly high incidence of bad stumps. Hardly any attempt is made to rehabilitate the patient, who is allowed to return to his village to fend for himself. Contractures at knee or hip often follow and if and when an opportunity for limb fitting presents itself, a poorly shaped, badly scarred stump with contractures at proximal joints makes the task difficult or impossible.

This situation has to change. The surgeon needs to get out of the confines of the urban hospital setting and spend a little more time in trying to understand his society, its stratified nature, its traditions and culture, its rural and urban divide and a willingness to assume a leadership role in developing and organizing alternate strategies to improve the lot of amputees.

Combining the available scientific knowledge from the west with the recent thinking on appropriate technology (Reddy, 1981) can offer a range of options to widen the choices currently available in most developing countries (Sethi, 1989).

Epidemiology

No reliable data is available regarding the prevalence of causes leading to amputations in India. The only source currently available stems from patients who report to the scanty number of limb fitting centres in India. A study of 1000 consecutive amputees (Patni, 1979; Sethi and Patni, 1980) led to the following general conclusions:

(1) The incidence of amputations due to major occlusive vascular disease in the elderly is distinctly uncommon. Buerger's disease in younger adults is frequently encountered but the majority of these lead to transtibial amputations. Diabetes is a frequent cause of partial foot amputations because of distal occlusive vascular involvement; a major transtibial amputation is usually a result of neglected spread of infection.

(2) Almost 80% of amputees lose their limbs following trauma. There is a disturbingly high incidence of people falling from overcrowded railway trains. This overcrowding, a relatively recent phenomenon, is clearly related to rural poverty which compels an increasing number of villagers to commute daily to urban areas or

during harvesting seasons to distant states in search of work. Such victims are usually young adults of a wage earning group. The majority have transtibial amputations, often bilateral (Fig. 30.1). It is possible that commuters with a traumatic amputation through the thigh may succumb to the injury before medical aid is available at an inaccessible site of accident. The vascular tree of these amputees is otherwise normal.

(3) Leprosy, leading to an anaesthetic, disorganized infected foot remains an important reason for transtibial amputations.

(4) Mycetoma foot continues to account for a small proportion of transtibial amputations in any large series.

(5) There has been a distinct decline in the incidence of amputations due to tubercular osteoarticular lesions with multiple sinuses or for intractable chronic osteomyelitis since the advent and availability of effective chemotherapy and antibiotics.

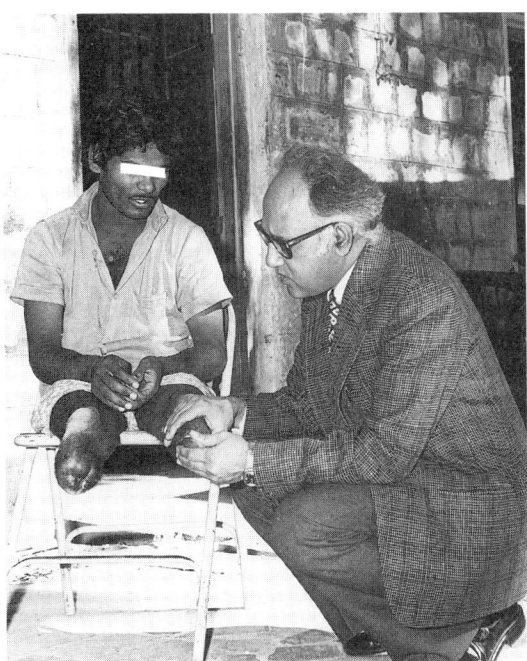

Fig. 30.1. Fall from overcrowded trains is a common cause of post-traumatic amputations. Most victims reaching limb-fitting centres are young adults with transtibial amputations, often bilateral.

The preponderance of young transtibial amputees with a normal vasculature presents a simpler problem than the elderly, dysvascular transfemoral amputees commonly seen in the west. Any strategy which can offer functional prostheses to such an amputee population is a worthwhile effort, converting dependent individuals into socially productive members of society.

Persuading patients for amputations

The most appropriate amputation at the right time is a rarity in an Indian scenario. Most patients have witnessed the fate of amputees and are understandably reluctant to be reduced to a crutch walking/crawling existence. Surgeons are inclined to be impatient when their advice is turned down, not realizing that their western education and non-familiarity with the local dialect, idiom and beliefs make it difficult for them to effectively communicate with an average rural patient. The services of an intelligent trained ward orderly can be utilized to leisurely talk the matter over with the patient and his family, using a language and idiom which is mutually understood, after the hurried ward round is over. This takes away many misgivings and is a strategy strongly recommended to get a timely consent.

By arranging a dialogue with some successfully rehabilitated amputees, the apprehensions about an uncertain future are allayed. Mere promises of arranging for a prosthesis are not good enough. Recruiting a few locally residing amputees to help the surgeon in this endeavour is very rewarding.

Using such simple tactics can offer a better chance of performing an amputation at an earlier stage with all the advantages which accrue from a timely intervention.

Level of amputation

The issues involving levels of amputations have been discussed at length by Murdoch (1970).

With each higher level of amputation more joints are lost, and there is less power and less leverage to control a prosthesis which, in turn, becomes heavier and more complex at each succeeding level. Nevertheless, it must be appreciated that a good amputation at any level is always preferable to a poor amputation at a lower level.

Selection of level of amputation, in the Indian context, often involves extraneous consideration. A bargaining duel is often witnessed between the family and the surgeon. The former are simplistic in their approach and would fight for every inch of the limb. It requires some considerable effort on the surgeon's part to get them to understand why the level of amputation should be where it ought to be.

At the same time, specially in extensive trauma, the surgeon is often tempted to save as much of the limb as possible. What one should be really concerned about is the saving of functional units rather than just saving viable tissues. Zealously applied surgical technology can salvage unusable tissues and parts. If as a result, the residual limb is stiff, painful or insensate in any part, the use of the rest of the limb would be compromised.

Salvage without function is a technical fraud. The operation may succeed but the patient will fail.

It is of utmost importance that a definitive amputation should heal without infection. Since the majority of post-traumatic amputations are required to be performed under adverse conditions, usually in remotely situated small hospitals nearest to the site of an accident by a relatively inexperienced doctor, there is every chance that primary wound healing may not be achieved. Often, by the time a decision for amputation is arrived at, infection has already set in. It would be rash in such a situation to amputate high up at the site of election, close the skin and then suffer post-operative wound infection. A wiser course would be to remove the limb as low down as feasible and fashion skin flaps to allow coverage of bare bone with sutures tied over a pack of non-adherent gauze emerging out on either side to permit adequate drainage. The surgeon should learn to break free from orthodox fashioning of anterior and posterior flaps and realize that modern amputation surgery permits medial and lateral flaps or skewed flaps to be able to preserve adequate skin coverage for a transtibial amputation. Even a guillotine amputation can be followed by skin traction to pull down the available skin to cover a substantial area of the raw stump – a technique successfully used in World War II but now almost forgotten. The resultant stump may not conform to the stringent standards of modern amputation practice, but a subsequent replacement of the terminal scar with a local or cross-leg flap may allow the knee mechanism to be preserved. Utilizing some unorthodox strategies may make all the difference between a transfemoral versus a transtibial amputation.

In undergraduate medical teaching, a much greater curriculum time needs to be allocated to teach the principles and practice of primary, emergency amputations and their management. Definitive surgery at a later date by trained surgeons then becomes so much more effective.

Evaluation of viability of tissues

In recent years we have witnessed the emergence of an entire range of sophisticated investigations to determine the viability of skin or muscles in deciding the correct level of amputation, especially in dysvascular patients. None of these are foolproof and in any case are unavailable even in teaching hospitals. A careful clinical examination, estimating the skin colour, temperature and texture, presence or absence of hair, the condition of toes with attrition of the soft tissues, trophic changes in the toe nails and of course a systolic blood pressure above and below the knee, can provide a reasonably accurate estimate. The surgeon should be willing to revise his decision at the time of operation by observing the vascularity of the skin and the colour and behaviour of muscles while pinching or cutting them. A necessary informed consent to amputate at a higher level if deemed necessary should be taken from the patient prior to surgery.

Transfemoral versus transtibial amputations

It is almost a cliché that every effort should be made to preserve the knee joint. This is even more important in the Indian context where the bulk of amputees residing in rural areas find it difficult to traverse the rugged terrain of our countryside if a transfemoral prosthesis is used. In the absence of a simple, inexpensive and durable knee mechanism the transfemoral prosthesis is used with the knee joint locked. The gait becomes laborious and it is no easy matter to swing the locked prosthesis over obstacles strewn over the rural landscape. The majority of transfemoral rural amputees soon discard the prosthesis and revert to crutches. There is a desperate need for an inexpensive, efficient and durable knee mechanism to be designed.

In this context, a disarticulation at the knee is so much easier to manage and even though not often feasible, it has proved to be an excellent operation. The superior control of the long stump by the intact thigh musculature allows a free knee joint to be used and rural amputees handle the prosthesis well.

Site of election and long transtibial stumps

Having been brought up on the classical concept of 'sites of election', surgeons insist on the conversion of long transtibial stumps to the shorter 'site of election level'. Most Indian amputees refuse to be re-amputated at a higher level. Instead of refusing to provide them with a prosthesis, these were fitted

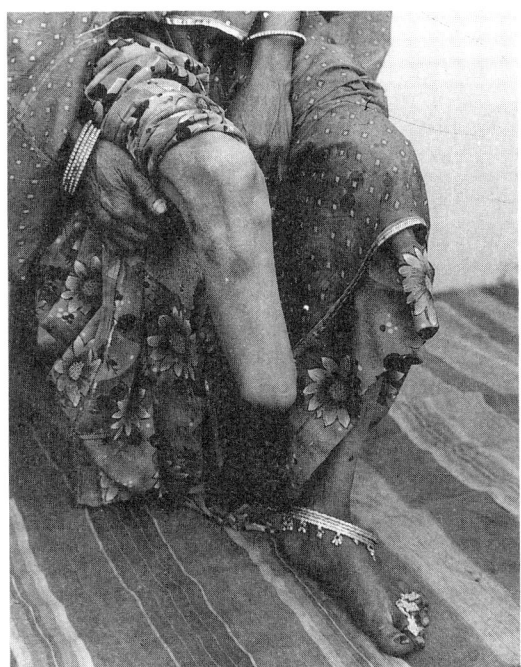

Fig. 30.2. Amputees with long transtibial stumps usually refuse a revision amputation at the 'site of election'. These can be successfully fitted with a Jaipur limb without skin breakdowns.

'under protest', as it were. The anticipated skin breakdowns, surprisingly, were not encountered. The normal vascular tree of the majority of young amputees residing in a warm climate could be an obvious explanation for this experience. In fact, a longer stump often provides a more effective lever arm to control the prostheses.

One wonders whether such a fuss should be made about the site of election in amputees with a normal vascular tree. The long tapering stump undoubtedly is ill suited for a total contact socket but a socket design which is a hybrid of a patellar tendon bearing socket and the old fashioned plug fit seems to work well in these cases (Sethi, 1989) (Fig. 30.2).

Ankle disarticulation (Syme) and partial foot amputations

End-bearing stumps, preserving limb length, allow amputees to ambulate without a prosthesis, and this is a feature which cannot be ignored when prosthetic services are scarce. Cosmetic considerations play a secondary role. It is conceded that situations permitting a Syme amputation are few and the technique of surgery is demanding in providing a durable stump, but when executed well, the performance level of a Syme amputee is superior to a transtibial alternative.

Mid-tarsal or Chopart's amputations are often frowned upon. When an equinus deformity causes weight bearing on the sensitive and vulnerable terminal end of the stump. It can usually be resolved by performing a Lambrinudi type of fusion to provide a plantargrade foot and several such cases have personally been followed-up by the author for nearly 20 years without any skin breakdown. These operations clearly have no place when the plantar skin is anaesthetic, as in leprosy.

Elective amputation surgery

Amputation surgery is reconstructive in its intentions, creating a terminal end organ for obtaining the most effective control of the prosthesis and the most satisfactory man-machine interface. Viewed in this light, amputation surgery is technically demanding. This message has still not percolated to the average Indian surgeon who continues to regard such operations as ablative, with an accent on merely getting rid of an undesirable part. Such an attitude brings with it an element of ruthlessness with muscles slashed wildly and allowed to retract away to a state of uselessness and a rough handling of tissues leading to painful neuromas, projecting bony spurs, infection ending up in an adherent scar and casual post-operative care.

Good surgery is the first step towards successful rehabilitation. The skills of a plastic surgeon are required. Instead of relegating amputations to the end of a long operating list to be performed by a junior assistant, the status of the event has to be elevated and only an experienced surgeon should undertake such work. Amputation surgery, in other words, requires to be upgraded. Considerable work of late has been done with regard to decision making and the craft of amputation surgery and the recently published report of the ISPO Consensus Conference on Amputation Surgery (Murdoch *et al.*, 1992) should become compulsory reading for all Indian surgeons.

Continuing education courses for surgeons who can interact with amputees and prosthetists are absolutely vital towards this end. However, unless the existing prosthetic services can be quickly augmented, the subject will never be taken seriously. One enters into a kind of circular argument where the surgeon sees little point in spending more time to fashion a good stump when prosthetic services are hardly available, and on the other hand the prosthetists lament at the alarming number of bad stumps they are required to fit. This stalemate has to be broken.

Post-operative dressing

The two major requirements are prevention of oedema which interferes with good wound healing and the prevention of contractures. Several alternatives are in vogue which do work well when handled by experienced surgeons. Elastic compression bandage, properly applied, is very effective in controlling oedema but badly applied it can be harmful. It has to be reapplied several times a day to maintain a firm pressure gradient diminishing proximally and this requires constant vigilance.

A rigid total contact plaster cast over a thin padding, is very effective in controlling oedema, diminishing post-operative pain by splinting the

limb and preventing contractures. There is a very narrow margin of safety, however, and it cannot be recommended for general use. 'Safety with excellence' is a motto to be followed.

The combination of a soft, bulky dressing of cotton wool held firmly in place by a thin covering of plaster bandage offers the virtues of both methods. This is the kind of dressing which can be applied by any surgeon and which can be unreservedly advocated.

Early post-operative fitting

Immediate post-surgical fitting which took the surgical world by storm when Marian Weiss (1969) described it is too hazardous to be recommended for general use. However, an early temporary fitting after the wound has healed has a special place in the Indian context. If an Indian amputee is allowed to return to his village with instructions to apply an elastic compression bandage to shape and shrink the stump and return a couple of months later for a definitive limb fitting, either the bandaging would certainly be managed incorrectly or there would be no certainty of the amputee coming back.

A well applied plaster socket, with a PVC pipe as a thermoplastic pylon which can be heat moulded to secure proper alignment has been found to be a very worthwhile tactic to ensure a speedy maturation of stump, and a much faster rehabilitation after definitive fitting. With his 'umbilical cord' still attached to the hospital, his return is nearly assured. Expensive devices are neither available nor would ever be allowed to be taken out of the hospital premises. A PVC pipe, in this respect, serves as a practical alternative to alignment devices; the only equipment needed is a heat gun for spot heating to adjust alignment.

Prosthetic considerations

All refinements in amputation surgery become irrelevant if a prosthesis cannot be provided to the amputee.

The view held by the academic pundits is that the design of a prosthesis is essentially a matter of biomechanical substitution of the lost member of a limb. With a better understanding of the mechanism of locomotion, the use of new materials, many of which are the product of aerospace research, limbs can be made which are increasingly efficient, lighter and stronger. In short, the design of a prosthesis is held to be primarily a bioengineering and technological issue.

This seems to be a rather simplistic understanding of a complicated problem. The lower limbs are not meant merely for walking on level ground. The database acquired on the level walkways of expensive gait analysis laboratories, useful in many ways, is incapable of reproducing the ground reactions and difficulties encountered in a real life situation with amputees walking in farms, marshy lands, deserts or mountainous terrains. In the ultimate analysis, the acceptance of a technical solution by an actual user should determine its suitability and though subjective in nature, his feedback should not be scorned. After all, it is the 'user' who has to be the ultimate judge.

Lower limbs, in addition, also have to have a capacity to lower the body to the floor, to permit sitting on haunches (squatting) or cross-legged, a posture customarily adopted by the majority of the floor sitting population in most developing countries. This involves a degree of mobility at the hip, knee and ankle which is not needed for the chair sitting people in the western world. Designs adequate for the chair sitting culture in the west are clearly unacceptable to the floor sitting people of the third world (Fig. 30.3)

Prosthetic design is not, and cannot be a simple technological issue. It has to take into account the climate, the terrain, the life style and the economic background of the consumer. Failure to understand this and continuing to imitate the west, as has been done so far, would continue to cater for the top 10% of the urban affluent population who are in a position to take to a western style of living. The needs of 90% of our rural and urban poor cannot be ignored. This is only achieved if alternative designs, locally available materials and a technology which can be handled by locally trained manpower recruited from amongst readily available traditional craftsmen are given serious consideration (Sethi, 1986) (Fig. 30.4).

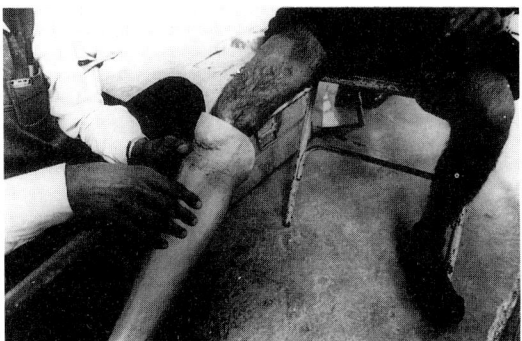

Fig. 30.4. 'Rapid-fit' limb technology, using simple equipment and traditional craftsmen, permits, a large and quick turnover of work.

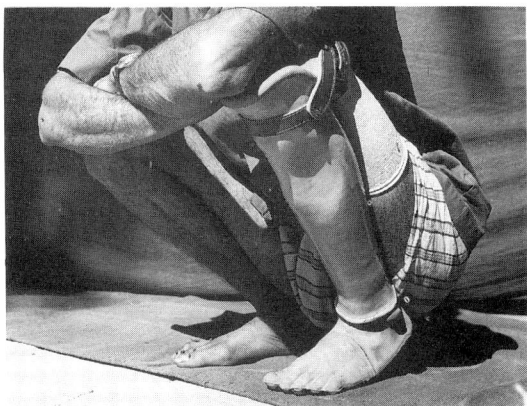

Fig. 30.3. Squatting, a common posture, demands adequate dorsiflexion in the footpiece. Specially designed suspension strap permits full knee flexion.

Of the foregoing, primacy should be given to the design of the prosthetic foot. A controlled mobility in the foot piece to dampen ground reactions, permit transverse rotation of the shank and allow postures adopted in daily life is the key issue (Fig. 30.5). Cosmetic appearance and a durable, waterproof exterior to permit barefoot walking are secondary but nevertheless important considerations which allow an amputee to return to his village and resume his earlier vocation, instead of migrating to an urban area, taking a sedentary job under some welfare scheme and uprooted from his family and friends. A prosthesis which permits this has an element of 'built-in rehabilitation'. This is what 'true rehabilitation' is really all about (Sethi, 1988) (Figs 30.6 and 30.7)

Total contact sockets have been shown to be much superior to the earlier 'plug-fits'. There are two difficulties which ought to be aired and discussed freely. Firstly, it would require a very

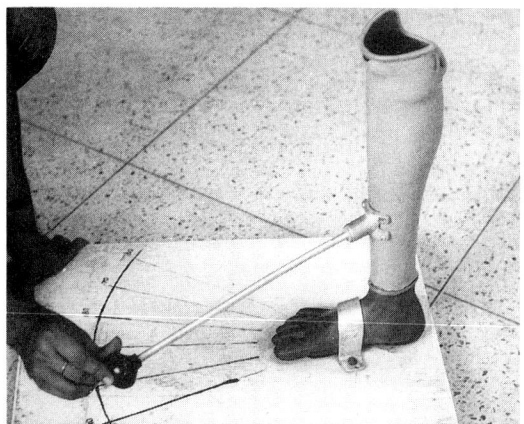

Fig. 30.5. Transverse rotation of the shank is built into the design of the Jaipur foot.

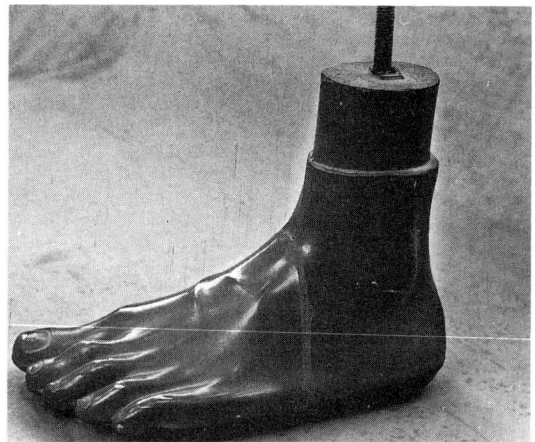

Fig. 30.6. A footpiece suitable for barefoot walking.

(a)

(b)

Fig. 30.7(a,b). A durable, waterproof limb suitable for walking on uneven surfaces.

accomplished prosthetist to successfully and safely fit a poorly crafted scarred stump in a total contact socket (Fig. 30.8). Such experts are just not available. Secondly, even with good stumps, the hot and humid climate in tropical countries makes a total contact socket unbearably stifling. The amputees demand an airy socket system. It is perfectly feasible to use an open ended socket, but utilizing the latest techniques of handling pressure tolerant and pressure sensitive areas to get the maximum biomechanical advantage. In several thousands of young, post-traumatic transtibial amputees, the problem of distal oedema has not been encountered. In the few dysvascular or anaesthetic stumps, a total contact system can and ought to be used.

In short, there has to be a compromise between what is biomechanically superior versus what is comfortable. The analogy of a biomechanically superior laced up closed footwear and a less efficient but more airy open sandal in a hot climate is not out of place.

The choice of materials is secondary. Availability, cost and familiarity of local craftsmen with traditional materials allows one to provide limbs to a much larger number of amputees than an insistence on systems currently fashionable in London or New York.

The number of amputees in developing countries moving around on crutches is so large that an insistence on superior but sophisticated and expensive limbs leads to preposterous situations. How can one justify an amputee being placed on a waiting list and asked to report back for limb fitting after 20 years? This is precisely what the

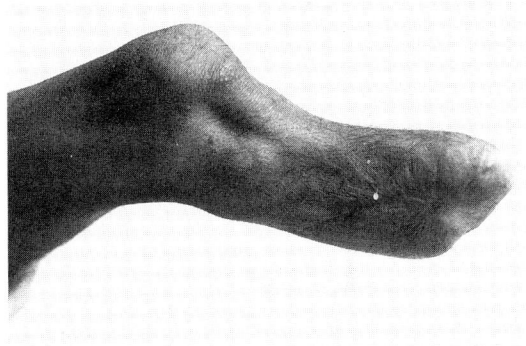

Fig. 30.8. Providing total-contact sockets for such stumps requires considerable expertise. Jaipur limb easily takes care of such amputees.

author witnessed at an extremely well equipped workshop in one of the largest teaching hospitals in a neighbouring country.

Civil wars, landmines and traffic accidents are leading to an unprecedented density of lower limb amputees. Alternative strategies for meeting such situations have to be evolved. There is no going back to peg legs, however. An open ended, flexible approach using the advances made in the west but tailoring them to devise designs and delivery systems which allow 'reaching the unreached' poses a tough challenge and one must not shy away in despair from facing the Indian reality (Ring and Sethi, 1981).

The person who needs to be sensitized to the complexities of this problem is the surgeon. Not only has he to give up his contempt for amputation surgery, but he alone is in a position, by virtue of his social and professional status, to play a leadership role in encouraging and devising models of prosthetic services.

References

Murdoch, G. (1970) Indications, Levels and Limiting Factors in Amputation. In: *Prosthetic and Orthotic Practice* (ed. G. Murdoch), Edward Arnold London, pp. 7–13.

Murdoch, G., Jacobs, N.A. and Wilson A.B. Jr (eds.) (1992) *Report of ISPO Consensus Conference on Amputation Surgery.* ISPO, Copenhagen, Denmark.

Patni, P. (1979) *Amputee Census.* M.S. Thesis. University of Rajasthan. Jaipur. India.

Reddy, A.K.N. (1981) Appropriate technology for rural development. In: *Appropriate Technology for Primary Health Care.* Indian Council of Medical Research, New Delhi, pp. 15–18.

Ring, N.D. and Sethi, P.K. (1981) A rapid-fit limb using alternative technology in India. *J. Biomed. Eng.,* **3,** 318.

Sethi, P.K. (1986) How to make the best use of local means and resources when designing and producing technical aids. *ICTA/AHRTAG Seminar on Appropriate Technical Aids for Disabled People,* Stockholm. pp. 29–34.

Sethi, P.K. (1988) Jaipur foot revisited. In: *Recent Advances in Surgery 2* (ed. R.L. Gupta), Jaypee Bros., New Delhi, pp. 307–321.

Sethi, P.K. (1989) Technological choices in prosthetics and orthotics for developing countries. The Knud Jansen Lecture. *Prosthet. Orthot. Int.,* **13.3,** 117–124.

Sethi, P.K. and Patni, P. (1980) *An Amputee Census.* Paper presented at Annual Conference of Association of Surgeons of India. Calcutta.

Weiss, M. (1969) Physiological amputation, immediate prosthetic and early ambulation. *Prosthet. Int.,* **3(8),** 38–44.

Section 14

The neuropathic foot

31

The neuropathic foot

J. H. Bowker and P. D. Poonekar

Introduction

Diabetes mellitus is the condition most often linked, in the developed world, with injury to the feet due to loss of protective sensation. In the United States alone, estimates of prevalence range to 14 million (Reiber, 1993). In much of the developing world, however, insensitivity due to leprosy (Hansen's disease) is seen more frequently. Other fairly common causes of insensate feet, seen worldwide, are alcoholism and myelomeningocele. In this chapter, the authors will describe both specific and shared features of foot neuropathy and its treatment as related to aetiology from diabetic mellitus or leprosy. To begin, a concise review of leprosy is given.

Leprosy, though predominantly a problem in developing countries, stands as a global public health concern. Estimates in 1988 (WHO, 1988) pointed to 12 million people infected by leprosy calling for a world-wide leprosy control programme and energetic rehabilitation measures. With the introduction of multidrug therapy (MDT) as advocated by the World Health Organization (WHO), cases dropped to 5.5 million in 1991 (Noordeen et al., 1992; Smith, 1992). Leprosy patients with residual deformity are estimated to be in the range of 2–3 million (Noordeen et al., 1992). The long latent period of the infection, usually in the range of 2–3 years with extremes of 1–20 years, makes the problem difficult to control. The accompanying disabling effects of the disease are further compounded by the associated social prejudices and isolation, making the rehabilitation of leprosy patients a herculean task.

Leprosy conventionally has been defined as 'a chronic disease of man resulting from infection with *Mycobacterium leprae*, affecting primarily the nerves, skin, and/or mucosa of the upper respiratory tract' (Pannikar, 1992). Due to the wide clinical spectrum of its manifestations, the elusive ideal experimental model and the epidemiological overtones, the Sixth WHO Expert Committee on Leprosy in 1988 defined a leper as 'a person showing clinical signs of leprosy, with or without bacteriological confirmation of the diagnosis and requiring chemotherapy' (WHO Expert Committee on Leprosy, 1988). Even to this day, the bacillus cannot be cultivated except in the mouse footpad (Shepard, 1960), the nine-banded armadillo (Kirchheimer and Storrs, 1970) and the sooty mangabey monkey (Wolf et al., 1985). Recent cytochemistry studies regarding the immunogenicity of *M. leprae* have resulted in a tremendous increase in the basic understanding of this bacillus (Salame et al., 1983; Jacobs et al., 1987; Rawlinson et al., 1988; Colston and Lamb, 1989; Seghal et al., 1989 a, b; Ross et al., 1992; Bottasso et al., 1993; Turk et al., 1993). This, in turn, is reflected in newer frontiers in serodiagnosis (Smith, 1992) as well as chemotherapy and vaccines (Antia and Birdi, 1988; Gupte, 1992) for leprosy.

The portal of entry is predominantly the upper respiratory tract. A small percentage of cases do result from direct contact by susceptible persons, often greatly facilitated by breaks in the skin from trauma, including tattoo sites (Rao *et al.*, 1987; Seghal and Joginder, 1989). Approximately 95% of individuals are immune to *M. leprae* with only a small percentage of contacts ever resulting in leprosy.

Though leprosy stands as a single disease, the clinical picture is essentially dictated by the host's immunological response to the bacillus. That is, the particular manifestation reflects the position of the patient in the immunopathological spectrum. With tuberculoid leprosy (TT) on one end of the spectrum and lepromatous leprosy (LL) on the other, there are also the intermediate forms: borderline-tuberculoid (BT), borderline/midborderline (BB) and border-line-lepromatous (BL) (Fig. 31.1). The exact diagnosis is based on the degree and extent of skin and nerve involvement, along with the histology and bacteriological index (Ridley-Jopling classification, 1966) (Seghal, Koranne, Nayyar *et al.*, 1980; Ridley and Ridley, 1986). The type of leprosy decides the choice of recommended chemotherapy in terms of the drugs, their dosage and duration of use (WHO Expert Committee on Leprosy, 1988; Fennstra, 1990; Ji and Grosset, 1990; WHO model prescribing information, 1991). Initial multidrug therapy is given for a period of 6 months to 3 years, depending on the type of leprosy and is closely correlated with the histological picture (WHO Expert Committee on Leprosy, 1988).

Tuberculoid leprosy (TT) is seen in a host with mildly impaired immunity to *M. leprae*. Typically, TT is restricted to skin nerves and regional lymph nodes. Clinically, it presents as hypopigmented patches on dark-skinned patients or erythematous ones in the light-skinned, with loss of sensation and dry, scaly macular lesions associated with thickened peripheral nerves. These are best felt where they are subcutaneous (ulnar at the elbow, common peroneal at the fibular neck and auricular over the mastoid) with distal sensorimotor loss. Nerve involvement includes 'stocking and glove' sensory loss, with associated motor impairment, resulting in Charcot neuroarthropathy of the feet.

On the other hand, lepromatous leprosy (LL) is typically seen in a host with very little or no immunity to *M. leprae*. This results in an aggressive form which presents as a systemic disease involving the skin, nerves, lymph nodes, nose, mouth, eyes and internal organs. The clinical findings in LL include loss of eyebrows and eyelashes, macular, papular and nodular lesions all over the body, including the face (leonine face) as well as destruction of nasal cartilage (saddle-nose). Nerve involvement is relatively late. Though

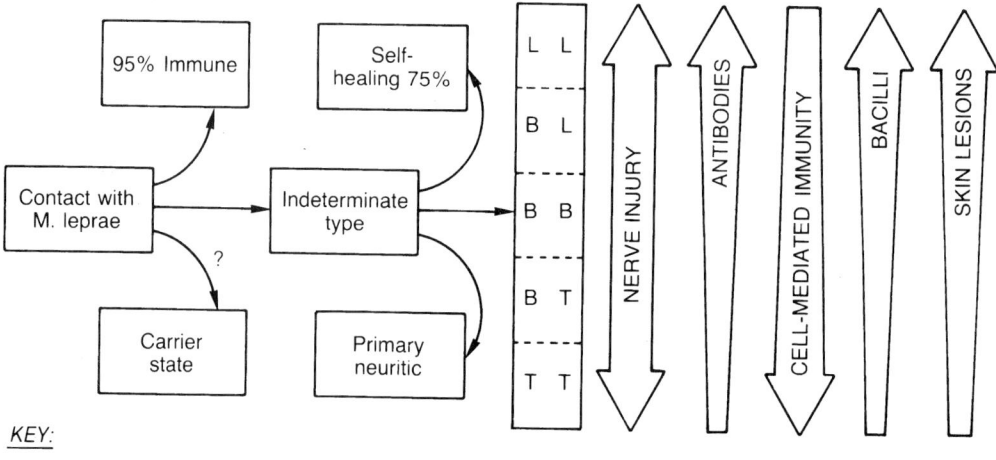

KEY:
LL – Lepromatous leprosy
BL – Borderline lepromatous
BB – Mid-borderline leprosy
BT – Borderline tuberculoid
TT – Tuberculoid leprosy

Fig. 31.1. Profiles of leprosy: response pathways following contact with *M. leprae* and types of clinical manifestation.

internal organs seldom show significant functional derangement, involvement of the testes is common, resulting in testicular atrophy, sterility and altered secondary sexual characteristics.

The neuropathic foot

When evaluating the causation and prevention of foot ulceration, with its great possibility of partial or complete limb loss, one must first consider the risk factors involved. Primary among these is peripheral neuropathy, which is manifested by interference with normal transmission of signals in peripheral nerves. Neuropathy can be of sensory, motor and autonomic types. Most health care professionals are acutely aware of *sensory* neuropathy in which loss of protective sensation results in damage to the feet.

Leprosy is considered to be the commonest cause of severe peripheral neuropathy in developing countries (WHO Peripheral Neuropathies, 1980). In most types of leprosy, nerve involvement is a common denominator. This is partly explained by the bacillus' predilection for cooler areas (36°C/96°F), with subcutaneous nerves fulfilling the requirement. Moreover, the unique immunological protection bestowed on the bacilli by the Schwann cell basement membrane, the multilayering of perineurium, the absence of lymphocyte recirculation within the fascicles and the natural blood–nerve barrier, results in a secure state for the bacilli in the nerve (Shetty and Antia, 1988). They can thus freely multiply locally, resulting in a higher bacillary load in the nerve as compared to the skin (Antia and Pandya, 1976). Nerve lesions may often be seen without a skin lesion. This has resulted in an additional subgroup called 'primary neuritic' leprosy (Seghal 1984; Seghal *et al.*, 1989; Talwar *et al.*, 1992). The predilection for nerve involvement has a bearing on therapy and prognosis, in that chemotherapy often leads to rapid clearance of bacilli from the skin and lymph nodes, while some may persist in the nerves causing relapse at a later stage when chemotherapy has been stopped (Shetty *et al.*, 1992). In one study, some relapses occurred only in the nerves (Pereira *et al.*, 1991). Nerve involvement may start before, during and/or after drug therapy (Ridley, 1969). Nerve damage is clinically manifested as thickening and tenderness of the nerves, and may progress

to abscess formation. Nerve involvement leads to distal sensory loss, motor loss and autonomic dysfunction of the leg with predominant effects on the plantar aspect of the foot.

In *diabetes mellitus*, sensory neuropathy appears to be directly related to the duration of the diabetic condition (Boulton *et al.*, 1986). It may also be related to the strictness of diabetic control over time in that poor control of blood sugar leads to overall glycosolation of body tissues. In diabetics, peripheral sensory neuropathy begins at the end of the toes and progresses proximally.

Motor neuropathy may affect the foot dorsiflexors, permitting a rapid uncontrolled descent of the forefoot following heel strike. The slapping gait may result in damage to the skin and to the metatarsal heads. Aggravation of this effect can occur following paralysis of the foot intrinsic muscles, leading to shift of the metatarsal fat pad distally as the toes go into a clawed position, leaving the skin under the metatarsal heads with little protective fat padding and the foot without its normal resilience. Paralysis of the foot evertors may lead to an unstable subtalar joint, allowing ankle sprains and fractures. Major motor nerve involvement appears to be more common in leprosy than in diabetes mellitus. *Autonomic* neuropathy, manifested by dryness of the skin due to lack of normal sweat production, will lead to fissuring, creating sites of entry for bacteria on the plantar aspect of the foot (Bowker, 1992) (Fig. 31.2).

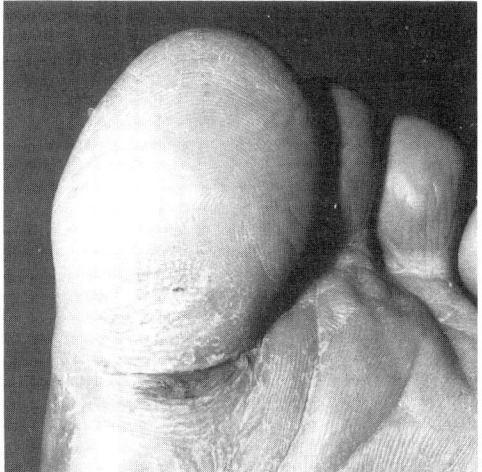

Fig. 31.2. Dry fissured skin on plantar aspect of great toe of 57-year-old diabetic with autonomic neuropathy.

The aetiopathogenesis of plantar ulcers in the neuropathic foot, whether due to leprosy or diabetes mellitus, is very similar. The combined sensorimotor impairment described above results in disruption of the normal gait cycle with an altered loading pattern on the foot. Prolonged and excessively high pressures occur, especially over the bony prominences on the plantar aspect of the foot, leading to localized callus formation. As noted above, this is further compounded by autonomic dysfunction with loss of normal sweating, resulting in the plantar skin becoming dry and brittle, impairing its ability to accommodate the axial and shearing forces imposed on the foot. Ultimately, ulceration at high pressure (bony) areas on the plantar surface of the foot will occur with the risk of infection and partial or complete amputation of the foot (Figs 31.3 and 31.4). Carcinomatous changes have been reported in long-standing trophic ulcers in leprosy, though rarely (Fleury and Opromolla, 1984; Kumar and Patil, 1985; Reddy *et al.*, 1985; Arora and Mukherji, 1987; Pfaltzgraff, 1989; Furuta *et al.*, 1990; Smith and Richardus, 1991).

Plantar pressure distortion is further related to changing bony architecture, both in the bone structure itself and in the arch patterns. Infection due directly to the leprosy bacillus may also lead to

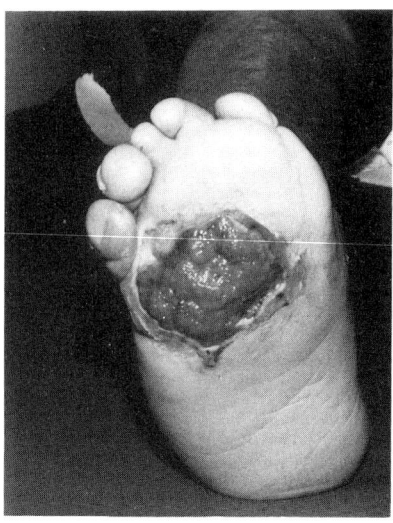

Fig. 31.4. Severe ulceration of plantar aspect of right foot in leprosy patient with neuropathic arthropathy.

primary lesions, although this accounts for only about 4% of foot infections (Selvapandian and Sundararaj, 1991). Bone involvement (Charcot neuroarthropathy) leads to varying degrees of periostitis, osteoporosis and microfractures leading to tarsal disintegration and bone resorption distally (Fig. 31.5). These changes in turn result in progressive foot deformity with gradual collapse of

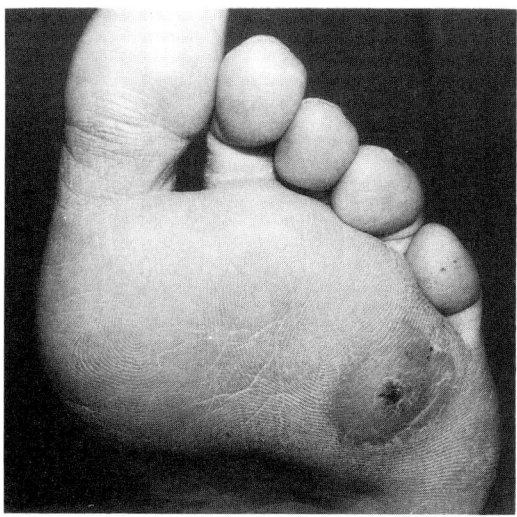

Fig. 31.3. Localized callus and healing ulcer beneath prominent fourth metatarsal head in 51-year-old diabetic. After several ulcer recurrences, removal of plantar one-third of metatarsal head was curative.

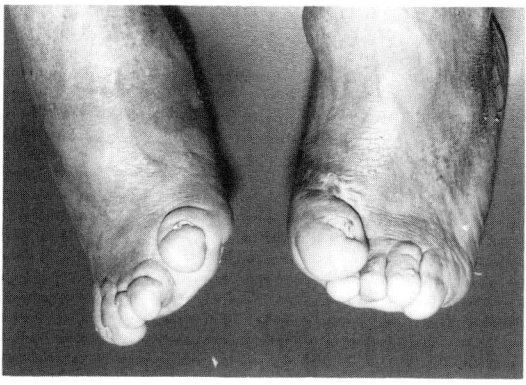

Fig. 31.5. Feet of leprosy patient, showing absorption of phalanges resulting in shortening of foot.

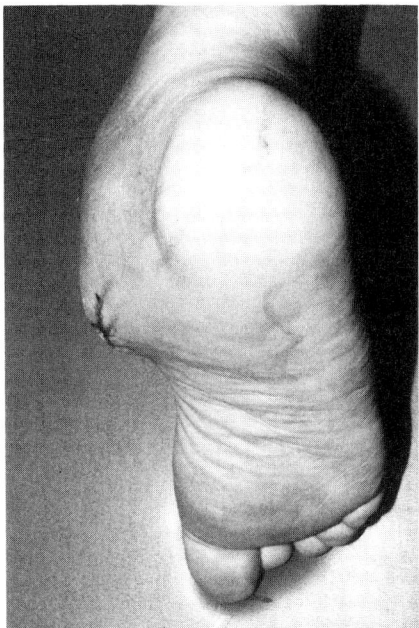

Fig. 31.6. Plantar photograph of diabetic neuroarthropathic foot with severe hindfoot destruction. Note lateral displacement of entire foot with ulceration and abscess formation medially. The patient underwent successful two-stage Syme ankle disarticulation.

the arches of the foot leading to a 'rocker bottom' or 'boat-shaped' foot. In severe cases, the foot may be reduced to a 'bag of bones' with the distal end of the tibia bearing weight beneath the heel pad (Fig. 31.6).

Treatment of the neuropathic foot and its complications

The objectives in treatment of the neuropathic foot and its complications are:

(1) Healing of neuropathic ulcers.
(3) Healing of the disintegrating foot architecture.
(4) Correction of deformity,
 (a) skeletal elements,
 (b) neuromuscular elements.
(5) Optimal pressure distribution on the foot.
(6) Rehabilitation of the patient as a whole.

Acute management

As in the management of any disease process, the management of lesions in the insensate foot is expedited by a classification methodology that has therapeutic significance. For this reason, we manage our insensate foot lesions by the method of Meggitt and Wagner (1978). This system is based on the presumption that most of these lesions, other than dry gangrenous changes from proximal arterial occlusion, are penetrating lesions, either an ulcer or a direct puncture of the foot. There are six grades, each related to lesion depth and involvement of soft tissue and bone (Fig. 31.7).

Grade 0 – bony prominence and callus, but skin intact
Grade 1 – superficial ulcer
Grade 2 – ulcer penetrating skin
Grade 3 – deep ulcer penetrating bone and/or joint
Grade 4 – gangrene of part or all of forefoot
Grade 5 – gangrene of entire foot

Grade 0, with a bony prominence but without skin breakdown, represents a foot at risk due to insensitivity. When the skin is squeezed between a prominent bone internally and an unyielding, hard walking surface and increasingly stiff callus externally, it will necrose beneath the callus, producing a superficial ulceration or grade 1 lesion. This can be prevented by use of proper shoewear with custom-moulded inserts as well as trimming of the callus to allow the skin to be more compliant to direct pressure and shear forces (Young *et al.*, 1992). If these measures are not instituted, a grade 1 lesion will occur with progression to a grade 2 lesion, which is a deep ulcer penetrating to the tendon or joint capsule, but not involving the bone or joint. Although frequently colonized with bacteria, these lesions are not overtly infected and may be treated by non-weight bearing with crutches or walker. A great many patients, however, are reluctant to use these devices because of denial related to lack of pain. If the patient remains non-compliant, a series of healing casts may be offered to allow healing (Boulton *et al.*, 1986). They are extremely effective in evenly distributing weight-bearing forces over the entire plantar surface, thus reducing them over the bony prominence beneath the ulcer. The cast is closely conforming, but never tight. Depending on the size of the ulcer, casting

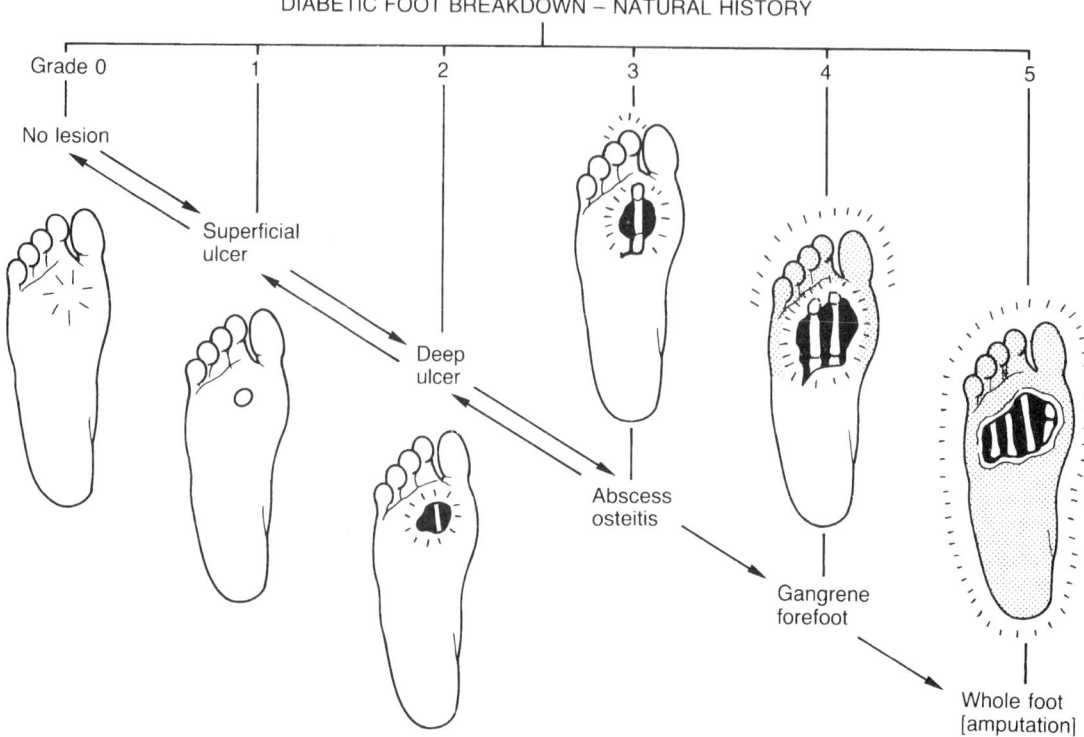

Fig. 31.7. Meggitt–Wagner diabetic foot lesion grading system. Reprinted from Resident Training Manual, courtesy Rancho Los Amigos Medical Center, Downey, California.

can take anywhere from 3 to 8 weeks to produce healing, with weekly and biweekly changes. More recently, special half-shoes, which bear all or most weight on the heel, have been introduced to help with the healing of forefoot lesions (Figs. 31.8 a, b). These are obviously not suited for midfoot or hindfoot lesions for which healing casts are the best option. Once healing of a grade 1 or 2 ulcer has been achieved, protective footwear is prescribed. Over the last few years

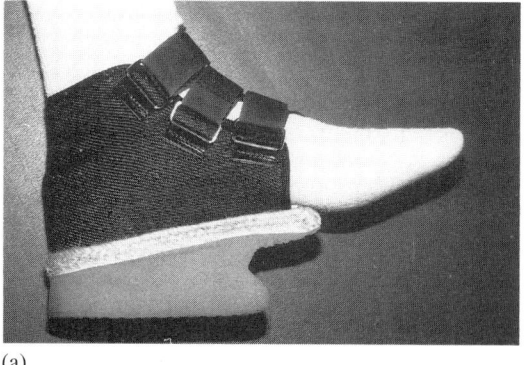

(a)

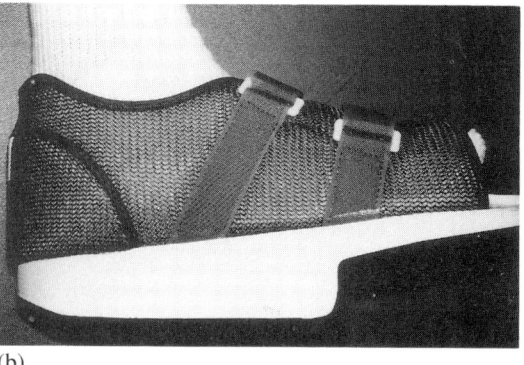

(b)

Fig. 31.8.(a, b) Two types of hindfoot shoe for ambulatory management of grade 1 and 2 forefoot ulcers. These convenient shoes are suitable only for the *compliant* patient.

several studies have indicated the usefulness of various topical agents and dressing materials for healing neuropathic ulcers in leprosy (Sugita, 1992). This has been a direct extrapolation of the studies and findings in treatment of chronic ulcers including diabetic foot ulcers (Alvarez *et al.*, 1993). Leprosy being predominantly a problem of developing countries, the availability and high cost of these dressing materials will be deciding factors determining their use in neuropathic ulcers in leprosy patients on a regular long-term basis.

Grade 3 lesions, which penetrate into bone and/or joint, can easily be diagnosed by probing of the wound (Fig. 31.9). If bone is exposed in the depths of the wound, osteomyelitis is usually present. Plain radiographs will determine the extent of bony involvement. Bone scans are not necessary in the usual penetrating lesion. Occasionally, however, Charcot neuroarthropathy and a penetrating ulcer may coexist, leading to an erroneous diagnosis of osteomyelitis. Magnetic resonance imaging, rather than bone scans, is best used to make the distinction (Wang, *et al.*, 1990).

Grade 3, 4 and 5 lesions require surgical management. Prior to surgery, the circulatory status of the grade 3 or 4 foot must be evaluated to

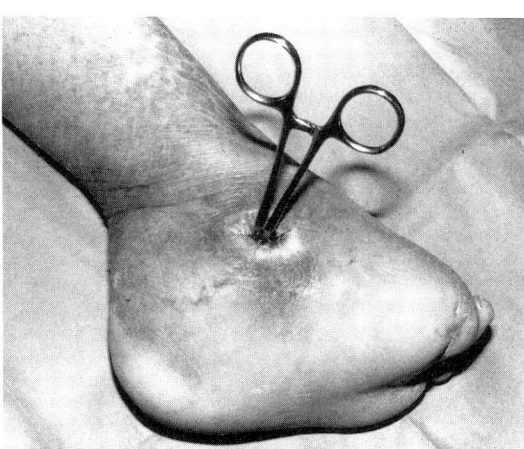

Fig. 31.9. Left foot of 53-year-old leper. Note absorption of the bony structure of the forefoot and the probe in a deep sinus, leading to an area of chronic osteomyelitis in the mid- and hindfoot. A transtibial amputation was required. Reprinted with permission from *AAOS Atlas of Limb Prosthetics*, 2nd Edition. J.H. Bowker and J.W. Michael, Editors, Mosby Year Book, St. Louis, 1992.

help determine what surgical approach may be feasible. This can be done with simple Doppler evaluation of the systolic blood pressure at the ankle and distal forefoot or by use of other methods such as trans-cutaneous oxygen determination (Haward *et al.*, 1985). There is a strong correlation between premature onset of atherosclerosis and peripheral arterial calcification in diabetics, making dysvascularity a major factor in management of infection and gangrene in diabetics (Levin, 1993) as compared to lepers. If the circulatory status is poor, consultation with a vascular surgeon should be obtained regarding possible vessel reconstruction or by-pass procedures. For a detailed discussion of this subject, please see the contribution by P. E. Holstein.

Nutritional status and immunocompetence should also be assessed to give the surgeon confidence that healing is likely to occur at a given level. Nutritional status is considered adequate with a serum albumin level of 3.5 gm/dl or above while a total lymphocyte count of at least 1500 mm^3 is considered evidence of immunocompetence (Dickhaut *et al.*, 1984).

While it is highly desirable to have blood sugar levels under the best possible control to enhance healing in diabetics, it is understood that good control may not be possible until the infection has been controlled. All 'surgical grade' lesions should be cultured aerobically and anaerobically. Due to the polymicrobal nature of most diabetic foot infections, broad spectrum antibiotics are given pending the results of sensitivity testing (Louie *et al.*, 1976). Diabetic control should be initiated simultaneously. Antibiotic treatment of the infection may be only partially effective because of the inhibitory effect of hyperglycaemia on leucocyte functions (Tan *et al.*, 1975). The interdependence of these factors reinforces the case for early excision of necrotic and infected tissue. In lepers, in the rare cases of nerve abscess, incision and drainage is done. Intraneural granulomas respond well to low-dose steroid therapy (Bourrel, 1991).

Prior to any *acute* surgical procedure, it is extremely important to discuss all possible as well as probable procedures and outcomes with the patient and family, i.e., to obtain truly informed consent for treatment. In the operating room, the most important aspect of the procedure is the removal of all infected and necrotic tissue. The surgeon must be prepared to take the person back a second time, if necessary, for further debridement. Once a clean wound has been achieved, whether it is closed over a continuous flow-through irrigation system (Kritter, 1973) or left open, the patient must voluntarily limit weight bearing until the foot is entirely healed over a period of many weeks.

Grade 3 lesions will often require a toe or ray amputation, while grade 4 lesions, characterized by gangrene of part or all of the forefoot, will require additional tissue removal. Although amputation is required, its severity can often be mitigated by acting on a careful consideration of the vascular condition of the foot, coupled with an operation which includes a conservative debridement of infected and necrotic tissue only, leaving all useful normal tissue for immediate or later reconstruction. Longitudinal amputations such as single or multiple toes or ray (toe and metatarsal) procedures are preferable to transverse amputations, such as trans-metatarsal amputations, because the length of the foot is preserved, enhancing functional restoration (Figs 31.10 a, b, c). If both the first and second rays must be removed, however, a trans-metatarsal amputation is recommended. With a more proximal foot infection, a Syme ankle disarticulation is often preferable to a Lisfranc or Chopart procedure since, if either of the latter two fail to control the infection, the chance for a Syme procedure is gone and a transtibial amputation is inevitable. Syme ankle disarticulations hold up well in insensate patients with good vascularity and good prosthetic fitting. Partial foot amputation in leprosy patients is much less common due to their better vascularity and greater resistance to secondary infection than diabetics.

An alternative to surgical ablation of gangrenous portions of a diabetic's foot is 'autoamputation' or natural separation of the necrotic parts over time. Jernberger (Starkhammer and Jernberger, 1989) has recommended this as a means of preserving secure muscle attachments, hence maximum foot function. He recommends placing the foot in a plaster of Paris cast with a fenestration over the necrotic part to allow dressing changes. Over a mean duration of nine months, he observed a healing rate of 49% of 51 feet. The present authors have restricted the use of autoamputation to cases of dry gangrene with insufficient blood flow to meet the additional metabolic needs of tissues attempting to heal after the trauma of surgery (Figs 31.11 a, b).

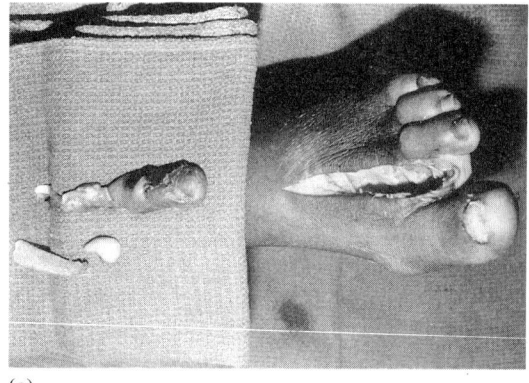

(a)

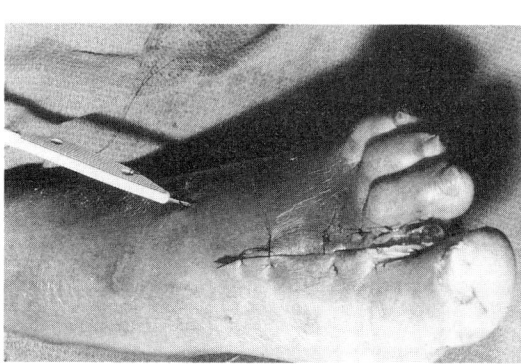

(b)

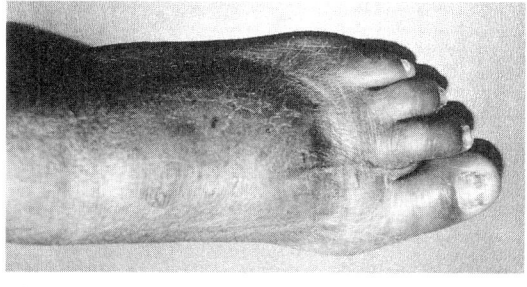

(c)

Fig. 31.10.(a, b, c) Series of intraoperative photographs of second ray resection for osteomyelitis secondary to penetrating (grade 3) diabetic ulcer. (**a**) Operative wound with removed toe and metatarsal. Note infective dissolution of metatarsal neck. (**b**) Wound closed over flow-through irrigation system. Note widely spaced sutures which allow egress of irrigation fluid into absorptive cotton gauze dressing. (**c**) Same foot showing good cosmetic result. (c) reprinted with permission from *AAOS Atlas of Limb Prosthetics*, 2nd Edition. J.H. Bowker and J.W. Michael, Editors. Mosby Year Book, St. Louis, 1992.

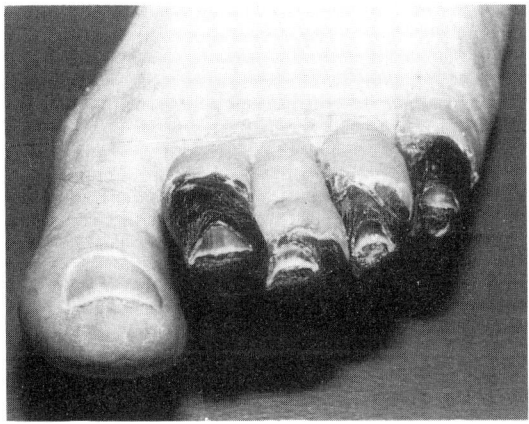

(a)

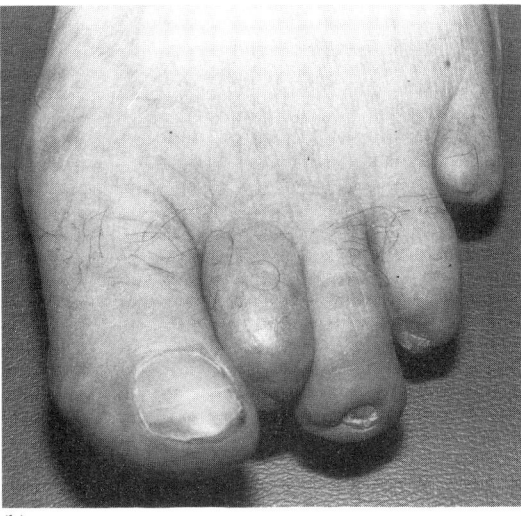

(b)

Fig. 31.11.(a, b) Left foot of diabetic with peripheral vascular disease resulting in dry gangrene of parts of toes 2 through 5. (**a**) Photographic showing apparently extensive involvement of these toes. (**b**) Six months later, autoamputation had resulted in separation of all necrotic tissue, leaving healed wounds with greater toe length than would have been achieved by early surgical debridement.

Charcot neuroarthropathy

Immobilization of the foot remains an important form of therapy for Charcot neuroarthropathy. This is most efficiently achieved by application of a well-moulded short-leg plaster cast which is carefully padded over all bony prominences. The total-contact non-weight-bearing cast provides rest to the foot facilitating the slow consolidation of the disintegrating bony architecture. Early healing is evidenced by a rapid reduction in swelling, warmth and pain in the foot. This may require cast replacement weekly for a time. Casting with regular changes is carried out for a period of 6–9 months. Touch-down foot contact may be started at 12–16 weeks, gradually progressing to partial weight bearing. Regular radiographic follow-up helps in deciding the duration of cast immobilization and the permissible amount of weight bearing.

Following a variable period of cast immobilization, the patient is provided with a 'clam-shell' ankle foot orthosis with well padded plantar contact and rocker-bottom sole (Fig. 31.12). The orthosis allows ambulation with restriction of movements at the ankle, subtalar and midtarsal joints, at the same time easing forward progression during stance phase by the rocker-bottom configuration. After a period of bracing, custom shoes are

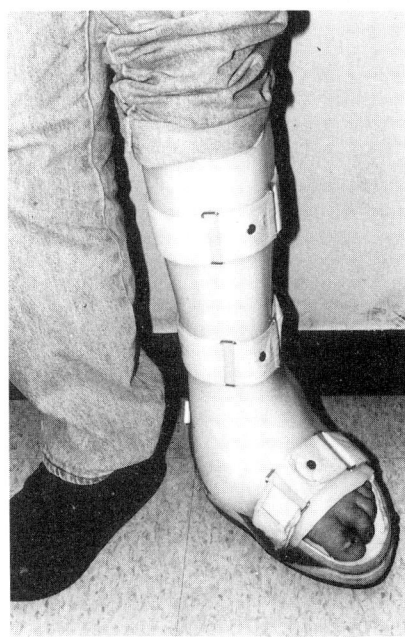

Fig. 31.12. Two piece 'clam shell' orthosis with rocker bottom applied after definitive bony consolidation of neuroarthropathic foot fractures had been achieved by casting. The orthosis will be worn for several months before advancing to a custom-made shoe.

fitted with inserts made of various combinations of thermomoulding materials such as polyethylene (plastazote), synthetic cork and urethane polymer (PPT). This results in greatly decreasing the pressure and shearing forces imposed on the foot during the gait cycle. The entire process of consolidation may require 12–14 months.

Role of reconstructive surgery

In spite of many possible footwear modifications utilizing advanced materials, surgery may still be required to reduce high pressure areas and/or correct major foot deformity in the neuropathic foot. Neuropathic ulcer *per se* in leprosy seldom calls for surgery. In the face of infection, however, local debridement with excision of infected bone and or joint may be required. If localized skin pressure beneath a bony prominence with a sharp apex cannot be mitigated by shoewear modification, a prophylactic osteotomy is done to reduce or remove the offending prominence (Fig. 31.13). Care is taken to avoid disrupting the architectural stability of the foot.

Surgery may be required to stabilize the foot either for proper weight transmission or to prevent further deterioration. Stabilizing procedures include subtalar arthrodesis, ankle arthrodesis and triple arthrodesis. Arthrodesis to correct major deformity in neuropathic feet should be undertaken with caution, since many patients with insensate

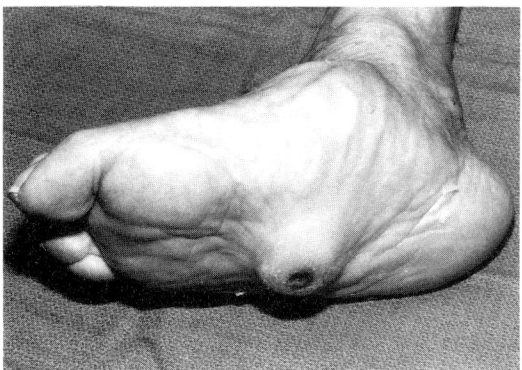

Fig. 31.13. Neuroarthropathic foot with ulcer beneath displaced midfoot bony fragment. Removal of prominence will prevent recurrence of ulcer and ease shoe-fitting.

feet will be poorly compliant with the limited weight bearing regimen required for bony healing, due to lack of protective sensation.

Motor involvement of the posterior tibial nerve may result in claw-toe deformity, calling for release of the flexor tendons and associated soft tissues of the toes. If associated with dislocation/subluxation of the metatarsophalangeal or interphalangeal joints, appropriate skeletal correction and fixation with Kirschner wires may be required. Involvement of the peroneal nerve results in foot drop. If identified at an early state, with regular physiotherapy and bracing, a foot drop deformity (triceps surae contracture) can be prevented or reduced. If surgery for foot drop correction is not indicated or desired for any reason, a moulded plaster ankle foot orthosis (AFO), which can be worn inside almost any footwear, can be prescribed. Alternatively, a modified toe pick-up strap may be used, depending on local resources.

In leprosy cases, posterior tibial tendon transfer may be undertaken (Bourrel, 1991). Our technique involves rerouting the posterior tibial tendon via the interosseous route or circumtibial route and reinserting it in the dorsum of the first cuneiform. In another technique, the posterior tibial tendon is split longitudinally into two slips and passed anteriorly through the interosseous membrane. One slip is sutured to the tendon of the tibialis anterior medially and the other to the peroneus brevis or tertius laterally (Weber *et al.*, 1992). In the presence of a recently healed foot ulcer, 'double tendon transfer' may be performed exclusively in the leg (Carayon *et al.*, 1953). The tibialis posterior tendon is sutured to the tibialis anterior tendon to correct varus and foot drop. The flexor digitorum communis tendon is attached to both the extensor hallucis tendon and extensor digitorum communis tendons. This completes correction of varus and foot drop, also assisting eversion of the foot. If equinus contracture is present, percutaneous fractional Achilles tendon lengthening is done at the time of tendon transfer. If the foot has a tendency to go into supination due to peroneal weakness/paralysis, and adequate tibialis anterior muscle power is present, a split anterior tibial tendon transfer (SPLATT) procedure can be undertaken (Waters *et al.*, 1978). During tendon transfer procedures, tendon anchorage is performed while the foot is held in the desired amount of dorsiflexion and eversion. Post-operatively, the foot is immobilized in a short leg cast for about 6 weeks, followed by regular physiotherapy and gait training. A protective AFO will be required temporarily or permanently to prevent recurrence of the deformity. Surgery is undertaken when the pattern of paralysis has stabilized, ideally before joint stiffness or muscle contracture has set in. With respect to multidrug therapy (MDT), surgery is usually performed 6 months after beginning of MDT, in a phase clear of any reaction (Bourrel, 1991).

With better understanding of the pathophysiology of leprosy, there are increasing efforts to restore plantar sensory function and sweat secretion by surgical decompression of the posterior tibial neurovascular bundle (Bourrel, 1991; Rao and Siddalinga Swamy, 1989). In one study it resulted in increased sweating in 50% of the cases (Pallen and McDermott, 1986). For optimal results, decompression should be done before nerve damage becomes definitive (Bottasso *et al.*, 1993; Bourrel, 1991). Recently, microsurgery has been employed to repair peripheral nerves using denatured autogenous muscle grafts (Pereira *et al.*, 1991).

Preventive measures

There are a large number of preventive instructions regarding foot care that have been given to patients with insensate feet over the years, but they can be reduced to relatively few:

(1) Proper shoes. as described above, should be used at all times.
(2) Feet and shoes should be checked daily by direct observation and palpation to be sure there are no lesions on the feet or foreign bodies in the shoes.
(3) Never, under any circumstances, walk barefoot except when getting into the shower or tub.
(4) Avoid all possibility of thermal injury from excessive heat or cold.
(5) Never use keratolytic substances, such as salicylic acid, to remove calluses.
(6) Avoid all 'bathroom surgery'; instead have calluses removed by professionals.
(7) Limit walking, if feet are insensitive.
(8) Cease smoking.

Footwear

Considerable work has been done throughout the world to obtain ideal footwear for the patient with insensate feet (Coleman, 1993; Janisse, 1993). For the leper, this effort has been largely governed by the constraints of locally available material and climatic factors. In spite of material variations, the general principles are the same. Often a cast of the deformed foot is taken to make an insole for a custom moulded shoe. The insole is essentially a multilayered structure of resilient material formed to match closely the plantar aspect of the foot, thereby ensuring total contact while providing maximum cushioning of the foot. A rigid rocker-bottom sole greatly facilitates forward progression. Mild valgus or varus deformity is corrected for ideal plantar weight transmission by providing a medial or lateral heel wedge, respectively. The wedge may be extended to the sole distally depending on the angle of foot–ground contact. If there is a more marked tendency of the foot to go into valgus/varus deformity with instability of the ankle, medial/lateral flared heels are provided. These may also be extended into the sole. Climatic conditions in the tropics often dictate that the footwear be of open sandal-type. When straps are used, care is taken to round off their edges. In an effort to provide leprosy patients with proper footwear, projects have been undertaken to make them on a mass scale using moulded plastics enhanced with custom-built insoles (Antia, 1990). These are gradually replacing the conventional wooden clogs which have long been known for their durability, ease of maintenance, sculpted total plantar contact and rocker bottom. The replacement is welcomed essentially due to the lightweight, waterproof properties, ease of repair and feasible mass production of plastic footwear. The uppers of the plastic footwear need regular checks to avoid sharp edges cutting into the dorsum of the insensate foot.

Barring neuroarthropathic changes, most diabetics throughout the course of their disease have feet of normal appearance in gross terms, except perhaps for toe clawing and metatarsal head prominence secondary to paralysis of the intrinsic muscles of the foot. Therefore, they try to fit their feet into various common types of footwear, disregarding the fact that they do not have normal protective sensation. With progression of neuropathy, patients become unsure that properly-sized footwear really fit and will tend to purchase shoes that are much too tight, helping to induce ulcer formation. Recommendations for footwear for diabetics include avoiding a number of features common to ordinary shoewear such as tightness of the throat (shoe entry) and toe box, non-compliant materials (plastic uppers as opposed to leather), high heels associated with pointed toes and seams which protrude into the shoes. Conversely, favoured features include a roomy throat and toe box, compliant leather uppers and padded tongues and insoles. If the metatarsal heads are prominent, a custom-moulded insole may be of considerable value in reducing callus formation which would eventually lead to ulcers. It is well known that foot ulcers usually heal given sufficient time and rest from weight bearing; the major problem is to prevent their recurrence by the consistent use of protective footwear. The patients may see no harm in walking barefoot, at least in the house, but in doing so thereby expose themselves to penetrating wounds from various objects left on the floor. Walking barefoot on hot surfaces, common in the tropics, can also produce damage. With loss of normal pain sensation and proprioception, both acute and chronic trauma to the foot can occur without the awareness of the individual.

Management of the neuropathic foot patient need no longer be a formidable health problem as in the past. Best results are obtained using an interdisciplinary approach to encompass patient education, drug therapy, surgery (as indicated), physiotherapy, proper footwear, progressive vocational rehabilitation and social welfare programmes. By this team approach, not only are the health and social issues addressed but the patient is reintegrated as a useful and productive member of society (Lenon, 1988; Loretti, 1989; Levin, 1993).

References

Alvarez, O.M., Gilson, G. and Auletta, M.J. (1993) Local aspects of diabetic foot ulcer care. In: *The Diabetic Foot*, 5th Edition (eds M.E. Levin, L.W. O'Neal and J.H. Bowker), Mosby Year Book, St. Louis, pp. 259–281.

Antia, N.H. (1990) Plaster footwear for leprosy. *Leprosy Rev.*, **61**, 73–78.

Antia, N.H. and Birdi T.J. (1988) Leprosy vaccine – a reappraisal. *Int. J. Leprosy*, **55**, 657–666.

Antia, N.H. and Pandya, N.J. (1976) Qualitative histology and quantitative bacteriology in various tissues of 50 leprosy patients. *Leprosy Rev.*, **47**, 175–183.

Arora, S.K. and Mukherji, R.D. (1987) Malignant melanoma over trophic ulcers. *Int. J. Leprosy*, **59**, 414–415.

Bottasso, O., Besuschio, S., Merlin, V. *et al.* (1983) Lepromatous leprosy treated with recombinant interferon gamma-cutaneous histologic changes. *Int. J. Dermatol.*, **31**, 813–817.

Boulton, A.J.M., Bowker, J.H., Gadia, M. *et al.* (1986) Use of plaster casts in the management of diabetic neuropathic foot ulcers. *Diabet. Care*, **9**, 149–152.

Boulton, A.J.M., Kubrusly, D.B., Bowker, J.H., *et al.* (1986) Impaired vibratory perception and diabetic foot infection. *Diabet. Med.*, **3**, 335–337.

Bourrel, P. (1991) Surgical rehabilitation. *Leprosy Rev.*, **62**, 241–254.

Bowker, J.H. (1992) The choice between limb salvage and amputation: Infection. In 'AAOS Atlas of Limb Prosthetics. 2nd Edition (Eds J.H. Bowker and J.W. Michael), Mosby Year Book, St. Louis, pp. 39–43.

Carayon, A., Chippaux-Maths, J. and Megh, E. (1953) Nouvelle intervention palliative pour pied equin paralytique. *Rev. Med. Chir. A Ex-Orient*, **1**, 24–28.

Coleman, W.C. (1993) Footwear in a management programme for injury prevention. In: *The Diabetic Foot*, 5th Edition (eds M.E. Levin, L.W. O'Neal and J.H. Bowker), Mosby Year Book, St. Louis, pp. 531–547.

Colston, M.J. and Lamb, F.I. (1989) Molecular biology of the mycobacterium. *Leprosy Rev.*, **60**, 89–93.

Dickhaut, S.C., DeLee, J.C. and Page, C.R. (1984) Nutritional studies: importance in predicting wound-healing after amputation. *J. Bone Joint Surg.*, **66A**, 71–75.

Fennstra, P. (1990) Basic requirements for implantation of multidrug therapy – ILEP Medical Bulletin. *Leprosy Rev.*, **61**, 381–390.

Fleury, R.N. and Opromolla, D.V.A. (1984) Carcinoma in plantar ulcers in leprosy. *Leprosy Rev.*, **55**, 369–378.

Furuta, M., Obara, A., Ishida, Y., *et al.* (1990) Leprosy and malignancy – autopsy findings of 252 leprosy patients. *Int. J. Leprosy*, **58**, 697–703.

Gupte, M.D. (1992) The relevance of future leprosy vaccine to disease control. *Leprosy Rev.*, **63**, Supplement 1, 99s–105s.

Haward, T.R.S., Volay, R., Golbranson, F. *et al.* (1985) Oxygen inhalation-induced transcutaneous PO_2 changes as a predictor of amputation level. *J. Vas. Surg.*, **2**, 220–227.

Jacobs, W.R. Jr., Tuckman, M. and Bloom, B.R. (1987) Introduction of foreign DNA into mycobateria using a shuttle plasmid. *Nature*, **327**, 532–535.

Janisse, D.J. (1993) Pedorthic care of the diabetic foot. In: *The Diabetic Foot*, 5th Edition (eds M.E. Levin, L.W. O'Neal and J.H. Bowker), Mosby Year Book, St. Louis, pp. 549–576.

Ji, B. and Grosset, J.H. (1990) Recent advances in chemotherapy of leprosy. *Leprosy Rev.*, **61**, 313–329.

Kirchheimer, W.F. and Storrs, E.E. (1970) Attempts to establish the armadillo (*Dasypus novemcinctus*) as a model for the study of leprosy. Report of lepromatoid leprosy in an experimentally infected armadillo. *Int. J. Leprosy*, **39**, 693–702.

Kritter, A.E. (1973) A technique for salvage of the infected diabetic gangrenous foot. *Orthop. Clin. N. Am.*, **4**, 21–30.

Kumar, A.G. and Patil, S.G. (1985) Squamous cell carcinoma developing in trophic ulcer in leprosy – a case report. *Ind. J. Leprosy*, **57**, 879–882.

Lenon, J.L. (1988) A review of health education in leprosy. *Leprosy Rev.*, **56**, 611–618.

Levin, M.E. (1993) Pathogenesis and management of diabetic foot lesions. In: *The Diabetic Foot* 5th Edition (eds M.E. Levin, L.W. O'Neal and J.H. Bowker), Mosby Year Book, St. Louis, pp. 17–60.

Loretti, A. (1989) Leprosy control – the rationale of integration. *Leprosy Rev.*, **60**, 306–316.

Louie, T.J., Bartlett, J.G., Tally, F.P. *et al.* (1976) Aerobic and anaerobic bacteria in diabetic foot ulcers. *Ann. Int. Med.*, **85**, 461–463.

Noordeen, S.K., Bravo, L.L. and Sundaresan, T.K. (1992) Estimated number of leprosy cases in the world. *Leprosy Rev.*, **63**, 282–287.

Pallen, M.J. and McDermott, R.D. (1986) How might *Mycobacterium leprae* enter the body? *Leprosy Rev.*, **57**, 289–297.

Pannikar, V.K. (1992) Defining a case of leprosy. *Leprosy Rev.*, **63**, Supplement 1, 61s–65s.

Pereira, J.H., Palande, D.D. and Gschmeissner, S.E. (1991) Mycobacteria in nerve trunks of long-term treated leprosy patients. *Leprosy Rev.*, **62**, 134–142.

Pereira, J.H., Palande, D.D., Subramanian *et al.* (1991) Denatured autologous muscle graft in leprosy. *Lancet*, **338**, 1239–1240.

Pfaltzgraff, R.E. (1989) Carcinoma in plantar ulcers of leprosy patients – a report of four cases from Turkey. *Leprosy Rev.*, **60**, 160–161.

Rao, K.S. and Siddalinga Swamy, M.K. (1989) Sensory recovery in plantar aspect of the foot after surgical decompression of posterior tibial nerve. *Leprosy Rev.*, **60**, 283–287.

Rao, K.S., Balakrishnan, S., Oommen, P.K., et al (1987). Restoration of plantar sweat secretion in feet of leprosy patient. *Int. J. Leprosy*, **59**, 442–449.

Rawlinson, W.D., Batson, A., Britton, W.J. et al. (1988) Leprosy and immunity: Genetics and immune function in multiple case families. *Immunol. Cell Biol.*, **66**, (part 1), 9–21.

Reddy, N.B.B., Srinivasan, T., Krishna, S.A.R. *et al.* (1985) Malignancy in chronic ulcer in leprosy – a report of 5 cases from northern Nigeria. *Leprosy Rev.*, **56**, 249–253.

Reiber, G.E. (1993) Epidemiology of the diabetic foot. In *The Diabetic Foot*, 5th Edition (eds M.E. Levine, L.W. O'Neal and J.H. Bowker) Mosby Year Book, St Louis, pp. 1–15.

Ridley, D.S. (1969) Reaction in leprosy. *Leprosy Rev.*, **40**, 77–81.

Ridley, D.S. and Jopling, W.H. (1966) Classification of leprosy according to immunity; a five group system. *Int. J. Leprosy*, **34**, 255–273.

Ridley, D.S. and Ridley, M.J. (1986) Classification of nerve is modified by delayed recognition of Mycobacterium leprae. *Int. J. Leprosy*, **54**, 595–606.

Ross, M., Barr, R.J. and Bocachia, J.H. (1992) A rapid method for the cytodiagnosis of multibacillary leprosy. *Int. J. Dermatol.*, **30**, 632–634.

Salame, P.R., Mahaderan, P.R. and Antia, N.H. (1983) Mechanism of immunosuppression in leprosy – presence of suppression factor(s) from macrophage of leprosy patients. *Infec. Immun.*, **40**, 1119–1126.

Seghal, V.N. (1984) Evolution of classification in leprosy. *Int. J. Dermatol.*, **55**, 424–426.

Seghal, V.N., Bhattacharya, S.N., Shah, Y. *et al.* (1989b) Reaction in leprosy – acute phase reactant response during and after remission. *Int. J. Dermatol.*, **28**, 632–634.

Seghal, V.N., Jain, M.K. and Srivastava, G. (1989a) Evolution of the classification of leprosy. *Int. J. Dermatol.*, **28**, 161–167.

Seghal, V.N., Koranne, R.V., Nayyar, M. *et al.* (1980) Application of clinical and histological classification of leprosy. *Dermatologica*, **161**, 93–96.

Seghal, V.N. and Joginder (1989) Tuberculoid (TT) leprosy – localization on a tattoo. *Leprosy Rev.*, **60**, 241–242.

Selvapandian, A.J. and Sundararaj, G.D. (1991) Infections of the foot and ankle, including leprosy, mycetoma and yaws. In: *Disorders of the Foot and Ankle – Medical and Surgical Management* (ed. M.H. Jahss), W.B. Saunders, Philadelphia, pp. 1958–2001.

Shepard, C.C. (1960) The experimental disease that follows injection of human leprosy bacilli into foot pad of mice. *J. Exp. Med.*, **112**, 445–454.

Shetty, V.P. and Antia, N.H. (1988) Nerve damage in leprosy. *Int. J. Dermatol.*, **56**, 619–621.

Shetty, V.P., Suchitra, K., Uplekar, M.W. *et al.* (1992) Persistence of Mycobacterium leprae in the peripheral nerve as compared to skin of multidrug treated leprosy patients. *Leprosy Rev.*, **63**, 329–336.

Smith, P.G. (1992) Revised estimate of global leprosy numbers. *Leprosy Rev.*, **62**, 317–318.

Smith, P.G. (1992) Serodiagnosis in leprosy. *Leprosy Rev.*, **63**, 97–100.

Smith, T.C. and Richardus, J.H. (1991) Squamous cell carcinoma in chronic ulcer leprosy – a review of 38 consecutive cases. *Leprosy Rev.*, **62**, 381–388.

Starkhammer, A. and Jernberger, A. (1989) Diabetes foten LIC-ortopedi Sweden, pp. 44–49.

Sugita, Y. (1992) Rapid healing of a chronic wound surrounded by hyperkeratosis in a leprosy patient after hydrocolloid occlusive dressing. *Leprosy Rev.*, **63**, 379–380.

Talwar, S., Jha, P.K. and Tiwari, V.D. (1992) Neuritic leprosy – epidemiology and therapeutic responsiveness. *Leprosy Rev.*, **62**, 263–268.

Tan, J.S., Anderson, J.L., Watanakunakorn, C. *et al.* (1975) Neutrophil dysfunction in diabetes mellitus. *J. Lab. Clin. Med.*, **85**, 26–33.

Turk, J.L., Curtis, J. and DeBlaquiere, G. (1993) Immunopathology of nerve involvement in leprosy. *Leprosy Rev.*, **64**, 1–6.

Wagner, F.W., Jr. (1978) Orthopaedic rehabilitation of the dysvascular limb. *Orthop. Clin. N. Am.*, **9**, 325–350.

Wang, A., Weinstein, D., Greenfield, L. *et al.* (1990) MRI and diabetic foot infections. *Magnetic Resonance Imaging*, **8**, 805–809.

Waters, R.L., Perry, J. and Garland, D. (1978) Surgical Correction of gait abnormalities following stroke. *Clin. Orthop. Rel. Res.*, **131**, 54–63.

Weber, M.W., Van Soet, A., Neff, G. *et al.* (1992) Results of surgical procedures for correction of foot drop and of lagophthalmos due to leprosy. *Leprosy Rev.*, **63**, 255–262.

WHO (1980) *Peripheral Neuropathies*. Technical Report Series No. 654.

WHO (1988) *A Guide to Leprosy Control*. WHO Geneva.

WHO Expert Committee on Leprosy. (1988) *Technical Report Series No. 768*. WHO Geneva.

WHO (1991) *Model Prescribing Information: Drugs Used in Mycobacterial Diseases*. WHO, Geneva.

Wolf, R.H., Gormus, B.J. and Martin, L.N. (1985) Experimental leprosy in three species of monkeys. *Science*, **227**, 529–531.

Young, M.J., Cavanagh, P.R., Thomas, G. *et al.* (1992) The effect of callus removal on dynamic plantar fat pressures in diabetic patients. *Diabet. Med.*, **9**, 55–57.

32

Minor amputations after revascularization for gangrene in diabetics

P. E. Holstein

Introduction

Limb salvage by vascular reconstruction or angioplasty can be obtained in an increasing number of diabetic patients. In the presence of gangrene, local revision or a minor amputation of the foot often becomes necessary. Careful attention to detail is required because failure to achieve healing will result in a major amputation, and the revascularization procedure will have been wasted, possibly even complicating the major amputation. Whether or not a minor amputation is related to vascular surgery, healing is dependent not only on the arterial supply but on surgical technique, postoperative treatment and follow-up care.

Arterial supply

The crucial condition for healing is adequate arterial supply. Segmental blood pressure measurements, in particular the toe blood pressure, is valuable in predicting healing of skin lesions and small amputations of the feet (Holstein and Lassen, 1980; Ramsey *et al.*, 1983; Holstein, 1984; Larsen *et al.*, 1989; Apelqvist *et al.*, 1989). Ankle pressures are less reliable due to medial sclerosis often causing the pressures to be falsely high. In small amputations performed after vascular reconstruction it should be noted that the distal circulation is not stable. In contrast to ankle pressures which immediately increase, the toe pressures

representing the most distal – and the most relevant – circulation rises more slowly (Fig. 32.1). After 24 h only about 30% of the increase has been obtained. After 1 week the increase is 60% with a stable situation being recorded after 1 month. These results, obtained by sequential recordings after angioplasty as well as after arterial reconstruction (Noer *et al.*, 1980), explain why minor amputations done immediately after the reconstructive procedure are often followed by skin necrosis. Our routine postponement of the amputation for about 1 week is justified by these measurements.

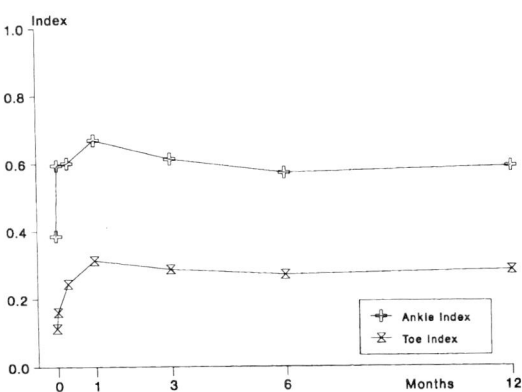

Fig. 32.1. Ankle and toe index after angioplasty in 50 diabetic patients (iliac 17, femoro-popliteal 33). From Holstein, Jorgensen and Larsen (1991). With permission from the *European Journal of Vascular Surgery.*

Due to the relatively slow improvement of the distal circulation it is difficult to identify amputations that will fail. None the less, toe pressure recordings are valuable to document the results. Failed small amputations at adequate blood pressure ought to be rare events.

In the case of severe invasive infection it may be necessary to drain and revise the lesion immediately after the vascular procedure which should then be carried out as an emergency procedure. The therapy for invasive infection in an ischaemic diabetic foot is difficult and must be designed for the individual patient according to the state with regard to septicaemia, ketoacidosis, nutritional status, tissue damage, pus formation, the degree of ischaemia and any technical problems in revascularization.

Operative procedure

Antibiotics

Antibiotics are always given intravenously during the operation. In the case of primary wound closure oral antibiotics are continued for 3–5 days. When the wound is left open for secondary healing, the antibiotics are maintained until bones and tendons are covered with granulation tissue. A combination of dicloxacillin and ampicillin is used but dependent on wound cultures this standard regimen may be adjusted, in most cases due to the presence of Gram negative organisms resistant to ampicillin. In some cases of primary wound closure, local antibiotics (gentamicin beads, Gentacol) may be considered appropriate.

Amputation level

There is no objective method for measuring the blood supply of the foot which can distinguish the suitability of different toe and foot amputation levels. When the circulation has been found to be adequate for a local amputation the level is determined by the extension of the tissue necrosis.

Although it would be preferable, one cannot plan exactly the amputation level preoperatively. The extension of the tissue damage can only be assessed during the operative procedure. For this reason, pre-operative information to the patients must be given with some reservations. It also takes an experienced surgeon to judge the local viability. Primary suture requires healthy tissues. If doubt exists, the wound should be left open for closure by second intention.

Local revisions and free-style amputations

Amputations can in most cases be restricted to levels distally on the feet (Fig. 32.2). There is no reason to place the incision far away from the gangrene, since the ischaemia has been eliminated by the revascularization. Thus free-style amputations rather than strictly prescribed amputation levels can frequently save important weight bearing areas (Fig. 32.3). In fact, the procedure can often be confined to debridement of dead tissue with bone resection sufficient to allow granulation tissue and epidermis to fill and cover (Fig. 32.4).

Primary closure of the wound will shorten the time of healing but carries the risk of necrosis and infection; moreover, weight bearing parts of the foot skeleton should never be sacrificed simply to obtain sufficient skin flaps for primary wound closure. Flaps which cover the foot on the plantar

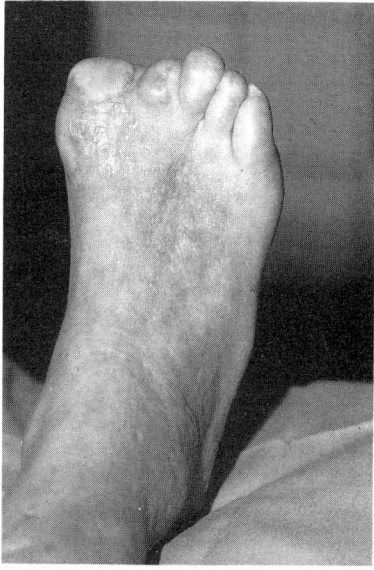

Fig. 32.2. Multiple toe gangrene after free-style revision. With permission from the *European Journal of Vascular Surgery.*

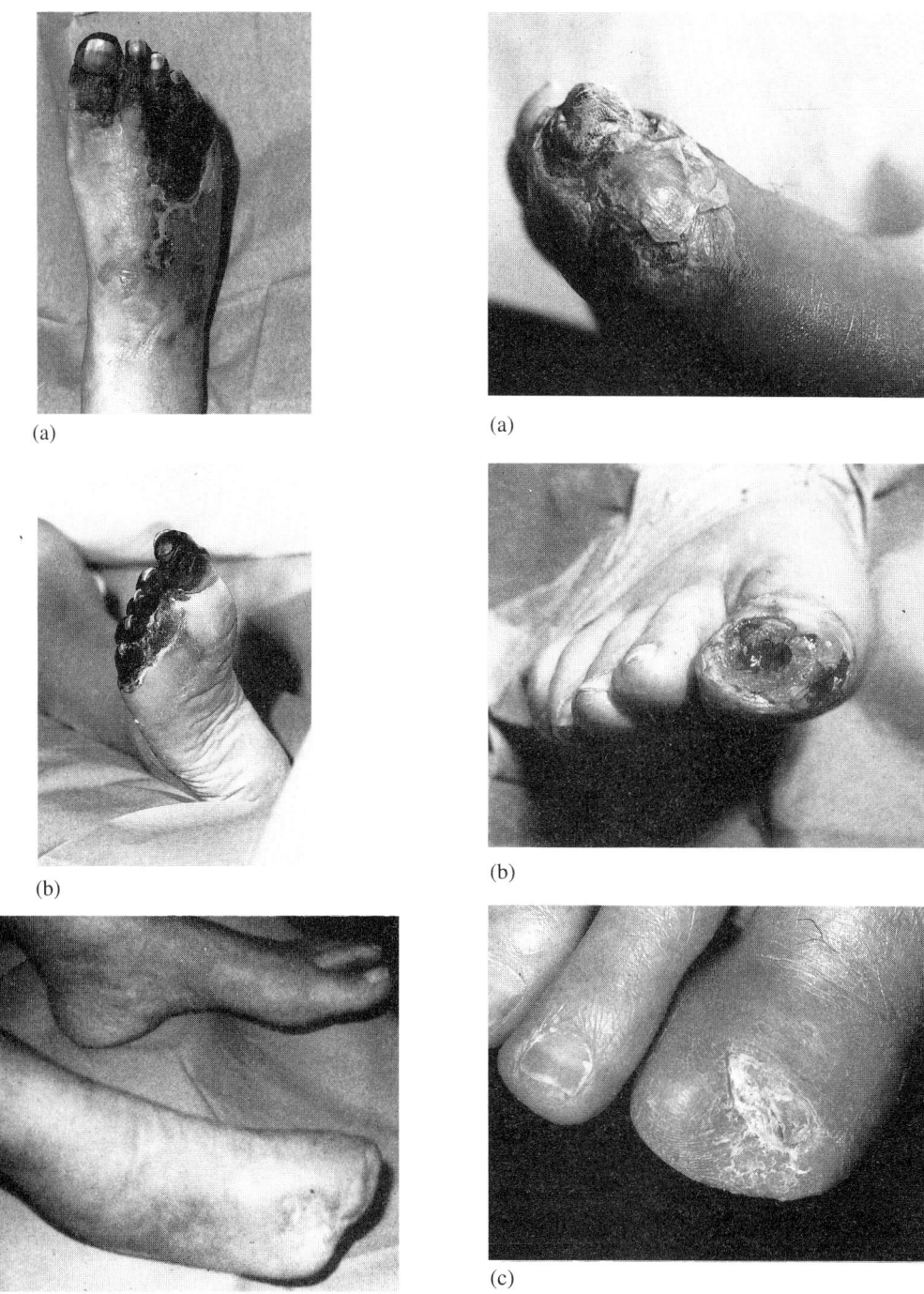

(a)

(b)

(c)

(a)

(b)

(c)

Fig. 32.3. Major gangrene of the forefoot and toes. Free-style revision for saving the metatarsal heads is a better solution than a straightforward forefoot amputation.

Fig. 32.4. Gangrene of the hallux. Revision 1 week after arterial reconstruction revealed that a ray resection, which had been planned, was not necessary (a,b,c). With permission from the *European Journal of Vascular Surgery.*

surface and provide cover distally will be sufficient with any remaining dorsal defects covered by split skin grafts.

Exposed cartilage should never be left in an open wound. Removal of the cartilage allows granulation tissue to penetrate from the bony surface.

Auto-amputation, i.e. spontaneous demarcation and rejection, is seldom used because of the risk of the invasion of infection during the long healing process.

Ray amputations

When a toe lesion extends a small distance beyond the metatarsal-phalangeal crease, ray resection (toe and metatarsal) is probably the procedure of choice. Usually no more than two rays should be resected. If more resections are required a trans-metatarsal amputation should be considered as an alternative. The most important technical point is to preserve healthy tissue and ensure covering of the neighbouring metatarso-phalangeal joints. If this fails, the neighbouring toe will soon be lost.

Forefoot (trans-metatarsal) amputation

This procedure results in a nicely shaped foot stump, but wherever possible we try to resect distal to this level, preferring if need be somewhat irregular minor resections. In this way the important weight bearing metatarsal heads may be saved (Fig. 32.3). Nevertheless, the trans-metatarsal amputation is useful when indicated (Fig. 32.5).

Tarsal amputations and Syme ankle disarticulation

The tarsal levels are often abandoned due to the threat of an equino-varus deformity of the foot remnant. In the elderly patient with limited walking function the tarsal levels are, however, valuable in the case of extensive necrosis of the forefoot. Handmade shoes are adequate for walking function.

The Syme ankle disarticulation was popularized by Wagner (1977) 2 decades ago. This regime requires a normal or close to normal perfusion from the posterior tibial artery and when this is

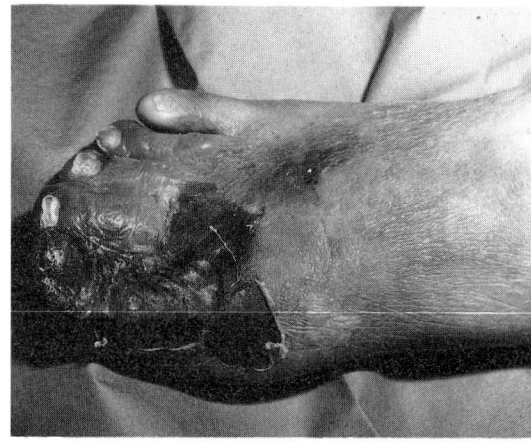

(a)

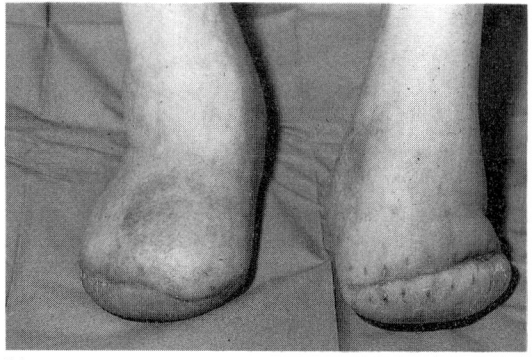

(b)

Fig. 32.5. Forefoot amputation in the right foot following revascularization by femoro-crural bypass. The left foot was salvaged in a similar manner after angioplasty of the superficial femoral artery. With permission from the *British Journal of Surgery*.

present we would prefer more distal amputations. However, it may be the solution of choice when more distal levels cannot be chosen because of extensive skin damage or when a more distal procedure on the foot has failed to heal.

Postoperative treatment

Weightbearing

It is of crucial importance that no weightbearing is allowed until a primarily sutured wound has healed, usually 3–4 weeks after the amputation. In

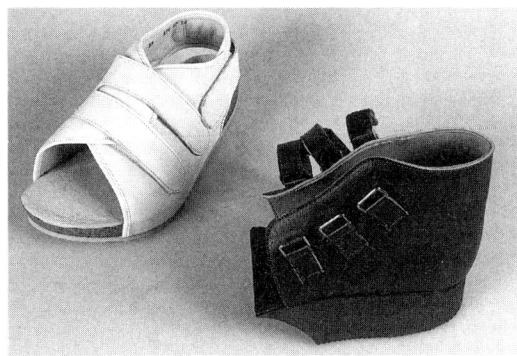

Fig. 32.6. Therapeutic shoes suitable for mobilization after minor amputations on the feet.

case of wound healing by second intention, the foot should not bear weight until bones and tendons are covered by healthy granulation tissue. Earlier weight bearing can, however, be tolerated if the patient is able to walk with a 'heel-shoe' (Fig. 32.6), which prevents load on wounds of the toes and forefoot. The device is very useful but it should not be used in patients with contralateral amputations or in blind patients.

We have also tried to mobilize patients in plaster casts but in most cases they proved uncomfortable with the risk of pressure sores under the plaster and even skin lesions on the other leg.

Oedema

Oedema is a significant problem, particularly after extensive femoro-popliteal reconstruction. Aggressive treatment with diuretics, compression bandages and elevation facilitates wound healing.

Follow-up

All patients undergoing amputations are examined by specialists in shoe and insole fitting. In the majority of the patients custom-made shoes are prescribed. It is required not only because of the amputation but often because of pre-existing deformities on the other foot.

Diabetic patients require regular foot care. After amputations on the feet there is a propensity to secondary deformities and life-long surveillance should be maintained.

Results

During a 3-year period, from 1.1.88 to 31.12.90, 89 patients had digital amputations following revascularization in 95 legs (median age 74 years, range 39–94 years) (Albrektsen, Henriksen and Holstein, 1994) (Table 32.1). Forty-three were in diabetics and 52 in non-diabetics. The preceding revascularizations were aorto-iliac 6; distal to the inguinal ligament 85; combined 4. Forty distal reconstructions were to tibial or pedal arteries. Two out of the six aorto-iliac and 22 out of the 85 distal revascularizations were balloon angioplasties.

Table 32.1 Minor amputations after revascularization, primary and final level. Numbers in brackets are diabetic cases. In survivors 71% (69% in diabetics) had final levels distal to the trans-metatarsal level.

Amputation	Primary level	Final level
Partial digital	25	20 (7)
Digital disarticulation	3	3 (2)
Ray resection	47	34 (16)
Ray resection + partial digital	6	5 (2)
Trans-metatarsal (forefoot)	12	13 (4)
Tarsal	2	4 (3)
Symes	0	3 (2)
Transtibial	0	4 (2)
Transfemoral	0	1 (1)
Death before healing	0	8 (4)
Total	95	95 (43)

Eight patients, four diabetics and four non-diabetics, died (8%). Five legs were revised to major amputation, three in diabetics, two in non-diabetics. These were transtibial in four and transfemoral in one. Thus, the healing results in survivors were distal to the trans-metatarsal level. Independent walking was obtained in all survivors but four, who all had contralateral major amputations. In these patients the salvaged limb, however, was important in transfers and for balance. Table 32.2 shows the healing results in relation to toe and ankle blood pressures. The four cases of major amputation occurring with toe pressures below 30 mmHg were caused by graft occlusion or reocclusion of the balloon dilated arterial segment. The one case of major amputation despite adequate arterial supply was

Table 32.2   Toe pressures (upper panel) and ankle pressures (lower panel) before and after vascular reconstruction procedure in survivors.

	Before	After	Major amp.
Toe BP			
< 20 mmHg	58	7	4
20–29 mmHg	16	8	0
> 30 mmHg	7	42	1
Not measured	6	30	
Total	87	87	5
Ankle BP			
< 50 mmHg	26	3	2
50–99 mmHg	51	22	1
> 100 mmHg	4	52	1
Not measured	6	10	1
Total	87	87	5

caused probably by deep tissue necrosis due to infection prior to antibiotic treatment and revascularization.

The figures demonstrate that the presence of gangrene requiring local amputation after the revascularization procedure does not seriously compromise the results and the outcome in diabetics is not different with non-diabetics. It should, however, be emphasized that the above results were obtained by strictly following the principles as outlined above.

References

Albrektsen, S.B., Henriksen, B.M. and Holstein, P. (1994) Minor amputations on the feet for gangrene following arterial reconstruction. *Br. J. Surg.*, **81**, 1596–1599.

Apelqvist, J., Castenfors, J., Larsson, J., Stenstroem, A. and Agardh, C.-D. (1989) The prognostic value of systolic ankle and toe blood pressure levels in the outcome of diabetic foot ulcers. *Diabetes Care*, **12**, 373–378.

Holstein, P. (1984) The distal blood pressure predicts healing of amputations on the feet. *Acta Orthop. Scand.*, **55**, 227–233.

Holstein, P. and Lassen, N.A. (1980) Healing of ulcers on the feet correlated with distal blood pressure measurements in occlusive arterial disease. *Acta Orthop. Scand.*, **51**, 995–1006.

Holstein, P., Jorgensen, B. and Larsen, K. (1991) Minor amputations on the feet in diabetics (with a note on major amputations). In: *The Diabetic foot*. Proceedings of the first international symposium of the diabetic foot. The Netherlands, 1991 (eds K. Bakker and A.C. Nieuwenhuijzen Kruseman), Excerpta Medica, Amsterdam, pp. 125–136.

Larsen, K., Holstein, P. and Deckert, T. (1989) Limb salvage in diabetics with foot ulcers. *Prosthet Orthot Int.*, **13**, 100–103.

Noer, I., Tonnesen, K.H. and Sager, Ph. (1980) Minimal distal pressure rise after reconstructive arterial surgery in patients with multiple obstructive arteriosclerosis. *Acta Chir. Scand.*, **146**, 105–107.

Ramsey, D.E., Manke, D.A. and Sumner, D.S. (1983) Toe blood pressure. A valuable adjunct to ankle pressure measurement for assessing peripheral arterial disease. *J. Cardiovasc. Surg.*, **24**, 43–49.

Wagner, F.W. (1977) Amputation of the foot and ankle. *Clin. Orthop.*, **122**, 62–69.

Section 15

Problem amputations

33

The problem amputation stump

K. P. Robinson

Introduction

No matter how expertly the surgery is performed the patient is left with a healed large and complex wound at the end of the shortened extremity in close relationship to a portion of the skeleton that was never designed to be a terminal structure. When one considers that this healed wound is also expected to be the interface between the amputee and the artificial limb, the importance of problems that may affect the amputation stump are obvious.

With the advances in surgical practice and prosthetic design, the features that constitute an ideal amputation stump are well recognized and clearly defined. The stump should be pain free and comfortable within the prosthesis, which should provide function as close to the normal as can be achieved. To attain this end, the lever length from the distal joint must be carefully judged. The musculature must be well developed, innervated and firmly attached to maintain function. The skin should be fully sensitive with a well placed linear scar, mobile over the bone end and deep tissues. There should be adequate soft tissue covering the bone end. The functioning muscle groups should be well balanced and where possible myoplasty and myodesis should maintain their function. The bone end should be smooth and rounded (Fig. 33.1). The blood supply of the whole stump should be adequate. Venous return should be unimpeded and the transected nerves protected from stress. The whole stump should have an aesthetically acceptable shape with parallel sides and a rounded

end to be compatible with the prosthesis that will be provided. While these ideals are well recognized and frequently achieved, the causal condition may dictate some compromise which in the long term of the patient's remaining lifespan may prove to cause problems requiring treatment (Baumgartner, 1983; Bernd *et al.*, 1991).

The age at which the amputation is performed may have an important effect on such problems that the stump may produce. Limb deficiency at birth or amputation in the growth period will produce problems arising from continuing bone growth and the degree of activity of the extremity following amputation with inevitable effects on the growth and development of the musculo-skeletal

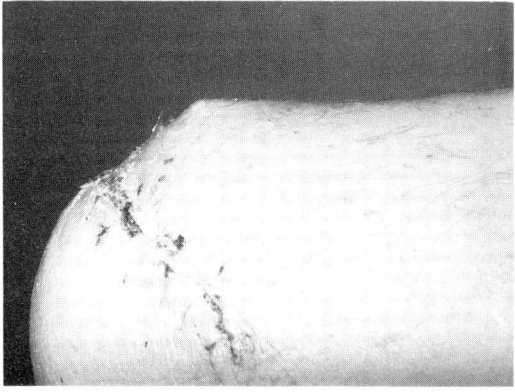

Fig. 33.1. Subcutaneous sharp ridge of bone due to badly shaped tibia – transtibial amputation treated by refashioning and reshaping of the bone end.

system. The smaller blood vessels and reduced growth of the hemi-pelvis following lower limb amputation in childhood is well recognized but rarely causes a serious problem. The continuing bone growth may require repeated revisions of the amputation unless a well judged epiphyseal stapling or bone capping procedure is carried out. If there is growth failure of the proximal bone in the limb a bone lengthening procedure can have a place in equalizing femoral length.

Those who have their amputation in mid-life can encounter problems as increasing age brings the onset of disease processes that can affect any limb and the amputated limb is not excluded. Deterioration of the arteries (Khokhol, 1979), the onset of diabetes and degenerative joint disease will affect the previously satisfactory amputation stump and produce increasing problems (Khokhol, 1979; Bunt, 1990; Paes, 1990). In addition to these factors, an alteration in the amputee's lifestyle may also affect the amputation stump. A change in occupation may alter the time scale and stresses to which the stump is exposed and may require a modification of prosthesis or activity (Bowker, 1990). Exposure to extremes of heat, humidity or cold may produce ulceration of the skin and difficulties in healing of the lesions produced. Pregnancy and obesity may increase the loading of the stump and result in the development of a skin breakdown, adventitious bursae and the onset of pain. A change from an athletic outlook to a sedentary existence will dramatically change prosthetic usage and may result in a loss of muscle strength and change in volume of the stump itself (Fernie, 1978).

The design and technique of the original amputation whether arising from developmental abnormality, accidental damage, war-time injury or disease processes, may have determined the length, character and site of the amputation encompassing to some degree of compromise such that the stump from the outset is not totally ideal and has inherent problems which may subsequently cause difficulty (Baumgartner, 1983; Gottschalk et al., 1989). The length of the amputation stump is frequently controversial and changes in prosthetic practice produce variation in what is considered to be the optimum length for a particular amputation site. While, in general, the longer an amputation is below the functioning joint the better, some prosthetic considerations dictate a limitation to the lever length, particularly in the transfemoral,

transtibial, transhumeral and transradial amputation sites. Increasingly in female patients cosmetic considerations can rate highly enough to justify re-amputation provided good prosthetic function can also be achieved. The shank size on a long transtibial amputation may demand a wide ankle in the prosthesis. The wide knee of the knee disarticulation prosthesis may elicit a request for revision to a transfemoral level, as may an overlong transfemoral amputation. At the transfemoral site, shortening is sometimes necessary to allow a sophisticated knee joint mechanism to be achieved without displacing the knee joint axis to an abnormally low level. In particular, many athletic patients find restricted knee joint function a great disadvantage and welcome shortening to allow 12 cm between the end of the stump and the knee joint axis line. Where a very short residue of femur has been retained at hip joint level, this may add to the problems of prosthetic fitting and require excision of the bone residue, with significant improvement.

Conversely, a stump that is too short may also cause problems and while expert prosthetic practice can accommodate very short stumps below the useable joint there is a limit to what can be achieved. Until recently, the alternative has been to resort to a higher level of amputation but surgical techniques have become available which may provide a better option, particularly in young patients. Reference can be made to the Ilizarov (1971) technique of bone lengthening which can be applied to the amputation stump or the bone proximal to the amputation stump producing significant leg or stump lengthening.

Problems of amputation stump length

The problem of excessive length

Overgrowth of the stump skeleton is a major complication, associated with epiphyseal activity in patients prior to epiphyseal fusion. It is seen most often in the humerus, fibula, tibia and femur, in that order (Aitken, 1963), sometimes to the extent that the bone end may ulcerate through the soft tissue and skin interfering with prosthetic fitting. The younger the child at the time of amputation, the more frequent is the complication and treatment for overgrowth may require to be repeated on more than one occasion. A 'capping' of the stump using an autogenous cartilage bone graft to take the weight on the bone end and protect the skin was advocated by Marquhart (1976). The procedure of bone capping has been successfully performed on both the upper and lower limbs (Bernd *et al.*, 1991; Pfiel, 1991). The soft tissue is displaced from the bone end at operation and the bone end resected, leaving two flaps of periosteum and muscle. The projecting bone is split longitudinally using bone from the iliac crest or head of the fibula if it is available. The widened bone marrow cavity is filled at its lower end with a cancellous bone graft and the hole covered with the periosteum securing the grafted bone through small drill holes. If the cap is unstable, a screw or cross pins may be required which are removed after 2–6 months. Other techniques use silicon implants, relying on an intramedullary stem for stability.

While much has been written on the necessity for the maximum length of bone to be retained in the amputation stump and the concept of the ideal amputation site for different levels has been largely discredited, the result is that many amputations, particularly in those performed for trauma, are initially at levels that leave a stump of such a length that becomes troublesome and prevents the best use of a prosthesis. This is a problem encountered in both upper and lower limb amputations and certainly where the initial amputation was essentially a debridement. This can be argued to be good advice and the need for a secondary, definitive amputation can be accepted in most cases in the industrial world. However, Kaphingst and Heim (1988) draw attention to the difficulties in the concept of secondary amputation, in many races in whom an intact body is an important religious attribute and consent may be withheld for the re-amputation procedure, leaving the patient with an unsatisfactory limb. It is not unusual in the western world to encounter a psychological resistance to the idea of a further operation once the patient has a healed amputation stump. While there is no suggestion that the emergency amputation should necessarily be the definitive procedure, at least it should be at a level that is compatible with reasonable function. The long transtibial amputation stump requires a wide shank to the prosthesis with an unsatisfactory cosmetic appearance and

this may be a problem for the female patient. The knee disarticulation amputation may also be unacceptable to the female patient, who finds the width and the displacement of the knee joint axis unacceptable. Similar considerations apply to the supracondylar and Gritti–Stokes amputation. An unduly long transfemoral amputation stump may prevent the use of a prosthesis incorporating a sophisticated knee joint mechanism and the patient, may request a re-amputation as referred to above.

In the upper limb, controversy continues with regard to the decisions required regarding long transradial stumps or wrist disarticulation and similarly regarding long transhumeral stumps or elbow disarticulation. Better strength, sensation and prosthetic suspension are the attributes of disarticulation but with less than satisfactory prosthetic cosmesis. On the other hand, wrist disarticulation may prevent the use of a satisfactory rotator for the prosthetic hand and, at elbow level, sufficient room should be allowed for the prosthetic elbow joint, usually 5 cm, and the same for the wrist.

The surgical technique of reamputation differs not at all from the technique of the primary amputation at that level with a tourniquet in place, antibiotic prophylaxis and gentle surgical technique. It requires very careful and detailed dissection of the stump structures to provide the best possible stump.

The stump that is too short

The surgeon quite properly seeks to provide the longest and most suitable amputation that can be preserved. In the final analysis, this length is determined by the length of stump that can be accommodated sufficiently securely in the socket to allow a full range of useful movement providing adequate lever length within the socket, with an acceptable method of suspension and control. At transtibial level, this is determined by the length of stump protruding below the hamstring tendons when the knee is flexed and, at the transfemoral level, the amount of stable stump between the ischial tuberosity and the perineum. Until recently, the problem of inadequate length could only be solved by a re-amputation to a higher level with considerable loss of function, particularly when a transtibial amputation had to be revised to a transfemoral level or a transradial to a trans-humeral level.

The short stump may be deprived of soft tissue cover, leaving a bone end covered only with skin which may be tight and threatening to ulcerate or there may be adequate skin and soft tissue cover but with insufficient bone for a useable stump. There are now plastic surgery techniques available to remedy the soft tissue deficiency and ortho-paedic procedures to overcome the deficiency in stump length.

Techniques to overcome soft tissue deficiency

It has long been recognized that the use of pedicle and crossleg flaps have severe disadvan-tages in the patient with an amputation stump, principally due to the prolonged immobilization and the involvement of the contralateral limb which may not always be normal. Re-amputation had always been the preferred option. However, other soft tissue transfer techniques are now available. Skin and subcutaneous tissue may be progressively expanded by the use of a subcuta-neous implanted silicon chamber which is pro-gressively expanded to stretch the skin, giving an increase in volume of between 30% and 50% over a period of 4–6 weeks. At the removal of the expander, the now redundant skin can be used to cover areas of deficiency not readily grafted by any other technique (Frykman, 1987; Comtet *et al.*, 1989; Jennings, 1991). Alterna-tively, a transfer of well-vascularized muscle may be utilized to form a bed for a subsequent split thickness skin graft or, the preferred option, of transfer of a well-vascularized muscle carry-ing a related island of skin receiving its blood supply through the blood supply to the muscle. The reverse flow saphenous island flap has been used successfully to bring soft tissue to a defi-cient transtibial amputation stump (Torri *et al.*, 1989), while at transfemoral level an area of skin can be carried on the pedicle of the rectus abdominus muscle and brought to the mid and upper thigh to reconstitute deficient soft tissue in that region (Muguti, 1990). The ultimate soft tissue transfer is the transference of muscle, subcutaneous tissue and skin complete with arte-rial and venous blood supply, transferred as a free graft with a microvascular anastomosis of the supply vessel and the local circulation (Fryk-man, 1987). The most practical technique for this soft tissue reconstruction is the use of the latissimus dorsi muscle with its overlying skin and subcutaneous fat carried on the subscapular artery and vein which can be anastomosed directly to the popliteal or femoral vessels by a microvascular anastomosis.

Techniques to overcome a deficiency in bone length

A deficiency of bone length remains the most serious problem and attempts to improve the bone length by capping or local bone graft procedures are largely unsuccessful. The situa-tion is now improved by the successful treatment of a number of patients using the Ilizarov tech-nique (Latimer *et al.*, 1990). While this tech-nique is very successful in improving the bone length, it does not itself expand or increase the soft tissue which must be adequate prior to use of this technique or improved by one of the preceding procedures ahead of the bone length-ening operation. The principle of the Ilizarov technique is to perform a subperiosteal corti-cotomy in the proximal portion of the bone, taking care not to disturb the medullary bone at this site. The bone is secured in the Ilizarov fixator which consists of three parallel rings in a spaced frame, allowing the bone fragments to be traversed by securing wires maintained under tension. The distance between the proximal ring and its wires, representing the fixed point, is gradually increased, while the two remaining rings and wires maintain the alignment and position of the distal fragment. The process of distraction is continued for 3–4 months; one report has indicated that lengthening of the tibia below the knee from 7 to 15 cm can be achieved during this period. This procedure is, therefore, of the greatest value in the younger patient but would have to be considered with great caution in the more elderly patient and is certainly contra-indicated in the elderly geriatric amputee who would find great difficulty in cooperating with the stringent regime. The technique has also been used with success in upper limb amputations.

Problem joints

In amputations performed for congenital disorders, abnormally developed joint or joints in abnormal positions may add considerably to the problems of successful functional rehabilitation and have to be taken into consideration in the planning of the original procedure. Sophisticated orthopaedic procedures such as the Van Nes (1950) Rotationplasty where the ankle joint takes on the function of a 'knee' may be indicated. Less severe abnormalities can be treated by suitable orthoses or, incorporating the principles of the orthotics in constructing an 'orthoprosthesis'. The more commonplace problems affecting the joints in amputee patients are those in which the joint sustained an injury in a traumatic amputation, where the joint is already diseased or where disease has developed subsequent to the amputation in the form of degenerative osteoarthritis.

An unstable joint at hip or knee level can be accommodated by the use of external support, e.g. transtibial prosthesis fitted with side steels and a lace-up thigh corset or a transfemoral prosthesis incorporating a hinge and rigid pelvic band to stabilize the hip joint. However, in a younger patient these options would be weighed against the improved function that might be achieved by orthopaedic surgical procedures on the knee or hip joint such as replacement arthroplasty. The most common problem to be encountered is that of the rapidly acquired flexion deformity resulting from either delay before limb fitting or a pre-existing deformity. While this flexion is initially reversible and can be overcome by careful physiotherapy, if it is delayed a manipulation under anaesthetic may be required. Rarely tenotomies of contracted flexor tendons and perhaps capsular division can be considered. It is usual to obtain the maximum degree of improvement with physiotherapy and incorporate a modification in the prosthesis so that the socket is aligned to accommodate the flexion deformity in the expectation that increasing activity will gradually and progressively reduce the flexion contracture.

Skin problems in the amputation stump

The skin of the amputation stump can be affected by any dermatological abnormality but also by the abnormal micro-climate in the socket producing abnormal warmth and moisture resulting in septic lesions affecting the hair follicles, acneform pustules affecting the sebaceous glands, hydradenitis pustules affecting the sweat glands, occasionally progressing to a major abscess or carbuncle. Avoiding prosthetic usage, exposing the skin, draining and culturing any pus and instituting the appropriate antibiotic if the infection is deep seated will usually result in rapid resolution. Advice to the patient is frequently needed on the care of the stump, stump socks and the socket. However, there are patients in whom persistent sweating may be the underlying problem and sympathectomy can be considered. The upper limb sympathectomy can be achieved most readily by a trans-axillary laparoscopic diathermy coagulation of the upper two ganglia of the thoracic sympathetic chain maintaining the integrity of the first; alcohol injection at this level is more likely to produce a Horner's syndrome. In the lower limb a phenol-in-Myodil sympathetic block can achieve sympathetic denervation; results, however, are unpredictable and, for certainty of effect, a retro-peritoneal surgical exposure of the lumbar sympathetic chain and removal of the lower four ganglia preserving the first lumbar is usually successful in producing a warm, dry skin. For a transfemoral stump, the first ganglion should be included unless a bilateral procedure is required.

The most common skin problem encountered in the amputation stump, remains ulceration and is frequently due to excessive localized pressure or shear due to unsatisfactory prosthetic fitting, an increase in weight or the onset of diabetes producing increased stress on the skin. This is particularly likely to occur where the skin is anaesthetic, as in patients with spina bifida or neurological abnormalities. Diabetic patients are particularly at risk, as are patients in whom the blood supply is compromised by a disease process. Correction of the underlying cause will result in healing of the ulcerated skin. A period without prosthetic usage is usually required to achieve healing and when the underlying problem is corrected the ulceration should not recur. However, unless the causation is carefully analysed and fully corrected the ulcer may become a persistent problem. Not infrequently seen is the ulceration of a transfemoral amputation stump in an elderly patient in whom retraction of the soft tissues has allowed the skin to become stretched over the bared femur which painfully erodes and finally ulcerates the skin. This condition requires early recognition and shortening of the femur to avoid the severe and intractable pain associated with the process of ulceration.

Subcutaneous fat and muscles

The fat content of an amputation stump is extremely variable, generally conforming to the patients body configuration. Fat atrophy does not cause a problem but excess fat in an obese patient can cause great difficulties in limb fitting and soft tissue reduction may be necessary and may need to be repeated at intervals. Liposuction is a technique which may be preferable to direct surgical reduction of the redundant fat and skin and the attendant risk of an unfavourable scar (Kruger, 1990). The Iceross (Kristensson, 1993) type of silicone liner in the transtibial prosthesis by its total contact and 'second skin' characteristics appears to stabilize the fat and reduce the need for surgical intervention. A balanced functioning musculature achieved by myodesis or myoplasty and, in the case of the transtibial stump, osteomyoplasty of the Ertl (1949) type may become less effective as the muscles atrophy with increasing age. Where the technique of muscle attachment has not achieved long-term fixation the muscles may retract from the bone end leaving the bone end in a subcutaneous situation. This leaves a stump of an unacceptable outline, cosmetically unattractive and difficult to fit with a sophisticated prosthesis. Reference has been made to the situation with the transfemoral amputation where the femur may become subcutaneous and subsequently protrude. In a young patient, an attempt at refixation of the muscles by an improved technique with well chosen suture materials may be successful in restoring the muscle attachments. The author's experience has been disappointing with muscle retraction recurring after a period of months despite a myodesis employing drill holes in the bone end and strong non-absorbable sutures. The alternative is to shorten the bone to a length that allows the myodesis or myoplasty to be reconstructed with less tension.

Blood vessels (Fig. 33.2)

The arterial perfusion and venous return from the amputation stump quickly adapt and, although there is some overall reduction in size of the feeding vessels, the perfusion remains satisfactory. The occasional case has been described of an arteriovenous fistula forming at the site of a mass ligation of an artery and its venae comitans but there is no record of this occurring in an amputation stump (Campbell, 1989). Should it occur, a re-exploration and dissection with individual ligation would be required to correct the abnormality, the vessels in this site being terminal and too small to reconstruct. The amputation itself may have been performed on account of a vascular anomaly and a number of amputations have been performed for the Klippel–Trenaunay syndrome (Lindenauer, 1971) in childhood and adolescence to remove the enlarged and deformed limb. Invariably this seems to result in the abnormality persisting in both the amputation stump and the tissues above which may cause variable swelling in the stump and, in one case, chylous leakage from the skin.

The large majority of lower limb amputations at the present time are performed for peripheral vascular disease affecting the arteries of the lower limb and the level of amputation is selected at the most distal site where the arterial perfusion will allow primary healing. This implies that the blood

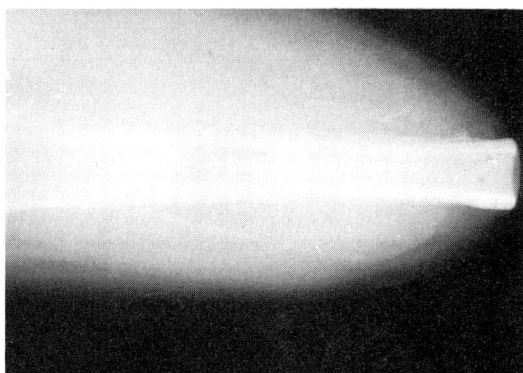

Fig. 33.2. Radiograph of transfemoral amputation stump following failed vascular surgery. Despite a myoplasty and a well rounded bone end, muscle retraction has occurred and resection of the distal 4 cm of the femur and refashioning of the myoplasty was required.

supply at this site is less than normal but adequate and this leaves the possibility that, with the progression of the peripheral vascular disease, the blood supply will deteriorate and the stump be affected by intermittent claudication (Setacci, 1989).

The effect of smoking is well recognised in the progression of atheroma and thrombo-angiitis obliterans. Where ischaemic changes threaten to compromise the amputation stump, the most important measure is to encourage the patient to stop smoking. In the smoker, revascularization is unlikely to succeed and continued deterioration can be expected (Lind, 1991). This frequently precedes any colour change in the skin and is not always associated with palpable temperature change and thus may be very difficult to diagnose, constituting one of the principal causes of intractable pain. This situation can also occur in a long-standing amputation performed for trauma in which the patient subsequently develops peripheral vascular disease and the amputation stump becomes affected by the generalized arterial stenosis presenting the same clinical picture. Transcutaneous oximetry has proved a useful tool in making this diagnosis and blood flow studies using isotope wash-out techniques such as Xenon-133, I-131 or I-125 tagged Iodoantipyrine are also helpful. However, to determine the treatment, injection arteriogram usually by the transfemoral route from the opposite side is required, although digital subtraction angiography and magnetic resonance imaging can be helpful. Where it is available, a duplex ultrasound image can also be utilized. While it is unlikely that the progression of vascular disease will permit any opportunity for revascularizing the stump where vascular disease has supervened, there is the possibility of a balloon angioplasty, a bypass graft or profundaplasty to revascularize the amputation stump. The place of lumbar sympathectomy remains controversial but could be used in the case of borderline ischaemic changes. An axillo-profunda, ilio-profunda and femoro-femoral crossover grafts have all been used to revascularize an ischaemic transfemoral stump.

Venous abnormalities may result in troublesome swelling of the amputation stump when amputation is at the site of a pre-existing or subsequent deep vein thrombosis. While the occasional patient may benefit from a veno-venous bypass graft at groin or iliac level these cases are very exceptional and the situation will usually have to be accepted by the patient with the use of multiple stump socks or an adjustable socket. At the present time, the efforts to graft venous valves and reconstruct the deep venous system in the lower limb have had very little success.

The nerves in the amputation stump (Fig. 33.3)

Every transected nerve in the amputation stump, being composed of fully medullated white nerve fibres in a neurolemmal coating, will attempt to regenerate by the outgrowth of neurofibrils when transected. In the absence of the distal segment in which to regenerate, the neurofibrils form a disorganized mass in relation to the cut end associated with a variable amount of fibrous tissue and constitutes the amputation stump neuroma (Martini and From, 1989). While it is well recognized that the major anatomical nerves will form macroscopically obvious neuromata, it is often forgotten that the cutaneous nerves will behave in the same way and their neuromata become incorporated in the terminal scar tissue of the stump. The neurofibril outgrowth may be more diffuse without the appearance of a detectable neuroma. The neuromata do not appear to be inherently harmful and, unless compressed or subjected to shear stress, will cause very little problem but, if the neuroma is allowed to develop at a site that is subjected to stress then pain will be the result. This may be a trigger point the patient

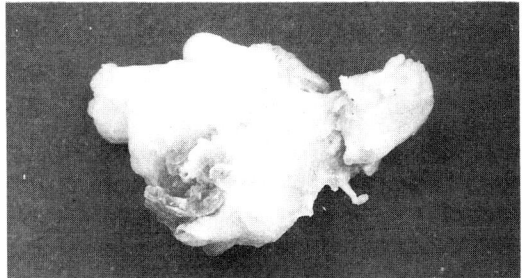

Fig. 33.3. A small amputation neuroma integral with fibrous tissue dissected at a site of exquisite tenderness in a transtibial amputation stump. Originating from the peroneal nerve it had become incorporated in the terminal scar.

can recognize or can be found on careful palpation of the stump. Where the neuroma is detectable at a point of acute tenderness, then it is likely that resection of the neuroma and the new neuroma allowed to develop at a more favourable site, will result in an improvement in the patient's tenderness and pain. This problem is particularly troublesome when the neuroma is subjected to pressure by the prosthesis and surgical treatment may be required if the prosthesis cannot be modified to off-load the pressure from the tender point. However, the wholesale removal of neuromata in a quest for a painless stump is deplored and the results of poorly considered neuroma resection are bad. In detection of a neuroma, ultrasound scanning has not proved to be effective (Hurvitz *et al.*, 1989). CAT scanning and magnetic resonance imaging (Donnal, 1989) will both detect neuromata but the decision to resect the neuroma must remain a matter of clinical judgement and not be based on scanning information which reveal neuromata be they troublesome or not.

The bone in the amputation stump

With the exception of disarticulations, most amputation levels require the transection of a long bone with its compact cortical structure and cancellous medullary cavity. The transection of the long bone through its medullary cavity produces a disorganization of the vascularity of the bone and although this is minimized if the bone end is closed off. Even when the bone end is covered with an acceptable degree of soft tissue, the transected bone end will remain in the distal part of the stump where it is exposed to most of the stresses involved in prosthetic usage and will carry some degree of the patient's body weight however small. A healthy bone in this situation should be painless and the amount of demineralization and osteoporosis will be determined by the blood supply and degree of physical activity. While the surgeon performing the amputation will take every care to provide a well rounded, smooth bone end to ride comfortably in the soft tissue structure of the stump, many are seen in which the bone transection remains a simple saw cut with sharp edges around the circumference of the transected bone. Myositis ossificans in the adjacent musculature is a not uncommon finding and produces a hard bulky mass at the bone end which of itself does not require any surgical treatment.

The bone end may develop local protuberances and sharp thorn-like osteophytes and cause particularly troublesome pain when wearing a prosthesis. Almost invariably an adventitious bursa forms over the site of exceptional bone pressure and inflammation in the bursa may add to the patient's problems. While the bursa may be readily removed surgically, unless the underlying bony abnormality is also corrected a recurrence is to be expected. Where an adventitious bursa is overlying an osteophyte it is treated by excision of the adventitious bursa and the osteophyte itself is best removed and shaped by using a turbine dental drill with a cylindrical burr and adequate irrigation.

Rarely, the residual bone in an amputation stump may be subjected to the problems of any bone – fracture from trauma, pathological fracture, Paget's disease, metastatic bone disease or osteoporosis. It is, therefore, important that unexpected pain in an amputation stump should always be investigated by plain X-rays of the bone structure as well as biochemical analysis including particularly prostate specific antigen and alkaline phosphatase estimations. Isotope bone scanning and isotope tagged white cell scans may be of assistance.

Osteomyelitis in the transected bone is not uncommon and left untreated may form a sequestrum, typically a ring, but often a smaller fragment, that proves to be a persistent focus of infection in the amputation stump. That infection may in turn result in a sinus or an ulcerated area that persists until the sequestrum is separated and extruded – a debilitating, time-consuming and painful process for the patient who may be unable to wear a prosthesis during this stage.

When osteomyelitis is detected, this should be treated by exploration under antibiotic cover, sequestrectomy and reshaping of the bone end if the inflammation is quiescent. The application of gentamycin beads at the site of the affected bone is particularly effective.

There remain a number of amputation stumps in which a complex of these local problems cause pain, discomfort and prosthetic difficulty (Melzack and Wall, 1965; Setacci, 1989). In the case of the patient with an adherent scar, contracted muscles, a sharp and subcutaneous bone end with one or more tender points of nerve entrapment and possibly areas of skin ulceration, all contributing to a generally unsatisfactory stump be it upper or lower limb, an elective and calculated revision is clearly worthwhile. The procedure consists essentially of excising the original scar, all unsatisfactory soft tissue, remodelling the bone ends and reconstructing the amputation stump. Performed under a tourniquet and an epidural anaesthetic, this can be a relatively atraumatic procedure in patients who might otherwise be unfit for major surgery and the benefits are extremely worthwhile. The original scar is completely excised. New flaps are formed. The muscles are dissected to form a myoplasty or myodesis. An osteomyoplasty can be performed at transtibial level to improve the shape of the bone end. Some bone shortening is usually required to avoid tension in the refashioned soft tissue. Skin closure is with fine interrupted sutures, the fat and fascia being closed with fine absorbable sutures. A suction tube drain is always inserted. The new stump gives the patient the opportunity for normal prosthetic fitting and restoration of the expected level of function.

The ulcerated amputation stump (Figs 33.4–33.6)

Perhaps one of the most common complications of the mature amputation stump is the stump which becomes ulcerated, a problem for which there are so many causes that the first step in management is to establish the exact diagnosis so that the correct management can be instituted. The ulcerated stump is perhaps most troublesome in patients who have had an amputation for spina bifida abnormalities resulting in a totally anaesthetic stump. In common with other patients with anaesthetic stump skin, they are at risk from unnoticed minor injuries which, with continued trauma, will result in skin ulceration. Some research may be required to determine the nature of the trauma as this may not be necessarily due to the socket and could be inflicted by a chair, an orthosis or even the posture adopted in bed. However, once the cause is identified and the pressure or stress removed from the affected site, healing can usually be expected. This generally means that the prosthesis may have to be set aside for a period of days until healing is complete.

The fully sensate stump is not immune from similar problems. Many patients expecting some level of discomfort from their amputation stump endure unnecessary discomfort for long periods which may result in destruction of a localized area of skin in relation to a point of excessive pressure. This may be related to the underlying bone, or to faulty limb fitting with excess pressure generated by abnormal surface loads. A common problem is found in the transtibial amputee wearing a patellar tendon bearing socket when stump volume reduces

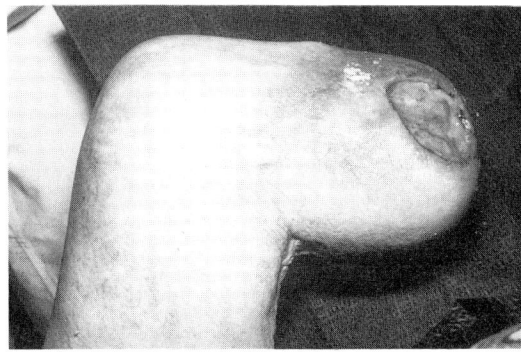

Fig. 33.4. Neglected ulceration in a transtibial amputation stump in a diabetic patient, treated by shortening and refashioning with a wedge resection technique.

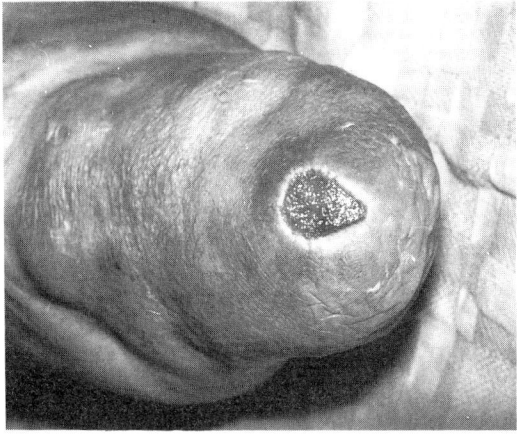

Fig. 33.5. Ulceration of a transtibial amputation stump at the site of pressure from the prosthesis. This was treated by not wearing the prosthesis until healing had occurred and then refitting the socket to obtain an optimal pressure distribution.

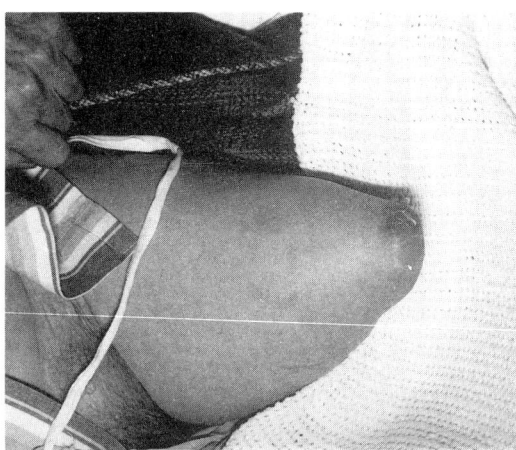

Fig. 33.6. A transfemoral amputation stump with protrusion of the femur ulcerating the skin and causing intractable pain. In this case, a myoplasty or myodesis had not been performed and the bone end remained sharp.

and permits the stump to descend too far into the socket with the result that the anterior tibial border at its lower end impinges on the socket and becomes ulcerated. A revision of the socket or the provision of an extra sock may solve this problem without resort to surgical treatment. However, if there is a prominent osteophyte or a badly shaped bone end, repeated ulceration may be an indication for surgical treatment.

With the predominance of amputations for vascular disease, the commonest cause of ulceration is a deficient arterial perfusion of the stump in whole or in part which is subject to excessive pressure first with the onset of ischaemic rest pain and ultimately ulceration affecting both skin and deeper structures. Almost inevitably, the chronic nature of the lesion leads to infection resulting in enlargement and deepening of the ulceration which may progress to destruction of the entire end of the stump and exposure of the bone end. In the early stages of deficient arterial perfusion, the situation may not be apparent and a temporary improvement may result from control of the infection with antibiotics, antibacterial dressings and avoiding use of the prosthesis for a period. However, unless the arterial ischaemia is recognized the treatment fails repeatedly and the ulcer fails to improve or becomes worse despite all local management. Unfortunately, investigation by tissue oxymetry,

isotope clearance studies or arterial imaging often reveal that the problem is obliteration of the distal vessels leaving little scope for an arterial reconstruction or a bypass to revascularise the stump (Lind, 1991). As indicated above this situation is frequently encountered in patients who continue to smoke and the improvement that results when they are persuaded to stop is frequently sufficient to enable the ulcerated area to heal or at least allow some chance of success with surgical treatment. If the amputation stump has not been successfully revascularised by stopping smoking the remaining alternative is a re-amputation. However, this option may be avoided or delayed if it is possible to do an effective excision of the ischaemic tissue and the wedge resection, advocated by Murdoch (1977) and reviewed by Hadden et al. (1987), is a useful procedure to consider before accepting the need for a re-amputation. In the reported cases, the procedure proved successful in 37 out of 42 patients who had their transtibial amputation performed for peripheral vascular disease by a long posterior flap amputation. The principle of the wedge resection is for the original scar to be completely excised including the ulcerated area and any skin of doubtful arterial perfusion in the vicinity by incisions extending to the full diameter of the stump. In addition, the muscles and soft tissue must also be excised from this ellipse down to the bone end which must be shortened to form the apex of the excision of a wedge of tissue. Tension is avoided by resection of 1.5–2 cm of bone and the flaps of soft tissue can then be approximated without tension. The bone end must be carefully filed and revised to a uniformly smooth contour and the fibula shortened and smoothed to an appropriate level. The principle embodied in this procedure can be applied to transfemoral amputations, Symes ankle disarticulations and to upper limb amputations.

The swollen amputation stump

Increase in volume of an amputation stump may cause considerable prosthetic difficulties and may require successive refitting of the socket. The usual and most obvious cause is the patient increasing in body weight. However, the stump is also susceptible to swelling as is the natural limb with venous disease as the predominant cause.

Deep vein thrombosis is a not uncommon accompaniment of lower limb amputations and when this occurs in the post-operative period, dramatic stump swelling may be encountered (Frost, 1991). A deep vein thrombosis may, however, occur spontaneously at a later date and produce the same effect. If the swelling involves the thigh as well as the lower leg, the possibility of an iliac vein thrombosis must be considered and if seen in the acute stage requires urgent treatment with venous imaging by doppler duplex scan or, if necessary, a venogram which would have to be obtained by direct femoral puncture or per trochanteric injections. The demonstration of thrombus in the iliac veins requires energetic treatment with systemic heparin, or alternatively, streptokinase, urokinase or tissue plasminogen activator. The earlier the treatment is started, the more effective the thrombolysis is likely to be and the smaller the chance of long-term swelling or pulmonary embolus. However, if the stump swelling is encountered after the acute event, venous imaging may reveal an occlusion or stenosis of the femoral or iliac veins but only rarely is this amenable to surgical treatment. Fortunately, collateral compensation eventually becomes established with some improvement. It is important to make sure that the transfemoral prosthesis does not press on the subcutaneous collateral veins that frequently develop in the inguinal region and cross to the opposite side to re-establish drainage from the affected leg.

Where an amputation has been performed for malignant disease or where recurrent cellulitis has resulted in inflammatory obliteration of the lymphatics, secondary lymphoedema may become a problem. Lymphoedema may also be encountered if the patient has hypoplasia or aplasia of the lymphatics. This may be associated with a Klippel–Trenaunay type of congenital abnormality. It is uncommon to find a lymphatic abnormality that lends itself to a lympho-venous anastomosis but this possibility remains if the lymphoedema is massive and persistent.

Stump pain

When all these situations have been assessed and investigated, the predominant problem remains that of stump pain. It is still most difficult to treat. While there is medication that is now more successful than any that has been used in the past, there are a number of patients who may still require full neurophysiological investigation and rarely neurosurgery. It is questionable whether any amputation is comfortable in the sense that a normal extremity is completely without any intrusion into the patient's awareness. However, the point at which discomfort or pain in the amputation stump interferes with the patient's well-being and ability to use the amputated limb is the time when the problem has to be acknowledged. It has been recognized since the earliest times (Keil, 1990) that there are many amputation stumps in which recurrent or persistent pain occurs for which no apparent cause can be found. The pain is frequently severe, spasmodic in nature and may be accompanied by involuntary muscular movements, jactitations and may be, to a variable degree, accompanied by the patient's awareness of residual pain in the absent part of the limb, the 'phantom phenomenon'. The phantom phenomenon is probably experienced by every amputee, presumably the result of the major nerves being divided at the site of the amputation wound and providing inappropriate sensory stimuli to the central nervous system. Whether the 'phantom' is painful is to some extent determined by the amount of pain that was experienced in the limb prior to the amputation operation (Fisher and Meller, 1991) and perhaps even more significant to the amount of pain that is experienced from the stump wound in the early healing period. If this is allowed to predominate over all other sensations, the central nervous system becomes set in a pattern whereby predominance of painful signals is given precedence in processing through the sensory gates, thus making pain the dominant sensation felt in the phantom. Such a pattern can easily become set into a long-term problem. As other sensations take predominance over the phantom sensation, the degree of pain becomes progressively less. It is our practice to foster the application of warmth, massage and comfort to the stump with the maximum of non-painful contact to gradually over-ride the pain element in the phantom. Effective medication may also go a long way to minimizing the element of pain so that eventually, although the phantom sensation remains, the pain is gradually eliminated. If the amount of pain in the phantom is allowed to dominate the patient, the most extreme effect is the development of a Sudek type of sympathetic dystrophy syndrome which will reduce the patient to a total invalid incapable of mobility or prosthetic usage (Erdmann, 1992).

In the pre-operative period, adequate analgesia is essential and we have advocated the use of a continuous epidural analgesia for up to 48 h prior to the surgical operation and for the same period afterwards wherever possible. Where this cannot be used, the self-regulated intravenous morphine system can be applied and it is certainly helpful if the major nerves are injected with a long-acting local anaesthetic at the time of surgery. Early activity, early handling of the limb by the patient and frequent handling of the limb by everyone concerned in the patient's care with the application of warmth and soft dressings wherever possible is recommended. There should be a cheerful acknowledgement of the phantom sensation, accepting it as a natural phenomenon, a curiosity rather than a threat, while not dismissing it or making references to it in a too light-hearted manner. Any reference to the phantom sensation as a problem is studiously avoided. We have observed on more than one occasion that if a patient who is a long-term sufferer with 'phantom pains' is in contact with other patients who have an amputation during their period in hospital, then they too will have 'phantom pain', while at other times the problem is simply labelled as 'phantom sensation' and not regarded with any special concern. There is one report of phantom pain being triggered by an epidural anaesthetic (Bulder and Smelt, 1991) and there is also a report of phantom pain being triggered by the patient having a magnetic resonance imaging scan (Paes, 1990). It seems likely that a much wider observation would be necessary before these effects could be considered significant.

Various medical agents have been used to treat the limb pain – beta blockers and propanolol have been used. Anticonvulsant agents by acting on the spinal cord may diminish the facilitatory circuits at segmental level. Carbamazepine is the most frequently used and largely successful agent; phenytoin and baclofen are alternatives. Chlorpromazine appears to have a largely sedating effect. There is no doubt that many of the patients do have an endogenous depression and benefit from antidepressant medication – amitriptyline, imipramine and trazodone have been used, as have benzodiapine and clonazepam. Nevertheless, many patients progressively move through the range of specific analgesic agents and many reach the opioid alkaloids to which they quickly become habituated, in some cases, dependent and, in fewer cases, addicted.

To rationalize the management of the patient with intractable post-amputation pain, a period of in-patient assessment is usually invaluable. This gives the opportunity for a full investigation to exclude an organic cause for the pain such as ischaemia (Setacci, 1989), bone disease or neuroma entrapment which can be appropriately treated. A cautious reduction in the analgesic dosage can be achieved. A psychiatric assessment can assess the degree of depressive illness and the opportunity can be taken to solve any prosthetic problems, advise on stump management, particularly to increase the amount of sensory input from the stump that is not painful and a routine application of infrared radiation, warm bathing, massage and the well tried technique of trans-cutaneous nerve stimulation at sites remote from the amputation stump. When combined with increased physical activity (Cohen, 1991) and widening of the patient's interests and horizons, there is nearly always an improvement while the patient is in the abnormal in-patient environment but the likelihood of relapse when the patient returns to their previous way of life is disappointingly high.

The place of neurosurgical treatment remains difficult to define. Simple nerve neurectomy produces only a temporary improvement and we have noted on the occasions when this has been performed that it frequently results in a hyperaesthesia and increased pain awareness in the zone around the numb area where the divided nerve has been. Spinal tractotomy, thalamic tractotomy, subcortical neurectomy and even lobectomy have been applied in the past but with disappointing results and it seems that the key to management is to keep the level of surgical intervention to a minimum and to provide the maximum support while minimizing the adverse factors in the patient's lifestyle that may be exacerbating the problem.

Conclusion

The problem amputation stump may have causes that can be trivial and easily remedied or may

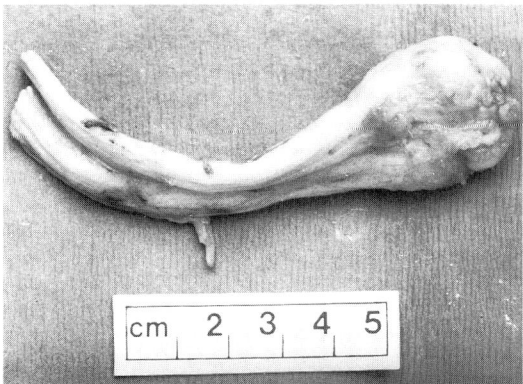

Fig. 33.7. A resected sciatic nerve neuroma which was causing pain as it was exposed to pressure from the socket and when shortened the problem was eliminated.

require months or years of attention (Figs 33.7–33.10). Every patient will hope that their problem can be solved by changes in the prosthesis or possibly by physiotherapy or medication rather than surgical intervention. While this is clearly the first objective in the management, there are many patients who spend months or years with a problem that could be readily resolved by surgery. It is

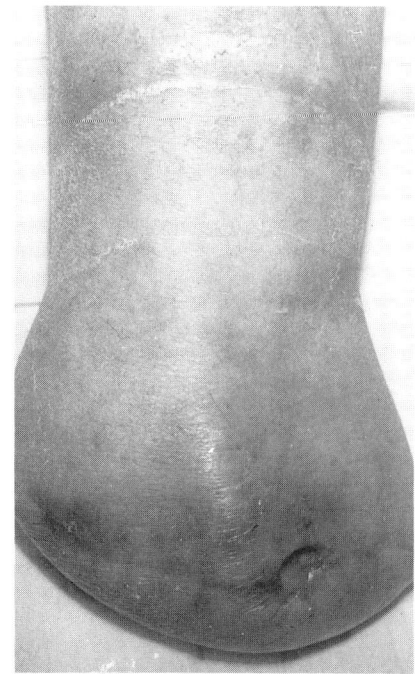

Fig. 33.9. Transtibial amputation stump which had failed to heal satisfactorily due to wrong level selection. In addition, the conformation of the stump would produce considerable problems in prosthetic fitting and on this patient reamputation was performed to transfemoral level.

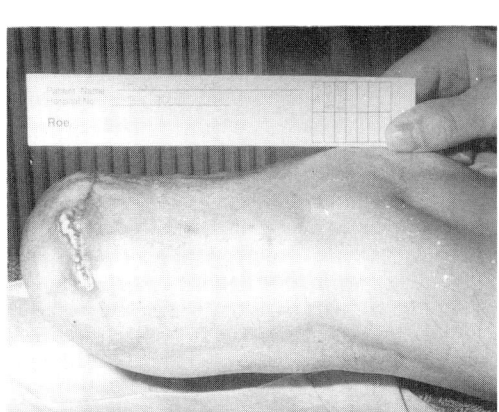

Fig. 33.8. Transtibial amputation stump with a chronic low grade breakdown along the suture line due to progression of the peripheral vascular disease. The patient was a smoker, but when he was able to stop smoking spontaneous healing occurred. Ulceration due to stresses imposed by the prosthetic fitting again relieved by a period of rest without use of the prosthesis followed by represcription and manufacture of a new prosthesis.

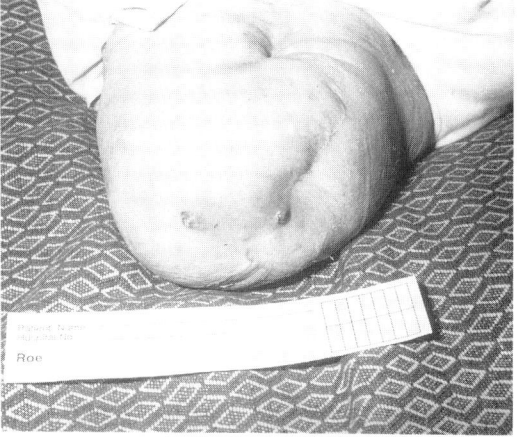

Fig. 33.10. A transfemoral amputation stump following vascular surgery showing a chronic sinus communicating with the residual arterial graft. Once this had been removed, stump healing was complete and prosthetic use was resumed.

clearly important that a sensible balance is achieved and where a cause of pain or a problem is clearly defined and surgery is needed there is little benefit in postponing this treatment.

References and further reading

Aitken, G.T. (1963) Surgical amputation in children. *J. Bone Joint Surg.*, **45A**, 1735–1741.

Baumgartner, R. (1983) Failures in through knee amputation. *Prosthet. Orthot. Int.*, **7(2)**, 116–8.

Bernd, L., Blasius, K., Lukuscher, M. and Lucke, R. (1991) The autologous stumplasty. Treatment for bony overgrowth in juvenile amputees. *J. Bone Joint Surg.*, **73(2)B**, 203–206.

Bowker, J. (1990) Surgical techniques for conserving tissue and function in lower limb amputations for trauma, infection and vascular disease. *Inst. Course Lectures AAOS 39*, pp. 355–360.

Bulder, E.R. and Smelt, W.L.H. (1991) Onset of amputation stump pain associated with epidural anaesthesia. *Anesth. Analg.*, **72(3)**, 394–396.

Bunt, T.J. (1990) Iatrogenic tibial pseudo aneurysm following below knee amputation. *Am. Surg.*, **56(9)**, 546–547.

Campbell, W.B. (1989) Patients who have undergone amputations (letter). *Ann. R. Coll. Surg. Engl.*, **7(4)** suppl), 58.

Cohen, L.G. (1991) Motor reorganisation after upper limb amputations in man. A study with focal magnetic stimulation. *Brain*, **114(Pt 1B)**, 615–627.

Comtet, J.J., Saint Cast, Y., Remy, G., Herzberg, G. and Michel, A. (1989) Emergency knee joint salvage utilising a free musculo-fascio cutaneous flap based on the anterior tibial artery. *Microsurgery*, **10(4)**, 302–309.

Donnal, J.F. (1989) MR imaging of stump neuroma. *Am. J. Phys. M.R.*, **68(5)**, 234–239.

Erdmann, M.W. (1992) Familial reflex sympathetic dystrophy syndrome and amputation. *Injury*, **23(2)**, 136–138.

Ertl, J. (1949) About amputation stumps. *Chirurg.*, **20**, 218–223.

Fisher, A. and Meller, Y. (1991) Continuous post-operative regional analgesia by nerve sheath block for amputation surgery – a pilot study. *Anaesth. Anal.*, **73(3)**, 300–303.

Frost, F.S. (1991) High resolution real tissue ultrasound for the diagnosis of venous thrombosis in the rehabilitation setting. *Am. J. Phys. Med. Rehabil.*, **70(1)**, 3–4.

Frykman, G.K. (1987) Amputation salvage with microvascular free flap from the amputated extremity. *J. Trauma*, **27(3)**, 326–329.

Gottschalk, F., Kourosh, S., Stills, M., Mcclennan, B. and Roberts, J. (1989) Does socket configuration influence the position of the femur in above knee amputation? *J. Prosthet. Orthot.*, **2**, 94–102.

Hadden, W., Marks, R., Murdoch, G. and Stewart, C. (1987) Wedge resection of amputation stumps; a valuable salvage procedure. *J. Bone Joint Surg.*, **69(2)B**, 306–308.

Harrison, G. (1956) Phantom limb pain occurring during spinal analgesia. *Anaesthesia*, **11**, 249–251.

Hurvitz, E.A., Ellenberg, M., Lerner, A.M., Pope, S. and Wirthlan, L. (1989) Ultrasound imaging of residual limbs; a new use for an old technique. *Arch. Phys. Med. Rehabil.*, **70(7)**, 556–558.

Ilizarov, G.A. (1971) Osnovnye printsipy chreskostonogo kompressionnogo i distraktsionnogo osteosinteza. *Ortop. Travmatol. Protez.*, **32**, 7–15.

Jennings, J.F. (1991) Amputation stump salvage using a 'banked' free tissue transfer. *Am. Plast. Surg.*, **27(4)**, 361–363.

Kaphingst, W. and Heim, S. (1988) Cultural considerations and appropriate technology in orthopaedics for developing countries. In: *Amputation Surgery and Lower Limb Prosthetics* (ed. G. Murdoch), Edinburgh: Blackwell, pp. 384–389.

Kasabian, A.K. (1991) The role of microvascular free flaps in salvaging below knee amputation stumps; a review of 22 cases. *J. Trauma*, **31(4)**, 495–500.

Keil, G. (1990) So-called initial description of phantom pain by Ambroise Paré. *Fortschr. Med.*, **10;108(4)**, 62–66.

Khokhol, M.I. (1979) State of blood circulation in the leg stump depending on the characteristics of mechanical stress in the prosthesis. *Orthop. Traumatol. Protez.*, **(11)**, 20–24.

Kristensson, O. (1993) The ICEROSS concept: a discussion of a philosophy. *Prosthet. Orthot. Int.*, **17**, 49–55.

Kruger, L.M. (1990) Suction assisted lipectomy – an adjunct to orthopaedic treatment. *J. Pediatr. Orthop.*, **10(1)**, 53–57.

Latimer, H.A. *et al.* (1990) Lengthening of below the knee amputation stumps using the Ilizarov technique. *J. Orthop. Trauma*, **4(4)**, 411–414.

Lind, J. (1991) The influence of smoking on complications after primary amputations of the lower extremity. *Clin. Orthop.*, **267**, 211–217.

Lindenauer, S.M. (1971) Congenital arterio-venous fistula and the Klippel-Trenaunay Syndrome. *Ann. Surg.*, **174(2)**, 248–263.

Martini, A. and Fromm, B. (1989) A new operation for the prevention and treatment of amputation neuromas. *J. Bone Joint Surg.*, **71(3)B**, 379–382.

Marquardt, E. (1976) Plastiche operationen bei drohender knochendurchspieBung am kindlichen Oberarmstumpf. *Eine Vorlanfige Z. Orthop.*, **114**, 711–714.

Melzack, R. and Wall, P.D. (1965) Pain mechanisms – a new theory. *Science*, **150**, 971–979.

Muguti, G.I. (1990) The deep inferior epigastric artery. Local musculocutaneous flap; a method of preserving sensation. *Br. J. Plast. Surg.*, **43(2)**, 236–40.

Murdoch, G. (1977) Amputation surgery in the lower extremity – part II. *Prosthet Orthot Int.* 183–192.

Paes, E.H. (1990) Late vascular damage after unilateral leg amputation. *Z. Unfallchir. Versicherungsmed.*, **83(4)**, 227–236.

Pfiel, J. (1991) The stump capping procedure to prevent or treat terminal osseous overgrowth. *Prosthet. Orthot. Int.*, **15(2)**, 96–99.

Setacci, C. (1989) Post amputation pain in patients with

vascular diseases. *Angiologia* (Spanish). Sup. **41(5),** 194–196.

Van Nes, C.P. (1950) Rotation-plasty for congenital defects of the femur: making use of ankle of the shortened limb to control the knee joint of a prosthesis. *J. Bone Joint Surg.*, **32B,** 12–16.

Yukh, W.T. (1992) Phantom limb pain induced in amputee by strong magnetic fields. *J. Magn. Imaging*, **2(2),** 187–189.

34

The problem stump: prosthetic solutions

N. A. Govan

Every amputee who is fitted with an artificial limb presents his or her own unique individual range of problems to be overcome. Each stump is unique and demands a socket tailor-made to its particular shape and properties. The lifestyle and expectations of the amputee have an influence on prescription decisions that are made by the rehabilitation team. The personality of the individual cannot be separated from the prosthetic possibilities. It has to be recognized that, in fitting a socket to a stump and fitting a prosthesis to an individual, we attempt something quite unnatural, as, for example, asking the amputee to take pressure in areas not designed by nature for that specific task. By experience and by biomechanical analysis ways have been found to overcome the problems presented.

In all of this confusion prosthetists come to recognize different groups of amputees in which similar patterns and circumstances arise and that for the most part amputees within each such group can be treated similarly. But every so often an amputee will be encountered with a stump which presents problems of such a nature that they can be considered to be outside of these common groups. Another way of considering this spectrum of disability would be to regard such amputees as a group so small as to be considered a rarity. These are the amputees with problem stumps. The unique quality of each case makes it difficult to lay down rules for solution of the different problems presented. As it happens, prosthetists, by and large, use weapons already in their armoury to find solutions.

Some of the problems encountered include extremely short stumps in which the pressure on the stump will be necessarily high, flexion contractures of hip or knee, scarring or skin grafts in areas where the socket would normally exert pressure, skin problems (Levy, 1983), hypersensitive or insensate stumps, problems relating to neuromata, stumps with bone spurs and prominent bone ends and scar tissue adherent to bone. Stump problems encountered can be only part of the problem and the whole picture can be complicated by the presence of other disabilities, health problems, poor general fitness, deformities of the affected or other limb, multiple limb deficiencies or simply having a difficult personality.

The first thing to be recognized is that such amputees require a higher level of care than normal. They will have to be seen more often and often for a longer time than is normal. An accurate and full record of the attempts to overcome a problem must be kept in order that the rehabilitation team does not chase round in circles looking for a solution. Team work becomes more important than ever when these problems have to be overcome. The amputee may have to be carefully instructed in matters such as stump care including inspection, hygiene and the correct use of stump socks.

The prescription for amputees with problems must be carefully developed and the possibility of future changes to the prescription must be faced. Just as particular attention would be paid to the prescription of a prosthesis for a young athlete who wished to pursue athletic activities, so we must consider the special needs of someone who may need to sit more often than to stand or walk, or of someone with a stump problem whose plea is simply to be able to walk a very few yards. Solutions to problems may have to be found from other amputation levels. Examples are the ankle disarticulation amputee who requires proximal socket loading, the ankle disarticulation amputee with a paralysis who requires a thigh corset, the transtibial amputee with a painful stump who needs an ischial bearing thigh corset (Fig. 34.1), the transtibial amputee who requires not only a thigh corset but the additional security and control of a pelvic band and the amputee with an extremely short transfemoral stump who may benefit from the fitting of a modified Canadian hip disarticulation prosthesis.

The choice of materials for the socket is significant. In some cases, a flexible socket will be preferred to a rigid one and soft inserts or liners may prove beneficial at any amputation level. Silicone inserts are sometimes used as distal pads in sockets but there is no need to restrict these inserts to the distal socket. Very occasionally there may be a need to turn the clock back and try a leather or wooden socket in place of a modern plastic one. Such thinking can only derive from an open minded attitude to options available.

Fig. 34.1. The addition of a thigh corset to a patellar-tendon-bearing prosthesis.

Decisions on the shape of the socket are of prime importance in the presence of stump problems. First of all come general considerations such as whether an ischial containment or a quadrilateral socket is preferred in a particular circumstance. More particularly, decisions about the precise shape of the socket can be made when working with modern casting, rectification and manufacturing techniques. It should be remembered that an area which seems to need relief from pressure does not necessarily require the application of a build-up in rectification. The desired effect may be achieved by pressure being applied in some other area. The prosthetist will have an understanding of where the forces will be applied between socket and stump and during rectification of the model should be asking questions like 'What would happen if . . .?'

The use of transparent check sockets is of particular importance when dealing with problem stumps. Again full documentation is required, especially if a series of these sockets is anticipated. When trying the sockets, the influence of alignment on socket pressures should not be ignored and occasionally an alignment which is not ideal from an aesthetic point of view may be required to create the pressure pattern required.

While the prosthetist may be able to fit almost any stump he is presented with there can be no doubt that results will be better when the stump is closest to ideal in features such as length, quality of tissue cover and mobility of the knee. The prosthetist must always be aware of the help available from other members of the clinic team. A stump with excessive distal soft tissue can be surgically revised. A flexion contracture may be reduced by appropriate physiotherapy.

All of this is really a plea for quality amputee care from the rehabilitation team but so often constraints are cited on, for example, the prosthetist's time or even on the ability of the team to get together. If the team works properly at its job and if the amputee's cooperation is achieved and determination applied it becomes extremely difficult to set down limitations on what is possible.

Reference

Levy, W.S. (1983) *Skin Problems of the Amputee*. St Louis: Warner H. Green Inc.

Section 16

The arm and the hand

35

Amputation surgery of the arm in adults

G. A. Hunter

Introduction

More than 90% of all upper extremity amputations are a result of accidents, and most of these occur in young males (see Table 35.1).

The loss of an upper extremity has more devastating consequences than the loss of a lower extremity. In the UK, arm amputations account for only 3% of the total amputations referred for prosthetic rehabilitation (Day, 1990). In the USA, in contrast, an estimated 12 000 persons lose a hand or arm each year.

Most patients are seen after industrial accidents and accordingly the average surgeon has little, if any, experience with the management of this problem. The operation therefore consists of wound debridement with primary or secondary closure at the most distal level, thereby 'conserving all length possible' to which one must add 'consistent with good surgical judgement and eventual prosthetic fitting and rehabilitation'. Loss of one or both hands results in lack of grasp and sensation. The higher the level of amputation, the greater the functional loss. If the dominant limb has been lost within a few months, the patient can be trained to use the contralateral hand for most day to day activities. The ultimate goal after amputation of the upper extremity is the successful acceptance and operation of a prosthesis, so that the patient will be able to return to modified work and resume household and recreational activities.

Surgical principles

There are two basic types of amputation or disarticulation.

Provisional (open)   Used when primary healing is unlikely because of infection, ischaemia or inadequate wound debridement.

Definitive (closed)   Used after provisional amputation or for elective surgery.

One should save all possible viable length around the shoulder and below the elbow joint, bearing in mind the problems of subsequent prosthetic fitting. There is little to be gained by preserving a useless hand, with one or two stiff fingers, often covered by an insensitive, ugly free flap.

The risk of infection will be reduced by gentle handling of all tissues, irrigation with at least 6–9 l of normal saline using a pulsed lavage system (+/– antibiotics), adequate wound debridement, appropriate anti-tetanus measures, i.v. antibiotic therapy for 48 h combined with adequate haemostasis. The amputation should usually be left open, and secondary wound closure with suction drainage planned within a few days, provided there is a healthy wound. Split skin grafts may be used at this stage to preserve joints and length in a non

Table 35.1. Aetiology of upper limb amputation – adults.

CONGENITAL	LATE RESULTS OF CONGENITAL DEFORMITIES
ACQUIRED TRAUMA	INDUSTRIAL ACCIDENTS KNIFE & GUN WOUNDS HIGH PRESSURE INJECTION INJURIES FAILED REVASCULARIZATION LATE PROBLEMS AFTER FRACTURES BURNS (THERMAL, ELECTRICAL OR CHEMICAL)
NEOPLASMS	PRIMARY TUMOURS OF BONE OR SOFT TISSUES POST-IRRADIATION PROBLEMS
INFECTION	GAS GANGRENE HUMAN BITES
ISCHAEMIC	ARTERIAL THROMBOSIS OR EMBOLISM FROST BITE INTRA-ARTERIAL INJECTION (DRUG ADDICTS)
IATROGENIC	ARTERIAL CANNULISATION (DIAGNOSTIC OR THERAPEUTIC)
NEUROPATHIC	BRACHIAL PLEXUS INJURIES

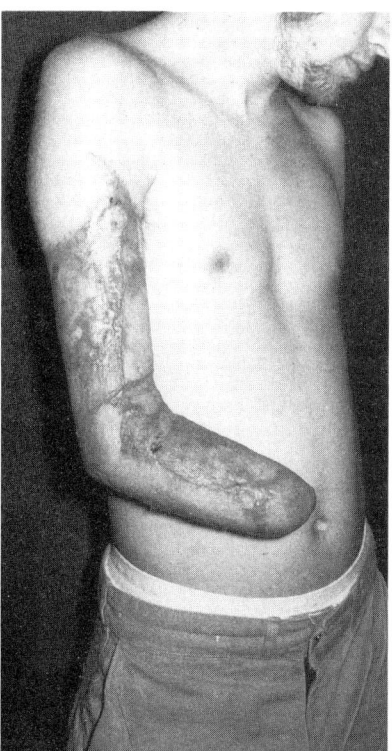

Fig. 35.1. Use of split skin grafts to preserve the elbow joint and allow prosthetic fitting.

weight bearing limb (see Fig. 35.1). The revision rate after skin grafts used in the early care of the upper limb amputee is low (Wood *et al.*, 1987) (29%), but the use of a pedicle flap with inadequate sensation is to be questioned (see Fig. 35.2).

Except in the management of tumours, a tourniquet is usually unnecessary in upper limb amputations.

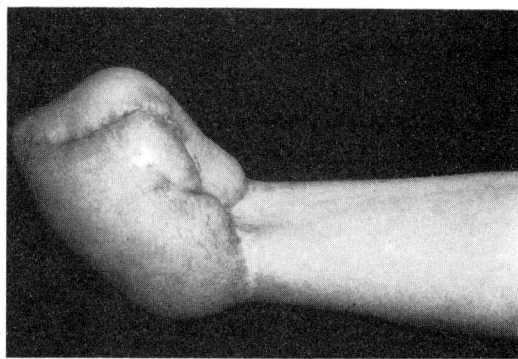

Fig. 35.2. Triumph of technique over reason, requiring transradial amputation to fit a prosthesis.

Handling of tissues

Skin

The skin flaps should measure slightly more than one half the diameter of the limb at the level of proposed bone section. They may be either anterior and posterior or medial and lateral depending on the viability of local damaged skin.

Muscle

Muscles and tendons are divided distal to the site of bone section and opposing muscle groups are sutured loosely over the bone to each other (myoplasty).

Bone and joint

Bone is cut cleanly across at the proposed level for subsequent skin and muscle cover. If a disarticulation is performed, the articular cartilage should be preserved. In transradial amputees, the radius and ulna should be sectioned at the same level where possible. Bone chips should be meticulously irrigated from the wound.

In children, epiphyseal growth lines must be preserved whenever possible. Their growth potential is not equal. In the humerus, it is the proximal and in the forearm, the distal epiphyseal lines that are much more important (Bowker and Michael, 1992).

Nerves

A neuroma is an inevitable consequence of nerve resection. It appears preferable to isolate the nerves and after gentle traction, divide the nerves with a sharp blade, allowing the nerves to retract proximally into the soft tissues away from the scar line and prosthetic pressure points.

Either at the time of the initial elective surgery, or when the wound is closed, care should be taken to avoid adherence of the skin to underlying bone (see Fig. 35.3). Redundant soft tissues, especially common above the elbow, are to be avoided (see Fig. 35.4), and suction drainage is used routinely for 1 or 2 days.

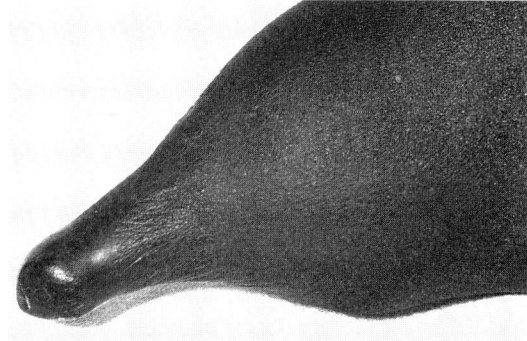

Fig. 35.3. Excessive retraction of soft tissues in transhumeral amputation.

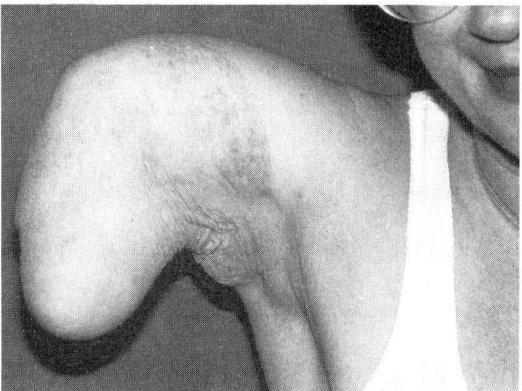

Fig. 35.4. Redundant soft tissues after transhumeral amputation.

The use of wool and a light plaster of Paris dressing is desirable, as is early post-surgical prosthetic fitting (2–3 weeks) to control oedema, reduce post-operative pain and to aid in the psychological adjustment of the patient's disability. The joint above the amputation site must be mobilized by early active stump exercises. A good stump should be obtained within 2–3 weeks to allow prosthetic fitting; the patient is naturally very reluctant to undergo further shortening of the upper limb to facilitate prosthetic fitting. The patient will ask 'Why did the surgeon not get it right the first time?'

A good stump should have the following qualities:

(1) Freedom from pain, which implies good skin cover, a non-tender well placed scar and adequate soft tissue padding with intact sensation at sites of socket stabilization.
(2) Adequate range of motion in residual joints – 135° of elbow motion and 45° of forward elevation of the shoulder are considered adequate to provide good prosthetic function.
(3) Sufficient muscle power to position, stabilize and activate a functional terminal device.

Sites of election

The level of amputation is determined by the extent of the injury; the surgeon has no control of the aetiological trauma, but may be able to influence the success of prosthetic fitting by careful technique and a knowledge of prosthetic components available.

It should be again stressed that the average surgeon has little if any knowledge of amputation of the upper limb, and no knowledge of the existing commercially available prosthetic components.

The site of election is the most distal point in the upper extremity, where sound surgical principles will permit the formation of a satisfactory amputation stump. The appropriate levels described for prosthetic fitting are shown in Table 35.2.

Wrist disarticulation

Wrist disarticulation is popular, since the operation is said to conserve natural supination – pronation of the forearm provides for prosthetic suspension and creates a force tolerant distal end of the limb.

I prefer a long transradial amputation for the following reasons:

(1) The theoretical aim of preserving pronation and supination by maintaining stump length is not realistic; close fitting of the standard hook limits rotation by 50% and some rotation is possible by incorporating rotation within the prosthesis and by using flexible hinges at the elbow. A myoelectric prosthesis allows both active and passive rotation within a Muenster socket.

Table 35.2. Levels of amputation in the upper limb.

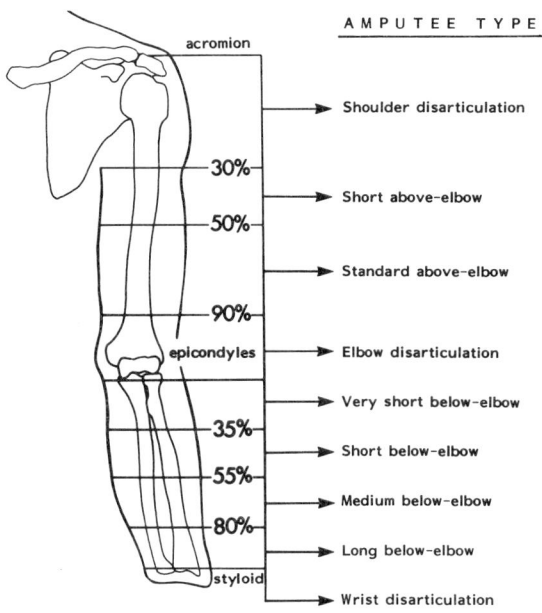

(2) Prosthetic fitting is more difficult, and it is distressing for the patient to notice the artificial limb longer than the normal hand, particularly when the elbow is flexed.

(3) I feel that the importance of rotation to a unilateral transradial amputee has been overstated in the past.

(4) Personal observation over a period of years indicates that rotation is often more limited at the wrist disarticulation level than at the transradial level, possibly because of associated or iatrogenic injury to the inferior radioulnar joint and associated triangular fibrocartilage disc.

Transradial level

A minimum of 6 cm above the wrist joint is required to install an adequate wrist unit, allowing for interchange of the various terminal devices (see Fig. 35.5 abcd). At this level, 70% of rotation is usually preserved. The range of rotation decreases as the length of the stump decreases; when 60% or more of the forearm is lost, minimal rotation is possible. Conserve all possible length below the elbow joint; a minimum of 10 cm below the lateral epicondyle of the humerus is preferred. However, it is worth preserving even 3 cm of stump, combined with section of the biceps tendon, when

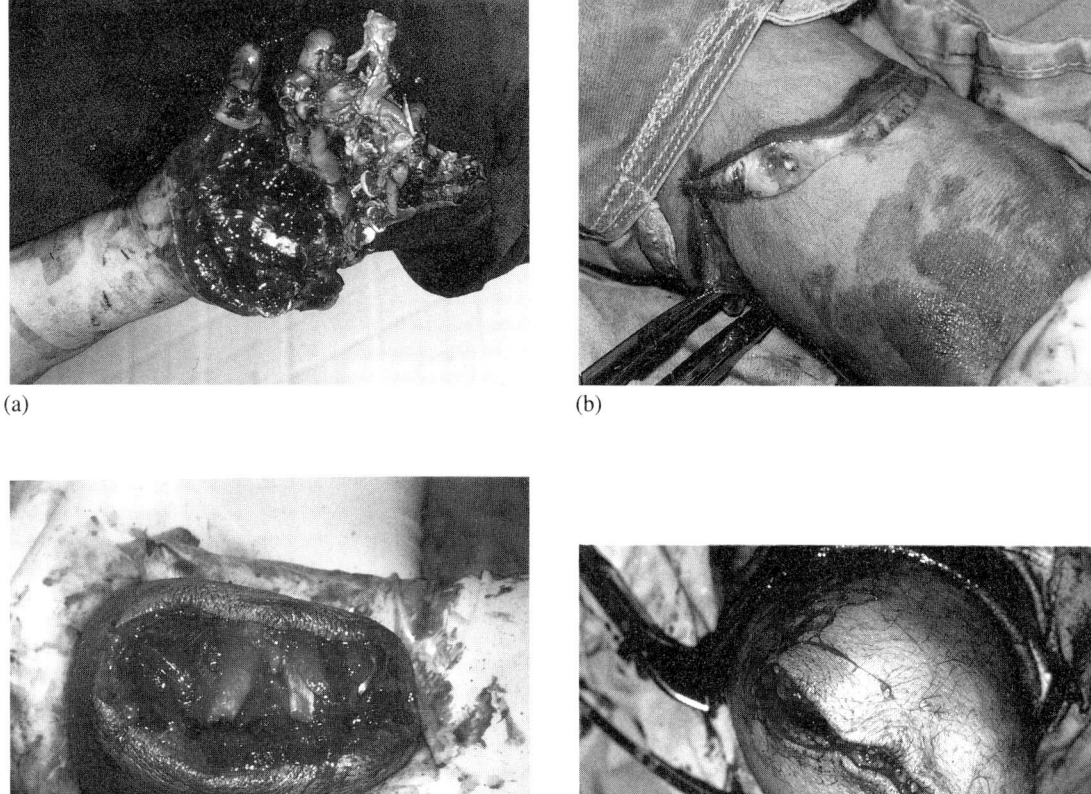

(a)

(b)

(c)

(d)

Fig. 35.5 (a–d). Open transradial amputation after punch press injury.

brachialis is intact, in order to add functional length and facilitate prosthetic fitting. The value of preserving the elbow joint whenever possible cannot be over-emphasized.

Elbow disarticulation

Elbow disarticulation has the advantage of the condylar flair to aid prosthetic suspension, and humeral rotation is transmitted to the prosthesis, avoiding the need for a mechanical turntable for rotation required in above elbow prostheses. The longer lever arm may improve strength and provide a force tolerant stump.

However, I prefer transhumeral amputation to elbow disarticulation because:

(1) After elbow disarticulation, the standard cable operated prosthesis requires outside locking hinges, which are bulky and easily damage clothing.
(2) The standard elbow disarticulation prosthesis has a reduced number of forearm locking positions.
(3) The cosmetic appearance is poor.
(4) Elbow disarticulation, as of today, does not allow for the fitting of a myoelectric prosthesis.

Transhumeral level

The amputation should be at least 6 cm above the lateral epicondyle to allow sufficient space for the turntable multiple locking elbow unit, which allows motion from 5 to 135° of flexion and has eleven locking positions in this range.

A stump of 10 cm length can be fitted with a conventional transhumeral prosthesis. If the stump is shorter, one may use a forearm spring lift assist device incorporated into the prosthetic elbow joint.

Angulation osteotomy of long above-elbow stumps in children and adults may be considered on an elective basis, because this operation improves mechanical coupling between the humerus and prosthesis, so that humeral rotation of the prosthesis is readily controlled by natural humeral rotation (Marquardt and Neff, 1974).

Amputation around the shoulder

When the transhumeral stump measures less than 10 cm from the axilla to the end of the stump, the patient should be fitted as a shoulder disarticulation. It is essential to preserve as much of the upper end of the humerus as possible, and to avoid disarticulation of the shoulder to improve the cosmetic appearance, and make for easier prosthetic fitting. The reader should refer to the classic article by Littlewood (1922) for details of the posterior approach to inter-scapulo-thoracic amputation.

At this level, early prosthetic fitting is delayed, and the patient is first provided with a light shoulder cap to avoid pressure on the chest wall, and to provide a filler for the shoulder of the clothing to give a normal outline, and to allow women to carry the shoulder straps of undergarments.

To summarize, the numbers 6 and 10 should be remembered:

Transradial amputation – preferred level is 6 cm above the wrist joint and at least 10 cm below the elbow joint.

Transhumeral amputation – preferred level is 6 cm above the elbow joint and at least 10 cm below the shoulder joint.

Post-surgical management

The amputee should be referred to a specialized centre 10–14 days after the amputation. The amputee team will supervise prosthetic fitting and discuss the emotional and financial problems resulting from the amputation. Vocational counselling and retraining, where necessary, should aim to return the upper limb amputee to work within 3 months of the injury. The longer the fitting is delayed after surgery, the less the chance of useful prosthetic function.

It is the author's practice to start all new transradial and transhumeral amputees with a standard cable operated prosthesis with a hook. A functional hand is provided for social reasons. A myoelectric prosthesis is usually offered to the transradial amputee, and a hybrid prosthesis (cable controlled elbow and myoelectric hand) to the transhumeral amputee between 6 and 9 months after the injury. Most patients use both a standard body powered and a myoelectric prosthesis alternately at work and home. For further details of prosthetic fitting and results, the reader is referred to recent publications (Hunter, 1984; Heger *et al.*, 1985; Millstein *et al.*, 1986; Atkins and Meier, 1989; Bowker and Michael, 1992; Dykes, this volume).

Krukenberg procedure

This reconstructive procedure is performed on a below-elbow stump to provide a crude pinching mechanism with sensation (Swanson and Swanson, 1980). The objective is achieved by splitting the below elbow stump longitudinally at the interosseous membrane into radial and ulnar rays, widely separating the ends of the bone and closing the skin deficit between the rays by the use of skin grafts. The stump is unsightly but may be covered by a cosmetic prosthesis. It does have a place in underdeveloped countries where prosthetic fitting is unavailable (Mathur *et al.*, 1981; Jain, 1995). However, in developed countries, it should be reserved for blind bilateral upper limb amputees, and should be performed on the dominant side only to improve sensory feedback. It is interesting to note that all blind bilateral amputees in Britain have refused the operation (Vitali *et al.*, 1986).

Bilateral upper limb amputees

Unlike children with severe upper limb deficiencies present at birth, the ability to manipulate the feet is not helpful in the adult bilateral acquired upper extremity amputee. All limb lengths should be preserved, provided surgical judgement is exercised. In children, the epiphysis should be preserved to conserve bone growth. Serious consideration should be given in bilateral high upper limb amputees to provide electrically powered prostheses (muscle or switch controlled) including wrist rotation, or to consider a hybrid system of body and electric powered prostheses. The patient must be transferred to a specialized unit for these fittings.

Complications specific to upper limb amputees

(1) Skin problems around the shoulder produced by the harness.
(2) 'Overuse syndromes' in the soft tissues and joints of the contralateral upper limb.
(3) Redundant soft tissues especially above the elbow, preventing adequate prosthetic fitting.
(4) Problems with daily living and self care, which increase tenfold with bilateral upper limb amputees.
(5) Loss of acquired skills at work, at home and in recreational activities.
(6) Associated defects of vision increase the handicap of the patient.
(7) Loss of sensory feedback with an inability to express emotions.
(8) Frustration, lack of independence and severe depression especially in bilateral upper limb amputees.

Replantation or amputation

Although revascularization and replantation are now possible in specialized centres, considerable surgical judgement is necessary to avoid a technical triumph that leads to poor function, multiple operations, severe emotional and often financial problems and eventually the need for a late amputation.

Informed consent is not possible under such tragic circumstances. Factors to consider when advising replantation include the following:

(1) Age of the patient. Replantation is almost always indicated in children in whom good recovery of repaired tendons and nerves is expected, and in whom further growth of the replanted limb can be expected. On the other hand, a patient aged over 50 years with associated medical problems would not be a suitable candidate.
(2) Technical facilities and a skilled surgical team must be available to perform the replantation, and rehabilitation services should be available to provide expert therapy after the operation.
(3) Associated life threatening injuries, and a time interval between injury and replantation of more than 6–10 h preclude replantation.

(4) Function of the opposite dominant or non-dominant limb.
(5) Mechanism of injury. Was it a slice, crush or avulsion injury? Is there extensive wound contamination? What is the level of injury, and are there associated injuries proximal and distal to the amputation?
(6) Mental status of the patient preventing adequate rehabilitation. Was this a suicide attempt? Is the patient using non-prescription drugs?

The rate of success of this type of surgery has increased, and regularly receives worldwide press endorsement. However, little progress has been made over the years in preventing adhesions of flexor tendons or improving nerve repair to improve sensation, both important factors that determine the degree of functional recovery.

Acknowledgements

The author thanks Churchill Livingstone for permission to reproduce Tables 35.1 and 35.2 and Figs 35.1 and 35.3 from *Adult Orthopaedics* (1984), Vol. 2, 1137–1159 (eds R.L. Cruess and W.R.J. Rennie)

References

Atkins, D.J. and Meier, R.H.III (1989) *Comprehensive Management of the Upper Limb Amputee*. Springer New York.

Bowker, J.H. and Michael, J.W. (1992) *Atlas of Limb Prosthetics*. American Academy of Orthopaedic Surgeons, 2nd Edition, Mosby Year Book Inc., St Louis.

Day, H.J.B. (1990) Amputation Surgery in the Upper Limb. In: *Report of ISPO Consensus Conference on Amputation Surgery* (eds. G. Murdoch, N.A. Jacobs and A.B. Wilson), University of Strathclyde, Scotland, October 1990.

Heger, H., Millstein, S. and Hunter, G.A. (1985) Electrically powered prostheses for the adult with an upper limb amputation. *J. Bone Joint Surg.*, **67**, 278–281.

Hunter, G.A. (1984) Amputation and prosthetic fitting of the upper extremity. In: *Adult Orthopaedics*. Vol. 2 (eds R.L. Cruess and W.R.J. Rennie), Churchill Livingstone, New York, pp. 1137–1159.

Jain, S.K. (1995) This volume.

Littlewood, H. (1922) Amputations at the shoulder and at the hip. *Br. Med. J.*, **1**, 381.

Marquardt, E. and Neff, G. (1974) The angulation osteotomy of above elbow stumps. *Clin. Orthop.*, **104**, 232–238.

Mathur, B.P., Narang, I.C., Piplani, C.L. and Majid, M.A. (1981) Rehabilitation of the bilateral below elbow amputee by the Krukenberg procedure. *Prosthet. Orthot. Int.*, **10**, 27–34.

Swanson, A.B. and Swanson, G.D. (1980). The Krukenberg procedure in the juvenile amputee. *Clin. Orthop.*, **148**, 55–61.

Vitali, M., Robinson, K.P., Andrews, B.G., Harris, E.E. and Redhead, R.G. (1986) *Amputations and Prostheses*, 2nd ed. Bailliere Tindall, London, p. 109.

Wood, M.R., Hunter, G.A. and Millstein, S.G. (1987) The value of stump skin grafting following amputation for trauma in adult upper and lower limb amputees. *Prosthet. Orthot. Int.*, **11**, 71.

36

The Krukenberg operation

S. K. Jain

The hand is one of the most intricately designed mechanisms of the human body. The loss of one hand causes a severe degree of handicap, and, inevitably, with the loss of both hands the handicap is multiplied. In contrast, it has to be recognized that the function of the arm and hand can never be matched by the functions achieved by the prosthetic hand.

In addition to the fine controlled movements of the hand, proprioception and the sense of touch are of vital importance. Both these sensory functions together act as a third eye for an individual; thus, even a blind person can recognize an object by its shape, density, surface, etc. We may hope to develop a prosthesis with such sensory functions in the next century but up until now all we have is a hand prosthesis exhibiting only slight movements and moderate cosmesis.

Loss of vision in addition to the loss of both hands is, thankfully, rare and usually takes place when a device explodes while handling. Thrasher accidents, frost-bite, electric burn injuries and machine (industrial) accidents are other causes of amputation of both the hands.

The Krukenberg procedure of converting two non-functional below-elbow stumps into sensitive and functional organs makes a double hand amputee into a totally independent individual. In such an amputee who has also lost his vision the procedure is of vital importance. This procedure was devised by Krukenberg (1917) with the aim to make a double hand amputee totally independent. However, the procedure fell into disrepute due to its unsightly appearance.

The Krukenberg procedure has been regularly performed at the Artificial Limb Centre, Pune since the late 1950s, with an average of eight operations per year. Out of a total of 250 operations, the author's contribution is 103 in 64 patients.

Principle

There are two sets of movements in the forearm, flexion/extension and supination/pronation. Flexion/extension takes place at the elbow joint whereas supination and pronation take place between the superior and inferior radio-ulnar joints. In the transradial stump or in the Krukenberg operation where the inferior radio-ulnar joint no longer exists the supination and pronation movements can be remoulded and re-educated to opening and closing movements of the radial and ulnar stumps in any position of the elbow. The supinators open the stumps while the pronators close them. Therefore during the re-education programme special attention must be given to the strengthening of both supinators and pronators.

The ideal Krukenberg stump

Krukenberg stump

The Krukenberg stump can be likened to a pair of forceps. The two prongs are formed by the radial and ulnar stumps and their base is the proximal forearm where the bones have not been separated.

Movements

Movements of the elbow, i.e. flexion and extension, should exhibit a full free range and should also demonstrate adequate strength. The opening and closing movements of the stumps should also have adequate strength and gripping force. The distance between the tips of the stumps, when fully opened should be 4 inches (10 cm).

Length of the base

The length of the base from the elbow crease should be about 3 inches (7–8 cm) (Fig. 36.1) If the length is less than 3 inches the pronator teres and/or its nerve supply is likely to be damaged, resulting in a poor functional result.

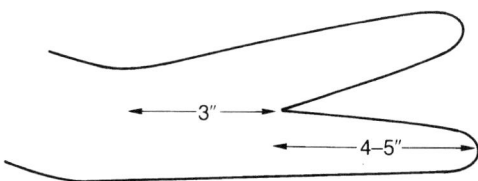

Fig. 36.1. Ideal Krukenberg stump.

Length of the stumps

The ideal length of the stumps is 4–5 inches (10–13 cm). Both the stumps should be of equal length. When there is no choice in a short transradial stump, a shorter radial and ulnar stump length is acceptable. Stumps longer than 4–5 inches have a post-operative problem in terms of healing because of poor vascularity. In addition, excessive length will weaken the gripping force at the tip.

Skin cover

In the Krukenberg procedure it is the radial stump which is required to rotate over the ulnar stump. It is advisable to cover both the stumps with normal skin. In both wrist disarticulation and the long transradial stump, there is always sufficient skin to cover both stumps. If it is not possible to cover the stumps with normal skin then the radial stump should be covered with such skin that is available and the ulnar with a split skin graft on its non-opposing surface. Whatever scar results should be placed over the non-opposing surfaces of the stumps.

It is not always possible to create ideal Krukenberg stumps, but one must strive to do so. Very often a compromise is preferable to a conventional amputation and whatever prosthesis is available.

Indications

Precise indications for a Krukenberg procedure cannot be outlined but the various issues involved are worth exploring.

Blind bilateral hand amputee

Loss of vision in addition to amputation of both hands is an absolute indication for the Krukenberg operation. With the help of the stumps which have pincer action and normal sensations the blind person will be able to recognize an object, which he would not have done otherwise. The operation must be performed on both sides in these patients given that the original stumps are long enough.

Bilateral hand amputee

Though it is not an absolute indication, the procedure should be performed on both sides in the bilateral hand amputee, for total independence. The functional achievements are so gratifying that the patient sets aside the unsightly appearance. The appearance of the Krukenberg stumps can satisfy social requirements by covering them with cosmetic gloves or with a prosthesis.

Unilateral hand amputee

In unilateral cases the indications are based on the individual need, since the remaining normal hand can perform almost all the functions and a prosthesis on amputated side can act as a supporting hand. Rehabilitation of an amputee who has lost the dominant hand is improved by this procedure on the amputated side. When the non-dominant hand is lost, adequate rehabilitation may be achieved by a prosthesis. In manual workers it is perhaps better to perform the operation irrespective of which side is involved.

The Krukenberg procedure has not been popular in the past due to its unsightly appearance. This handicap can now be overcome by providing a prosthesis or gloves to cover both radial and ulnar stumps and achieve a good cosmetic appearance. When performed the operation should be done on both sides or at least on the dominant side.

Surgical technique

Assessment and planning

Pre-operatively, the amputated limb should be examined in detail for adequacy of length, movement at the elbow and radio-ulnar joints, the availability of skin, such requirement there is for a skin graft, the vascularity and the presence of painful or hypersensitive areas for the general planning of the operation. The operation should be performed under a pneumatic tourniquet.

Skin incision

Anteriorly the skin incision is placed on the ulnar side, i.e. $\frac{1}{2}$ inch (1–1.5 cm) medial to the mid line. At its proximal end the incision turns sharply laterally towards the radial side (Fig. 36.2a). On the posterior aspect the incision is made in the mid line or slightly towards the radial side and at its proximal end it turns sharply to the medial side (Fig. 36.2b). Distally, both the incisions meet at the tip of the stump.

The incision is deepened and the deep fascia is divided in the same line but without separating it from the skin flaps. This is essential in order to preserve the vascularity of the skin flaps, since the base of the flaps is narrow compared with its length.

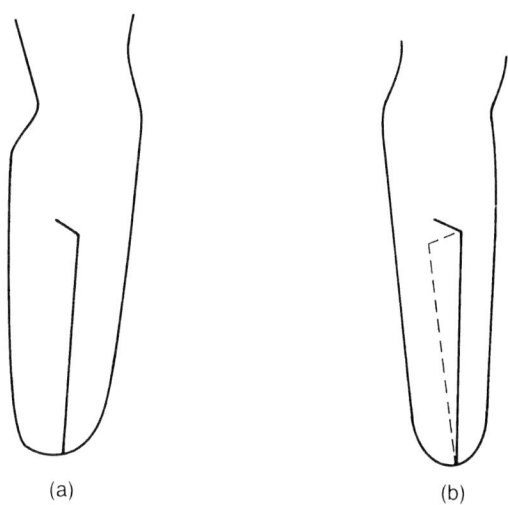

(a)   (b)

Fig. 36.2.(a) Incision on anterior aspect of the forearm located on the ulnar side and turns laterally at its proximal end. (b) Incision on posterior surface lying on the radial side and its proximal edge turns towards the medial side.

Management of muscles

Both the radial and ulnar stumps should be well covered with muscles. It is necessary to be careful to avoid damage to the pronator teres and its nerve, which is vulnerable during the separation of radial and ulnar stumps.

The muscles are separated into two equal halves of flexor digitorum sublimis and flexor digitorum profundus by blunt dissection as far as the interosseous membrane. This procedure is carried out on the posterior aspect in a similar fashion again as far as the interosseous membrane.

It can be tempting to excise the muscles to help closure of the skin without a skin graft. However, it has been our experience that when the superficial muscles are excised the distal skin suffers ischaemia and necrosis takes place, and in the same way excision of the deeper muscles leads to sequestration of the terminal ends of the bones. When the muscles are retained, wound closure may be difficult but healing is better. If the muscles are not excised there remains a raw uncovered area but split skin grafting should complete stump coverage.

Interosseous membrane

The interosseous vessels should be carefully isolated to avoid damage, as damage to them would effect the nutrition of the stumps. The interosseous membrane should be divided away from the vessels, leaving at least a small fringe attached to the bone. The division is then extended proximally under vision. Both the bones are separated gently by supinating, pronating and separating movements resulting in the opening out to 4–5 inches (10–12 cm) at the tips of the bones. Care should be taken to avoid damage to the superior radio-ulnar joint.

Bones

Both the bones must be divided at the same level with minimal possible periosteal stripping. Unequal bone lengths hamper the function of holding small objects by the tip of the stumps. Excessive periosteal stripping leads to terminal bone necrosis due to the already compromised circulation. The periosteum is divided by an encircling incision $\frac{3}{4}$ inch (2 cm) distal to the proposed site of bone section; the periosteum is then reflected proximally up to a point just 1–2 mm proximal to the bone section. The margins of the bones are smoothly sculptured and the periosteal sleeve is closed, covering the end of the bone.

The muscles are then sutured over the end of the bone and in passing placing a few secure sutures through the periosteum, thus keeping the muscles at their original tension to achieve optimal function.

Nerves

Both median and ulnar nerves are divided as they emerge from the muscle mass. It is best managed if they are gently pulled down, ligated and divided, so that the cut ends of the nerves retract and are buried in the muscles.

Blood vessels

The radial, ulnar and interosseous vessels are ligated as distally as possible, avoiding damage to the smaller vessels. The tourniquet is then removed and complete haemostasis ensured.

Wound closure

The proximal medial tip of the radial flap on the anterior aspect is drawn posteriorly between the bones and sutured at the proximal medial corner on the ulnar stump. Similarly, the posterior proximal tip of the ulnar flap is pulled between the bones and sutured at the proximal anterior corner of the radial stump.

Radial and ulnar stumps are then closed. If it is not possible to close the wound without tension, a split skin graft should be applied over the non-opposing surface of the ulnar stump. The skin in the web space between the stumps is also sutured together (Fig. 36.3).

Suction drains are placed separately in both stumps as well as at the base of the stumps.

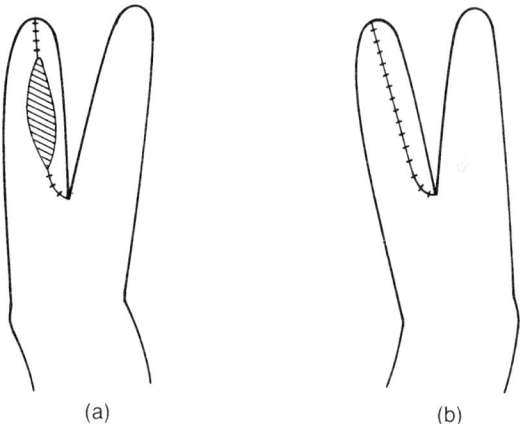

Fig. 36.3.(a) Ulnar stump with split skin graft. (b) Radial stump without skin graft.

Post-operative management

The closed wound is covered with a thin layer of vaseline gauze and then padded well with cotton wool followed by dressing with an elastic bandage. The stumps should be kept separated by introducing cotton wool pads between them. The stumps are elevated with the help of a sling to prevent oedema.

The first dressing is changed usually on the fifth day permitting examination of both the wound and any split skin graft. The second dressing is carried out on the tenth post-operative day and skin sutures are removed after 2 weeks. It is essential to keep the radial and ulnar stumps separated with the help of cotton wool padding throughout this period, otherwise the stumps would tend to close.

Complications

All efforts must be made to avoid the range of possible complications. The following complications are typical.

Skin flap necrosis

This is normally due to ischaemia in long flaps with a short base. Tight closure of the wound, damage to blood vessels during operation, undue excision of muscles and rough tissue handling are all typical precipitating factors.

Bone necrosis

The tip of the bone may sometimes undergo avascular necrosis as a result of undue periosteal stripping. Division of the interosseous membrane overly close to the bone may also contribute to the problem.

The necrosis of skin flaps or bone impose special problems. If these problems do occur one should wait for the line of demarcation to develop and then excise the dead portion. It is true that dead bone at the tip of the bone will separate gradually, but will take a very long time. In order to maintain equal length of the stumps it may be necessary to remove a portion of normal bone from the tip of the other stump.

Failure of pincer action

Failure of pincer action may be due to damage to the nerve to pronator teres, or too wide separation of radial and ulnar stumps, leading to undue stripping of pronator teres muscle from its insertion. Temporary loss of pincer action can also take place due to prolonged use of the tourniquet.

Training and rehabilitation

Training should be started immediately after wound healing and the removal of sutures. Pronation, supination, flexion and extension of the elbow are performed passively at the same time as the patient is encouraged to do these movements actively.

It is the radius which rotates over the ulnar and brings about the pincer effect; thus the pronators are responsible, in effect, for the closing action whereas the supinators open the stumps. Pronator function can be improved by electrical stimulation of the pronator teres muscle.

All patients are given training in the activities of daily living and for other relevant activities with the help of various appliances. On an individual basis they are also given an 'on the job' training. Bilateral transradial amputees become totally independent not only in activities of daily living but also in recreational activities and can even earn their livelihood. If they are blind they may manage well enough to be financially independent. Figures 36.4a and 36.4b demonstrate the level of activities performed by a Krukenberg amputee.

For cosmetic purposes a suitable prosthesis or a cosmetic glove can be worn over the Krukenberg stumps.

The results have been found to be so gratifying that the patients disregard the alleged poor appearance and most patients ask to have the Krukenberg procedure.

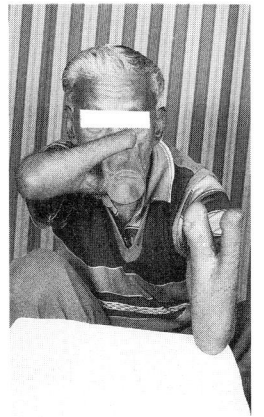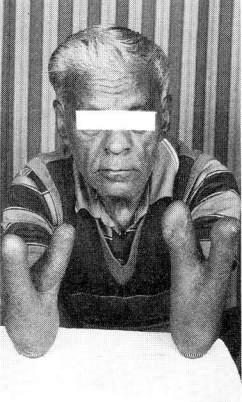

Fig. 36.4.(a, b) Showing functional achievements of a Krukenberg amputee.

In children with wrist disarticulation amputation the distal radial and ulnar epiphyses must be preserved. Throughout the growth period in the case of transradial amputees the bones should never be shortened.

The success of the Krukenberg operation depends on pre-operative planning, rigid adherence to the principles of Krukenberg surgery, immaculate post-operative care and dedicated post-operative training. It usually requires a period of some 3–4 months after operation to use the stumps effectively.

This operation is popular in India and it is felt that even the most sophisticated prosthesis cannot match the functional advantages of the Krukenberg procedure which is commonly known as 'ALC scissors operation' by our patients.

Reference

Krukenberg, H. (1917) *Über die plastiche Umvertung von Amputionsstümpfen*. Enke, Stuttgart.

37

The autogenous stump capping procedure

E. G. Marquardt

Introduction: The problem of terminal bony overgrowth

'The major complication of amputation surgery in children is bony overgrowth . . .'. The cause of this phenomenon, even now, remains obscure. It is not a complication of disarticulation surgery, but only follows metaphyseal or diaphyseal transection. Histologically, it is appositional bone growth from the end of the skeletal stump. Implantation of a metal marker at the time of amputation and serial follow-up by x-ray examination have conclusively shown that it is an additive phenomenon, not the *vis-a-tergo* of the proximal epiphysis (Kruger, 1992). Usually there is a bursa formation between the bone and the soft tissue of the stump. As the overgrowth progresses, the soft tissue cushion is invaded and the bursa becomes subcutaneous and increasingly painful. In neglected cases, the skin is eroded and a low-grade superficial infection develops (Aitken and Pellicore, 1981).

This phenomenon of terminal bony overgrowth, however, is not only the major complication in acquired amputations in children, but is also found in both transverse metaphyseal and diaphyseal deficiencies discovered at birth (Pellicore *et al.*, 1974), and especially in children exhibiting the amniotic band syndrome (Lovett, 1987). However, a hypoplastic stump resulting from underdeveloped and consequently a less strained growth plate, as in the middle third of the upper arm, but yet possessing condyles and thus cartilage at its distal end and have the shape of a diaphyseal stump, will never develop terminal bony overgrowth.

The problems of a painful bursa including the perforation of the spike through the skin occur in transhumeral and transtibial stumps much more frequently than in transfemoral stumps, even if the femoral stump demonstrates the same kind of terminal bony overgrowth. Therefore, the epiphyseal contribution of the more active epiphyses of the proximal humerus (80–81%), tibia (55–57%) and fibula (60%), and the distal femur (69–70%) compared to the epiphysis of the proximal femur (30–31%) or of the distal humerus (19–20%) cannot be ignored (Aitken, 1968; Pellicore *et al.*, 1974).

The spiky bone stump end in contrast to the large bony and cartilaginous surface of a disarticulation stump, is unable to stimulate the skin or, indeed, the muscles in a myoplasty for an equivalent physiological growth. Romano and Burgess (1966) hold responsible a further cause: '. . . the circulation of the epiphysis remains intact after the amputation and the bone grows normally, but the soft tissue loses its potential to grow because of muscle atrophy and diminished blood supply.' In summary, the perforation of the skin is preprogrammed.

Proximal epiphysiodesis (Vom Saal, 1939, 1943) cannot be the answer because of the fact that bony overgrowth is the result of terminal bony apposition (Aitken, 1968). Even after an epiphysiodesis, the bony stump end remains thin and spiky.

Even if stump revision of either a transhumeral or transtibial amputation is carried out with skill so 'that (1) the soft tissue flaps be left longer in children than in the adult amputee; (2) the fibula be cut short and that the bone be handled carefully . . .' (Romano and Burgess, 1966), the development of a spiky and thin bony stump cannot be prevented.

Accordingly, many other surgical techniques have been developed such as the Silicone Rubber Implant of A.B. Swanson (1969, 1972) or the porous polyethylene cap of L.C. Meyer and B.W. Sauer (1975). Even so the following statements reflect opinion today: 'The present state of the art is such that in the human there is no certain way of preventing this complication . . .' 'Treatment is the revision of the stump, removal of the bursa and overgrowth bone, and then prosthetic reapplication when healing is secure.' (Aitken and Pellicore, 1981).

Stimulated by the Silicone-Rubber-Implant of A.B. Swanson (1972) and by the homogenous or heterogenous cartilage-bone-transplants of O.A. Buchtiarow (1973) as well as by G.T. Aitken's statement that bony overgrowth 'is not a complication of disarticulation surgery', the author carried out autogenous cartilage-bone transplants in children first in 1974 with the goal of converting a metaphyseal or diaphyseal transverse deficiency amputation stump into a kind of disarticulation stump (Marquardt, 1976).

For his first and second patients the author used autogenous cartilage–bone transplants, but in the following 19 stumps homogenous cartilage–bone material from the bone bank was employed in order to minimize the extent of the operation. Those experiences from the very beginning, however, showed the superiority of the autogenous transplants. This had been corroborated very clearly by the examinations of the data by M. van der Goten (1989) and L. Bernd et al., 1991).

Consequently, it is recommended to use autogenous transplants in stump capping procedures if at all possible (Kruger, 1988); otherwise skin traction (Freidmann and Freidmann, 1985), angulation osteotomy for long stumps only (Marquardt, 1972, 1980, 1981, 1992; Marquardt and Neff, 1974) or the titanium-polyethylene implant (Marquardt, 1987) could be alternatives to the exclusive resection of the spike.

The autogenous stump capping procedure: surgical techniques

The autogenous cartilage–bone graft should possess the shape of a condyle or have an epiphysis with a convexity at its end; it should be not bigger than the taut elastic bursa before its perforation but preferably before surgery. Available as the transplants (Marquardt, 1989: Pfeil et al., 1991) are:

(a) **for the transhumeral stump:** the posterior iliac spine or, in the case of a multimembral deficient child, an epiphysis which cannot provide any function or an equivalent condyle with well shaped cartilage and with its cancellous bone attached and removed as an integral part of a reconstructive procedure.

(b) **for the transtibial stump:** the head of the fibula.

(c) **for the acquired amputation** (transhumeral, transfemoral, or transtibial) **to prevent terminal bony overgrowth:** a well shaped and non-injured epiphysis or condyle from the amputated part of the limb.

The extraction of the transplant as well as the transplantation itself must be carried out under strictly sterile conditions. The extraction and preparation of the stump cap is the first step in the whole procedure. The second step of the appropriate surgical procedure depends upon the condition of the bony stump be it humeral, femoral or transtibial and especially as to whether the skin of the stump is healthy or not. Scars at the end of the stump are responsible for most failures during the growth period.

There are two surgical techniques, which require to be described:

(a) The Autogenous Stump Capping Procedure (ASTCP) of the Pillar-Type. This type is indicated in the case of a distally directed long spike, if marked diminution of the diameter of the tubular bone begins significantly proximal to the bursa (Fig. 37.1).

In such cases the histological findings demonstrate additive bone at the end as Aitken described, but, in addition, a process of reduction of the cortical bone proximal to the amputation.

(b) The autogenous stump capping procedure (ASTCP) of the trans-section planar type. This

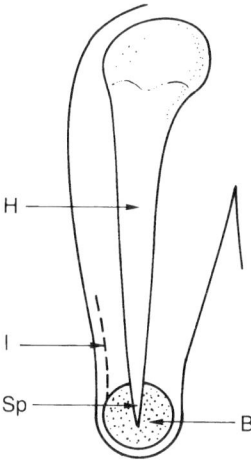

H is HUMERUS
Sp is SPIKE
B is BURSA
I   is INCISION

Fig. 37.1. Terminal bony overgrowth of a transverse deficiency discovered at birth affecting the middle third.

Fig. 37.2. Transverse and longitudinal osteotomies: transverse to remove the spike, longitudinal to prepare two pillars. MPF are the two musculo-periosteal flaps.

type is indicated in the case of a short spike, distally directed and tubular bone of almost normal diameter or in the case of an acquired amputation.

Autogenous stump capping procedure (ASTCP) – Pillar Type

The ASTCP is indicated before the spike has perforated bursa and skin.

The incision should not cross the end of the stump; it should be made as proximal as possible to ensure that the end of the stump is free of scars.

After the incision of the skin and of the bursa, the spike will be exposed. Figure 37.2 shows the situation after trans-section and removal of the spike, after the elevation of two musculo-periosteal flaps and after the longitudinal osteotomy of the distal end of the humeral stump. The distance for osteotomy from the end of the stump bone depends on the length of the remaining humerus but must be long enough for the split to provide two pillars, normally between 2 and 5 centimetres.

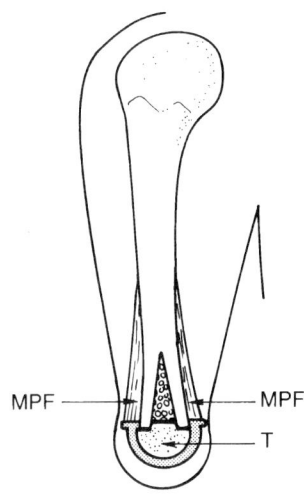

MPF = musculoperiosteal flap
T   = the autogenous cartilage-bone transplant, (stump cap)

Fig. 37.3. Autogenous cancellous bone chips between pillars.

Figure 37.3 shows the finished procedure: the distal humeral stump is separated into two pillars, which are set into two prepared grooves of the cancellous bony part of the stump cap (T). The cartilage is directed distally and the fixation is secured by a screw or with Kirschner wires. Then the musculo-periosteal flaps are sutured to the transplant and in the space between the pillars and in between the periosteum and the pillars are placed small particles of autogenous cancellous bone. There should be no separation between periosteum and muscles.

If at all possible the bursa is then closed over the cartilaginous part of the transplant; the liquid produced by the bursa is essentially similar in composition to that of synovial liquid.

The skin sutures should be lightly applied to ensure accurate abutment. A sterile dressing and bandage is applied. Immobilization is generally not necessary provided fixation of the transplant is secure.

In a case of doubt, however, skin traction built into the inside of a plastic therapy socket fabricated in the Munster technique, will reduce the tension, reduce and control postoperative oedema and will favourably influence the healing of the wound. The immobilization of the stump with a heavy plaster of Paris socket is more troublesome than useful. In contrast, the plastic therapy socket is light, washable and can be removed as often as the dressing has to be changed. It should be prefabricated before surgery and designed to accommodate the postoperative dressing.

Case report

The patient was a boy with multiple limb deficiencies: left upper arm normal, right lower limb longitudinal deficiencies: rays 3, 4, 5 total, fibula total, tibia extremely hypoplastic in fusion with a subtotal deficient femur inside the soft tissues of the pelvis, left lower limb longitudinal deficiencies, rays 4 and 5 total, fibula total, femur total (Fig. 37.4).

On the right transhumeral remnant there was imminent perforation of the skin by a bony spike. Accordingly ASTCP was indicated before the spike would penetrate the skin. An operation of the 'pillar' type was carried out on 2 December 1980. The first step was extraction of the cartilage–bone transplant and of cancellous bone chips from the right femoral tibial rudiment; with temporary preservation in sterile Ringer's solution. The second step was the ASTCP as shown in Figure 37.3. Fixation of the transplant was secured with 3 K-wires, the protruding ends of which were shortened and bent over the surface of the cartilage (Fig. 37.5).

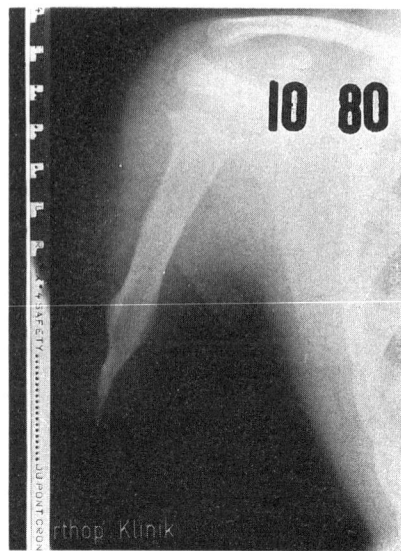

Fig. 37.4. X-ray of a transverse deficiency, upper arm middle third, right with terminal bony overgrowth in a 5-year-old boy (X-ray October 1980). Two-thirds of the spike penetrates a large, painful bursa.
The length of this humeral stump is 9 cm.

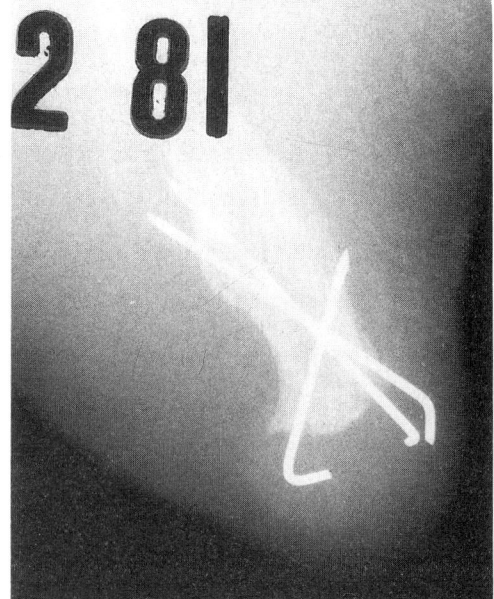

Fig. 37.5. X-ray of the same stump almost 3 months after surgery. The pillars are no longer visible; they have merged with the cancellous bone chips and consolidated with the transplant. The distance between the distal end of the bony stump and the bend of the K-wires indicates the thickness of the cartilage.

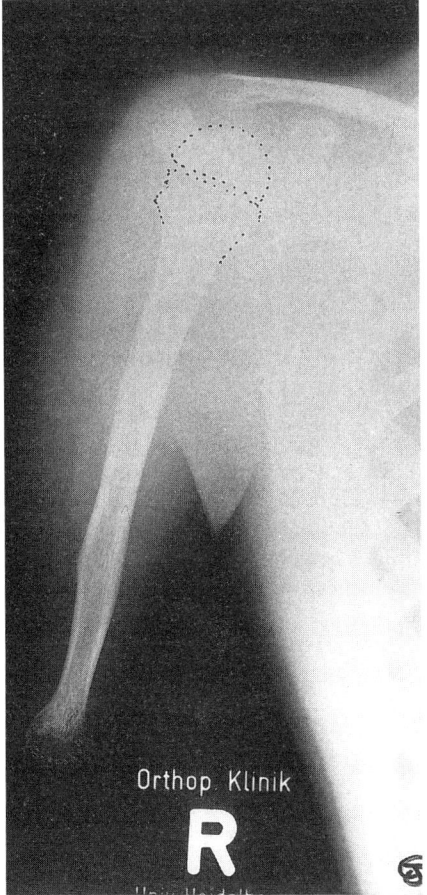

Fig. 37.6. X-ray of the same stump, February, 1984. The length of the humerus stump measures 15.2 cm, excepting the marked cartilage cap (before surgery 9 cm, minus 1.5 cm resection of the spike). Consequently, the length of the stump has doubled in a period of 2 years and 2 months.

Stump stressing by end-bearing training as described by Marquardt (1976, 1981 and 1989) was applied thus stimulating longitudinal growth. Thickening of the bone was achieved in turn by bending stresses as in forward and backward lifting, abduction and adduction of the shoulder joint against resistance performed by the parents but supervised by a physical therapist.

The Kirschner wires were removed on 6 March 1981, 3 months after surgery (Fig. 37.6). Schumacher (1976) had found that the phase of healing requires a period of some 5 to 9 weeks in the case

of cancellous bone and about 12 weeks for cortical bone. In the 'pillar' type of ASTCP both cancellous and cortical bone were in the process of consolidation thus determining the interval from surgery to the start of end-bearing training.

The boy was demonstrated and examined by international experts at the ISPO Symposium on the Limb Deficient Child at Heidelberg in August 1988. He proved to have no pain and no stump problems. The stump was observed to be healthy, with no scar at the end-bearing area and good mobility of the skin over the cartilaginous surface of the stump cap. Painless end-bearing capacity was deemed to be half of the body weight. Confirmation of the condition of the boy was obtained by telephone on 26 October 1993 almost 13 years after the ASTCP. The right transhumeral stump of this patient remains healthy, is in direct use in all daily activities (no prosthesis) without pain and the end-bearing capacity is unchanged since 1981. The young man has been in the USA for one year, where, in addition to other activities, he has been trained for competition in international swimming matches for the handicapped. This example together with many others (L. Bernd *et al.*, 1991) demonstrates the advantage of the ASTCP compared to simple resection of the spike.

Autogenous stump capping procedure (ASTCP) – Transection Planar Type

In this procedure the transection plane of the tubular bone as well as the transplant itself should be a little larger than that of the host bone. The transplant itself should be at least of the same diameter: this is to ensure that the cartilage of the transplant will prevent the development of a new distally directed spike. Fixation of the transplant itself is secured first by Kirschner wires followed by a screw before final suturing of the musculoperiosteal flaps to the transplant. Closed suction drainage is employed and a lightweight rigid cast applied over the wound dressing.

The same technique can be used to bridge tibia and fibula in a transtibial amputation during the growth period and in the case of transverse deficiency discovered at birth. In this instance a bridge of periosteum and bone chips is established immediately proximal to the transplant thus increasing bone contact between stump and transplant. This procedure differs from that of Ertl (1939, 1949) and Dederich, (1963, 1987) in that a periosteal tube is not used.

In the case of a transfemoral stump whether it be acquired or present at birth the transplant should have a growth plate. In contrast to the active growth potential of the transhumeral and transtibial growth plates that of the transfemoral is comparatively poor. Nevertheless every centimetre of growth is fundamental to ensuring the best eventual stump length at the end of growth if a good prosthetic fit and related function is to be obtained. When the transplant comprises the head of the fibula (Marquardt *et al.*, 1983) it is vitally important to reconstruct the lateral ligament of the knee to preserve the integrity of the joint. In any event all shortening procedures should be avoided if at all possible even if it requires the transplant of a non-vascularized complete epiphysis.

The first non-vascularized autogenous stump capping procedure was performed by the author in 1980 on an eleven-year-old boy with a very short acquired transfemoral stump. Nine years later he was found to be extremely active using a total contact socket prosthesis with significant end-bearing capacity and no pain. Another non-vascularized autogenous transplant was performed on a six-year-old boy carried out during the primary procedure using the distal epiphysis of the tibia along with its growth plate and 1 mm of attached cancellous bone (Marquardt and Correll, 1984; Neff, 1988; Pfiel *et al.*, 1991).

Given optimal conditions, transplantation with an intact vascular supply can be performed by micro-surgery; the opportunity to do so is rarely possible.

The following criteria are important:

(1) the cancellous bone attached should be of 1 mm thickness from the transection plane to the growth plate.
(2) The normal precautions regarding overheating of the bone should be observed if a power saw is used.
(3) Once K-wires are removed and a prosthesis with total contact socket fitted, daily end-bearing training is imperative otherwise the growth plate will become inactive and an extremely hypoplastic stump will result.

Where a knee disarticulation amputation is considered to be contraindicated because of inadequate skin cover, consideration should be given to femoral shortening by osteotomies, provided the blood supply is preserved (Marquardt, 1987; Baumgartner and Botta, 1989). Additional plastic surgery may be required to obtain complete skin cover (Marquardt *et al.*, 1986, 1987). In this way growth is ensured and valuable end-bearing obtained.

Regrettably it is not possible to apply autogenous transplantation in each and every acquired and transverse limb deficiency even when clearly indicated. Because of the high failure rate (Bernd *et al.*, 1991) homogenous and heterogenous cartilage bone transplants cannot be recommended. There remains the hope of a successful review of the development of the Titanium-polyethylene implant designed to treat terminal osseous overgrowth (Marquardt, 1987); the cap requires to be removed at the end of the growth period.

The results of autogenous stump capping procedures (Bernd, *et al.*, 1991; Kruger, 1991; Watts, H.G. and Setoguchi, 1992) testify to the efficacy of the procedures. Virtually all lower limb procedures were successful – all patients were using their prostheses regularly and no secondary procedures were required. In transhumeral stumps there was an overall failure rate of some 10–15% requiring further operations.

Conclusions

The Autogenous Stump Capping Procedure (ASTCP) is clearly an important step forward in the prevention of terminal osseous overgrowth with the added bonus of at least a degree of end-bearing in most cases.

The success of ASTCP is, however, predicated on healthy soft tissue cover. A scar at the stump end acquired in initial trauma, as part of an acquired amputation, previous bone spike excision or a wrongly placed incision will become wider, thinner, vulnerable and painful. Skin traction may help and should certainly be used before spike resection is undertaken but the latter is necessary if spike perforation has occurred or is imminent. In these circumstances the more proximal longitudinal incisions for stump capping are clearly not appropriate:

- In the case of a perforation the infected spike and bursa has to be resected and ASTCP deferred until sterile conditions exist. Thereafter skin traction inside a therapy socket or prosthetic socket will promote healing (Friedmann and Friedmann, 1985).
- Where a scar already exists it should be excised and the stump along with spike, shortened as far as necessary to accommodate the transplant and permit closure of the wound without tension, employing skin traction if need be. The use of a tissue expander for four weeks prior to the operation may provide a sufficiency of skin where the scarring is extensive.
- Where these measures are doomed to failure and with microsurgical facilities available, the large expanse of wound exposed after resection can be covered by a neurovascular, musculocutaneous latissimus dorsi 'island' flap. (Marquardt, Martini, von Bazan, 1983).

The extensive and comprehensive experience recorded here makes it clear that patients requiring treatment of terminal osseous overgrowth should be referred to a specialized clinical and surgical team. Otherwise – and this is true for many regions in both industrial and developing countries – the better and safer answer is revision of the stump and removal of the bursa followed by a new prosthetic fitting (Bowker *et al.*, 1992; Tooms, 1992).

References

Aitken, G.T. (1962) Overgrowth of the amputation stump. *Inter-Clin. Info. Bull.*, **1**, 1–8.

Aitken, G.T. (1968) The child with an acquired amputation. *Inter-Clin. Info. Bull.*, **7(8)**, 1–15.

Aitken, G.T. (1969) Osseous overgrowth in amputation in children. In: *Limb Development and Deformity: Problems of Evaluation and Rehabilitation* (Ed. C.A. Swinyard), Thomas, Springfield.

Aitken, G.T. and Pellicore, R.J. (1981) Introduction to the child amputee. In: *Atlas of Limb Prosthetics: Surgical and Prosthetic Principles*. American Academy of Orthopedic Surgeons, C.V. Mosby, St Louis, pp. 493–500.

Baumgartner, R., Botta, P. (1989) *Amputation und Prosthesen-Versorgung der unteren extremitat*. Enke, Stuttgart p. 137.

Bernd, L., Blasius, K., Lukoschek, M., Lucke, R. (1991) The autologous stump plasty. *J. Bone Joint Surg. (Br)*, **73B**, 203–206.

Bowker, J.H., Keagy, R.D., Poonekar, P.D. (1992) Musculoskeletal complications in amputees: their prevention and management. In: *Atlas of Limb Prosthetics, Surgical, Pros-thetic and Rehabilitation Principles*. 2nd edn (eds J.H. Bowker and J.W. Michael. American Academy of Orthopedic Surgeons, Mosby Year Book, St Louis, pp. 665–680. Bony overgrowth in children, pp. 677–678.

Buchtiarow, O.A. (1973) Rekonstructions-plastiche operationen an den Extremitatstumpfen mit der verwendung der knocken-Knorpeligen Hono- und Hetero-transplantate in Verbindung mit der Prothetik. *Proc. 1st International Congress on Prosthetic Techniques and Functional Rehabilitation, Vienna*, Vol. 1, pp. 45–46.

Dederich, R. (1963) Plastic treatment of the muscles and bone in amputation surgery. *J. Bone Jt. Surg.*, **45B**, 60.

Dederich, R. (1987) *Amputationen der Gliedmaszen*. Georg Thieme, Stuttgart. New York There: BK Amputation pp. 124–134. Amputation in Children pp. 39–42.

Ertl, J. (1939) *Regeneration Ihre Anwendung in der Chirurgie*. Barth, Leipzig.

Ertl, J. (1949) Uber amputationstumpe. *Chirurg.*, **20.**, 218.

Friedmann, L.W., and Friedmann, L. (1985) The conservative treatment of the bony overgrowth problem in the juvenile amputee. *Inter-Clin. Info. Bull.*, **20(2)**, 17–23.

van der Goten, M. (1989) *Die Stumpfkappenplastik nach Marquardt: postoperative verlauf und derzeitiger befund*. Inaugural dissertation der Ruprecht-Karls-Universitat, Heidelberg, pp. 64–75.

Kruger, L.M. (1988) Surgical management of lower limb deficiencies. In: *Amputation Surgery and Lower Limb Prosthetics* (ed. G. Murdoch), Blackwell Scientific Publications, Oxford, pp. 279–283.

Kruger, L.M. (1992) Lower limb deficiencies, surgical management. In: *Atlas of Limb Prosthetics Surgical, Prosthetic and Rehabilitation Principles*. 2nd edn, (eds J.H. Bowker and J.W. Michael.) American Academy of Orthopedic Surgeons, Mosby Year Book, St. Louis, pp. 795–834.

Lovett, R.J. (1987) Osseous overgrowth in congenital limb deficient children. *Journal of the Association of Children's Prosthetic-Orthotic Clinics*, **22**, No. 2, 26.

Marquardt, E. (1972) Steigerung der Effektivat von oberarm prosthesen durch winkelosteotomie. *Rehabilitation*, **11**, 244.

Marquardt, E. and Neff, G. (1974) The angulation osteotomy of above elbow stumps. *Clin. Ortho.*, **104**, 232.

Marquardt, E. (1976) Plastiche Operationen bei drohender Knockendurich spieburg am kinlichen oberarmstumpf. *Z. Orthop.*, **114**, 711–714.

Marquardt, E. and Martini, A.K. (1979) Gesichts Punkte der Amputations Chirurgie der Aberen Extremiten. *Z. Orthot.*, **117**, 622–631.

Marquardt, E. (1980) The multiple limb deficient child. In: *Atlas of Limb Prosthetics: Surgical and Prosthetic Principles*, American Academy of Orthopedic Surgeons, C.V. Mosby, St. Louis, pp. 595–641.

Marquardt, E. (1980) The Knud Jansen Lecture. The operative treatment of congenital limb malformation – Part I. *Prosthetics and Orthotics Internat.*, **4**, 135–144.

Marquardt, E. (1981) Part II. Case study. *Prosthetics and Orthotics Internat.*, **5**, 2–6.

Marquardt, E. (1981) The multiple limb-deficient child. In:

Atlas of Limb Prosthetics: Surgical and Prosthetic Principles, American Academy of Orthopedic Surgeons, C.V. Mosby, St. Louis, pp. 595–641.

Marquardt, E. and Puhl, W. (1981) Die Knorpel-Knochen-Transplantation zur Behandlung Drohender Durchspiessung am Amputationsstumpf. In: *Implantate und Transplantate in der Plastischen und Wiederherstellungs Chirurgie.* (eds H. Cotta and A.K. Martini), Springer, Berlin, Heidelberg, pp. 292–301.

Marquardt, E. *et al.* (1983) In: *Regionale Plastiche und rekonstructive Chirurgie un Kindesalter* (ed. C. Naumann) Springer, Berlin, Heidelberg, pp. 190–199.

Marquardt, E., Martini, A.K. and von Bazan (1983) Stumpfkapponplastik and stumpfverlangerung bei traumatischen amputationen. In: *Regional Plastiche und Reconstructivie Chirurgie in Kindesalter* (eds W. von Kley and C. Naumann) Springer, Berlin, Heidelberg pp. 190–199.

Marquardt, E. and Correll, Y. (1984) Amputations and Prostheses for the Lower Limb. *Internat. Orthopaedics (SICOT)*, **8**, 139–146.

Marquardt, E. (1987) Indikation und spezielle Amputationstechnik am Oberschenkel. In: *Hefte zur Unfallheilkunde, Heft 189.* (ed. A. Pannike) Springer, Berlin, Heidelberg, pp. 809–819.

Marquardt, E. (1987) Titanium polyethylene implant to prevent terminal osseous overgrowth. *JACPOC*, **22(2)**, 26.

Marquardt, E. (1989) The Heidelberg experience. In: *Comprehensive Management of the Upper-Limb Amputee.* (eds. D.Y. Atkins and R. H. Meier III), Springer, New York, pp. 240–252.

Marquardt, E. (1992) The multiple limb-deficient child. In: *Atlas of Limb Prosthetics: Surgical and Prosthetic Principles.* 2nd edn, (eds J.H. Bowker and J.W. Michael), American Academy of Orthopedic Surgeons, C.V. Mosby, St. Louis, pp. 839–854.

Meyer, L.C. and Sauer, B.W. (1975) The use of porous high density polyethylene caps in the prevention of appositional bone growth in the juvenile amputee: A preliminary report,. *Inter-Clin. Info. Bull.* **14**, 1–4.

Neff, G. (1988) Surgery in Above-Knee Amputation In: *Amputation Surgery and Lower Limb Prosthetics.* (eds G. Murdoch and R.G. Donovan.) Blackwell Scientific Publications, Oxford, pp. 117–129.

Pellicore, R.Y., Sciora, Y., Lambert, C.N., *et al.* (1974) Incidence of bone overgrowth in the juvenile amputee population. *Inter-Clin. Info. Bull.* **13**, 1–8.

Pfeil, J. *et al.* (1991) The Stump Capping Procedure to prevent or to treat terminal osseous overgrowth. *Prosthetics and Orthotics Internat.* **15**, 96–99.

Romano, R.L. and Burgess, E.M. (1966) Extremity growth and overgrowth following amputations in children. *Inter. Clin. Inform. Bull.*, **5(4)**, 11–12.

vom Saal, F. (1939) Epiphysiodesis combined with Amputation. *J. Bone Jt. Surg.*, **21**, 442–443.

vom Saal, F. (1943) Amputations in children. *Surg. Gynec. Obstet.*, **76**, 708–710.

Schumacher, G. (1976) *Das verhalten von autologen kompakta- und spongiosa Transplantaten im Tierexperiment.* Habilitation Schrift der Rupert-Karls Universitat, Heidelberg.

Swanson, A.B. (1969) Bone overgrowth in the juvenile amputee and its control by the use of silicone rubber implants. *Inter-Clin. Info. Bull.*, **8**, 9–16.

Swanson, A.B. (1972) Silicone-rubber implants to control the overgrowth phenomenon in the juvenile amputee. *Inter.-Clin. Inform. Bull.*, **11(9)**, 5–8.

Tooms, R.E. (1992) Acquired amputations in children. In: *Atlas of Limb Prosthetics*, 2nd edn. (eds J.H. Bowker and J.W. Michael), *American Academy of Orthopedic Surgeons*, Mosby Year Book, St. Louis pp. 735–741.

Watts, H.G. and Setoguchi, Y. (1992) Marquardt stump capping for overgrowth. *Journal of the Association of Children's Prosthetic-Orthot. Clinics.*, **27**, No. 2, 61.

38

Angulation osteotomy of transhumeral stumps

E. G. Marquardt

This procedure, first suggested by the author at the International Congress of Paediatricians in Vienna in 1971 provides much improved prosthetic function in transhumeral amputations and also serves to prevent bone overgrowth. Angulation osteotomy is applicable in both children and adults.

The *first* method is for children with long transhumeral stumps or in cases of bone overgrowth. Once the bone is exposed via a lateral incision the periosteum is opened longitudinally on its anterior aspect and the cortical bone transected 3 to 5 cm from stump end. The thicker the soft tissues the longer the distal fragment required for good prosthetic control. The bone is bent over the fulcrum of the index finger to the desired angle of 75° to 90° the posterior periosteum remaining intact. The position is secured by means of a Kirshner wire retained normally for some six weeks.

The *second* method is again for use in children and especially for those with middle length transhumeral stumps. The bone is approached via a postero–lateral incision. The periosteum is incised longitudinally and by virtue of two oblique incisions, distal and proximal flaps are formed and secured by holding sutures. The bone freed of

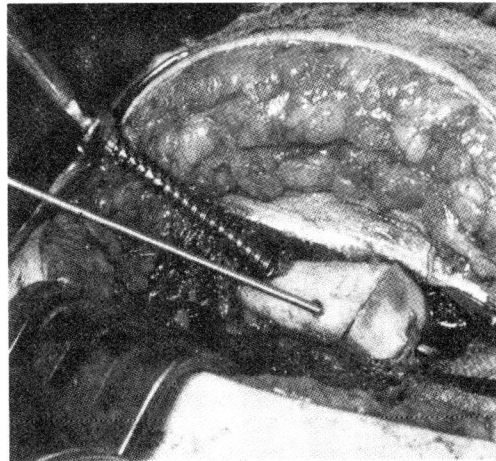
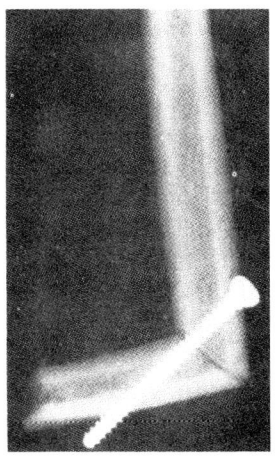

(a)   (b)

Fig. 38.1.(a) Fixation with AO compression screw with temporary Kirschner wire keeping osteotomy in position; (b) X-ray 4 weeks post-operatively.

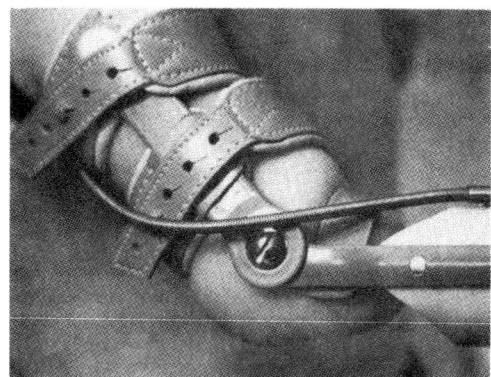

Fig. 38.2. Open splint construction.

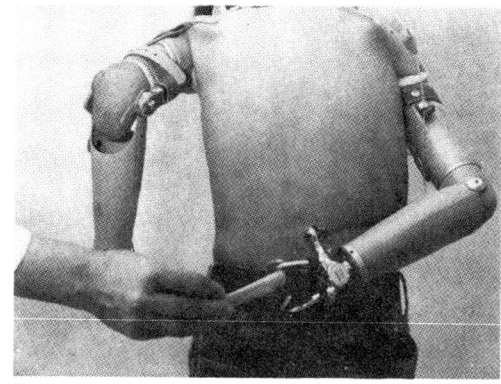

(a)

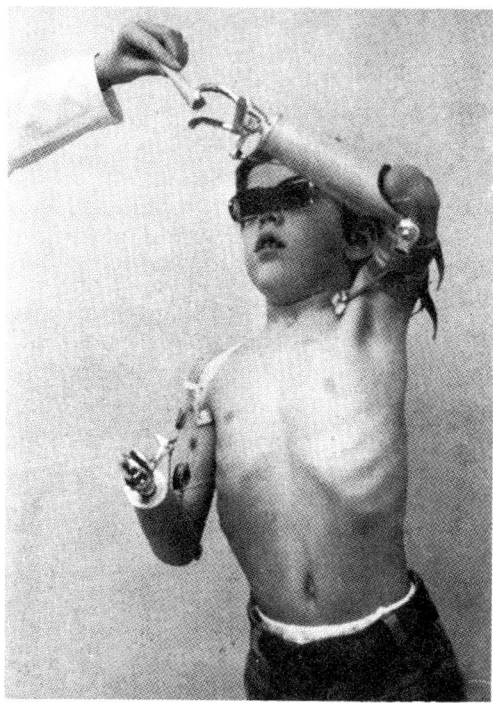

(b)

Fig. 38.3. Bilateral angulation osteotomy; **(a)** open splint construction (with free stump for touch sensation); **(b)** with plastic socket. Angulation osteotomy and light prosthetic replacement compensate for wrist rotation permitting prehension function in front and behind trunk.

periosteum is osteomized posteriorly, an incomplete division being sufficient thus preserving the anterior cortex. The angle is formed easily bending over a fulcrum and the angle secured with a Kirshner wire for some 8–10 weeks. The dorsal bone defect is covered by the two periosteal flaps to allow ossification to proceed. Immobilization in a cast is maintained for 4 to 5 weeks prior to prosthetic fitting.

The *third* method designed for adults requires removal of an anteriorly based wedge of between 70° and 90° dependent on stump length and sited some 6–7 cm from the stump end. Stump length also determines whether the whole or half cross section is incorporated in the wedge. The bone fragment must not be separated in any way from the soft tissues and careful secure positioning is required to permit fixation with a compression screw (Fig. 38.1). After wound closure a light cast is applied to control swelling and permit early prosthetic fitting.

In long transhumeral stumps an open splint prosthetic construction (Fig. 38.2) with an outside lock and active hook and hand is employed thus emancipating the shoulder. The prosthesis is fitted closely on the end of the stump so that any shoulder movement is transferred directly to it. The prosthesis is suspended on the stump end.

In middle length stumps, the shoulder is again left free and careful close fitting of the cap socket to the anterior aspect of the angle and a small posterior window and velcro strap ensure secure suspension. A conventional elbow unit with cable control is fitted and similar control of the terminal

device used. Myoelectric technology can also be exploited.

The combination of the angulation at the end of the stump and the simple light prosthetic replace-

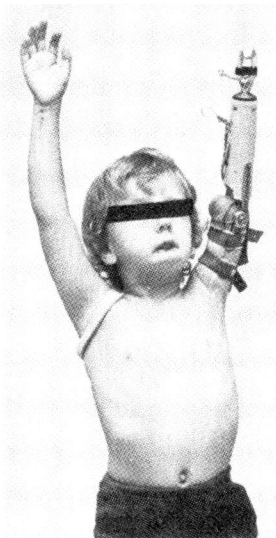

Fig. 38.4. Free elevation.

ment permitting free shoulder movement compensate to a high degree for wrist rotation, permits prehensive function in front and behind the trunk (Fig. 38.3) and free elevation (Fig. 38.4).

Further reading

Marquardt, E. (1972) Steigerung der Effektivitat von Oberarmprosthesen durch Winkel-osteotomie. *Rehabilitation* 11:244–248.

Marquardt, E. and Neff, G. (1974) The angulation-osteotomy of above-elbow stumps. *Clin. Orthop.* 104, 232–238.

Tooms, R.E. and Snell R.R. (1972) Prosthetic Principles – conventional upper limb prostheses. In *The Child with an Acquired Amputation* (ed. George T. Aitken), National Academy of Sciences, Washington.

Tooms, R.E. (1981) Acquired amputations in children. In *Atlas of Limb Prosthetics: surgical and prosthetic principles* (ed. American Academy of Orthopaedic Surgeons), C.V. Mosby: St. Louis pp. 553–559.

39

Partial hand amputation

G. E. Omer Jr

Amputation is among the most ancient of surgical procedures and continues to be a common operation in spite of increasing proficiency in replantation of amputated parts and imaginative reconstructive procedures (Omer, 1991).

An amputation is utilized to preserve life or improve limb function. Indications for amputation include tumours, uncontrollable infection, revision of a congenital deformity, extensive paralysis, peripheral vascular disease and trauma. Congenital deformities are significant under 5 years of age and peripheral vascular problems are important in the elderly, but trauma predominates the indications for amputation. Industrial machinery, vehicular accidents, burns and explosives often result in traumatic loss of parts that are avulsed, crushed or jagged and are poor candidates for replantation. Almost any amputated part can be successfully revascularized by an experienced microsurgery team (Wood, 1992; Louis, 1993) However, survival of the injured part should not be misconstrued as useful function and amputation may be the preferable salvage procedure (Goldner et al., 1990; Engel et al., 1991).

The basic components of hand function are grasp, hook and pinch. Grasp is the action of holding an object against the palm of the opposed thumb and fingers. Hook utilizes the fingers and is used in the transfer of objects in space. Pinch is the manipulation of objects within the grasp, either at fingertip level or palmar level. The goals of amputation surgery in the hand should be: preservation of functional length, preservation of useful sensibility, prevention of symptomatic neuromas or painful syndromes, prevention of adjacent joint contractures, short morbidity, early prosthetic fitting where applicable and early return of the patients to work or play (Louis, 1993).

Finger tip amputations

These injuries are the most common type of amputation in the hand. There is general agreement that all length in the thumb should be maintained, but there is less agreement concerning the necessity for maintaining the length of other digits following digital tip amputations. Split-thickness skin grafts are the most popular technique for closure of injuries involving only skin or pulp loss (Flatt, 1972), but conservative management gives appropriate results in adults as well as children (Louis, 1985). Conolly and Goulston (1973) studied amputations distal to the distal interphalangeal joint and found that grafts and flaps were associated with a higher morbidity than primary suture or allowing healing by secondary intention.

When bone is exposed, one must consider shortening the bone or utilizing special flaps such as volar V-Y flaps, lateral V-Y flaps, volar flap advancement for only the thumb, cross-finger pedicle flap, or the thenar flap for only the index and long fingers (Louis, 1993). In an adult with a fingertip injury and pulp loss, there is 30–50% cold intolerance and aberration of sensibility regardless of the technique utilized (Louis, 1993).

Amputations through fingers

Even when amputation is indicated, one should consider if parts of the finger may be useful for reconstructive procedures. Skin may be utilized as a free graft and skin, well supported by one or more neurovascular bundles, may be used as an island graft. Nerve segments are useful as autogenous grafts. Bones may be used as peg grafts or as free grafts.

The classic description of skin flaps in finger amputations is a long palmar and short dorsal. Unfortunately, classical skin flaps are rare when the amputation is the result of trauma and the presence or absence of an adequate circulation determines the length and position. Skin flaps should never be sutured under tension or the result will be a tender scar at the suture line and occasionally tissue breakdown. The bone is rongeured 0.5 cm proximal to the surrounding tissues. If the amputation is through the interphalangeal joints, the proximal phalanx is contoured by rongeuring off the volar and lateral condylar prominences. Flexor and extensor tendons should be drawn gently distally, divided and allowed to retract proximally. Tendons are never sutured over the bone end because they will limit the excursion of related tendons and result in a flexion contracture of the finger.

The digital nerves should be meticulously dissected from their volar beds and resected at least 0.5 cm proximal to the end of the skin flaps. The nerves are sharply divided and ligatures are not indicated. The end of the digital nerve should not lie at the same level as a palmar flexion crease but should be located in thick and healthy subcutaneous tissue. Neuromas are considered inevitable but they should be allowed to develop only in areas well padded with soft tissue (Omer, 1981, 1991). Gorkisch, Boese-Landgraf and Vaubel (1984) advocate a proximal transection of the fascicles and then suture the proximal stumps to each other, creating an interposed graft; this frustrates the process of neuroma formation. The digital arteries should be ligated with 7–0 or 8–0 sutures. Before flaps are closed, the tourniquet should be released and haemostasis obtained. Skin flaps are closed with small interrupted sutures.

Finger ray amputations (Figs 39.1, 39.2)

Ray amputation is most frequently undertaken as an elective procedure consequent to a disability resulting from a previous injury to a digit that renders it either functionally impaired or useless. This procedure may also be undertaken in the treatment of tumours or infections. Ray amputation is rarely indicated at the time of initial trauma.

If the index finger is to be amputated at its proximal interphalangeal joint or at a more proximal level, the remaining stump may prove to be awkward and in particular may hinder pinch between thumb and long finger. This has not been our general experience and it is acceptable to allow the patient to use the hand with the shortened index finger before deciding it is useless. In manual workers, the remaining index finger often participates in power grip.

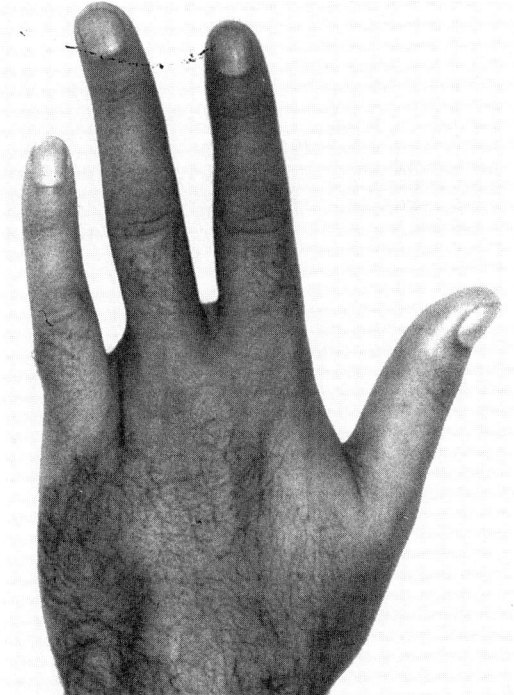

Fig. 39.1. Appearance of hand after excision of 4th ray and transposition of 5th ray to the base of the fourth metacarpal bone. The patient has a normal range of motion.

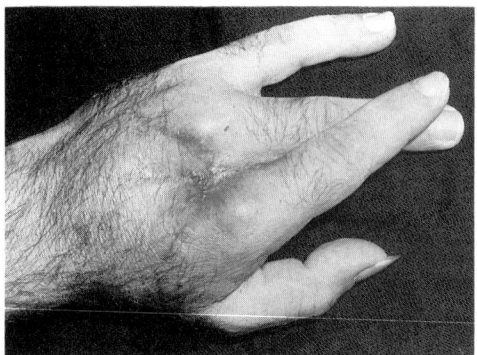

Fig. 39.2. Appearance of hand after excision of 3rd ray. In addition to the 'crossed fingers' there is limited flexion. The patient required transposition of the 2nd ray to the base of the third metacarpal bone and release of the tendons to the missing long finger.

The absence of either the long or ring finger creates a hole in the transverse palmar arch through which small objects can drop when the hand is used as a cup or scoop. In some patients, the remaining fingers deviate to the midline on flexion. If these problems are significant and persistent, a ray resection can be done by sub-periosteal dissection of the involved metacarpal without disturbing muscle arrangements. However, it should be recognized that excising the third metacarpal shaft weakens pinch by removing the stable origin for the adductor pollicis muscle. Patients should be aware that strength for grip will be diminished following these procedures (Murray *et al.*, 1977).

The little finger has an undeserved rating as the least important digit. Indeed the little finger is more important than the index finger for grasp and hook function (Omer, 1991). A short finger stump at the level of the proximal phalanx may be unsightly and awkward but also may contribute to functional strength. Even an amputation through the metacarpo-phalangeal joint contributes breadth for grasp and the metacarpal head should be bevelled for cosmetic reasons or because it may become painful.

The thumb

The thumb accounts for 40–50% of hand function and the longer the sensible amputation stump, the better the result. Local advancement flaps are

indicated as primary or secondary procedures to retain all possible length. The volar advancement flap has provided the best sensibility in our cases (Omer, 1991). There exists a wide range of microsurgical procedures for reconstruction (Steichen and Weiss, 1992; Weiss and Steichen, 1992). If pollicization is indicated, the index finger is the best digit for this procedure. (Brunelli and Brunelli, 1992). Neither replant nor revision thumbs have normal function (Wood, 1992).

Multiple finger amputation

A rudimentary grasping mechanism requires a cleft between two opposing poles that are rotated to be opposite one another or can adduct parallel to each other. Such a cleft can be developed by rotational osteotomies of metacarpals or by resecting a shortened second or fourth metacarpal. Ingenuity and imagination are required to provide motor power and preserve sensibility.

Prosthetic prescription

In amputations through the fingers, no prosthesis should be used if there are two remaining digits that have opposition and sensation. If only one digit remains, an orthosis can provide a flat surface for opposition; this is especially useful if the thumb remains and the fingers are amputated across the base of the palm. A partial hand with weak prehension may be fitted with a prosthesis for heavy work. All prosthetic devices must have a specific purpose and should be designed according to the patient's injury and the specific activity requirement. The cosmetic effect of the normal hand at rest may be simulated with a cosmetic glove (Pillet, 1981). A cosmetic hand does fill sleeves, can push and pull, provides two-handed holding and may be psychologically satisfying; however, most patients will select function over cosmesis.

References

Brunelli, G.A. and Brunelli, G.R. (1992) Reconstruction of traumatic absence of the thumb in the adult by pollicization. *Hand Clinics*, **8**, 41–55.
Conolly, W.B. and Goulston, E. (1973) Problems of digital amputations: a clinical review of 260 patients and 301

amputations. *Aust. NZ J. Surg.*, 43, 118–123.

Engel, J., Luboshitz, S., Jaffe, B. and Rotstein, Z. (1991) To trim or replant: a matter of cost. *World J. Surg.*, **15**, 486–492.

Flatt, A.E. (1972) *The Care of Minor Hand Injuries*, 3rd ed. C.V. Mosby, St. Louis, pp. 137–163.

Goldner, R.D., Howson, M.P., Nunley, J.A., Fitch, R.D., Belding, N.R. and Urbaniak, J.R. (1990) One hundred eleven thumb amputations: replantation vs revision. *Microsurgery*, **11**, 243–250.

Gorkisch, K., Boese-Landgraf, J. and Vaubel, E. (1984) Treatment and prevention of amputation neuromas in hand surgery. *Plast. Reconstr. Surg.*, **73**, 293–299.

Louis, D.S. (1985) To graft or not to graft – the fingertip. *J. Hand Surg.*, **10A**, 439–440.

Louis, D.S. (1993) Amputation. In: *Operative Hand Surgery*, 3rd edition (ed. D.P. Green), Churchill Livingstone, New York, pp. 53–98.

Murray, J.F., Carman, W. and MacKenzie, J.K. (1977) Trans-metacarpal amputation of the index finger: a clinical assessment of hand strength and complications. *J. Hand Surg.*, **2**, 471–481.

Omer, G.E., Jr (1981) Nerve, neuroma and pain problems related to upper limb amputations. *Orthop. Clin. N. Am.*, **12**, 751–762.

Omer, G.E., Jr (1991) Upper extremity amputation. In: *Flynn's Hand Surgery*, 4th edition (ed. J.B. Jupiter), Williams and Wilkins, Baltimore, pp. 543–558.

Pillet, J. (1981). The aesthetic hand prosthesis. *Orthop. Clin. N. Am.*, **12**, 961–969.

Steichen, J.B. and Weiss, A.B. (1992) Reconstruction of traumatic absence of the thumb by microvascular free tissue transfer from the foot. *Hand Clinics*, **8**, 17–32.

Weiss, A.P. and Steichen, J.B. (1992) Reconstruction of traumatic absence of the thumb by alternative microsurgical methods of reconstruction. *Hand Clinics*, **8**, 33–39.

Wood, M.B. (1992) Finger and hand replantation. *Hand Clinics*, **8**, 397–408.

40

Biomechanics and prosthetics

W. G. Dykes

Prosthetic replacement of the upper limb should be considered differently from the prosthetic replacement of the lower limb. Differences occur in patient motivation, functional replacement and cosmetic replacement. The lower limb amputee is motivated to wear a prosthesis by the considerable difficulty in ambulation when a prosthesis is not used. The use of crutches as a walking aid, while effective, prevents easy use of the hands and distinguishes the user from others. Therefore the motivation to wear a prosthesis to ambulate is high. Most upper limb amputees are unilateral and with experience can function very well; few activities of daily living are outside their capabilities. An upper limb prosthesis is not essential and they are unlikely to tolerate the device unless there are benefits. The lower limb prosthesis, in almost all instances, provides a reasonably efficient functional replacement for the missing limb, whereas the most advanced upper limb prosthesis available today can only be classed as a crude holding device which might assist the amputee in some simple tasks. From a cosmetic point of view the lower limb amputee once again benefits, as modern endoskeletal systems allow excellent prosthetic shaping and the prostheses can be hidden beneath clothing. For the upper limb the problems of providing a good cosmetic match in shape and colour to the missing limb are immense. The introduction of silicone rubber in the production of custom-made prosthetic cosmetic gloves has much improved the situation; however, mass produced silicone cosmetic gloves are presently only available to suit certain types and sizes of prosthetic hands.

While most of the gross movements of the upper limb prosthesis can be powered either by body power or external power the problems of providing sensation and positional feedback have yet to be overcome. The human hand in particular is a very complicated mechanism to attempt to replace and the limitations of the upper limb prosthesis must always be remembered. The human tolerance to gadgets is usually very low unless considerable advantage is obtained from their use and as a general principle, for the unilateral amputee, it is the simplest prosthesis which is the most successful.

Components

A large selection of components is available from manufacturers throughout the world, allowing the prescription team to match the requirements of the amputee as closely as possible. Consideration must be given to compatibility of components between manufacturers, e.g. does an electric hand from company A fit easily to a wrist unit from company B. The prosthetists should have sufficient information and knowledge to guide the prescription team.

When making decisions on components prescribers should consider carefully the needs of the amputee (Beasley, 1981). It is very easy to overload the amputee with apparently attractive technological developments. When making a decision on prosthetic prescription many factors need to be considered, including cultural and social acceptability. A prosthetic prehensor which is reported to be accepted and successful in one country may not necessarily be so in another.

Control of prosthetic components

Movement of prosthetic components can be accomplished either by body power or external power and a combination of both may be used within the same prosthesis. Body power is achieved by harnessing the movement of one body part relative to another; both the excursion and the force obtained is generally transferred to the prosthetic component through cables. Body power mechanisms are simple, reliable, low cost, easy to repair and work very well when used on the transradial prosthesis. On higher levels the system is less satisfactory because of more complicated cable runs, increased frictional losses and the number of control sites required.

Externally powered components commonly use electrical motors to drive the moving parts, although hydraulic and pneumatic systems have been used in the past. In the case of terminal devices this provides much higher prehension forces than can be achieved with body powered components. A number of methods are available to prosthetists to control electrically powered components. Commonly used methods include myoelectric control, micro switch control and touch sensitive switches.

Terminal devices

Terminal devices can be divided into two categories, passive and prehensile. Prehensile devices can be operated by body power or electrical power.

Body powered terminal devices are normally designed in voluntary opening or voluntary closing mode. The simpler and more popular voluntary opening method uses a positive force applied by the amputee through the operating system to open the terminal device (Fig. 40.1). The closing force is constant and provided by elastic bands or springs. The term 'voluntary opening' is now also applied to electrically powered devices which work on a

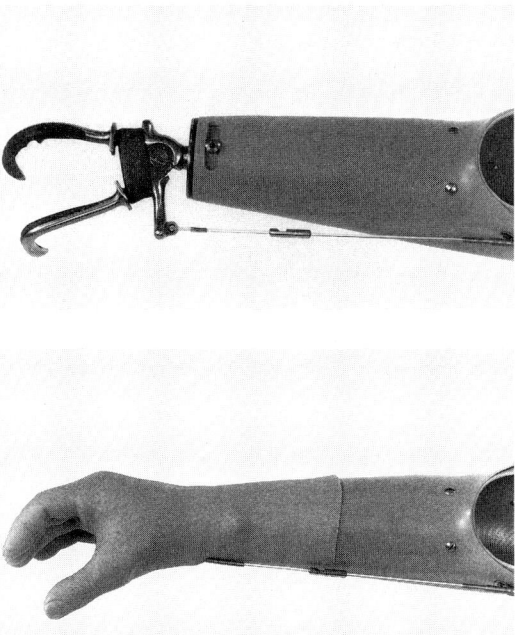

Fig. 40.1. Body powered split hook and body powered hand.

similar principle: a signal from a single site opens the device and keeps it open until the signal is released (Steeper SCAMP hand). When the signal is released the device closes under the tension of a coil spring. Alternatively, the device is closed by an electric motor with a current sensing circuit which switches off the power to the motor when the pinch force reaches a predetermined level (Variety Ability Systems Inc. VV series hands).

In body powered voluntary closing systems the force generated by the patient is applied to close the device, allowing direct control over the grasp force. To allow the grasp force to be maintained when body powered forces are relaxed, a locking system is built in.

Passive devices

The most common passive device is the lightweight foam hand, often referred to as a cosmetic hand. Typically, the hand is manufactured in a low density foam with malleable central finger wires to allow alteration of finger position. The foam is covered with a cosmetic cover (glove) manufactured in PVC or less commonly silicone rubber. Other passive devices may only resemble the shape of the human hand or be constructed to a special pattern for a particular activity.

Prehensile devices

Prehensile devices can be designed in the shape of the human hand or shaped purely for function. Attempting to replicate the shape of the human hand places constraints on the ability to achieve the best functional result.

Hooks

Acceptance of a hook-like device can be difficult for many amputees and for the parents of small children (Billock, 1986). Most experts would acknowledge the functional advantages of the body powered split hook. The split hook is generally much more mechanically efficient than a similar powered hand. It has the ability to grasp both large and very small objects, is light in weight and also importantly allows the operator to see the grasping surfaces without hindrance.

Active prosthetic hands

Both voluntary opening (the most common) and voluntary closing body powered prosthetic hands suffer from limited prehension force due to frictional losses and the resistance of cosmetic coverings. While the range of body powered hands available fulfil the needs of many unilateral amputees, the bilateral amputee will require a much stronger prehension force. Where appearance is important this can be accomplished by using an electrically powered hand giving a positive and stronger prehension force. Many bilateral amputees will sacrifice appearance for the added function available from a split hook.

Special and custom-made terminal devices

While the functional requirements of many amputees can be met with basic terminal devices such as prosthetic hands and hooks, some may have specific difficulties at work or with hobbies and pastimes. A number of off-the-shelf devices are available and special custom-made devices (Radocy, 1990) can be produced to suit particular problems (Fig. 40.2).

Fig. 40.2. Shoulder disarticulation amputee using a custom-made windsurfing attachment.

Wrist units

Three basic types of wrist units are in common use. These are the quick disconnect, which allows fast and easy interchange of the terminal device, the screw fitting still allowing interchange of the terminal device, and the permanent fixation which does not allow interchange of the terminal device. Wrist units for use with electric hands tend to be dedicated units, but manufacturers are now producing quick release units for electric hands with adaptors to allow the use of other types of terminal device. This is a very useful development for the amputee who prefers an electric hand for most tasks but wants to retain the capacity to use a body powered or specialized terminal device for occasional use, without additional expense or inconvenience of changing to a different prosthesis. Basic wrist units allow rotation of the terminal device and some have the ability to apply a positive lock, which may be useful to the amputee in accomplishing certain tasks. For the unilateral amputee the provision of wrist flexion/extension at the terminal device is considered an unnecessary complication and compensatory movements (e.g. humeral movements) can be made to alter the angle of the terminal device. Wrist flexion is more useful for the bilateral amputee and wrist flexion units such as those produced by the Hosmer Dorrance Corporation are available.

Elbow units and elbow hinges

A small selection of body powered elbow units, from different manufacturers are available for use with transhumeral amputations. Most provide similar functions: elbow flexion with locking in a number of positions of flexion, and internal/external rotation of the forearm as a substitute for rotation of the humerus. Locking of the elbow unit can be achieved either by a body powered cable or the contralateral hand. The space required to accommodate these mechanisms is typically from 8 cm for the larger sizes down to 4 cm for a child's friction elbow. Where space is limited or non-existent as in the case of the elbow disarticulation amputation, external elbow locking hinges must be used. These also provide flexion and extension with forearm locking but cannot make any provision for internal/external rotation of the forearm.

Externally powered elbow mechanisms are available for transhumeral amputations, providing electrically powered elbow locking with body powered elbow flexion or electrically powered forearm flexion/extension which lock automatically when the power is switched off.

Shoulder units

The use of a shoulder unit is not essential in the design of shoulder disarticulation and forequarter prostheses. The humeral section can simply be manufactured with the shoulder section in one piece. Most commercially available shoulder joints are simple devices providing passive movement under friction which allow limited movement to assist with dressing and positioning the prosthesis. Where a locking joint is required these tend to be manufactured on a custom basis, although a commercially produced unit from the MICA Corp., Longview, California, USA, has recently become available.

Amputation levels

Partial hand amputation

The restoration of the physical appearance following amputation of part of the hand or even part of a finger will be of no less importance than it is at a higher level. Cosmetic restoration can be achieved by supplying a PVC prosthesis selected to match the appearance of the hand from a range of standard sizes and colours. Alternatively, custom-made prostheses in silicon rubber (Pillet, 1981) can be obtained which provide extremely good matching. These are expensive to produce and require a high level of expertise in their manufacture. With most partial hand amputations it is very difficult to provide functional prehension devices with good cosmetic restoration. For this reason functional partial hand prostheses are often devices designed and worn for specific tasks (Fig. 40.3) and may be prescribed along with a prosthesis providing cosmetic restoration.

Wrist disarticulation

Amputation at the level of the wrist preserves pronation and supination and provides a stump capable of good force transmission. The distal shape may allow the use of a self suspending socket, although additional suspension straps may be required for heavy work. Problems usually arise where a wrist unit is provided to allow interchange

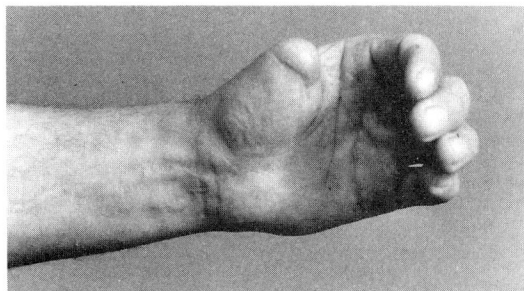

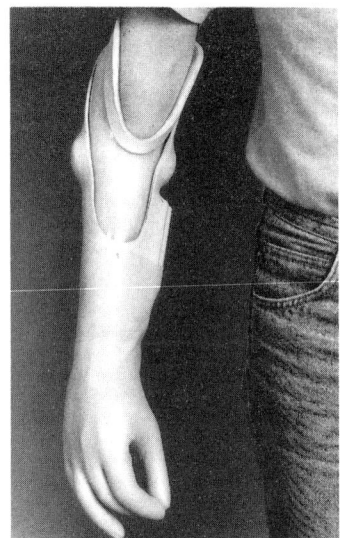

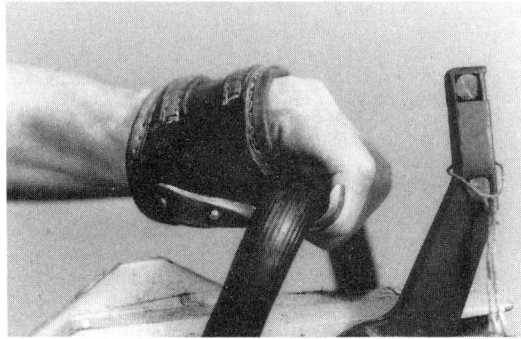

Fig. 40.3. A functional partial hand prosthesis.

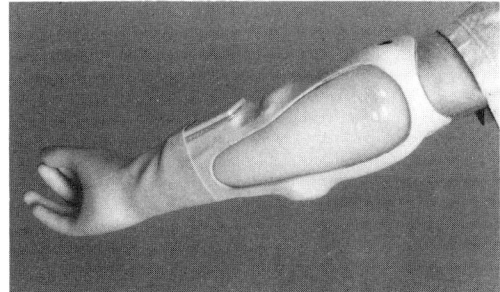

Fig. 40.4. Transradial (flexible socket with frame) self suspending myoelectric prosthesis.

of the terminal device. As with the case of most disarticulation amputation levels, it is not possible to place components distal to the amputation site and still maintain correct limb lengths. A typical adult size body powered hand or split hook and quick disconnect wrist unit requires 18 cm of space. A long prosthesis may be acceptable to the patient while functional terminal devices are in use, but will be very noticeable when a prosthetic hand is used. Where interchange of the terminal device is not required this problem may be overcome or at least improved by omitting the wrist unit and attaching the terminal device directly to the prosthetic socket.

Transradial

The preservation of even the shortest segment of the forearm and elbow joint gives the transradial amputee a distinct advantage over higher levels and with the added advantage of allowing the use of self suspending sockets. Recent developments in socket design and materials have made it possible to fit self suspending prostheses to almost

all transradial amputations. This gives considerable advantages to the amputee, particularly when used with externally powered (Fig. 40.4) terminal devices. The resulting self-contained prosthesis is easy to doff and don, less restrictive and more comfortable, due to the absence of harness and control straps. Even when used with body powered components the transradial self suspending prosthesis offers considerable advantage as harness straps are not required and control straps can be reduced to an absolute minimum (Fig 40.5). It is the opinion of the author that the self suspending socket should be the socket of choice for the transradial amputee and conventional harness suspension only considered where there are contra-indications. These may include immediate post-operative and early fittings, where activities dictate

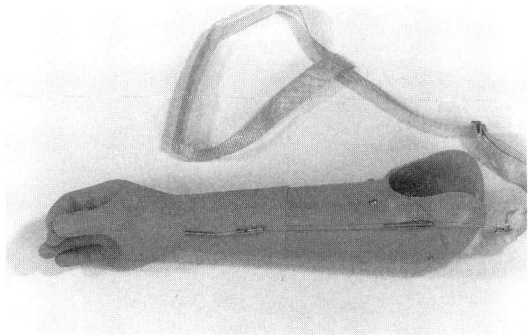

Fig. 40.5. Body powered transradial self suspending prosthesis with simple control cable.

the requirement for additional security by suspension straps, the fitting of the young child where suspension straps might be required to prevent the child removing the prosthesis. Where scar tissue or graft is present over the main socket suspension areas, suspension straps may be required, although the use of silicone as a socket liner (Fig. 40.6) can often overcome these problems.

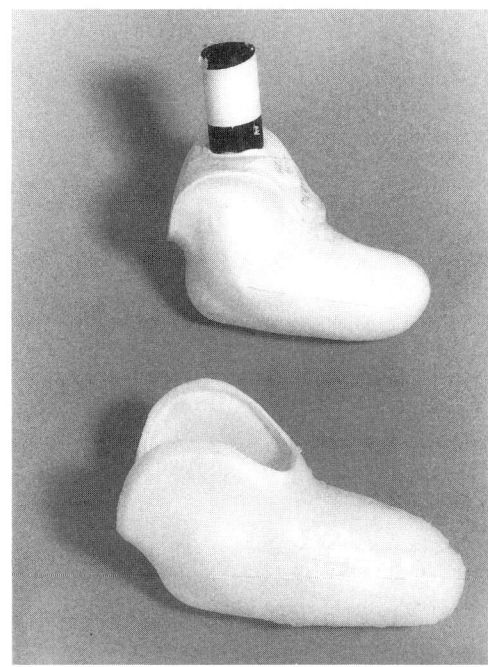

Fig. 40.6. Transradial silicone socket.

Elbow disarticulation

Careful consideration (Hunter *et al.*, 1981; McAuliffe, 1990) should be given to the patient's requirements before a decision to amputate at elbow disarticulation level is taken. While the distal shape of the stump will provide additional suspension and rotational stability over the transhumeral amputation, length limitations will dictate the use of external elbow hinges (Fig. 40.7). Hinges of this type allow only forearm flexion and extension with locking in a fixed number of positions. Internal/external movement of the prosthetic forearm must come from normal humeral rotation. The positioning of the hinges gives a very wide medio-lateral dimension at the elbow and the construction of the hinges is such that wear and tear on clothing tends to be high. The suspension obtained from the distal limb shape will usually be sufficient to retain a lightweight prosthesis, although additional suspension and control straps are likely to be required where a more functional prosthesis is fitted. It can be argued that in this situation the patient would benefit from a transhumeral amputation and the ability to use a more sophisticated elbow mechanism.

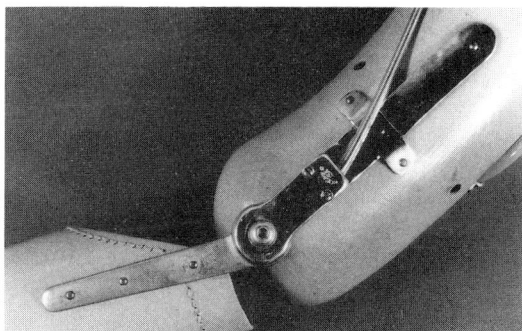

Fig. 40.7. Elbow disarticulation prosthesis with external joints.

Transhumeral amputation

Amputation through the distal third of the humerus will provide a stump length capable of good force transmission and yet still allow the use of production elbow mechanisms. At this level the use of body power to activate both elbow flexion and terminal device is usually possible but as stump length decreases body powered operation becomes increasingly more difficult and at proximal third level especially so. Limitations are due to the reduced cable excursion available and the low mechanical efficiency of the cable operation compared to the transradial system. Many transhumeral amputees, even those with amputations at the distal third level, will find operation of body powered prosthetic hands difficult and will resort to either the use of a cosmetic hand and perhaps a split hook, or look to external power (Sears *et al.*, 1989) to provide increased prehension. Usually there is enough force and excursion available to provide elbow flexion. Simple step-up systems (Fig. 40.8) can be used where excursion is limited and the provision of an electric hand or gripper gives much increased prehension forces. The increased distal weight of externally powered terminal devices does of course increase the forces required to flex the elbow. This situation can be alleviated by using the excellent forearm lift assist device available for use on Hosmer and US Manufacturing elbow mechanisms. This device, which is attached to the side of the elbow unit, allows the weight of the forearm and terminal device to be balanced out by spring tension. Where the patient prefers to retain body powered terminal device operation and the associated feedback, an electrically powered elbow mechanism can be used.

When considering the use of an electric elbow unit, cost, weight, speed of operation and maintenance need to be taken into account. Two basic types of above elbow socket are used for transhumeral amputation. The over the shoulder type where the proximal socket brim extends over the shoulder requires suspension forces to be taken directly through the socket, reducing harness forces. This type of socket will necessarily limit the amount of humeral abduction available and may not be suitable for some amputees. The second type has the proximal brim trimmed below the level of the acromion and relies solely on harness straps to carry the suspension forces. This type will allow a much fuller range of humeral abduction.

Self suspending sockets for transhumeral amputation, while possible, exhibit a number of problems, especially in donning and doffing using one hand and are not yet in common use.

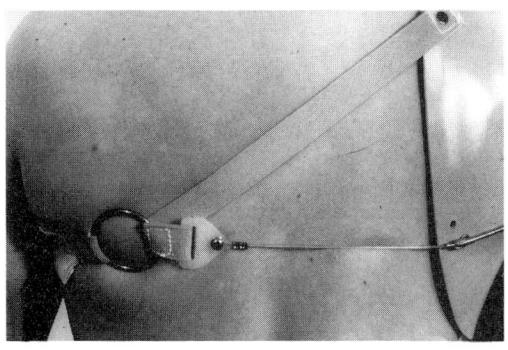

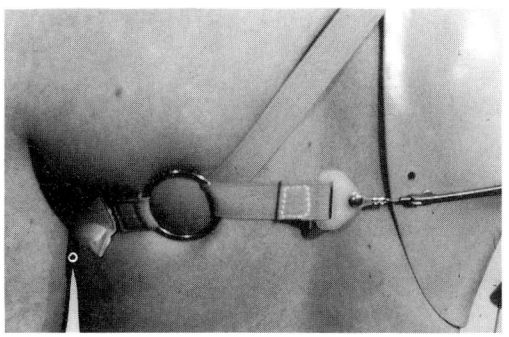

Fig. 40.8. Shoulder disarticulation prosthesis with simple forearm flexion step-up system.

Shoulder disarticulation amputation

The problems relating to body powered control of the elbow locking, forearm flexion and terminal device operation, already mentioned when considering the proximal third transhumeral amputation, also apply to the shoulder disarticulation. Similar component solutions can be applied. The amputee's expectations and requirements should be considered carefully when deciding on a prosthetic prescription. At this level motivation will feature strongly in the acceptance of a functional prosthesis. Successful fitting of the less motivated amputee will often be accomplished by prescribing a less sophisticated prosthesis with the emphasis on cosmetic restoration.

Forequarter amputation

At this level the provision of body powered operation is virtually impossible and where a functional prosthesis is required the use of externally powered components is essential. The resulting prosthesis will be heavy and still only of limited functional use and once again a high level of amputee motivation will be required for successful fitting. Many amputees will prefer a simple lightweight prosthesis with the main emphasis on restoring the profile of the amputated shoulder and cosmetic replacement. Indeed in some cases restoration of shoulder profile is the only requirement and a prosthetic shoulder cap (Fig. 40.9) ending at axilla level is sufficient.

Fig. 40.9. Forequarter amputation. Forequarter shoulder cap and forequarter endo-skeletal prosthesis.

References

Beasley, R.W. (1981) General Considerations in Managing Upper Limb Amputations. *Orthop. Clin. N. Am.*, 12(4).

Billock, J.N. (1986) Upper limb prosthetic terminal devices: hands versus hooks. *Clin. Prosthet. Orthop.*, 10 (No. 2),

Hunter, G.A., Kennard, A.B., Burt, W.F. *et al.* (1981) The upper limb amputee – experience of the Workers Compensation Board. *Amputation Surgery and Rehabilitation: the Toronto Experience*, pp. 161–166.

McAuliffe, J.A. (1990) Elbow disarticulation and transhumeral amputation: surgical principles. In: *Atlas of Limb Prosthetics*, 2nd Edition (eds J.H. Bowker and J.W. Michael). Mosby, pp. 251–253.

Pillet, J. (1981) The aesthetic hand prosthesis: *Orthop. Clin. N. Am.*, **12**, 961–970.

Radocy, B. (1990) Upper-limb prosthetic adaptations for sports and recreation. In: *Atlas of Limb Prosthetics*, 2nd Edition (eds J.H. Bowker and J.W. Michael). Mosby, pp. 325–344.

Sears, H.H. *et al.* (1989) Experience with the Utah arm, hand and terminal device. In: *Comprehensive Management of the Upper Limb Amputee* (eds D.J. Aitkens and R.H.H. Meier), pp. 194–210.

Sources

Otto Bock Orthopaedic (UK) Ltd.,
32 Parsonage Road,
Engelfield Green,
Egham TW20 0JW.

Hosmer Dorrance Corporation
561 Division Street,
P.O. Box 37,
Campbel, CA 95008

Hugh Steeper Ltd.,
Roehampton Disability Centre,
Queen Mary's University Hospital,
Roehampton Lane,
Roehampton,
London SW15 5PL.

VASI Inc.,
3701 Danforth Ave.,
Scarborough,
Ontario,
Canada M1N 2G2.

41

Prosthetic training

J. E. Edelstein

Introduction

Using an upper-limb prosthesis requires that the prosthesis fits properly and that the patient is able to don it quickly, control all mechanisms and incorporate the prosthesis in practical activities. In addition, the patient should be able to care for the skin, socks and prosthesis. Training should be conducted by a therapist who knows how the prosthesis functions and understands the patient from both psychological and physiological aspects and can thereby establish and maintain rapport. The programme should be carefully organized and specific to the individual so that there is constant challenge but not to a level that the patient is overwhelmed.

The training programme described here is designed for adults who have sustained a traumatic transradial or transhumeral amputation. The basic programme will require to be modified for children and those with bilateral or multiple amputations.

The key elements of training are:

(1) Orientation.
(2) Donning the prosthesis.
(3) Care of the amputation limb and the prosthesis.
(4) Control training.
(5) Daily activities training.
(6) Vocational and avocational training.

Orientation

Unlike lower-limb amputation which is often the culmination of years of peripheral vascular disease, amputation of the arm is usually sudden and the result of trauma. Consequently, the typical patient has little or no opportunity to become accustomed to the functional and cosmetic changes which amputation presents. Because upper-limb amputation is much less common, the patient usually lacks a role model. An ongoing orientation programme is necessary so that the individual and family can learn about the reality of amputation. Otherwise, the amputee may harbour unrealistically high expectations or, conversely, may be so discouraged as to forego any attempt to use a prosthesis. Key elements in orientation include:

(a) *Purpose of the prosthesis*. Although advances in technology may lead one to assume that the modern prosthesis has biological characteristics, the fact is that all upper-limb prostheses are essentially tools. The purpose of the prosthesis is to assist the remaining upper limb in the performance of activities. In addition, a prosthesis which has a hand restores some degree of cosmetic symmetry. The kinematic function of any prosthesis available today is far more limited than that of its anatomical counterpart. No prosthesis provides any tactile sensation. The extent of proprioception is limited to an appreciation of the force exerted on the harness if the terminal device is cable-controlled, or the extent of muscular contraction with a myoelectric terminal device.

(b) *Terminology.* By learning the technical names for the basic parts of the prosthesis, the patient gains insight into the operation of the device. Most people have never seen an upper-limb prosthesis and have no understanding as to how it works. The process of becoming familiar with the major components of the prosthesis also improves communication with all other members of the clinic team.

(c) *Training plan.* The patient should be aware that the therapist will retain the prosthesis until it can be donned and control of all components is mastered. By limiting access to the prosthesis for the first 10 or so days, the therapist can ensure that he or she does not develop contrived and awkward manoeuvres to achieve prosthetic activation. Once the patient can control the prosthesis reliably, the individual should be encouraged to wear the prosthesis full time and explore opportunities to use it.

At best training occurs on a daily basis and combines instruction, practice and rest. Most individuals fitted with a body-powered transradial prosthesis become proficient with donning and basic controls within 5 hours of instruction. Myoelectric prostheses typically require slightly more training time. Those with a transhumeral prosthesis typically require 10 hours of training to be able to perform daily activities. Vocational and avocational training times vary with the complexity of the movements required.

Donning

Unless the prosthesis can be donned independently and rapidly, it is most unlikely that the patient will use it.

(a) The body-powered (cable operated) prosthesis. The patient should wear a sock to shield the amputation stump from pressure from the edges of the prosthetic socket and to absorb perspiration. In addition, a soft cotton undershirt helps to distribute pressure from the

harness. The therapist should teach the patient to inspect the prosthesis to ascertain that all cables are secure in their retainers. If the harness is a figure of eight or figure of nine design, all straps should be buckled and should lie flat prior to donning. A prosthesis with a chest strap harness must have the chest strap unbuckled before the client dons it.

To apply the transradial prosthesis with figure of eight or figure of nine harness, the amputation stump is inserted into the socket, and the intact arm is slipped through the axillary loop of the harness. A transradial prosthesis with chest strap requires donning the socket first, leaning to the contralateral side so that the chest strap hangs vertically, then grasping the strap with the sound hand and securing it to its buckle. Some individuals hold the socket in place with the chin while reaching for the chest strap; others lean against a wall to prevent the socket from slipping while they reach and grasp for the chest strap.

The easiest way to don a transhumeral prosthesis with a figure of eight harness is to place the prosthesis on a table in an inverted 'U' pattern; with the terminal device near the patient, the socket opening is upward and the harness is untwisted with the axillary loop also near the client. The elbow unit is locked in partial flexion. The sound hand is placed in the axillary loop and grasps the terminal device; then the stump is eased into the socket. Donning a transhumeral prosthesis with chest strap is similar to the technique suggested for transradial prostheses.

(b) Myoelectrically controlled transradial prostheses are worn without any sock on the stump. If the prosthesis has a harness then an undershirt should be worn to protect the contralateral axilla. The patient should determine that all components of the prosthesis are in the proper place before donning.

To apply the prosthesis the stump is first encased in a stockinette tube, making certain that the proximal end of the tube is at the olecranon and the distal end of the tube is slightly longer than the amputation limb. Then the patient introduces the stump into the socket and pulls the stockinette completely out through the hole in the socket to draw superficial tissue into the socket. When properly oriented, the socket covers the amputation limb such that the embedded electrodes are in the optimum location.

Care of the stump and the prosthesis

(a) Stump care should be stressed from the first training session. Cleanliness is essential to prevent dermatitis and odour. The skin should be completely dry before the prosthesis is donned. A fresh sock should be used daily.

(b) Prosthesis care is an integral part of training. The patient should inspect the prosthesis daily to detect any problems, such as cable fraying or loose stitching and report these problems to the prosthetist promptly to avoid breakdown of the device. The socket should be wiped nightly with a damp cloth and allowed plenty of time to dry. Harnesses are usually made to be detached so they can be washed. The prosthetist or therapist may mark the buckle position so that the harness can be restored to its optimum length when it is reattached to the prosthesis. A list of instructions for preventing damage to all parts of the prosthesis, especially the cosmetic glove, is useful.

Controls training

The transradial prosthesis has only two components which require the wearer's control, namely the terminal device and the wrist unit. The transhumeral prosthesis also incorporates an elbow unit and a turntable enabling passive forearm rotation.

(a) A body powered terminal device, whether hand or hook, is activated by tension on the control attachment strap which passes behind the shoulder. Reaching forward by shoulder flexion applies tension to the cable which is attached to the strap. The cable transmits the tension to the terminal device. In the case of a voluntary-opening device, cable tension causes the device to open; relaxing tension allows the device to close. With a voluntary-closing terminal device, cable tension causes the device to close; relaxing tension allows the device to open. The patient should practice opening and closing first with the arm adducted and the elbow flexed 90°, the optimum position for the cable. Then practice should involve positioning the shoulder and elbow in various angles.

Once the patient understands how to operate the terminal device with shoulder flexion, the therapist should guide the individual into stabilizing the shoulder girdle without flexing the glenohumeral joint. Such a manoeuvre is important for controlling the terminal device in activities close to the body, such as buttoning a shirt.

A myoelectric terminal device is activated by muscular contraction at the appropriate electrode site. Prior to prosthetic fitting, the patient should practise isolated contraction of forearm flexors and extensors, with electrode(s) in place, while striving for a forceful reading on a voltmeter from one muscle group and relaxing the other group. With the complete prosthesis, practising muscle contraction for voluntary opening and closing continues and thereafter involves controlling the terminal device with the prosthesis overhead, to the side and extended.

(b) The wrist unit is usually controlled passively without cable or electric motor control. A form board is a very helpful training aid. The board consists of geometric objects of different sizes, shapes and textures. The patient practises positioning the wrist unit so as to grasp each object in the best possible manner.

(c) Elbow unit control should be taught after the patient has acquired a reasonable level of skill with the terminal device and wrist unit. Many practical activities can be accomplished without changing the elbow position; moreover, cable control of the elbow unit is usually more difficult for the patient to master. Elbow flexion is usually controlled by the same cable which controls the terminal device, hence the same shoulder motions.

Terminal device operation usually requires that the elbow be locked. Locking control should be taught in three stages. First, the patient should trigger the elbow lock with the prosthesis on the table, to appreciate the alternating mechanism. Next, while wearing the prosthesis, locking and unlocking is practised by a diagonal movement composed of shoulder girdle depression and shoulder extension, with the therapist supporting the prosthetic forearm. Finally, the patient learns to integrate elbow flexion by a combination of shoulder flexion and elbow locking via shoulder extension; thus locking must be done quickly to prevent the prosthetic forearm from dropping.

Daily activities training

Independence in dressing, grooming, eating and writing is fundamental. Initially, the therapist can suggest means of simplifying activities, such as choosing clothing which is easy to don and foods which can be managed readily with only a spoon or fork. Most adults with unilateral amputation learn to accomplish the full range of daily activities with little or no adaptation.

(a) Unimanual activities are ordinarily performed by the sound hand. For the individual who has lost the dominant hand, preliminary practice in change of dominance is desirable. The sound hand retains tactile and proprioceptive sensation and allows many more prehension patterns than does the prosthesis and in most instances, the sound hand also affords more forceful grasp.
(b) Bimanual activities require the use of the prosthesis to aid the sound hand. The patient should be guided to analyse a task to determine which element is relatively stationary or requires less force. For example, cutting meat with a knife and fork is most easily done with the fork held in the terminal device. After cutting, the patient may decide to transfer the fork to the sound hand and offer food to the mouth.

Vocational and avocational training

The usual hospital or rehabilitation centre does not have the facilities or time to retrain the client in specific vocational tasks. Nevertheless, return to work is a powerful motivation to persist with prosthetic training and wear. The therapist should explore the manual skills needed in any particular job and should be able to suggest ways of satisfying the requirements. A clerical worker, for example, may find that using a keyboard with the sound hand is preferable to using the prosthesis and the sound hand. Taking a telephone message is best performed if the prosthesis holds the receiver while the sound hand is used to write. Some individuals with amputation of the dominant hand find that writing with the prosthesis is easier than switching to the opposite hand.

Recreational pursuits should also be analysed to determine what modifications may be advantageous.

Conclusion

A carefully planned training programme designed to meet individual needs often means the difference between the individual's incorporating the prosthesis as an essential part of one's being or otherwise finding that the prosthesis is a nuisance which is best relegated to the closet. While a device cannot provide all the attributes of the missing arm, the goal of training nevertheless should help to determine how to make the most effective, efficient and meaningful use of the prosthesis.

Section 17

ISO/ISPO classification

42

The ISO/ISPO classification of congenital limb deficiency

H. J. B. Day

A logical system of classification and nomenclature is needed to facilitate scientific communication about congenital limb deficiency. The lack of a suitable system has allowed the use of the term 'congenital amputation', implying that a limb segment has been lost before birth, to be used for cases which are patently failures of formation. Furthermore, any classification should use simple words capable of translation into all languages. The use of terms derived from Greek or Latin roots may sound impressively scientific, but are both inaccurate and ambiguous, and are often misused, none more frequently than 'phocomelia' which is used to describe every level and type of deficiency.

The history of classifications devised since that of Frantz and O'Rahilly (1961), including those of Burtch (1966), Henkel and Willert (1969) and the work of the ISPO 'Kay' committee has been described previously by Kay, Swanson (1974, 1975, 1976) and Day (1988).

The ISPO system provided the framework which enabled the Working Group of ISO Technical committee 168 (Prosthetics and Orthotics) to set out a proposal for an International Standard. This has been accepted by the participating nations and has since been published as an International Standard.

8548–1:1989 'Method of describing limb deficiencies present at birth'

The standard is reproduced here with the permission of the International Organisation for Standardisation (ISO). Copies of this standard are available from the ISO Central Secretariat, Case Postale 56, CH-1211 Geneva 20, or from any ISO member body.

The standard has three constraints:

1. The classification is restricted to skeletal deficiencies and therefore the majority of such cases are due to a failure of formation of parts.
2. The deficiencies are described on anatomical and radiological bases only. No attempt is made to classify in terms of embryology, aetiology or epidemiology.
3. Classically derived terms such as hemimelia, peromelia, etc. are avoided because of their lack of precision and the difficulty of translation into languages which are not related to Greek.

Deficiencies are described as *transverse* and *longitudinal*.

The former resemble an amputation stump, in which the limb has developed normally to a particular level beyond which no skeletal elements are present. All other cases are classed as longitudinal, in which there is a reduction or absence of an element or elements within the long axis of the limb.

Method of description

Transverse

The limb has developed normally to a particular level beyond which no skeletal elements exist, though there may be digital buds. Such deficiencies are described by naming the segment at which the limb terminates, and then describing the level within the segment beyond which no skeletal elements exist (Fig. 42.1).

It is possible to use another descriptor in the phalangeal case to indicate a precise level of loss within the fingers.

Longitudinal

There is a reduction or absence of an element or elements within the long axis of the limb, and in this case there may be normal skeletal elements

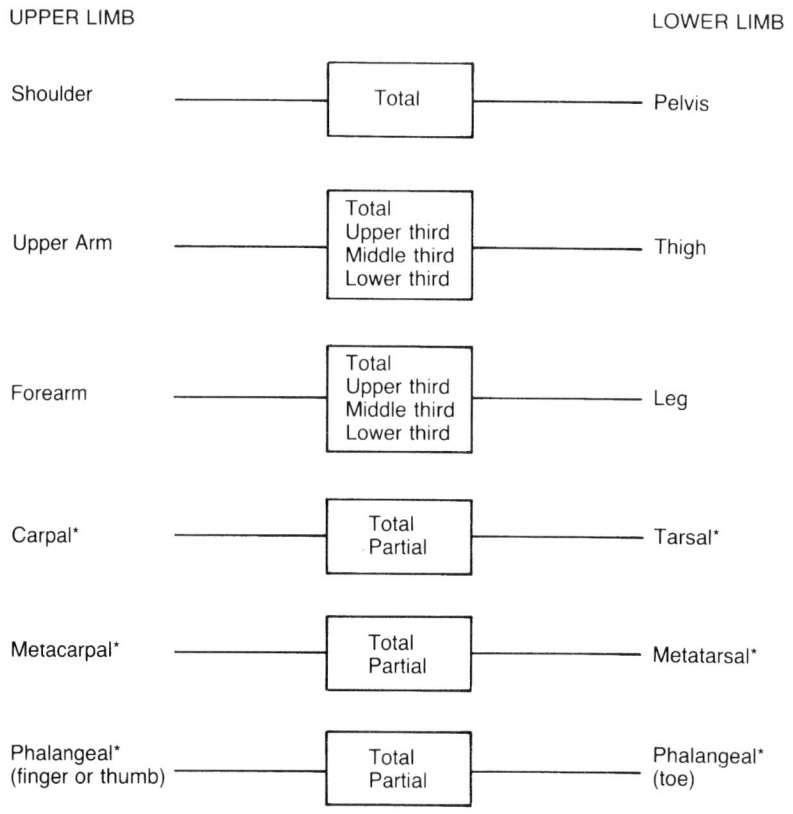

NOTE. – 1.   The skeletal elements marked * are used as adjectives in describing Transverse
deficiences, e.g. Transverse carpal total deficiency.
– 2.   Total absence of the shoulder or hemipelvis (and all distal elements) is a Transverse
deficiency. If only a portion of the shoulder or hemipelvis is absent, the
deficiency is of the Longitudinal type.

Fig. 42.1. Description of levels of transverse deficiencies of upper and lower limbs.

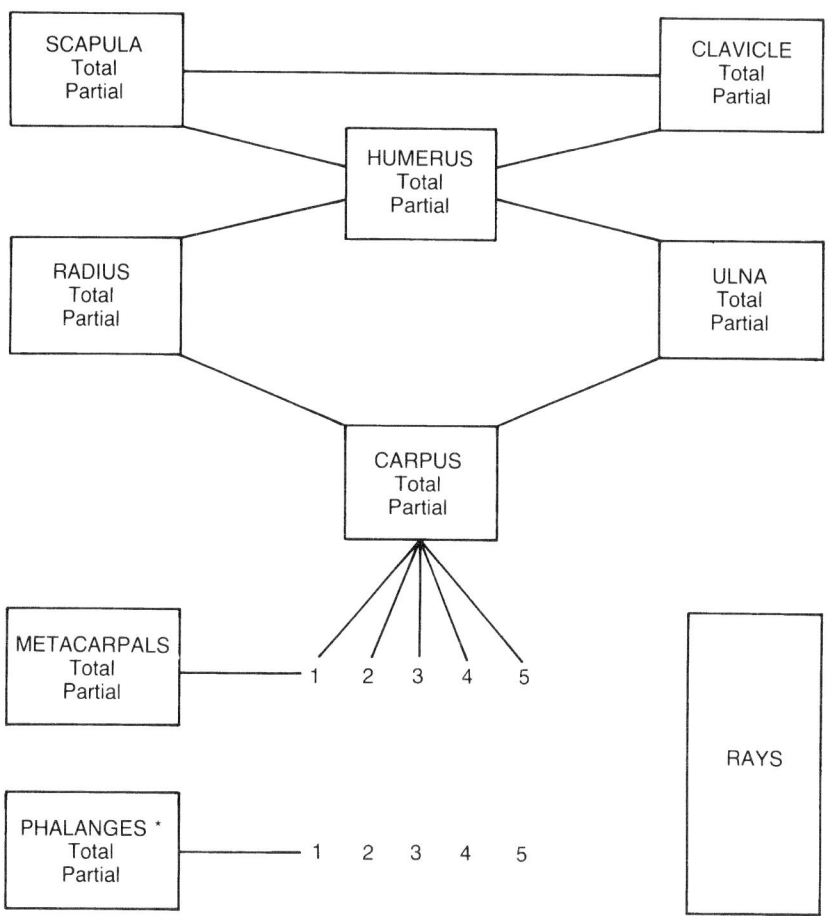

* The digits of the hand are sometimes referred to by name:- 1 = Thumb, 2 = Index, 3 = Middle, 4 = Ring, and 5 = Little (or small). For the purpose of this classification such naming is deprecated because it is not equally applicable to the foot.

Fig. 42.2. Description of longitudinal deficiencies of the upper limb.

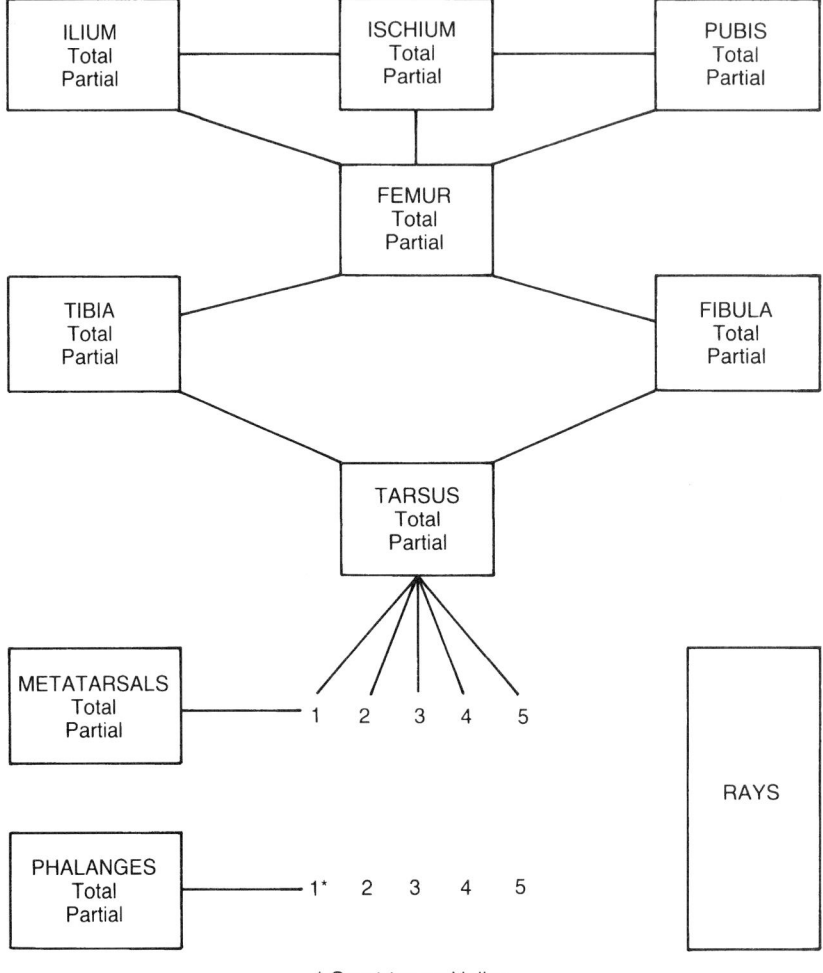

Fig. 42.3. Description of longitudinal deficiencies of the lower limb.

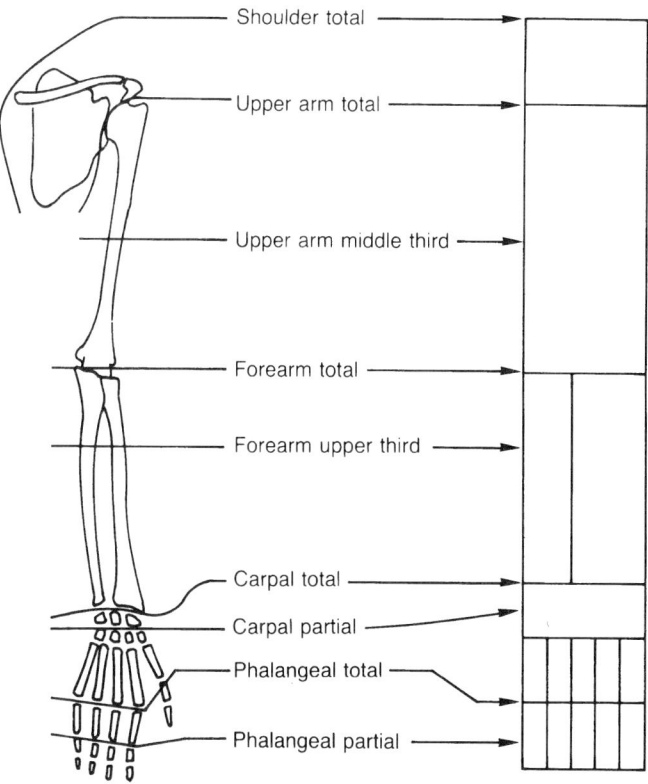

Fig. 42.4. Examples of transverse deficiencies at various levels, shown on the skeleton and as the author's stylized representation.

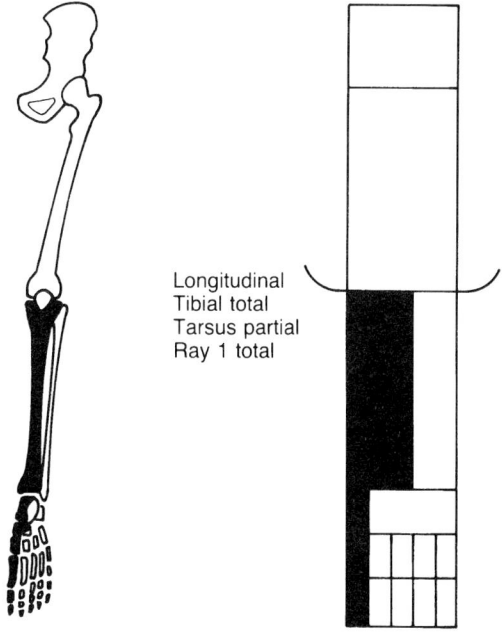

Longitudinal
Tibial total
Tarsus partial
Ray 1 total

Fig. 42.5. Example of a longitudinal deficiency shown on the skeleton and as the author's stylised representation, showing not only the original deficiency but also the treatment by knee disarticulation.

distal to the affected bone or bones. To describe such a deficiency (Figs 42.2 and 42.3):

1. Name the bones affected, in a proximo-distal sequence, using the name as a noun. Any bone not named is present and of normal form.
2. State whether each affected bone is totally or partially absent.
3. In the case of partial deficiencies the approximate fraction and the position of the absent part may be stated.
4. The number of the digit should be stated in relation to a metacarpal, metatarsal and phalanges, the numbering starting from the pre-axial, radial or tibial side.
5. The term 'ray' may be used to refer to a metacarpal or metatarsal and its corresponding phalanges.

Examples of transverse and longitudinal deficiencies are shown in Figs. 42.4 and 42.5, but it must be understood that the stylized representation of the limb which is used in these figures is neither part of the original ISPO 'Kay' committee work nor of the new International Standard. Nevertheless, the author has found it to be the most useful way of illustrating deficiencies in clinical notes and it can be used to indicate surgical treatment as well as the deficiency.

References

Burtch, R.L. (1966) Nomenclature for congenital skeletal limb deficiencies, a revision of the Frantz and O'Rahilly R. classification. *Artif. Limbs*, **10(1),** 24–25.

Day, H.J.B. (1988) Nomenclature and classification in congenital limb deficiency. In: *Amputation Surgery and Lower Limb Prosthetics* (ed G. Murdoch), Blackwell, Edinburgh, pp. 271–278.

Frantz, C.H. and O'Rahilly, R. (1961) Congenital skeletal limb deficiencies. *J. Bone Joint Surg.*, **43A,** 1202–1024.

Henkel, H.L. and Willert, H.G. (1969) Dysmelia, a classification and a pattern of malformation of congenital limb deficiencies. *J. Bone Joint Surg.*, **51B,** 399–414.

Kay H.W. (1974) A proposed international terminology for the classification of congenital limb deficiencies. *Inter-Clin. Inform. Bull.*, **13/7,** 1–16.

Kay, H.W. (1975) *The Proposed International Terminology for the Classification of Congenital Limb Deficiencies; the Recommendations of a Working Group of ISPO.* Spastics Int. Med. publications with W. Heinemann Medical Books Ltd., and J.P. Lippincott Co.

Swanson, A.B. (1976) A classification for congenital limb malformations. *J. Hand Surg.*, **1(1),** 8–22.

Section 18

Rehabilitation

43

Principles and philosophy

W. H. Eisma

Introduction

The Council of Europe (1989), founded in 1949, was the first European political organization under whose auspices cooperation on rehabilitation and resettlement of the disabled to achieve the aim of promoting the social integration of disabled people.

In the Netherlands, the USA and in many other countries, rehabilitation medicine is a medical specialty involved in the prevention, diagnosis, treatment and rehabilitation of patients with defects and physical incapacities, caused by illness, accidents or deformities present at birth.

The aim is to bring the patient to an optimum of physical, mental, emotional, social, vocational and economic efficiency.

The practice of rehabilitation requires that the responsibility of the physician does not end when medical treatment of the illness and any surgical efforts have met with their immediate and restricted objectives, but continues until the patient has been retrained for life with the more or less restricted possibilities that are left.

Rehabilitation endorses that part of medicine, which aims at early detection, diagnosis and treatment of impairments of organic nature and the related disabilities and handicaps.

International classification

The International Classification of Impairments, Disabilities and Handicaps of the WHO (1980) gives some definitions about impairment, disabil-ity and handicap. Trauma, illness and deformities at birth may lead to:

Impairment which is any loss or abnormality of psychological, physiological or anatomical struc-ture in terms of function. An impairment may be temporary or permanent.

Disability means any restriction or lack of ability to perform an activity considered normal for that human being the result of an impairment.

Handicap means disadvantage for a given individual, resulting from an impairment or disa-bility, that limits or prevents the fulfilment of a role that is normal for that individual. Handicap therefore represents the social and environmental consequences of impairments and disabilities.

One could say that the successive factors are:

disease – impairment – disability – handicap

The purpose of rehabilitation is to try and influence and if possible prevent the progress of these successive factors. Trauma and disease may result in spontaneous or man-induced recovery or death. Recovery may be incomplete and leave residual impairments, and if these interfere with the desired level of functional ability, some degree of disabil-ity is present.

Rehabilitation medicine

The methods of rehabilitation medicine are the same as in general medicine, namely objective observations, logically arranging the relative data, leading to the conclusion and evaluation of the treatment activities. However, there is an important difference between this way of thinking and that of more 'curative' centred medicine where in the first place more stress is laid on anatomical and physiological restoration, while in rehabilitation medicine the ultimate aim goes much further: the integration of man with his handicap in society.

The activities in rehabilitation medicine regarding diagnosis and prognosis involve not only examination of impairments but also determination of disabilities and handicaps. In practice, each of the classification dimensions can be assessed in the course of the rehabilitation process.

Impairment

Physical impairment may be assessed by measuring mobility, muscular strength, sensation, coordination, control of posture and movement, physical condition, circulatory function, respiratory function, bladder and rectal functionand also sexual function.

Consciousness, memory, intelligence and behaviour are criteria for assessing psychological impairments, while speech, language, hearing and sight are the measures of communication.

Measurement of disability focuses on everyday activities. The WHO Classification of disabilities appears to go somewhat further than the concept of disability itself and beyond scales measuring everyday activities such as those used by Katz, Barthel, Frenchay and the numerous dependence grids and scales used in rehabilitation medicine, public health and geriatrics.

Disability activities

The item list of disability activities consists of five categories:

- *Physical abilities*, such as walking, climbing, stairs, reading, lifting, stooping, manipulating, etc.
- *Activities of daily living*, such as eating, drinking, washing, bathing, dressing, etc.
- *Social activities*, such as transportation, job, family role, daily pursuits, hobbies, social contacts outside the family, etc.

- *Psychological abilities*, such as orientation, memory, concentration, behaviour, motivation, etc.
- *Communicative abilities*, such as understanding spoken language, speaking, hearing, seeing and writing.

Diagnosis

Eliciting the history and the examination of patients has to take place following this procedure. By assessing the patient against the background of these groupings, a rehabilitation diagnosis can be made. This is the first stage, problem analysis and orientation.

A trained eye can detect disturbances, but not precisely and not with reproducible accuracy. A clinician needs a reproducible and repetitive evaluation of motor activity. This may embrace sophisticated techniques such as electromyography and gait analysis and the measurement of disabilities in a realistic setting. It would be an enormous advantage if details of impairment and disability could be identified by movement analysis. Movement analysis techniques seem capable of assessing impairment, but not disability (Eisma, 1990). Movement analysis may be used in the assessment of the degree of rehabilitation of the patient and the rate of improvement to identify levels in response to treatment.

The results of motor analysis have to be presented in such a way that the clinician who is well acquainted with the techniques employed can readily understand the data and diagrams. Such evaluations are necessary before and after treatment and during follow-up.

Rehabilitation demands a methodical approach in which a system is built up in a logical, coordinated and purposeful sequence in order to reach the desired result.

Team approach

The treatment plan is conducted by a team which may include a surgeon, rehabilitation specialist, physiotherapist, occupational therapist, prosthetist and a psychologist. It seems logical that the whole procedure of, say, surgery and rehabilitation can be divided into a pre-and post-operative phase: that in both cases the team will vary dependent to which department the patient is admitted but it must be emphasized that in both cases the team needs one leader and that is usually a medical doctor.

However, some use a non-medical coordinator to good advantage (Ham *et al.*, 1987; Marks, 1987; ISPO Consensus Report, 1992).

The organization also depends on the country and related health care system. The ideal form of prosthetic training can be carried out in a limb fitting centre or in a rehabilitation centre either as an in-patient or on an out-patient basis.

The advantages are supervision by experts, the possibility for rapid and expert replacement of the prosthesis, intensive and frequent physiotherapy, occupational therapy and if necessary investigation and support for psycho-social problems.

Pre-operative management

Pre-operative management depends on the co-operation of the relevant specialists as a team and the special knowledge that the members of the team have at their disposal. If there is sufficient time prior to surgery the following assessment can be made and appropriate plans can be devised.

Important points also to be addressed are:

● The evaluation of the locomotor apparatus.
● The evaluation of the psychological condition of the amputee.
● To inform the patient on factors of aetiology.
● To locate the level of the amputation.
● To inform the patient of the kind of prosthesis that will be provided.
● To inform the patient on the expectations of the function of the prosthesis.
● To inform the patient when the prosthesis will be fitted.

Post-operative management

The aims of post-operative management are to:

● Provide for sound wound healing of the amputation stump.
● Maintain or improve the optimal condition of the locomotor apparatus of the patient.
● Train for independence in the activities of daily living (ADL).
● Support the patient in the event of any psycho-social problems.
● Prosthetic supply.

The integration of the amputee in the surroundings during the post-operative period is very important, not only from a human point of view but also in order to achieve an optimal rehabilitation process. The aim and goal of rehabilitation is to assist the patient in using his residual faculty to allow him to function in a way that is worthy of a human being and this aim can be achieved only by an integrated approach.

References

Council of Europe (1989) *The Use of the International Classification of Impairments, Disabilities and Handicaps (ICIDH) in Rehabilitation.* Publications and Documents Division, Strasbourg

Eisma, W.H. (1990) Rehabilitation medicine criteria for movement sciences. In *Proceedings of the Workshop CAMARC: Present and Future (Rome 1990)* (eds. V. Macellari and M. Torre), Instituto Superiore di Sanita, Rome, pp. 91–103.

Ham, R., Regan, J.M. and Roberts, V.C. (1987) Evaluation of introducing the team approach to the care of the amputee: the Dulwich study. *Prosthet. Orthot. Int.*, **11**, 25–30.

ISPO Consensus Report (1992) *Report of ISPO Consensus Conference on Amputation Surgery. Copenhagen.* ISBN 87–89809–00–9.

Marks, L. (1987) Lower limb amputees: advantages of the team approach. *Practitioner*, **23**, 1321–1324.

WHO (1980) *International Classification of Impairments. Disabilities and Handicaps: a Manual of Classification Relating to the Consequences of Disease.* World Health Organization, Geneva.

44

Early walking aids

L. J. Marks

The benefits of early mobilization of amputees have been recognized for over 30 years.

Berlemont (1961) first published the technique of immediate post-operative fitting, and the procedure received greater acclaim during the rest of the sixties. The technique involved applying a plaster cast at the end of the operation, to which was attached a prosthetic shin and foot. The procedure required close cooperation between the prosthetist and surgeon. The potential advantages were reported to be increased comfort; absence of pain; more rapid healing plus stump maturation, together with psychological benefits and early weight bearing. Despite Murdoch's (1969) conviction that the technique would become established practice, there was increasing evidence of problems, particularly in vascular amputees (Cohen et al., 1974) and enthusiasm for the technique has waned.

It was already known that the majority of amputations were being performed on elderly patients with peripheral vascular disease (Tunbridge Report, 1972) in whom prolonged immobilization is detrimental to successful rehabilitation. Elderly patients also have difficulty hopping on one leg, and this may be undesirable as the circulation in the remaining limb is likely to be impaired (Dickstein et al., 1982). A readily available, or rapidly fabricated walking aid is therefore required, and early reports of local initiatives (Devas, 1971; Hutton and Rothnie, 1977) showed what could be achieved and the benefits that accrued.

Several different early walking aids are now available and can be classified according to their construction. They vary from pneumatic and vacuum devices to a variety of aids using preformed or bespoke sockets mounted on a modular prosthetic stack.

Pneumatic devices

The Airleg

The Airleg was developed by Frances J. Bonner senior and colleagues, and is available in two forms, one for the transfemoral and the other for the transtibial amputee.

Both models have an inner air sac and an outer quadrilateral fibreglass shell, with a metal shank and prosthetic feet. Inflation is via a compressed air tank or manual or foot operated pump, and pressure is measured with a sphygmomanometer.

The inner air sac can be applied immediately after the operation, over a soft dressing, protecting the stump and controlling oedema. It is inflated to 12 mmHg initially and worn 24 h per day. When the amputee is ready to start weight bearing (usually 48 h post-op) the outer shell and foot are applied, with the pressure in the inner bag being increased to 25–30 mmHg. Because of its shock absorbing properties, weight bearing of greater than 9.0 kg can be achieved, and open or ulcerated stumps can be treated (Bonner and Green, 1982).

The device is light to wear and can be applied by staff with minimal training. It provides total stump contact and constant measurable pressure. The main problems are perspiration in warm weather (treated by using a thin sock over the stump), leakage of the air sac, and no immediate facility for adjusting height. The Airleg is supplied by: M.K. Henderson Industries Inc., P.O. Box 245, Radnor, PA 19087, USA.

The Jobst Air Splint

The Jobst Air Splint consists of a transparent plastic tube, with or without a foot, with a front opening zipper permitting easy application and removal. It is supplied with a foot pump and manometer. It is suitable for use with ankle disarticulation, transtibial and transfemoral amputees.

Like the Airleg, it can be applied immediately post-operatively over a soft dressing, but the variety with a foot causes problems with bed mobility. For use at rest, pressures of 25–30 mmHg are used but for ambulation the pressures should not exceed 40 mmHg (Monga et al., 1985). The same authors found that in the static position, mean weight bearing of 11 kg in transtibial amputees and 10 kg in transfemoral amputees could be achieved. Additionally, they commented that the Air Splint without a foot is more suitable for early ambulation in the parallel bars as deformation occurs later and excessive weight bearing is avoided. Once the amputee progresses to a frame the Air Splint with a foot may be more advantageous, as it allows more weight bearing, but this needs to be offset against the disadvantage of dragging the foot. Rausch and Khalili (1985) apply a metal cone over the splint for walking. The cone allows for increased weight bearing on the residual limb, and possibly aids stability.

The advantages of the Air Splint are primarily that it is light, simple to apply and equal pressure is maintained. However, it is bulky, hot, and suspension is a problem in transfemoral amputees who may require additional suspension. The Air Splint is supplied by: Jobst Institute Inc., Box 654, Toledo, Ohio 43694, USA.

PPAM Aid (Pneumatic Post Amputation Mobility Aid)

Little (1971) described an early walking device using a pneumatic air splint as a socket, supported in a metal frame with a prosthetic foot. This device was modified by the Biomechanical Research and Development Unit, Roehampton (Redhead, 1978) and more recently has undergone further modifications (Ham et al., 1989).

The equipment consists of a small cushion inner bag and larger transtibial and transfemoral bags. The bags are supported in an outer metal frame with a rubber rocker end (Fig. 44.1). The frame comes in three sizes: short, medium and long. The frames have a crucible sling support on their distal end, and there is a shoulder strap for suspension. Finally, there is a pump with a dial measuring pressure in mmHg.

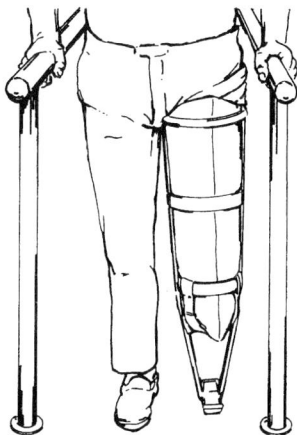

Fig. 44.1. PPAM Aid MK.II (Vessa Limited).

This device is suitable for use with transtibial, through knee, and transfemoral amputees. It can be used as early as 5 days post-operatively for non-vascular cases, but for vascular amputees it is preferable to defer its use till 7 or 10 days post-operatively. It can be applied over a soft dressing, stump support or plaster cast. It should not be used over an uncovered stump, as direct skin contact can cause excessive perspiration.

This device is applied with the patient half lying on a plinth. For transtibial and knee disarticulation amputees, the cushion bag (Fig. 44.2) and transtibial bag (Fig. 44.3) are used. For transfemoral amputees, the cushion bag and the transfemoral bag are used, the latter being distinguished by a collar ring at the proximal end with four vertical air

Fig. 44.2. Inner cushion bag for PPAM Aid (Vessa Limited).

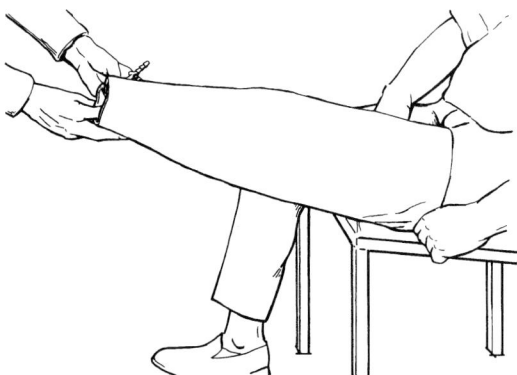

Fig. 44.3. Outer bag of PPAM Aid for transtibial amputation (Vessa Limited).

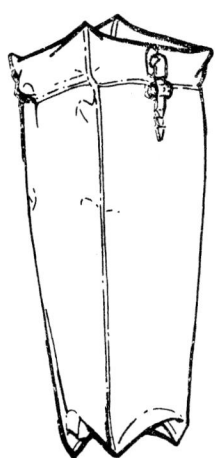

Fig. 44.4. Outer bag for transfemoral bag (Vessa Limited).

pockets (Fig. 44.4). The cushion bag is applied first, followed by the outer bag, and finally the frame, which should lie at least 8 cm below the top of the outer bag (Fig. 44.5).

On the first application the device should be applied for a 5-min trial period, at a pressure lower than the 40 mmHg recommended for walking (usually 15–25 mmHg). The device is then removed and the wound inspected. If there are no stump problems, the duration of use and the pressure are increased, until the patient can tolerate 40 mmHg. Standing exercises, followed by walking in the parallel bars, are then commenced.

It is important to note that the device is designed for use under professional supervision and is not recommended for home use. The stump *must* be inspected prior to, and following each application and any adverse problems reported. The device is designed for *partial weight bearing only* and problems may be encountered with younger, fitter patients who tend to apply more weight through the stump when walking.

Because of its immediate availability it can be used on several different patients during the course of the day, providing simple hygienic measures are employed. 'Minimal staff training' has been quoted as an advantage, but a study by Lein (1992) reported widespread variation in knowledge and use of the equipment.

Correctly used, it can be a useful device, reducing oedema and limiting stump volume change whilst not impairing wound healing (Redhead, 1983). It is supplied by: Vessa Limited, Paper Mill Lane, Alton, Hampshire GU34 2PY, England.

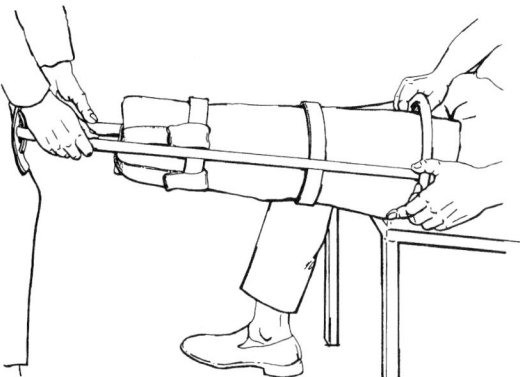

Fig. 44.5. Application of PPAM Aid frame (Vessa Limited).

Saarbrucken early fitting prosthesis

This device consists of a small pneumatic distal end closure bag, a pneumatic thigh sleeve, a distal end plate and a laminated resin frame socket. The lower leg is made up of Otto Bock Modular Components. The kit is supplied with a hand inflated manometer. The device is designed for use with transtibial amputees and knee disarticulation amputees and could possibly be used on long transfemoral stumps. It is presently available in Europe and USA but not the UK.

The device is used 7–10 days post-operatively. The outer frame and lower leg are initially applied over the stump to determine the correct length. They are then removed, and first the distal end closure bag and then the thigh sleeve are applied to the stump. The frame and lower leg are then reapplied and the thigh sleeve inflated to 40 mmHg and the distal end sleeve to approximately

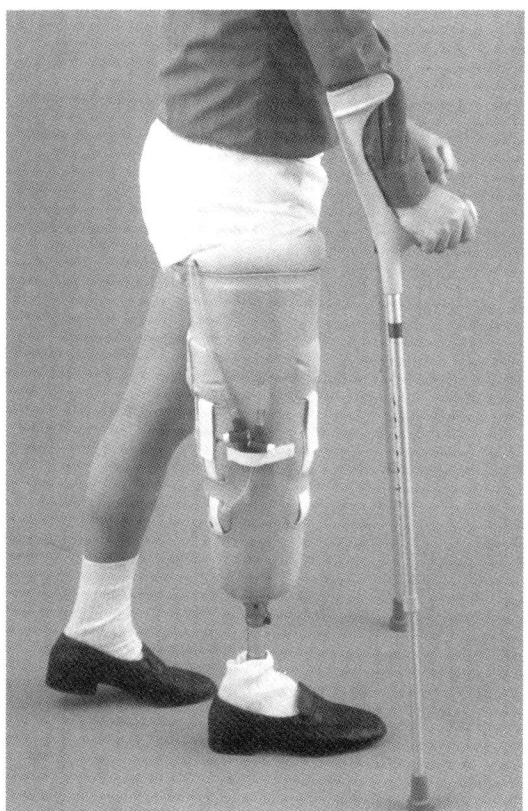

Fig. 44.7. Saarbrucken prosthesis: partial weight bearing only permitted.

20 mmHg. The device is aligned in the standing position (Figs 44.6, 44.7). The system should be used in a similar way to the PPAM aid, under supervision, and for partial weight bearing only. The main disadvantage is that it requires availability of modular prosthetic components and the skills of the prosthetist for optimum alignment. The pneumatic sleeves are also reported to have been unreliable (K. Harney, 1992, personal communication). It is supplied by: Otto Bock Orthopadische Industrie GmbH & Co. Postfach 12–60, D-3408 Duderstadt, Germany.

The Below Knee, Early Assessment Prosthesis (BK EAP)

This device is currently being developed by Rehabilitation Services Ltd., to provide an early walking aid for transtibial (TT) amputees.

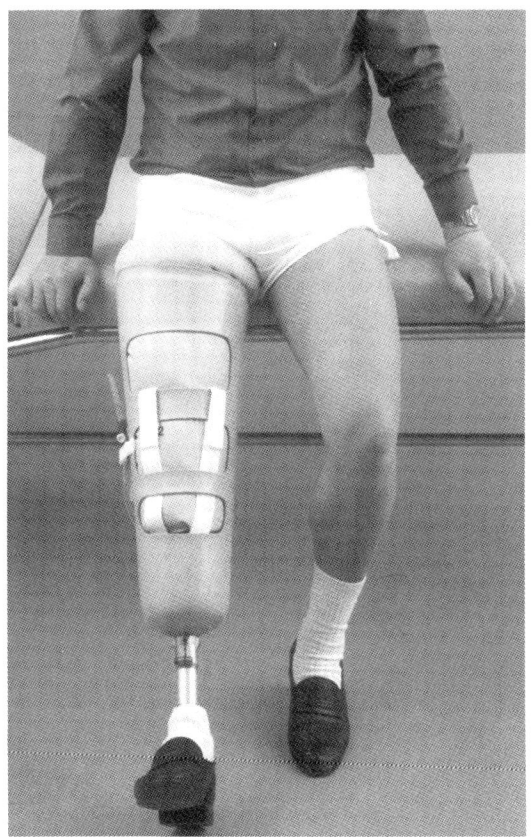

Fig. 44.6. Saarbruckens Early Fitting Prosthesis.

The system comprises a polypropylene outer TT socket inside which is a pneumatic sleeve. The TT socket is hinged to a polypropylene lower thigh frame into which is mounted a polythene thigh corset with rolled proximal edge, adjustable for height and closed with velcro straps. The lower leg is modular in construction with a SACH foot (Fig. 44.8). A pump is not routinely supplied at present, the manufacturers recommending use of the PPAM aid pump. Three sizes are currently being evaluated.

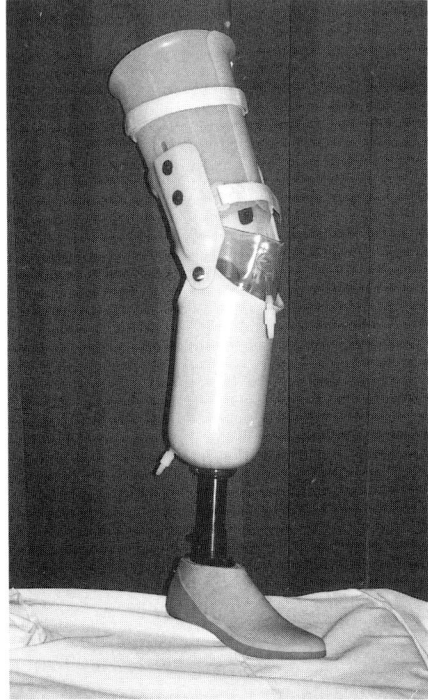

Fig. 44.8. The Below Knee, Early Assessment Prosthesis (BK EAP) permitting knee movement during early joint training.

The device can be used 7–10 days following amputation. Initially the pneumatic sleeve is pumped up to 20 mmHg and the amount of weight bearing through the stump is controlled by the position of the thigh corset. Pressure in the pneumatic sleeve is gradually increased to a maximum of 40 mmHg, and the thigh corset adjusted so that more or all weight is taken on the stump itself.

The main advantages of the system are that it allows the amputee to utilize knee movement during early gait training, and additionally up to 45° of knee flexion can be accommodated.

Stump oedema is also controlled by the pneumatic sleeve (G. Brown, 1992, personal communication). The main disadvantage is that the amputee will need to attend a prosthetic centre, serviced by RSL, for supply of the device. It is supplied by: Rehabilitation Services Ltd., Bunces Farm House, Lower Road, Tonge, Sittingbourne, Kent ME9 9BP.

Vacuum devices

Tulip Limb

The Tulip Limb was developed in Sweden by LIC in 1978 (Henriken *et al.*, 1978). It is suitable for use with transtibial amputees.

The device consists of an inner cloth sheath containing four vacuum packs filled with small plastic pellets in parallel connected sections, with velcro fastenings. The outer 'Tulip' frame is made up of four separate polypropylene sides or petals, joined at the base, and secured by velcro straps. The frame is mounted on an adjustable tube with a single axis 'universal' foot (size 40). A hand operated pump is supplied (Fig. 44.9).

The inner sheath is placed over the carefully bandaged stump, bandaging being vital for comfort when walking with the Tulip unit. The pellets in the vacuum packs are smoothed out by hand, and then the inner sheath is wrapped evenly around the stump. The lower end of the stump should be level with the 'tongues' at the base of the sheath. The tongues are then folded over the distal end of the stump. A second bandage is then applied over the inner sheath to ensure complete contact between the stump and sheath. The outer Tulip frame is then applied, and it is important that the distal end of the sheath sits firmly against the base of the frame. The frame is secured by the velcro straps, the lower of which has a patella pad which must sit relative to the patella tendon. Air is then *completely* evacuated from the system using the hand operated pump. Leg length is then adjusted by comparison with the contralateral leg, although it is usual to make the Tulip limb 1 cm shorter than the

Preformed sockets

Femurett

The Femurett was developed by LIC Orthopaedics in Sweden. It is available in two models, one without a knee mechanism (designed for transfemoral amputees and also for transtibial amputees who require tuber-bearing) and one with a manually lockable knee (for transfemoral amputees only).

The system without a knee mechanism consists of a laminate quadrilateral shaped open-ended socket, closed with velcro straps, and available in three sizes, right and left. The socket is fixed in 10° of flexion, but there is simple abduction/adduction adjustment. The tube mounting for the socket is telescopic to allow for length adjustment. The system is fitted with a universal prosthetic foot, size 40. A shoulder strap suspension is provided. For transtibial amputees there is a simple transverse strap, attached to the support tube, which is applied around the stump and helps prevent knee contracture.

The system with a knee mechanism is the same as above except that a single axis knee with extension spring and manual lock is placed in the tube mounting. The tube is adjustable above and below the mechanism so that the prosthetic knee joint will sit at the same level as the natural knee (Fig. 44.10).

The system can be used as soon as the stump is relatively free of pain, but stump oedema does not preclude usage and neither does incomplete healing. The stump is covered by a prosthetic sock and the system is applied in lying, ensuring that the ischial tuberosity is on the ischial seat. The socket is then firmly closed with the velcro straps. The shoulder strap is then supplied as a safety measure, as usually the device suspends on its own, the exceptions being when the stump is tender and the socket cannot be securely tightened, or when the stump is very conical (promotional data, P.I. Medical).

The main advantages of the system are that it can be used relatively early following amputation and without the need for prosthetic input, although the therapist does require some knowledge of prosthetic alignment. The hard socket with tuber seating provides greater support for transfemoral amputees than the PPAM aid (Parry and Morrison, 1989). The use of the device reduced average time

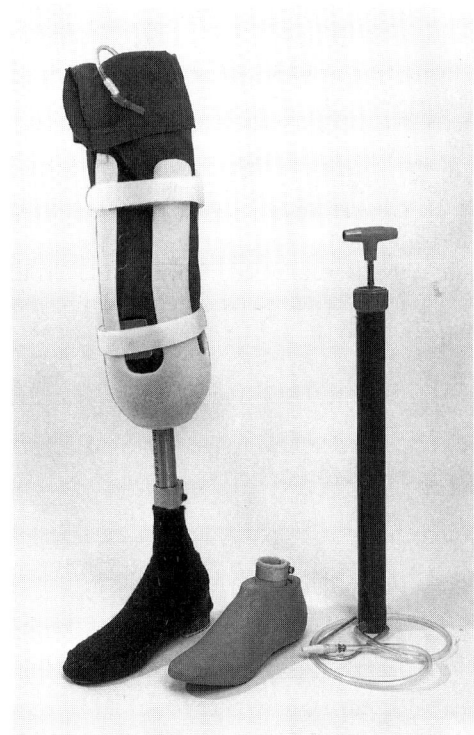

Fig. 44.9. Tulip Limb mounted on tube with 'universal' foot. For use 3 weeks post-operatively.

contralateral limb, to avoid catching the toe when walking with a stiff knee gait.

To use the Tulip limb the stump must be relatively insensitive to pain. However, swelling is not a problem, and full healing of the wound is not mandatory (P.I. Medical Promotional Video). Liedberg *et al.* (1983) advise delaying usage till 3 weeks after amputation, as earlier application can cause reddening and stretching of the suture line. The device is used under the supervision of a physiotherapist for 1–2 h per day.

The main advantages are that it is less bulky than the pneumatic devices, and can be used by the physiotherapist, without recourse to a prosthetist. The main disadvantage is that its use is best delayed for up to 3 weeks post-amputation. It is supplied by: P.I. Medical, Ab Box 67, S-751 03 Uppsala, Sweden, and R. Taylor & Son (Orthopaedic) Ltd., 49 Woodwards Road, Pleck, Walsall WS2 9RN.

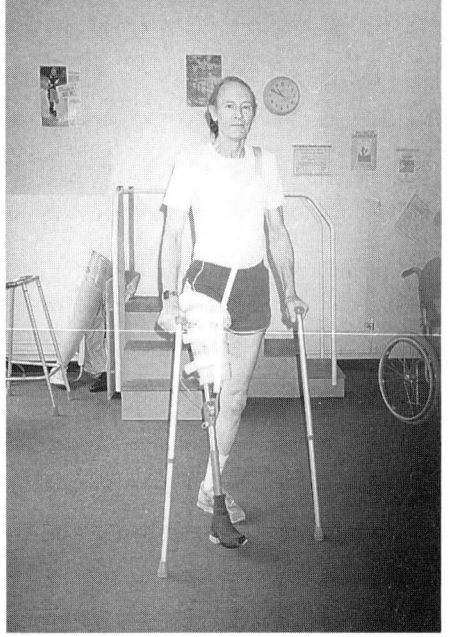

(a)

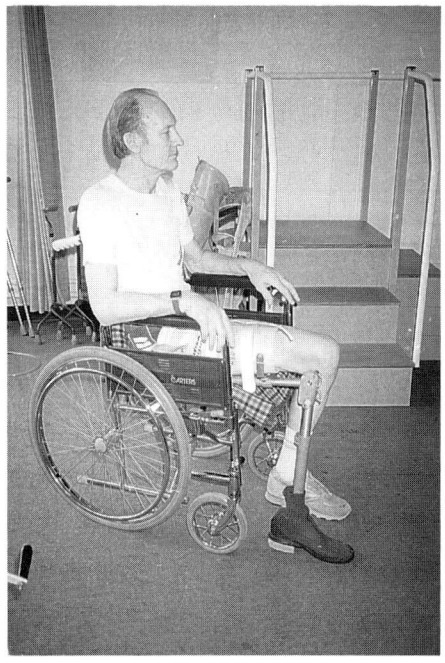

(b)

Fig. 44.10. Femurett System. **(a)** Ischial tuberosity bearing socket. **(b)** Model with knee mechanism.

to achieve independent gait in transfemoral amputees from 11.2 days to 7.16 days (Ramsay, 1988).

The main disadvantages for transfemoral amputees are the strong extension bias of the knee unit making free walking difficult, ineffective suspension, and variable socket discomfort. The device does not control oedema as well as some other devices, and the equipment is quite costly. Transtibial stumps with significant deformities cannot be accommodated. It is supplied by: P.I. Medical., Ab Box 67, S-751 03 Uppsala, Sweden and R. Taylor & Son (Orthopaedic) Ltd., 49 Woodwards Road, Pleck, Walsall WS2 9RN.

The Above Knee Early Assessment Prosthesis (AK EAP)

This device is being developed for transfemoral amputees. The philosophy behind its development being that early gait training is more meaningful if socket fit and alignment are optimal.

The system comprises a set of adjustable polypropylene ischial bearing sockets with plastic single axis hip joints, mounted on a modular endoskeletal assembly comprising a semi-automatic knee lock, a selection of height adjustable shin tubes and a right or left SACH foot. The system can also be fitted with a range of moving knee joints and a cosmetic cover can be added if required (G. Brown, 1992, personal communication) (Fig. 44.11).

This system offers full alignment potential and the ability to assess the amputees ability with a range of knee mechanisms. However, it requires the involvement of the prosthetist and attendance at a centre served by RSL. It is supplied by: Rehabilitation Services Ltd., Bunces Farm House, Lower Road, Tonge, Sittingbourne, Kent ME9 9BP.

The Lema prosthesis

This device was developed by LIC in Sweden for use with transtibial amputees who require tuber-bearing, and transfemoral amputees.

The equipment consists of preformed polypropylene open ended ischial bearing thigh 'sockets' which come in seven sizes, right and left, and are closed by velcro straps. The socket is riveted to simple side steels which can be supplied rigid or

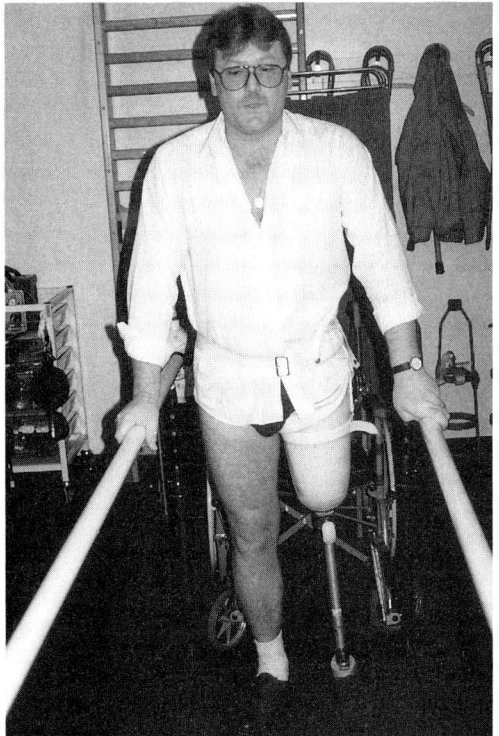

Fig. 44.11. Above Knee Early Assessment Prosthesis (AK EAP). Full alignment potential and permitting range of knee mechanisms. Prosthesis fitting only.

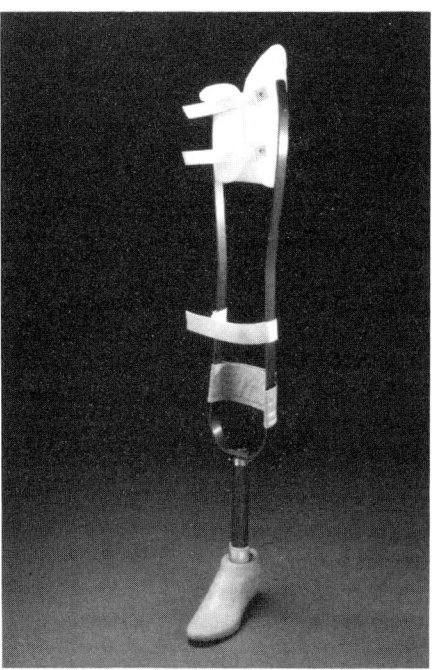

Fig. 44.12. Lema Prosthesis. Preformed ischial bearing thigh sockets in seven sizes RM.

with a manually lockable knee. The side steels are connected to a tube which is fitted with a prosthetic foot of the patient's choice. For transtibial amputees, there is an anterior strap (Fig. 44.12) above the knee joint and a posterior strap behind the stump, both of which help to prevent a flexion contracture.

The stump does not require any extra bandaging but the thigh must be protected by an extra stocking. The device is then applied in the supine position, ensuring that the ischial tuberosity is on the seating, before adjusting the velcro straps. The protective stocking is then pulled distally and tied to the lower end of the device (P.I. Medical Promotional Information).

The advantages of the system are that it is extremely light, and the patient can use their own footwear. The main disadvantages are that correct fitting and alignment require prosthetic skills, and there is no distal stump support.

The system has largely been superseded by the Tulip limb for transtibial amputees, and the Femurett for transfemoral amputees (B. Soderberg, 1992, personal communication). It is supplied by: P.I. Medical Ab Box 67, S-751 03 Uppsala, Sweden, and R. Taylor & Son (Orthopaedic) Ltd., 49 Woodwards Road, Plack, Walsall WS2 9RN.

Bespoke socket devices

Over the years, various materials have been used locally to form 'early' sockets. These include Hexcelite, Lightcast II, Scotcast and several others.

Two commercially available devices are described.

Halmsted prosthesis

This device was originally developed by LIC in Sweden for use on transtibial amputees. The socket is made of Sansplint and mounted on a modular

Otto Bock stack with foot. Suspension is assisted by a Juzo below-knee sleeve (Fig. 44.13).

The socket can be moulded directly onto the stump over the stump dressing, 5 days post-operatively. As stump shrinkage occurs, the socket can easily be changed in shape, and rarely more than two sockets are required before changing to a definitive prosthesis (B. Soderberg, 1992, personal communication). Some defer manufacturing the socket till 12–14 days post-operatively, and prefer to use a plaster cast to achieve socket shape, finding the sandsplint difficult to work on the patient (K. Harney, 1992, personal communication).

The prosthetic components are potentially re-usable but the system costs nearly as much as a definitive prosthesis. There is some concern that the Sansplint may deform with heat from domestic heating appliances (B. Soderberg, 1992, personal communication), and the appliance has not been formally approved for use in the UK. It is supplied by: Otto Bock Orthopadische Industrie GmbH & Co., Postfach 12–60, D-3408 Duderstadt, Germany and P.I. Medical, Ab Box 67, S-751 03 Uppsala, Sweden.

Habermann interim prosthesis

This system has been designed as an early walking aid for transfemoral amputees.

The system consists of a preformed Pedilon socket with installed valve ring which comes in

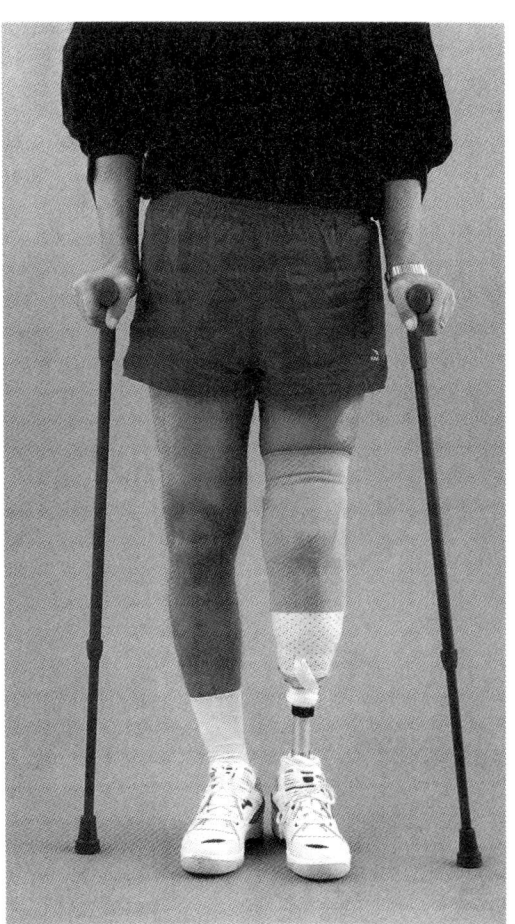

Fig. 44.13. Halmsted prosthesis (courtesy of Otto Bock). Custom-made socket in Sansplint or plaster of Paris.

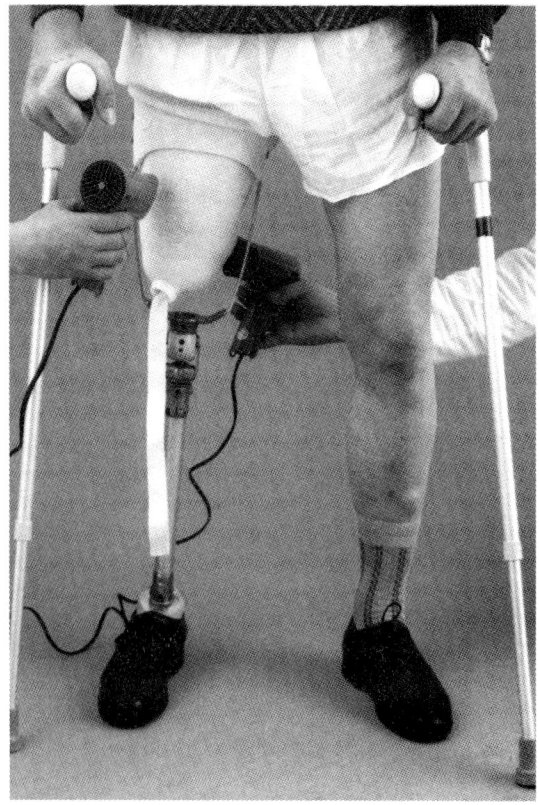

Fig. 44.14. Habermann Interim Prosthesis (courtesy of Otto Bock). Pedilon preformed socket being 'shrunk' onto stump.

two lengths, six circumferences and right and left. The socket is placed in an above knee brim (eight circumferences available) and mounted on aluminium side bars which are adjusted to the appropriate length, and degree of flexion/extension and abduction/adduction required. Heat protective stockinette is pulled over the stump and the prosthesis applied. The amputee is instructed to weightbear on the prosthesis, and the socket is 'shrunk' onto the stump using two hair dryers (Fig. 44.14). Shrinkage of the Pedilon starts at about 55° C. When the socket has been 'shrunk' to the appropriate dimensions, it is cooled with a moist cloth, and when completely cool, is removed from the patient. The brim and socket are secured in the frame. A distal end cushion can be fitted in the socket if required.

The stump must be sufficiently mature to tolerate 'pulling' into the socket, but clearly more correct socket dimensions can be achieved. The system is approximately half the cost of a definitive prosthesis, but requires considerable prosthetic time. At present the system is not approved for use in the UK (K. Harney, 1992, personal communication). It is supplied by: Otto Bock Orthopadische Industrie GmbH & Co., Postfach 12–60, D-3408 Duderstadt, Germany.

Conclusions

A variety of early walking aids are described, and it is established that their appropriate use undoubtedly speeds the rehabilitation process, as well as controlling post-operative oedema to a greater or lesser extent. Local factors will dictate which system(s) are used and whilst correct fit and optimum alignment are clearly beneficial to the amputee, it is the author's view that the systems which are not dependent on prosthetic skills, will generally be more favoured.

References

Berlemont, M. (1961) Notre experience de l'appareillage precoce des amputees des membres inferieures aux etablissements. Helio-Marins de Berck. *Ann. Med. Phys.*, **4**, 213.

Bonner, F.J. Jr and Green, R.F. (1982) Pneumatic Airleg prosthesis: report of 200 cases. *Arch. Phy. Med. Rehabil.*, **63**, 383–385.

Cohen, S.T., Goldman, L.D., Saltzman, E.W. *et al.* (1974) The deleterious effect of immediate post-operative prosthesis in below knee amputation for ischaemic disease. *Surgery*, **76**, 992–1001.

Devas, M.B. (1971) Early walking of geriatric amputees. *BMJ*, **1**, 391–396.

Dickstein R., Pillar, T., and Mannheim, N. (1982) The pneumatic postamputation aid in geriatric rehabilitation. *Scand. J. Rehabil. Med.*, **14**, 149–150.

Ham, R., Richardson, P. and Sweet A. (1989) A new look at the Vessa PPAM Aid. *Physiotherapy*, **75**, 493–495.

Henriken O., Marsch G. and Persson B.M. (1978) The Tulip prosthesis. *Acta Orthop. Scand.*, **49**, 107.

Hutton I.M. and Rothnie, N.G. (1977) The early mobilisation of the elderly amputee. *Br. J. Surg.*, **64**, 267–270.

Lein, S. (1992) How are physiotherapists using the Vessa Pneumatic Post Amputation Mobility Aid? *Physiotherapy*, **78**, (No. 5), 318–322.

Ledberg, E., Hommerberg, H. and Persson, B.M. (1983) Tolerance of early walking with total contact among below knee amputations – a randomized test. *Prosthet. Orthot. Int.*, **7**, 91–95.

Little, J.M., (1971) A pneumatic weight bearing temporary prosthesis for below knee amputees. *Lancet*, **1**, 272–273.

Monga, T.N., Symington, D.C., Lowe, P. and Elkin, N. (1985) Load bearing and suspension characteristics of airsplint as a temporary prosthesis. *Prosthet. Orthot. Int.*, **9**, 100–104.

Murdoch, G. (1969) Immediate post-surgical fitting: an editorial. *Prosthet. Int.*, **3(8)**, 2–7.

Parry, M. and Morrison, J.D. (1989) Use of the Femurett adjustable prosthesis in the assessment and walking training of the new above knee amputee. *Prosthet. Orthot. Int.*, **13**, 36–38.

Ramsay, E.M. (1988) A clinical evaluation of the LIC Femurett as an early training device for the primary above knee amputee. *Physiotherapy*, **74**, 598–601.

Rausch, R.W. and Khalili, A.A. (1985) Air splint in preprosthetic rehabilitation of lower extremity amputated limbs. *Phys. Ther.*, **65**, 912–914.

Redhead, R.G. (1983) The early rehabilitation of lower limb amputees using a pneumatic walking aid. *Prosthet. Orthot. Int.*, **7**, 88–90.

Redhead, R.G., Davis, B.C., Robinson, K.P. and Vitali, M. (1978) Post amputation pneumatic walking aid. *Br. J. Surg.*, **65**, 611–612.

Tunbridge Report (1972) *Rehabilitation. Report of a Subcommittee of the Standing Medical Advisory Committee.* HMSO, London.

45

Early management: the therapist's role

R. O. Ham

Introduction

Successful rehabilitation of the amputee patient has, for many years, been said to depend upon the 'team approach'. This is the efficient, coordination of many disciplines of staff working with the patient and their relatives, to maximize the patient's rehabilitation potential. Members of the amputee multidisciplinary team should include the following members of staff; surgeon, rehabilitation medicine specialist, nursing staff, physiotherapist, occupational therapist, medical social worker, clinical psychologist, prosthetist and community health care staff. Their roles will vary depending upon the individual patient's needs. The team leader is normally the surgeon or rehabilitation physician but the team's activity may be coordinated by any member of the team, such as the physiotherapist (Jamieson and Hill, 1976; Ham and Cotton, 1991).

The physiotherapist is said to be one of the 'tripod' members of the amputee rehabilitation team alongside the surgeon and the prosthetist (Burgess and Alexander, 1973). The physiotherapist spends a great deal of time with the patient and the family during the initial rehabilitation phase and develops a close professional relationship which allows time for the patient to express fears, concerns, health attitudes and share their goals for rehabilitation, all of which will play a part in the rehabilitation process.

The pre-amputation phase

Assessment

Today in the industrial world, the majority of amputees suffer from vascular disease, and most but not all will already be in a hospital ward. Once amputation has been decided to be the treatment of choice, the physiotherapist fully assesses the patient to gain as much useful information as possible to be able to set provisional goals and to contribute to the decision of amputation level selection. Information will also be collated by the team from a variety of sources; the vascular laboratory, the physiotherapist, occupational therapist, social worker, family and friends and the medical condition of the patient. This information is considered by the surgeon before the choice of amputation level is made (Ham and Cotton, 1991; Houghton *et al.*, 1992).

The therapists' assessment will cover three areas;

(i) *Physical aspects*: These include the lower and upper limb muscle power, joint range, skin sensation, proprioception, limb function, functional activities for movement, trunk mobility, balance and posture, sight, hearing, activities of daily living (ADL) for independence, the use of a wheelchair (if applicable), hand function and sensation, walking ability and such aids to independence that are in use will all be noted.

(ii) *Social aspects*: These will include the patient's present accommodation including; stairs, access to the outside, toilet and bathing facilities, lifestyle, family and friends, occupation, hobbies, social activities, the social support received from either the statutory or voluntary community services before admission, the means of transportation used and the patient's level of independence.

(iii) *Behavioural psychological aspects*: The cognitive state, pain behaviour, motivation, sociability, health attitude and psychological state will also be noted. Addresses, contact numbers of both relatives or carers and community staff are also recorded.

It is generally not possible to gain all of this information at one time, especially if the patient is seriously ill or toxic.

Examples of physiotherapy assessment forms are given elsewhere (Hilton, 1989; Ham and Cotton, 1991); the United Kingdom Chartered Society of Physiotherapy has also produced guidelines (CSP, 1992).

Treatment

Also in this pre-amputation phase, the physiotherapist whenever possible, continues to increase the patient's muscle strength and exercise tolerance through specific exercises and general functional activities, working also to reduce any joint contractures that are not fixed to increase the patient's rehabilitation and prosthetic potential. This is rarely possible in the elderly dysvascular patient.

From the beginning the physiotherapist will, along with other team members develop community links, engage family support and discuss the likely mobility of the patient on discharge.

Other aspects of rehabilitation are continually discussed and the patient and family informed of the likely steps in the rehabilitation process.

Information booklet

A lack of information has been found to lead to fear in the patient (Fletcher, 1973) and to an increase anxiety. Studies showing the benefit of giving patients information prior to surgery (Ridgeway and Mathews, 1982; Weinman, 1987) have been reported and information preoperatively has also been found to be useful for patients undergoing amputation (Ham and Kerfoot, 1986, 1987). The information must not include medical 'jargon' which has been found to be either meaningless or misinterpreted (Boyle, 1970; Korsch and Negrete, 1972) and it is important that the staff member checks that the information given is clearly understood by whoever is listening or receiving it (Ham, 1992).

Information booklets (Ham and Kerfoot, 1987) are handed out pre-operatively and are generally read by the family at this time and by the patient after the amputation. This written information complements verbal information and should never replace it.

Phantom limb sensation and pain

The phantom phenomenon is a universal phenomenon (Kolb, 1954) and can be either non-painful and described as *phantom sensation* or painful and described as *phantom pain* (Ribbers *et al.*, 1989). Virtually all amputees experience some phantom sensation which they feel as real as that in the opposite limb (Sternbach, 1968). The onset of phantom sensations can and usually does occur soon after the operation (Sternbach, 1968; Jensen *et al.*, 1984) and the patient may refuse to believe that the limb has been removed (Riding, 1976). In the majority of cases phantom sensation is experienced within 48 h of amputation. Unlike phantom sensation, phantom pain is experienced by only some amputees (Kolb, 1954; Postone, 1987).

When the sensations become intense enough for the amputee to define them as painful, they are known as phantom pain. The difference in the two appears to be one of intensity rather than a description of a new sensation. For example, warm sensations become painful when they are burning (Sherman, 1989).

The nature of the phantom is described to the patient and their family pre-operatively to reduce anxiety. It is important that phantom pain is not emphasized unduly but mentioned as a possible feature of the phantom. Feelings of the presence of the amputated limb, telescoping, the various sensations that can be experienced and the fading, are also described so that they do not feel 'insane if they feel a pain in a part no longer present' (Sherman and Sherman, 1985).

Post-operative management

Depending upon the speciality of the surgeon, the patient's stump will be managed post-operatively either in a plaster of Paris cast or a soft dressing. Bandaging should be avoided as it has been shown that even in experienced hands, pressures considerably greater than capillary pressure may be created (Isherwood *et al.*, 1975).

From the first post-operative day, therapy is begun. Initially this is on the ward bed but as the drips and drains are removed and the patient becomes more stable, therapy is extended to include the gymnasium, on or about the third post-operative day. A range of active exercises are given to the remaining limbs and general mobility is encouraged. Often the patient is scared to move but working with the nurses to control pain, mobility, independence and confidence are gradually increased. The physiotherapist ensures that a full range of movement is maintained in all joints.

During therapy sessions, the patient or relatives often need help coming to terms with the loss and in this the physiotherapist assists through the three core conditions of empathy, warmth and genuineness (Rogers, 1984). Also by responding in the therapy sessions through listening, reflection, paraphrasing and using open-ended questions, the physiotherapist will help to facilitate the patient's thought processes (Egan, 1982; Nelson-Jones, 1983).

Throughout the treatment sessions the physiotherapist observes the family through the bereavement process that follows any loss; notably numbness, despair and subsequent recovery (Ham and Cotton, 1991). If the process is seen as being atypical, it should be brought to the attention of the team for a referral to be made to a clinical psychologist or counseller.

Early home visit

It is useful if a member of the team visits the patient's home during the first 10 days post-operatively to meet with the family and carers on their 'own ground' and to assess and discuss the patients future needs (Ham and Cotton, 1991). This early home visit speeds up the patient's return to the community by setting in motion any plans that are made, ensuring that rehabilitation is continuous and that any social/health divide is contained.

This visit may also include statutory staff, for example in the UK this may include the Social Services who cover accommodation, home-help or assistance, and social needs, the primary or community health care team for continuing and future nursing and medical needs and employment links to negotiate for necessary adaptations, support or changes in the condition of service.

Once the patient's medical condition and stump wound are stable, the physiotherapist will begin to use one of the early walking aids described elsewhere (Chapter 44) (Fig. 45.1).

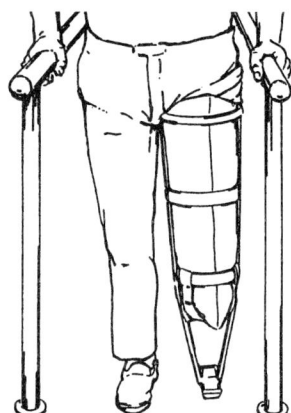

Fig. 45.1. PPAM early walking aid.

As time goes by the patient's rehabilitation programme will intensify and more time is spent in the therapy departments (Fig. 45.2).

Patients are encouraged to dress in day clothes from the beginning and will be provided with a wheelchair to increase their personal independence and sociability. Using a wheelchair is safer for the elderly patient and helps to avoid dependant oedema developing. Younger patients will be capable of using crutches safely but some may also need to use a wheelchair. Many patients will require a wheelchair at home to increase their mobility and to use when the prosthesis is not worn (Stewart and Jain, 1993).

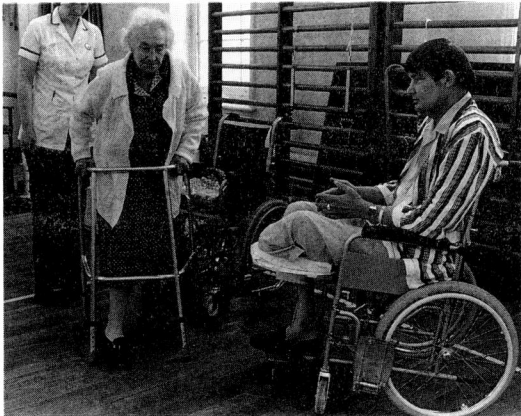

Fig. 45.2. Patients progress to therapy departments as condition stabilizes and early walking aids are introduced.

For all transtibial amputees it is necessary to maintain their knee extension to avoid stump oedema and flexion contractures. To reduce accidents in transferring, special stump boards have been designed, such as the King's stump board (Ham and Richardson, 1986) (Fig. 45.3).

The patient's mobility and strengthening programme will increase in length and intensity and they will be encouraged to become more and more independent. Exercise programmes are developed for each individual and will vary depending upon the patient's exercise tolerance, level of activity, age, motivation and lifestyle. Descriptions of such

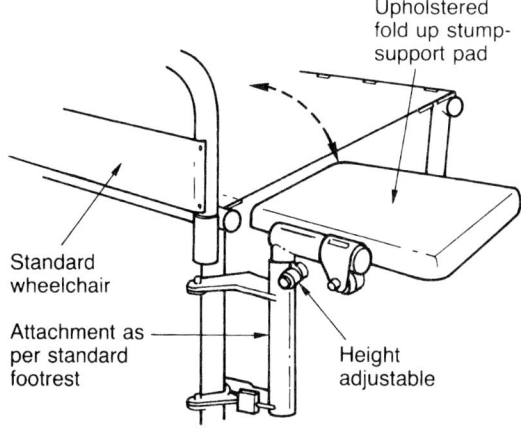

Fig. 45.3. Upholstered fold-up stump support pad fitted to standard wheelchair.

exercise programmes are given elsewhere (Engstrom and Van de Ven, 1985; Ham and Cotton, 1991).

Treatment sessions in groups provide patients with moral support and also encourages them to become more independent.

In some hospitals the patients remain as inpatients. In others the patients are discharged soon after the wound has healed and attend for therapy as outpatients on a frequent basis until they are fully rehabilitated in either a wheelchair or on a prosthesis. In the third example, patients are discharged home at an early stage and the hospital physiotherapist continues to treat them at home until rehabilitation is complete, providing the patient with continuity of care and avoiding long hospital stays. Support for each of the three methods are found in the literature (Ham *et al.*, 1987; AMRS, 1992; Stewart and Jain, 1993). However times are changing, health care is becoming more economically focused and acute hospital beds are becoming scarce. At first sight there is perhaps a reason to support the latter example, but costing the different protocols is extremely complex.

Prosthetic assessment

During therapy sessions, the patient will achieve some milestones of recovery which can be used to determine whether at this stage, they will be a prosthetic user. Such milestones are the ability to perform basic activities of daily living such as moving about the bed independently, washing, toileting, dressing both the upper and lower half of the body, transferring from different chairs and surfaces and the successful use of an early walking aid.

Also added to these milestones is the information gained at pre-operative assessment which includes the information regarding their mobility before admission and their previous lifestyle. If the patient is not medically capable or motivated enough to perform such activities, the decision to prescribe a prosthesis is postponed until these basic activities are achieved.

In England and Wales, the numbers of patients referred to prosthetic centres as suitable for prostheses was found to be 50% (Ham and Cotton, 1991) and this is in agreement with a more recent study (Houghton *et al.*, 1992). In the non-prosthetic users, full physiotherapy rehabilitation will continue, and directed towards wheelchair functioning but any changes observed in activity levels in the patient will be reported to the team. In another situation where a single team is responsible for the whole rehabilitation process from initial assessment, level selection and surgery through to prosthetic fitting and monitoring for life after discharge home a different picture emerges. Stewart and Jain (1993) report 81% of a total of 1846 primary amputees over a period of 25 years successfully fitted with a prosthesis. In the same report it is noted that 40% of all amputees were supplied with a wheelchair; of those fitted with a prosthesis a wheelchair was prescribed in 38%.

Prosthetic training

Once the patient has received the prosthesis, therapy training continues until fully rehabilitated or reach their maximal potential (Fig. 45.2). This will include donning and doffing the limb, learning to dress with the limb on, transferring, toileting, getting up off the floor, walking in and out of doors and up slopes, and more complicated activities depending on the patient's needs.

For the young trauma patient this requires full mobility in and out of doors, with advice on hobbies and occupational adaptation. For the elderly vascular case, this may only be walking a short distance indoors or using the prosthesis to assist with transfers from the wheelchair, for example, to the toilet safely. With daily practice training to a basic level of walking confidently takes on average 2 weeks. Patients reach their goals at different times depending upon their motivation, medical condition and cognitive state. Goals are constantly reviewed by the physiotherapist and the family or carers.

Patients are also taught about the care of prosthetic socks, the stump, skin care, the remaining foot, the importance of the shoe with prosthetic alignment and the hazards of walking with a prosthesis, especially on uneven surfaces. Patients practise getting up off the floor and climbing stairs. They are also taught the basic principles of a good prosthetic fit so they are aware for example, of having a stump that has either shrunk or swollen and the need for either fewer or more prosthetic

socks. In this way the physiotherapist works closely with the prosthetist to ensure that there is a good prosthetic fit and function. Details of this process are given elsewhere (Ham and Cotton, 1991; Barsby *et al.*, 1993).

The patient may initially be prescribed a less sophisticated prosthesis and this will be changed if performance merits it.

Non-walker or non-prosthetic user

The non-prosthetic user may be a bilateral amputee or a single amputee who is unable to or does not want to walk.

In the majority of these cases rehabilitation is a slower and longer process as their exercise tolerance is low, their general condition is generally poor, they may be depressed and they may be obese, often due to diabetes. They will, however, attend the gym when they are medically fit and continue an exercise programme with the other patients. Such patients require more encouragement and time spent with them by the staff and generally take longer to rehabilitate than prosthetic users. These patients are taught bed mobility, arm strengthening exercises to assist with the mobility, balance re-education, transferring activities, independence from the wheelchair, wheelchair safety and wheelchair manoeuvres.

Such patients are likely to get pressure sores, as their mobility is poor, and may have trouble with incontinence as it is difficult for them to reach the toilet in time or to access it. The nursing, occupational therapy and physiotherapy staff will work closely together to provide equipment aids, such as pressure relieving surfaces for the wheelchair and bed. Clothing adaptations can be made to assist with ADL and other equipment may be provided to assist with independence.

A final home visit takes place before discharge with the patient, hospital staff and ideally with relatives or carers. If the early home visit alerted the team to the need for major alterations, the staff concerned from the community Social Services agency or housing association should also be there to assist in speeding up the request procedures. The alterations, equipment, etc. are therefore requested and a discharge date is set. In this way time is not wasted and costs are kept to a minimum.

The clinic team

Throughout the rehabilitation process from comprehensive assessment after surgery, through the application of the individual skills of team members to the final level of locomotor ability and independence achieved the team monitors progress; formally once or twice a week during the in-patient period, re-setting goals when required and making arrangements for future review and any outpatient treatment sessions and community services. These might include financial benefit, telephones, transportation for the disabled, support groups, meals at home, care support and respite care, interest groups, holiday information, recreation and sport, automobile alterations.

The best chance of a successful rehabilitation for the amputee resides in the efficient and sympathetic operation of the clinic team.

References

Amputee Medical Rehabilitation Society (1992) *Amputee Rehabilitation. Recommended Standards and Guidelines.* Royal College of Physicians, London.

Barsby, P., Ham, R.O., Lumley, C. and Roberts, V.C. (1993) *Handbook of Amputee Management.* Kings College School of Medicine & Dentistry, London SE5.

Boyle, C.M. (1970) Differences between patients and doctors interpretation of some common medical terms. *BMJ*, **2**, 286–289.

Burgess, E.M. and Alexander, A.G. (1973) The expanding role of the physical therapist in the rehabilitation team. *Phys. Ther.*, **53**, 141–143.

CSP (1992) *Standards of Physiotherapy Practice for the Management of Patients with Amputation.* Chartered Society of Physiotherapy, London.

Egan, G. (1982) *The Skilled Helper.* Brooks Cole, Monterey, Cal.

Engstrom, B. and Van de Ven, C. (1985) *Physiotherapy for Amputees. The Roehampton Approach.* Churchill Livingstone, Edinburgh.

Fletcher, C.M. (1973) *Communication in Medicine.* Rock Carling Monograph Nuffield Hospital's Trust.

Ham, R.O. (1992) *The Vascular Amputee. From Aetiology to Community Resettlement.* Fellowship Thesis, Chartered Society of Physiotherapy, London.

Ham, R. and Cotton, L.T. (1991) *Limb Amputation; from Aetiology to Rehabilitation.* Chapman & Hall, London.

Ham, R.O. and Kerfoot, S. (1986) Considerations for staff involved in amputee rehabilitation. *Physiother. Practice*, **2(4)**, 161–165.

Ham, R. and Kerfoot, S. (1987) *Recovering from Amputation – The Dulwich Way.* Kings College School of Medicine & Dentistry, London SE5.

Ham, R. and McCreadie, M. (1992) Rehabilitation of elderly patients in the UK following lower limb amputation. *Topics Geriatr. Med.*, **8(1)**, 64–71.

Ham, R.O. and Richardson, P. (1986) The King's amputee stump board – Mark II. *Physiotherapy*, **72**, 124.

Ham, R.O., Regan, J.M. and Roberts, V.C. (1987) Evaluation of introducing the Team approach to the care of the amputee: the Dulwich Study. *Prosthet. Orthot. Int.*, **11**, 25–30.

Hilton, S. (1989) *A Review of Amputee Assessment Forms.* Postgraduate Diploma in Amputee Management project report, King's College School of Medicine & Dentistry, London SE5.

Houghton, A.D., Taylor, P.R. and Thurlow, S. *et al.* (1992) Success rates for rehabilitation of vascular amputees implications for pre operative assessment and amputation level. *Br. J. Surg.*, **79**, 753–755.

Isherwood, P.A., Robertson, J.C. and Rossi, A. (1975) Pressure measurements beneath below-knee amputation stump bandages, elastic bandaging, the Puddifoot dressing and pneumatic bandaging technique compared. *Br. J. Surg.*, **62**, 982–986.

Jamieson, C.W. and Hill, D. (1976) Amputation for vascular disease. *Br. J. Surg.*, **63**, 683–690.

Jensen, J.S., Krebs, B., Neison, J. and Rasmussen, P. (1984) Non-painful phantom limb pain in amputees: incidence, clinical characteristics and temporal course. *Acta Neurol. Scand.*, **70(1)**, 60–61.

Kolb, L. (1954) The painful phantom, psychological, physiological and treatment. *American Lecture Series* No. 235, Charles C. Thomas, Springfield.

Korsch, B.M. and Negrete, V.F. (1972) Doctor patient communication. *Sci. Am.*, **227**, 66–72.

Nelson-Jones, R. (1990) *The Theory and Practice of Counselling Psychology.* Cassell, London.

Postone, N. (1987) Phantom limb pain: a review. *Int. J. Psychiatry Med.*, **17:1**, 57–70.

Ribbers, G., Mveder, T. and Rijken, R. (1989) The phantom phenomenon: a critical review. *Int. J. Rehabil. Res.*, **12(2)**, 175–186.

Ridgeway, V. and Mathews, A. (1982) Psychological preparation for surgery. A comparison of methods. *Brit. J. Clin. Psychiatry*, **21**, 271–280.

Riding, J. (1976) Phantom limb: some theories. *Anaesthesia*, **31**, 102–106.

Rogers, C. (1984) *On Becoming a Person.* Constable, London.

Sherman, R.A. (1989) Stump and phantom limb pain. *Neurol. Clin.*, **7(2)**, 249–264.

Sherman, R.A. and Sherman, C.J. (1985) A comparison of phantom sensation among amputees whose amputations were of civilian and military origin. *Pain*, **21**, 91–97.

Sternbach, R.A. (1986) Phantom pain. In: *Psychology of Pain* (ed. R.A. Sternbach), Academic Press, New York.

Stewart, C.P. and Jain, A.S. (1993) Dundee revisited – 25 years of a total amputee service. *Prosthet. Orthot. Int.*, **17**, 14–19.

Weinman, J. (1987) *An Outline of Psychology as Applied to Medicine.* Wright, Bristol.

46

Later management: the therapist's role

J. E. Edelstein

Pre-prosthetic and non-prosthetic care

Rehabilitation of most individuals who sustained lower-limb amputation should be conducted under the aegis of a knowledgeable clinic team composed of physician, physical therapist and prosthetist. Many patients also benefit from the services of an occupational therapist, social worker, rehabilitation counsellor, and other specialists. The team should assess the patient to establish mutually satisfactory goals. A principal question which must be addressed is whether the individual is a good candidate for a prosthesis. Use of a temporary prosthesis is an excellent means of assessing the patient. The patient who returns home without a prosthesis, or who is not likely to gain from prosthetic use can be guided to achieve considerable personal, vocational and recreational independence without a prosthesis. The basic elements of rehabilitation apply regardless of level of amputation, although individuals with more distal amputations usually accomplish functional tasks more readily than do those with higher amputations (Narang *et al.*, 1984; Waters and Perry, 1989).

Goals of pre-prosthetic and non-prosthetic management

Chief goals of pre-prosthetic care are: (1) aiding the patient to become independent in self-care activities; (2) teaching the patient to become competent in caring for both the amputation and the contralateral limbs and (3) determining the individual's physical and emotional suitability for prosthetic use. A well coordinated, individualized programme of early care (Eisart and Tester, 1954; May, 1988) helps the individual adjust physically and psychologically to the loss of the limb.

Non-prosthetic self-care training

Whether or not the patient has a prosthesis, some activities, such as dressing, bathing, manoeuvring through short distances are performed without a prosthesis (Edelstein, 1992b). The individual with a unilateral amputation may have adequate balance to don undergarments while standing. Others, including those with bilateral amputations, find that sitting on the bed or a chair allows safer performance. Bathing in a tub is difficult for frail patients; they may find the tub slippery and may not be able to get out of the bathtub unassisted. A plastic seat secured to both sides of the tub enables some people to sit while bathing. Others find that a sturdy shower chair with rubber tipped legs provides adequate safety. Individuals with impaired circulation in the remaining foot should always test water temperature with the hand before turning on the showerhead to prevent scalding. Wall mounted bars help the client maintain balance, especially when transferring in and out of the shower or tub. Most individuals, however, simply sit on the shower floor to bathe.

Without a prosthesis, a young individual may be able to hop short distances. The older person with vascular disease, however, should be cautioned against hopping because the impact on the remaining foot may be damaging. Pivoting first on the heel, then on the forefoot, lessens the applied force, but still requires considerable agility.

Caring for the amputated and contralateral lower limbs

Early management should include attention to both the newly amputated limb and the intact, contralateral limb. For the amputation stump, wound healing, volume stabilization and skin care (Levy, 1983) are major concerns. An individually graduated programme of mat exercises will counteract the disuse atrophy which many dysvascular patients present, while improving cardiopulmonary function (Kegel *et al.*, 1981; Pitetti *et al.*, 1987).

The contralateral limb, particularly in the patient with peripheral vascular disease, should be evaluated periodically to determine what conservative and surgical management is appropriate to maintain limb function. Regardless of aetiology of amputation or the patient's age, the individual will place greater stress and reliance on the sound limb for it provides sensory feedback and more complex joint function than does any prosthetic substitute. The patient should learn how to inspect the sound foot with the use of a hand mirror to discover signs of incipient irritation. Bathing the foot in tepid water with mild soap and drying it gently are critical preventive measures. To protect the toes against abrasion, the staff should insist that the patient wears a clean sock and well-fitted shoe whenever standing or walking. An elastic sock can control dependent oedema and a custom-fitted foot orthosis is often necessary to compensate for the patient's altered balance.

Is the patient a candidate for a prosthesis?

Most individuals with leg amputation receive a prosthesis and benefit from wearing it. Prosthetic fitting may be instituted primarily to enhance the patient's self-esteem and sense of body image, with the wheelchair providing mobility. For others, the prosthesis enables them to walk and manoeuvre through narrow doorways, particularly in the bathroom, reserving wheelchair use for outdoor travel. Many young and middle-aged adults resume community ambulation and engage in sports while wearing a prosthesis. It is thus essential for the clinic team, the patient and family to predict the level of probable function in order to formulate a realistic rehabilitation plan which neither under- nor over-estimates the patient's capabilities.

A comprehensive history includes details of the present illness, particularly the patient's mobility prior to and following amputation. Past medical and psychological history provides clues regarding the individual's ability to cope with physical and emotional problems. A key point in examination is the patient's mentation, especially short-and long-term memory and ability to retain new information. Visual acuity, the extent of any arthritis, and cardiopulmonary and neuromuscular status have profound influence on one's likelihood of using a prosthesis. A major purpose of early management is to evaluate the patient's stress tolerance and adjustment to limb loss.

Using a temporary prosthesis

A temporary (preparatory) prosthesis introduces the patient to upright bipedal balance and gait. It also accustoms the individual to the pressures which a socket applies to the stump. Many designs of temporary prosthesis are available, including those which are custom-made from a cast of the stump (Hallam and Jull, 1988) and mass produced ones which the clinician adjusts to fit the specific patient (Bonner and Green, 1982; Monga *et al.*, 1985; Parry and Morrison, 1989). As a group, temporary prostheses offer a less expensive, readily adjustable means to institute prosthetic training. Temporary prostheses, however, usually are not as attractive as definitive prostheses and may also be heavier. For the patient who has questionable prospects for benefiting from a definitive prosthesis, such as the individual with multiple medical problems or bilateral amputations, use of a temporary prosthesis is an excellent way of determining the merit of prosthetic fitting.

Ideally, balance and gait training can be accomplished with the temporary prosthesis, so that the patient will require little additional care upon delivery of the permanent prosthesis.

Prosthetic care

Goals of care with a temporary or definitive prosthesis

These include teaching the patient to: (1) don the prosthesis and dress independently; (2) rise from various types of chair and sit in them; (3) walk in all directions safely with minimum extraneous movement; (4) climb stairs and similar surfaces and (5) take suitable care of the prosthesis. Many progress to advanced activities, such as kneeling, sitting on the floor and retrieving objects from the floor. For all individuals with amputation, the comprehensive goal of management is to assist the individual in returning to the community. Environmental adaptations at home and in the workplace may be required to achieve the goal. Amputation does not preclude engaging in a wide range of recreational endeavours. Peer and professional support groups are often effective in helping the patient and family achieve a satisfying lifestyle.

Basic training

Basic training for virtually all individuals who can be expected to use a prosthesis includes donning and dressing, transferring from chairs, balancing and walking (Humm, 1977; Banerjee, 1982; Engstrom and Van de Ven, 1985; Mensch and Ellis, 1986; Ham and Cotton, 1991). The techniques which follow suit individuals who wear a unilateral transtibial or transfemoral prosthesis. A skilful therapist, however, can adapt the procedures to benefit patients wearing bilateral prostheses (Adler *et al.*, 1987).

Dressing

Donning and dressing is fundamental to self-sufficiency. Indeed, if the client is unable to don the prosthesis independently with reasonable speed, then long-term prosthetic use is dubious. Most people find it easier to dress the prosthesis before applying it. Thus, the hose and shoe should be on the prosthetic foot. The patient should be wearing underwear prior to donning the prosthesis. For individuals wearing trousers or slacks, the prosthesis should be drawn through the trouser leg.

The sound limb should be inserted in its trouser leg. Prosthesis application begins with applying one or more socks to the stump while one is seated. Ankle disarticulation (Syme) and other prostheses which have a separate resilient liner are usually easier to don if one first removes the liner from the socket, dons the liner, and then inserts the stump into the socket. The transfemoral prosthesis whether suspended by partial suction or none also can be applied while the user sits, having first donned the necessary socks in the appropriate sequence. Once the prosthesis is situated on the stump, the client should rise from the chair to enable the stump to settle into the socket. Final closures can then be made.

Several methods enable one to don the transfemoral prosthesis which has total suction suspension. One can pull the stump into the socket with the aid of tubular cotton or nylon stockinette or elastic bandage which is applied to the thigh up to the groin. The patient passes the distal end of the stockinette or bandage through the valve hole. Most people prefer to stand while pulling downward on the fabric while alternately flexing and extending the contralateral hip and knee until the fabric can be pulled free from the socket. This is aided by lubricating the skin with a thin textured lotion. Once the stump is completely down into the socket, the patient inserts the valve.

Once the prosthesis is in place, the trousers are pulled up to the waist and fastened. A woman wishing to wear a skirt or dress would don it after putting on the prosthesis.

Standing and other transfer skills

To rise from a chair, the patient should put the sound foot behind the prosthetic one, bend forward slightly, then extend the hips and knees to cause the body to rise from the chair (Edelstein, 1992a). To sit, the patient should approach the chair, turn so that the chair is behind the individual, nudge the intact leg against the chair and transfer weight to that leg, then lower the torso to the seat. The prosthetic foot may be on line or slightly ahead of the intact foot. The beginner should practise with a sturdy armchair or a wheelchair with brakes secured. Eventually the patient should be able to transfer from a straight chair, the toilet seat, a bench and a car seat. Entering a car is influenced by the side of amputation. If the sound leg is closest to the car doorway, the patient will be able to enter somewhat more easily than if the prosthetic side is next to the door. In that instance, the patient may prefer to sit on the car seat with both feet outside the car, then lift the prosthesis with both hands to position it inside the car; finally, the sound leg is brought inside the car.

Balance and weight shifting

Control of the prosthesis requires that the wearer be able to shift weight onto and off it easily. Balance exercises may be performed in the parallel bars or alongside a sturdy table (Brunnstrom and Kerr, 1956). The patient should stand symmetrically with the feet 5–10 cm apart and shift weight to the prosthetic side, maintaining the body upright and both feet flat on the floor. This activity acquaints the patient with the contour of the socket and the degree of flexibility of the prosthetic foot. The therapist may increase the difficulty of the activity by placing a low stool ahead of the sound foot and asking the patient to step on it with the sound foot; in so doing, the patient must maintain good balance on the prosthesis (Gailey and McKenzie, 1989). The next stage is to shift to the sound side in a rhythmic manner. Forward and backward shifting is done in a comparable manner with respect to both foot and body position and with both knees kept extended. Next the patient should stand with the sound foot slightly ahead of the prosthetic one, and repeat the side-to-side and forward-backward shifting. Foot position is then reversed and shifting continued.

Control of the knee, whether anatomical or mechanical, requires that the patient associate hip flexion and extension with knee motion. The individual should alternate flexing and extending both knees first with the feet on a line, then with one foot ahead of the other.

Walking

Walking requires the patient to step forward with one leg while balancing on the other (Van Alste *et al.*, 1985; Summers *et al.*, 1986). To begin, direct the patient to step with the sound foot, thus requiring prosthetic control, an upright body and forward pelvic rotation. Next, the patient should step with the prosthesis, taking care to maintain equal step lengths. After the patient gains reasonable proficiency in forward walking in the parallel bars or next to a table, the therapist determines whether a cane held in the contralateral hand or a walking frame will be needed. In addition to forward walking, the patient should practise stepping backward, turning and walking on various terrains, such as carpet, grass and gravel. Agile individuals can be guided to step onto an escalator.

Gait deviations

Gait deviations indicate deficiency in the prosthetic fit, alignment or adjustment, or may suggest faulty movement patterns (Edelstein, 1988). Some common deviations from the gait of non-prosthetic wearers, such as lateral trunk bending and thrusting of the transtibial socket brim, are inevitable. One can expect some lateral trunk and socket shifting when the wearer is in the single support phase of gait. Because the prosthetic socket is fitted over soft tissue, skeletal fixation is reduced. The patient thus must shift the trunk toward the prosthetic side to maintain balance. Similarly, during midstance, the transtibial socket realigns itself on the wearer's leg, thrusting laterally. These deviations require attention when they are extreme in magnitude or cause the patient pain or undue fatigue. Using a cane obviates the problem.

A serious problem which may be observed when the individual wearing a transtibial prosthesis walks is excessive knee flexion during prosthetic stance phase. Such a deviation can upset the patient's balance and corrective measures should be employed. Causes range from wearing a shoe with a heel higher than appropriate for the prosthetic foot, a stiff heel cushion or posterior foot bumper, a foot malaligned in dorsiflexion, excessive socket flexion, excessive anterior displacement of the socket, a knee flexion contracture, or weak quadriceps.

The individual wearing a transfemoral prosthesis who walks with a wide base may have a prosthesis which is too long, or the stump may not be correctly situated in the socket. If the medial wall of the socket impinges on the perineum then the patient will abduct the prosthesis to reduce the discomfort. An abduction contracture or medial tissue redundancy will also force the patient to widen the walking base.

If the patient exhibits both excessive heel rise during early swing phase and impact at the conclusion of swing phase, consideration should be given to adjusting the friction mechanism on the knee unit to increase damping the shank motion. If, however, the patient displays only high heel rise then the likely cause is insufficiency in the extension aid. If the prosthetic foot rotates at heel contact, then softening the heel cushion will alleviate the problem. Some patients discover that if they 'vault', that is, exaggerate plantar flexion of the sound foot while the prosthesis is in swing phase, they can walk faster without the need to flex the prosthetic knee.

Climbing

Climbing stairs, curbs, ramps and, for some agile individuals, ladders enhances the patient's independence. Most people who wear transtibial prostheses can ascend and descend in the step-over-step manner, that is, climbing the first step with the right foot and the next with the left, regardless of prosthetic side. In contrast, those who require a transfemoral prosthesis must climb the first step with the sound foot, then raise the prosthesis to the first step. On subsequent steps, the sound foot is always the first to climb. Descending requires that the prosthesis is lowered first, then the sound foot is placed on to the same step, and continue in similar fashion. A few very confident individuals are able to descend stairs with the sound and the prosthetic foot on alternate steps. The manoeuvre requires that the individual bear weight on a flexed prosthetic knee unit while lowering the sound foot.

Prosthesis care

Caring for the prosthesis should be an integral part of prosthetic training so that the individual is not inconvenienced by being unable to use the prosthesis because of unforeseen repairs. The socket should be wiped with a damp cloth nightly and allowed to dry overnight. If the patient has to walk through puddles, the shoe and sock should be removed from the prosthetic foot so that the foot can dry thoroughly. One should anticipate rain storms and wear rubber overshoes to avoid immersing the foot, particularly if the foot has moving parts. A transfemoral prosthesis with a fluid-controlled knee mechanism should be placed in an upright position at night. The patient should inspect the prosthesis frequently to detect loosened parts or other mechanical defects. One should also examine the stump daily so that minor abrasions or other lesions can be addressed promptly. Wearing clean, well fitting sheaths of appropriate thickness and material is essential.

Advanced training

Advanced training includes teaching selected patients how to run, pick up objects from the floor, kneeling and sitting on the floor (Karacoloff *et al.*, 1992). Those who wear a transtibial prosthesis should be able to run with equal step lengths. With a transfemoral prosthesis, it may be easier to take two steps with the sound foot and then a quick hop on the prosthesis. Some individuals whose prosthesis includes a fluid-controlled knee mechanism are able to run by alternating left and right foot steps. Vaulting is another means of rapid movement. Whether wearing a transfemoral or transtibial prosthesis, the patient will have the least difficulty if the sound leg is placed under the individual's trunk and the prosthesis is kept somewhat posterior. Kneeling involves placing the sound foot slightly ahead and lowering oneself onto the prosthetic knee by controlled flexion of the sound hip and knee. The first attempt at kneeling should be done while the patient holds onto a sturdy chair. Sitting on the floor is a continuation of kneeling, in which the individual shifts onto the sound hip.

Reintegration

Reintegration into the community may be facilitated by environmental adaptations at home and at work. A professional home assessment conducted prior to discharge will reveal obstacles, such as loose rugs, narrow doorways and stairs lacking handrails. Simple alterations in the bathroom, such as handrails adjacent to the toilet, a secure shower seat or a tub seat, and strategically placed grab bars may contribute to the patient's self-sufficiency. Recent legislation in the United States mandates such accommodations in places of public access including offices, restaurants and schools. Those with amputation find employment in many occupations often require little or no adaptation.

Recreation

Recreation is an excellent and important component of prosthetic management. People with amputations engage in a wide range of sports, ranging from archery to weight lifting (Kegel, 1985). Some activities, such as swimming, are easier to accomplish without a prosthesis; others, such as bicycling are aided with modest adaptations of gear, for example a toe clip on the bicycle pedal (Fleiss and Rubin, 1983; McCormick, 1984; Pringle, 1987). A few sports require prosthetic adaptation (Banziger and Harding, 1991), such as a foot suited to an ice skate shoe. The individual may prefer to use a wheelchair for some sports. Many organizations provide instruction, companionship and competition for athletes with disabilities (Burgess and Rappoport, 1992).

Support

Support groups help the patient and the family adjust to the reality of amputation and the altered means of accomplishing activities. A new patient may find considerable encouragement from meeting someone who has become proficient with a prosthesis. Fears and misconceptions can be dispelled by individual and group encounters. Hospitals, rehabilitation centres and senior centres often sponsor such groups. A clinician can serve as a resource to respond to questions and foster rapport among participants.

Conclusion

Comprehensive rehabilitation for the patient with lower-limb amputation involves many elements, ideally conducted under the aegis of a rehabilitation team which takes full cognisance of the individual's physical, psychological and environmental characteristics.

References

Adler, J.C., Mazzarella, N., Puzsier, L. and Alba, A. (1987) Treadmill training programme for a bilateral below-knee amputee patient with cardiopulmonary disease. *Arch. Phys. Med. Rehabil.*, **68,** 858–861.

Banerjee, S.N. (ed.) (1982) *Rehabilitation Management of Amputees.* Williams and Wilkins, Baltimore, MD.

Banziger, E. and Harding, A. (1991) 'Aqua-Flex': a paediatric all-plastic prosthetic knee system. *J. Assoc. Child. Prosthet. Orthot. Clin.*, **26,** 53–55.

Bonner, F.J. and Green R.F. (1982) Pneumatic airleg prosthesis: report of 200 cases. *Arch. Phys. Med. Rehabil.*, **63,** 383–385.

Brunnstrom, S. and Kerr, D. (1956) *The Training of the Lower Extremity Amputee.* Charles C. Thomas, Springfield, IL.

Burgess, E.M. and Rappoport, A. (1992) *Physical Fitness: A Guide for Individuals with Lower Limb Loss.* Rehabilitation Research and Development Service, Department of Veterans Affairs, Washington, D.C.

Edelstein, J.E. (1988) Prosthetic assessment and management. In: *Physical Rehabilitation: Assessment and Treatment*, 2nd edn. (eds S.B. O'Sullivan and T.J. Schmitz), F.A. Davis, Philadelphia, PA.

Edelstein, J.E. (1992a) Functional activities for the amputee. In: *Manual for Functional Training*, 3rd edn. (eds M.L. Palmer and J.E. Toms), F.A. Davis, Philadelphia, PA.

Edelstein, J.E. (1992b) Rehabilitation without prostheses: functional skills training. In: *Atlas of Limb Prosthetics*, 2nd edn. (eds J.H. Bowker and J.W. Michael), C.V. Mosby, St. Louis, MO.

Eisert, O. and Tester, O.W. (1954) Dynamic exercises for lower extremity amputees. *Arch. Phys. Med. Rehabil.*, **35,** 695–704.

Engstrom, B and Van de Ven, C. (1985) *Physiotherapy for Amputees: The Roehampton Approach.* Churchill Livingstone, New York.

Fleiss, D. and Rubin, G. (1983) Devices to enable persons with amputation to participate in sports. *Arch. Phys. Med. Rehabil.*, **64,** 37–43.

Gailey, R.S. and McKenzie, A. (1989) *Prosthetic Gait Training Program for Lower Extremity Amputees.* University of Miami School of Medicine, Miami, FL.

Hallam, F.M. and Jull, G.A. (1988) Evaluation of a temporary prosthetic insert in the rehabilitation of elderly ischaemic below-knee amputees: a pilot study. *Aust. J. Physiother.*, **34,** 133–138.

Ham, R. and Cotton, L. (1991) *Limb Amputation: From Aetiology to Rehabilitation.* Chapman & Hall, London.

Humm, W. (1977) *Rehabilitation of the Lower Limb Amputee: For Nurses and Therapists*, 3rd edn. Bailliere Tindall, London.

Karacoloff, L.A., Hammersley, C.S. and Schneider, F.J. (1992) *Lower Extremity Amputation: A Guide to Functional Outcomes in Physical Therapy Management*, 2nd edn. Aspen, Gaithersburg, MD.

Kegel, B. (1985) Physical fitness: Sports and recreation for those with lower limb amputation or impairment. In: *Journal of Rehabilitation Research and Development Clinical Supplement No. 1.* Department of Medicine and Surgery, Washington, D.C.

Kegel, B., Burgess, E.M., Starr, T.W. and Daly, W.K. (1981) Effects of isometric muscle training on residual limb volume, strength and gait of below-knee amputees. *Phys. Ther.*, **61**, 1419–1426.

Levy, S.W. (1983) *Skin Problems of the Amputee.* Green, St. Louis, MO.

May, B.J. (1988) Preprosthetic management of lower extremity amputation. In: *Physical Rehabilitation: Assessment and Treatment*, 2nd edn, (eds. S.B. O'Sullivan and T.J. Schmitz), F.A. Davis, Philadelphia, PA.

McCormick, D.P. (1984) Handicapped skiing: a current review of downhill snow skiing for the disabled. *Phys. Occup. Ther. Pediat.*, **4**, 27–44.

Mensch, G. and Ellis, P.M. (1986) *Physical Therapy Management of Lower Extremity Amputations.* Aspen, Gaithersburg, MD.

Monga, T.N., Symington, D.C., Lowe, P. and Elkin, N. (1985) Load bearing and suspension characteristics of airsplint as a temporary prosthesis. *Prosthet. Orthot. Int.*, **9**, 100–104.

Narang, I.C., Mathur, B.P., Singh, P. and Jape, V.S. (1984) Functional capabilities of lower limb amputees. *Prosthet. Orthot. Int.*, **8**, 43–51.

Parry, M. and Morrison, J.D. (1989) Use of the Femurett adjustable prosthesis in the assessment and walking training of new above-knee amputees. *Prosthet. Orthot. Int.*, **13**, 36–38.

Pitetti, K.H., Snell, P.G., Stray-Gunderson, J. and Gottschalk, F.A. (1987) Aerobic training exercises for individuals who had amputation of the lower limb. *J. Bone Joint Surg.*, **69-A**, 914–921.

Pringle, D. (1987) Winter Sports for the amputee athlete. *Clin. Prosthet. Orthot.*, **11**, 114–117.

Summers, G.D., Morrison, J.D. and Cochrane, G.M. (1988) Amputee walking training: a preliminary study of biomechanical measurements of stance and balance. *Int. Disabil. Stud.*, **10**, 1–5.

Van Alste, J.A., la Haye, M.W., Huisman, K. *et al.* (1985) Exercise testing of leg amputees and the result of prosthetic training. *Int. Rehabil. Med.*, **7**, 93–98.

Waters, R.L. and Perry, J. (1989) Energy expenditure of amputee gait. In: *Lower Extremity Amputation* (eds. W.S. Moore and J.M. Malone), W.B. Saunders, Philadelphia, Pa.

Index